GENETICS, GENOMICS AND BREEDING OF TOMATO

Genetics, Genomics and Breeding of Crop Plants

Series Editor
Chittaranjan Kole
Department of Genetics and Biochemistry
Clemson University
Clemson, SC
USA

Books in this Series:

GENETICS, GENOMICS AND BREEDING OF TOMATO

Editors

Barbara E. Liedl
West Virginia State University
Gus R. Douglass Institute
Agricultural and Environmental Research Station
Institute, WV
USA

Joanne A. Labate
USDA-ARS Plant Genetic Resources Unit
Geneva, NY
USA

John R. Stommel
USDA-ARS Vegetable Laboratory
BARC-West
Beltsville, MD
USA

Ann Slade
Arcadia Bioscience
Seattle, WV
USA

Chittaranjan Kole
Department of Genetics and Biochemistry
Clemson University
Clemson, SC
USA

CRC Press
Taylor & Francis Group
Boca Raton   London   New York

CRC Press is an imprint of the
Taylor & Francis Group, an **informa** business

A SCIENCE PUBLISHERS BOOK

CRC Press
Taylor & Francis Group
6000 Broken Sound Parkway NW, Suite 300
Boca Raton, FL 33487-2742

© 2013 Copyright reserved
CRC Press is an imprint of Taylor & Francis Group, an Informa business

Cover photograph reproduced by kind courtesy of Barbara E. Liedl.

No claim to original U.S. Government works

Printed in the United States of America on acid-free paper
Version Date: 20120327

International Standard Book Number: 978-1-57808-804-1 (Hardback)

Library of Congress Cataloging-in-Publication Data

Genetics, genomics and breeding of tomato / editors, Barbara E.
Liedl ... [et al.]. -- 1st ed.
      p. cm. -- (Genetics, genomics and breeding of crop plants)
 ISBN 978-1-57808-804-1 (hardcover)
 1. Tomatoes--Genetics. 2. Tomatoes--Breeding. 3. Tomatoes-
-Genome mapping. I. Liedl, Barbara E. II. Series: Genetics,
genomics and breeding of crop plants.
 SB349.G426 2012
 635'.642--dc23

                                                          2012018332

Visit the Taylor & Francis Web site at
http://www.taylorandfrancis.com

CRC Press Web site at
http://www.crcpress.com

Science Publishers Web site at
http://www.scipub.net

Dedication

This book is dedicated to Professor Steven D. Tanksley, Liberty Hyde Bailey Professor of Plant Breeding Cornell University and CEO/Chief Scientific Officer of Nature Source Genetics for his innovative research to identify and study genes responsible for the variation found in tomato and a variety of other crops. His work with tomato began with his graduate work at the University of California-Davis under the guidance of Dr. Charles M. Rick, plant geneticist and botanist. The work in his early career to develop molecular markers and a saturated linkage map for tomato served as a critical resource for his and other laboratories to ask the larger questions about tomatoes and in other plant species. In 1993, scientists in his lab successfully cloned the first gene in tomato using a technique known as map-based cloning. His lab was also responsible for demonstrating that quantitatively inherited traits scattered across the genome can be dissected into Mendelian factors, called quantitative trait loci. This work has laid the ground work for understanding the inheritance of complex traits but opened up research avenues researchers in all areas of the life sciences. He was also involved in the international group responsible for sequencing the tomato genome. He has been recognized for his numerous accomplishments including being elected to the National Academy of Sciences and receiving awards such as Alexander von Humboldt Foundation Award, Martin Gibbs Medal of the ASPP, Wolf Foundation Prize in Agriculture and the Kumho Award in Plant Molecular Biology and Technology. His enduring legacy will not be only the bounty of research he has published and directed, but also the undergraduate, graduate and post graduate students from around the world as well as collaborators from all fields that have benefitted from his teaching, expertise, philosophy and passion for plants, particularly tomato.

Preface to the Series

Genetics, genomics and breeding has emerged as three overlapping and complimentary disciplines for comprehensive and fine-scale analysis of plant genomes and their precise and rapid improvement. While genetics and plant breeding have contributed enormously towards several new concepts and strategies for elucidation of plant genes and genomes as well as development of a huge number of crop varieties with desirable traits, genomics has depicted the chemical nature of genes, gene products and genomes and also provided additional resources for crop improvement.

In today's world, teaching, research, funding, regulation and utilization of plant genetics, genomics and breeding essentially require thorough understanding of their components including classical, biochemical, cytological and molecular genetics; and traditional, molecular, transgenic and genomics-assisted breeding. There are several book volumes and reviews available that cover individually or in combination of a few of these components for the major plants or plant groups; and also on the concepts and strategies for these individual components with examples drawn mainly from the major plants. Therefore, we planned to fill an existing gap with individual book volumes dedicated to the leading crop and model plants with comprehensive deliberations on all the classical, advanced and modern concepts of depiction and improvement of genomes. The success stories and limitations in the different plant species, crop or model, must vary; however, we have tried to include a more or less general outline of the contents of the chapters of the volumes to maintain uniformity as far as possible.

Often genetics, genomics and plant breeding and particularly their complimentary and supplementary disciplines are studied and practiced by people who do not have, and reasonably so, the basic understanding of biology of the plants for which they are contributing. A general description of the plants and their botany would surely instill more interest among them on the plant species they are working for and therefore we presented lucid details on the economic and/or academic importance of the plant(s); historical information on geographical origin and distribution; botanical origin and evolution; available germplasms and gene pools, and genetic and cytogenetic stocks as genetic, genomic and breeding resources; and basic information on taxonomy, habit, habitat, morphology, karyotype, ploidy level and genome size, etc.

Classical genetics and traditional breeding have contributed enormously even by employing the phenotype-to-genotype approach. We included detailed descriptions on these classical efforts such as genetic mapping using morphological, cytological and isozyme markers; and achievements of conventional breeding for desirable and against undesirable traits. Employment of the *in vitro* culture techniques such as micro- and megaspore culture, and somatic mutation and hybridization, has also been enumerated. In addition, an assessment of the achievements and limitations of the basic genetics and conventional breeding efforts has been presented.

It is a hard truth that in many instances we depend too much on a few advanced technologies, we are trained in, for creating and using novel or alien genes but forget the infinite wealth of desirable genes in the indigenous cultivars and wild allied species besides the available germplasms in national and international institutes or centers. Exploring as broad as possible natural genetic diversity not only provides information on availability of target donor genes but also on genetically divergent genotypes, botanical varieties, subspecies, species and even genera to be used as potential parents in crosses to realize optimum genetic polymorphism required for mapping and breeding. Genetic divergence has been evaluated using the available tools at a particular point of time. We included discussions on phenotype-based strategies employing morphological markers, genotype-based strategies employing molecular markers; the statistical procedures utilized; their utilities for evaluation of genetic divergence among genotypes, local landraces, species and genera; and also on the effects of breeding pedigrees and geographical locations on the degree of genetic diversity.

Association mapping using molecular markers is a recent strategy to utilize the natural genetic variability to detect marker-trait association and to validate the genomic locations of genes, particularly those controlling the quantitative traits. Association mapping has been employed effectively in genetic studies in human and other animal models and those have inspired the plant scientists to take advantage of this tool. We included examples of its use and implication in some of the volumes that devote to the plants for which this technique has been successfully employed for assessment of the degree of linkage disequilibrium related to a particular gene or genome, and for germplasm enhancement.

Genetic linkage mapping using molecular markers have been discussed in many books, reviews and book series. However, in this series, genetic mapping has been discussed at length with more elaborations and examples on diverse markers including the anonymous type 2 markers such as RFLPs, RAPDs, AFLPs, etc. and the gene-specific type 1 markers such as EST-SSRs, SNPs, etc.; various mapping populations including F_2, backcross, recombinant inbred, doubled haploid, near-isogenic and pseudotestcross; computer software including MapMaker, JoinMap, etc. used; and different types of genetic maps including preliminary, high-resolution, high-density, saturated, reference, consensus and integrated developed so far.

Mapping of simply inherited traits and quantitative traits controlled by oligogenes and polygenes, respectively has been deliberated in the earlier literature crop-wise or crop group-wise. However, more detailed information on mapping or tagging oligogenes by linkage mapping or bulked segregant analysis, mapping polygenes by QTL analysis, and different computer software employed such as MapMaker, JoinMap, QTL Cartographer, Map Manager, etc. for these purposes have been discussed at more depth in the present volumes.

The strategies and achievements of marker-assisted or molecular breeding have been discussed in a few books and reviews earlier. However, those mostly deliberated on the general aspects with examples drawn mainly from major plants. In this series, we included comprehensive descriptions on the use of molecular markers for germplasm characterization, detection and maintenance of distinctiveness, uniformity and stability of genotypes, introgression and pyramiding of genes. We have also included elucidations on the strategies and achievements of transgenic breeding for developing genotypes particularly with resistance to herbicide, biotic and abiotic stresses; for biofuel production, biopharming, phytoremediation; and also for producing resources for functional genomics.

A number of desirable genes and QTLs have been cloned in plants since 1992 and 2000, respectively using different strategies, mainly positional cloning and transposon tagging. We included enumeration of these and other strategies for isolation of genes and QTLs, testing of their expression and their effective utilization in the relevant volumes.

Physical maps and integrated physical-genetic maps are now available in most of the leading crop and model plants owing mainly to the BAC, YAC, EST and cDNA libraries. Similar libraries and other required genomic resources have also been developed for the remaining crops. We have devoted a section on the library development and sequencing of these resources; detection, validation and utilization of gene-based molecular markers; and impact of new generation sequencing technologies on structural genomics.

As mentioned earlier, whole genome sequencing has been completed in one model plant (Arabidopsis) and seven economic plants (rice, poplar, peach, papaya, grapes, soybean and sorghum) and is progressing in an array of model and economic plants. Advent of massively parallel DNA sequencing using 454-pyrosequencing, Solexa Genome Analyzer, SOLiD system, Heliscope and SMRT have facilitated whole genome sequencing in many other plants more rapidly, cheaply and precisely. We have included extensive coverage on the level (national or international) of collaboration and the strategies and status of whole genome sequencing in plants for which sequencing efforts have been completed or are progressing currently. We have also included critical assessment of the impact of these genome initiatives in the respective volumes.

Comparative genome mapping based on molecular markers and map positions of genes and QTLs practiced during the last two decades of the last century provided answers to many basic questions related to evolution, origin and phylogenetic relationship of close plant taxa. Enrichment of genomic resources has reinforced the study of genome homology and synteny of genes among plants not only in the same family but also of taxonomically distant families. Comparative genomics is not only delivering answers to the questions of academic interest but also providing many candidate genes for plant genetic improvement.

The 'central dogma' enunciated in 1958 provided a simple picture of gene function—gene to mRNA to transcripts to proteins (enzymes) to metabolites. The enormous amount of information generated on characterization of transcripts, proteins and metabolites now have led to the emergence of individual disciplines including functional genomics, transcriptomics, proteomics and metabolomics. Although all of them ultimately strengthen the analysis and improvement of a genome, they deserve individual deliberations for each plant species. For example, microarrays, SAGE, MPSS for transcriptome analysis; and 2D gel electrophoresis, MALDI, NMR, MS for proteomics and metabolomics studies require elaboration. Besides transcriptome, proteome or metabolome QTL mapping and application of transcriptomics, proteomics and metabolomics in genomics-assisted breeding are frontier fields now. We included discussions on them in the relevant volumes.

The databases for storage, search and utilization on the genomes, genes, gene products and their sequences are growing enormously in each second and they require robust bioinformatics tools plant-wise and purpose-wise. We included a section on databases on the gene and genomes, gene expression, comparative genomes, molecular marker and genetic maps, protein and metabolomes, and their integration.

Notwithstanding the progress made so far, each crop or model plant species requires more pragmatic retrospect. For the model plants we need to answer how much they have been utilized to answer the basic questions of genetics and genomics as compared to other wild and domesticated species. For the economic plants we need to answer as to whether they have been genetically tailored perfectly for expanded geographical regions and current requirements for green fuel, plant-based bioproducts and for improvements of ecology and environment. These futuristic explanations have been addressed finally in the volumes.

We are aware of exclusions of some plants for which we have comprehensive compilations on genetics, genomics and breeding in hard copy or digital format and also some other plants which will have enough achievements to claim for individual book volume only in distant future. However, we feel satisfied that we could present comprehensive deliberations on genetics, genomics and breeding of 30 model and economic plants, and their groups in a few cases, in this series. I personally feel also happy that I could work with many internationally celebrated scientists who edited the book volumes on the leading plants and plant groups and included chapters authored by many scientists reputed globally for their contributions on the concerned plant or plant group.

We paid serious attention to reviewing, revising and updating of the manuscripts of all the chapters of this book series, but some technical and formatting mistakes will remain for sure. As the series editor, I take complete responsibility for all these mistakes and will look forward to the readers for corrections of these mistakes and also for their suggestions for further improvement of the volumes and the series so that future editions can serve better the purposes of the students, scientists, industries, and the society of this and future generations.

Science publishers, Inc. has been serving the requirements of science and society for a long time with publications of books devoted to advanced concepts, strategies, tools, methodologies and achievements of various science disciplines. Myself as the editor and also on behalf of the volume editors, chapter authors and the ultimate beneficiaries of the volumes take this opportunity to acknowledge the publisher for presenting these books that could be useful for teaching, research and extension of genetics, genomics and breeding.

Chittaranjan Kole

Preface to the Volume

What is it about the lowly tomato that makes people wax poetic on eating tomatoes fresh from the garden? How have tomatoes inspired song writers, movie makers, artists, and yet risen to be the second most economically important vegetable in the world with more than 152 million tons produced in 2009. It all began in the New World where tomatoes originated, but they didn't begin to take over the world until they were distributed following the Spanish colonization of the Americas. Since then tomatoes have been used in variety of cuisines as both fresh and processed forms as well as being a good source of vitamins and antioxidants for our diet. Much of the success of tomato as a crop today is due to the dedicated research of the early geneticists and breeders in collecting land races and wild relatives, relating phenotypes to genotypes, and establishing linkage maps. With this firm foundation, in the last twenty years plant scientists working on tomato have seen an explosion in the fields of genetics and genomics elevating tomato to a model crop species as well as revolutionizing tomato breeding.

In this volume we combined the current research in both classical and molecular genetics and breeding on tomato and its wild relatives. The first chapter lays the ground work for the rest of the volume introducing tomato and its wild relatives in regards to their economic importance, botanical descriptions, taxonomy, diversity and domestication. Classical genetics and breeding is the subject of Chapter 2, which includes an updated classical linkage map, a list of major qualitative and quantitative traits incorporated into varieties and a historical record of critical breeding achievements. Diversity within cultivated tomato is addressed in Chapter 3 summarizing insights on the genes involved in fruit shape evolution. The various molecular marker systems are reviewed in Chapter 4 and their use in linkage disequilibrium analysis, association studies and identification of candidate genes. Mapping and tagging of simply inherited traits with emphasis on the fruit qualities of taste and color is covered in Chapter 5. The markers, statistical methods and software used to map complex trait with a wide variety of populations is presented in Chapter 6. Applying the work of the previous chapters is found in Chapter 7 on molecular breeding which also discusses the application of the new "omics" to the area. Positional cloning is discussed in Chapter 8 as an approach for functional analysis of genes from mutant phenotypes. Chapter 9 on structural genomics illustrates how the genotype contributes to phenotype. The work on tomato genome sequencing project is covered in Chapter 10 which lays the groundwork for Chapter 11 on the comparative genome sequencing with potato. The tools and methods used to characterize the tomato transcriptome are found in Chapter 12. A review of methods and application of proteomic and metabolomic studies on tomato fruit quality is discussed in Chapter 13. Finally the role of bioinformatics and a guide to available resources as a tool in research is presented in Chapter 14 to wrap up the volume. This book is geared to be of use to researchers not only working on tomato and other Solanaceous crops, but also those working with other crops.

Each chapter is written by authors who are experts, who have worked tirelessly compiling the research in their respective areas of expertise. We would like to acknowledge their effort and the time committed to this book.

Barbara E. Liedl
Joanne A. Labate
John R. Stommel
Ann Slade
Chittaranjan Kole

Contents

2. Classical Genetics and Traditional Breeding   37

John W. Scott, James R. Myers, Peter S. Boches,
Courtland G. Nichols and Frederic F. Angell

List of Contributors

Nazareno Acciarri
C.R.A. Agricultural Research Council, Research Unit #13 for Vegetable Crops, via Salaria 1, 63077 Monsampolo del Tronto (AP), Italy.
Email: _nazareno.acciarri@entecra.it_

Claire Anderson
Department of Horticulture and Crop Science, The Ohio State University/ Ohio Agricultural Research and Development Center, Wooster, OH 44691, USA.
Email: _andersonclaire3@gmail.com_

Frederic F. Angell
7281 Miller Ave, Gilroy CA 95020, USA.
Email: _ffangell@verizon.net_

Koh Aoki
Graduate School of Life and Environmental Sciences, Osaka Prefecture University, B4-215, 1-1 Gakuen-cho, Naka-ku, Sakai, 599-8531 Japan.
Email: _kaoki@plant.osakafu-u.ac.jp_

Hamid Ashrafi
Seed Biotechnology Center, University of California, 1 Shields Ave, Davis, CA, 95616, USA.
Email: _ashrafi@ucdavis.edu_

Cornelius S. Barry
Department of Horticulture, Michigan State University, East Lansing, Michigan 48824, USA.
Email: _barrycs@msu.edu_

Massimiliano Beretta
C.R.A. Agricultural Research Council, Research Unit #12 for Vegetable Crops, via Paullese 28, 26836 Montanaso Lombardo (LO), Italy.
Email: _beretta.massimiliano@gmail.com_

Peter S. Boches
Fall Creek Farm and Nursery, 39318 Jasper-Lowell Rd, Lowell, OR 97452,
USA.
Email: *peterb@fallcreeknursery.com*

Ana Caicedo
Biology Department, University of Massachusetts, 221 Morrill Science
Center, Amherst, MA 01003, USA.
Email: *caicedo@bio.umass.edu*

Mireille Faurobert
INRA, Unité Génétique et Amélioration des Fruits et Légumes, UR1052,
84000 Avignon, France.
Email: *Mireille.Faurobert@avignon.inra.fr*

Zhangjun Fei
Boyce Thompson Institute for Plant Research, Tower Rd., Cornell University
campus, Ithaca, NY 14853 USA
and
USDA-ARS Robert W. Holley Center for Agriculture and Health, Tower
Rd., Cornell University campus, Ithaca, NY 14853 USA.
Email: *zf25@cornell.edu*

Martin W. Ganal
Trait Genetics GmbH, Am Schwabeplan 1b, 06466 Gatersleben, Germany.
Email: *ganal@traitgenetics.de*

James J. Giovannoni
Boyce Thompson Institute for Plant Research, Tower Rd., Cornell University
campus, Ithaca, NY 14853 USA
and
USDA-ARS Robert W. Holley Center for Agriculture and Health, Tower
Rd., Cornell University campus, Ithaca, NY 14853 USA.
Email: *jjg33@cornell.edu*

Silvana Grandillo
CNR-IGV, Inst. of Plant Genetics, Research Division Portici, Via Università
133, 80055 Portici (Naples), Italy.
Email: *grandill@unina.it*

Theresa Hill
Seed Biotechnology Center, University of California, 1 Shields Ave, Davis,
CA, 95616, USA.
Email: *tahill@ucdavis.edu*

Sanwen Huang
Institute of Vegetables and Flowers, Chinese Academy of Agricultural Sciences, No. 12, Zhong Guan Cun Nan Da Jie, Beijing, 100081, China.
Email: *huangsanwen@caas.net.cn*

Yoko Iijima
Department of Nutrition and Life Science, Faculty of Applied Bioscience, 1030 Simo-ogino, Atsugi, Kanagawa, 243-0292 Japan.
Email: *iijima@bio.kanagawa-it.ac.jp*

Ilan Levin
Department of Vegetable Research, Institute of Plant Sciences, Agricultural Research Organization, the Volcani Center, P.O. Box 6, Bet Dagan 50250, Israel.
Email: *vclevini@volcani.agri.gov.il*

Antonio J. Matas
Department of Plant Biology, Cornell University, Ithaca, NY 14853, USA.
Email: *ajm75@cornell.edu*

Justyna Milc
University of Modena and Reggio Emilia, Department of Agricultural and Food Sciences, via Amendola 2, 42122 Reggio Emilia, Italy.
Email: *justynaanna.milc@unimore.it*

Lukas A. Mueller
Boyce Thompson Institute for Plant Research, Tower Road, Ithaca, NY 14853 USA.
Email: *LAM87@cornell.edu*

James R. Myers
Oregon State University, Dept. of Horticulture, ALS 4017, Corvallis, OR 97331-7304 USA.
Email: *myersja@hort.oregonstate.edu*

Courtland G. Nichols
1241Sunset Dr, Hollister, CA 95023, USA.
Email: *cgnichols@hotmail.com*

Nicola Pecchioni
University of Modena and Reggio Emilia, Department of Agricultural and Food Sciences, via Amendola 2, 42122 Reggio Emilia, Italy.
Email: *nicola.pecchioni@unimore.it*

Iris Peralta
Department of Agronomy, National University of Cuyo, Almirante Brown 500, 5505, Chacras de Coria, Luján, Mendoza, Argentina.
and
IADIZA-CONICET, C.C. 507, 5500, Mendoza, Argentina.
Email: *iperalta@fca.uncu.edu.ar*

Gustavo Rodriguez
Department of Horticulture and Crop Science, The Ohio State University/ Ohio Agricultural Research and Development Center, Wooster, OH 44691, USA.
Email: *grodrig@unr.edu.ar*

Catherine M. Ronning
Office of Biological and Environmental Research, SC-23.2/Germantown Building, U.S. Department of Energy, 1000 Independence Ave., SW, Washington, DC 20585 USA.
Email: *catherine.ronning@science.doe.gov*

Jocelyn K.C. Rose
Department of Plant Biology, Cornell University, Ithaca, NY 14853, USA.
Email: *jr286@cornell.edu*

Emidio Sabatini
C.R.A. Agricultural Research Council, Research Unit #13 for Vegetable Crops, via Salaria 1, 63077 Monsampolo del Tronto (AP), Italy.
Email: *emidio.sabatini@entecra.it*

Tea Sala
C.R.A. Agricultural Research Council, Research Unit #12 for Vegetable Crops, via Paullese 28, 26836 Montanaso Lombardo (LO), Italy.
Email: *tea.sala@entecra.it*

Arthur A. Schaffer
Department of Vegetable Research, Institute of Plant Sciences, Agricultural Research Organization, the Volcani Center, P.O. Box 6, Bet Dagan 50250, Israel.
Email: *vcaris@volcani.agri.gov.il*

John W. Scott
University of Florida, Gulf Coast Research & Education Center, 14625 CR 672, Wimauma, FL 33598, USA.
Email: *jwsc@ufl.edu*

Pasquale Termolino
CNR-IGV, Inst. of Plant Genetics, Research Division Portici, Via Università
133, 80055 Portici (Naples), Italy.
Email: *termolin@unina.it*

Esther van der Knaap
Department of Horticulture and Crop Science, The Ohio State University/
Ohio Agricultural Research and Development Center, Wooster, OH 44691,
USA.
Email: *vanderknaap.1@osu.edu*

Allen E. Van Deynze
Seed Biotechnology Center, University of California, 1 Shields Ave, Davis,
CA, 95616, USA.
Email: *avandeynze@ucdavis.edu*

Zhong-hua Zhang
Institute of Vegetables and Flowers, Chinese Academy of Agricultural
Sciences, No. 12, Zhong Guan Cun Nan Da Jie, Beijing, 100081, China
Email: *zhangzh.ivf@caas.net.cn*

Abbreviations

2TD	2-Tridecanone
AB	Advanced backcross
ABG	*Aubergine* gene
AB-QTL	Advanced backcross-QTL
AC	Ailsa Craig
AGP	Accessioned Golden Path
AGPase	ADP-glucose pyrophosphorylase
AFLP	Amplified fragment length polymorphism
AFT	*Anthocyanin fruit* gene
AGP	Accessioned golden path
AMV	Alfalfa mosaic virus
ANOVA	Analysis of variance
AOX	Antioxidant activity
APX	Ascorbate peroxidase activity
ARS	Agricultural Research Service
AS-PCR	Allele-specific polymerase chain reaction
AsA	Ascorbic acid content
ASC	*Alternaria* stem canker resistance
ATI	Anther tube index (ATL/ATW)
ATL	Anther tube length
atv	*Atroviolacium* gene
ATW	Anther tube width
AV	Aroma volatiles
AVRDC	Asian Vegetable Research and Development Center (Taiwan)
βLCY	*β-lycopene cyclase* gene
B	*Beta-carotene* gene
B	Soluble solids content (°Brix)
BAC	Bacterial artificial chromosome
BC	Backcross
BER	Blossom-end rot
BIL	Backcross inbred line
BIM	Bayesian interval mapping
BLAST	Basic Local Alignment Search Tool

BM	Biomass
BN	Total number of flower buds
bp	Base pair
BRN	Number of well developed branches
BSA	Bulked segregant analysis
BT	Flower bud type
BTI	Boyce Thompson Institute
BY	Brix x Total Yield
BYR	Brix x Red Yield
C	Cycle of recurrent selection
CA	Citric acid content
CaMV35S	Cauliflower mosaic virus 35S
CAPS	Cleaved amplified polymorphic sequence
CAR	β-carotene content
CAT	Catalase activity
CDA	Canadian Department of Agriculture
cDNA	Complementary DNA
Cf	*Cladosporium fulvum* gene
CF	Constriction of the fruit
CG	Candidate genes
CGEP	Center for Gene Expression Profiling
CI	Confidence interval
CI	Corolla indentation
CIM	Composite interval mapping
cM	centiMorgan
Cmm	*Clavibacter michiganensis* ssp. *michiganensis* gene
CMV	Cucumber mosaic virus
CNP	Cell number of the pericarp
COS	Conserved ortholog set
COS	Conserved orthologous sequences
COSII	Conserverd ortholog set II
COV	Cover
CP	Coat protein
CREDO	Cis regulatory element detection online
cRNA	Complementary RNA
CRTISO	*Carotenoid isomerase* gene
CTV	Curly top virus
CUL4	Cullin 4
CUTC	Cuticular conductance
Cy3	cyanine fluorochrome 3
Cy5	cyanine fluorochrome 5
CZP	Cell size of the pericarp
dbSTS	Sequence tagged sites database

DDB1	UV damaged DNA binding protein 1
DE	Days to seedling emergence
DEL	Delta gene
DET1	Deetiolated1 protein
DFCI	Dana-Farber Cancer Institute
DFL	Days to first flower
DFR	Days to first ripe fruit
DFS	Days to first set fruit
DH	Doubled haploid
DLW	Dry leaf weight
DM	Dry matter
DMCP	Dry matter content of the pericarp
DNA	Deoxyribonucleic acid
dNTP	Deoxy nucleotide triphosphate
DRW	Dry root weight
DSW	Dry stem weight
DTF	Flowering time
DTL	Days to first true leaf
DUS	Distinctness, uniformity and stability
DWP	Dry weight of the pericarp
εLCY	*ε-lycopene cyclase* gene
EA	Earliness
EBI	European Bioinformatics Institute
EGFR	Epidermal growth factor receptor
EGO	Eukaryotic gene orthologues
EI	Eccentricity index
ELA	Elasticity
ELISA	Enzyme-linked immunosorbent assay
EMBL	European Molecular Biology Laboratory
EMS	Ethyl methanesulfonate or Ethylenemethane sulfonate
ER	Epidermal reticulation
EST	Expressed sequence tag
ET	Expressed transcript
F	*Fusarium* wilt race 1 resistance
FC	Fruit colour
FCR	Fruit cracking
FD	Fruit equatorial diameter
FDW	Total dry matter weight of the fruit
FER	Fertility
FF	*Fusarium* wilt races 1&2 resistance
Fgr	Fructose to glucose ratio
FIR	Firmness
FISH	Fluorescence *in situ* hybridization

fl	Full-length
FL	Fruit polar diameter
FLA	Total flavonoid content
FLM	Flower morphology
FLZ	Flower size
FK	Fructokinase locus
FM	Functional marker
FN	Fruit number
FPC	Fingerprinted contigs
FRT	Fruitiness
FRU	Fructose content
FRW	Fresh root weight
FS	Fruit shape index (FL/FD)
FW	Fruit weight or fruit mass
FZ	Fruit size
g	Gram
GA	Glutamic acid
GBNV	Groundnut bud necrosis virus
Gbrowse	Generic genome browser
GBSSI	Granule bound starch synthase sequences
GCA	General combining ability
GDPC	Genomic diversity and phenotype connection
GEO	Gene Expression Omnibus
GLU	Glucose content
GMOD	Generic Model Organism Database project
GMM	Genic molecular marker
GO	Gene ontology
GRIN	Germplasm Resources Information Network
GRO	Growth
GRSV	Groundnut ringspot virus
GSS	Genome survey sequence
GWAS	Genome-wide association study
H4T	Height of the 4th truss
HA	Horticultural acceptability
HCA	Hierarchical cluster analysis
HI	Harvest index
HMM	Hidden Markov model
hp	High pigment genes
hp1	High-pigment 1 gene
HPLC	High-performance liquid chromatography
HR	Hypersensitive response
HTG	High throughput genotyping
HTGS	High Throughput Genome Sequence

HTP	High-throughput phenotyping
HYI	Hybrid incompatibility
IBC	Inbred backcross population
ICIM	Inclusive composite interval mapping
INL	Internode length
IL	Introgression line
ILH	Introgression line hybrid
IM	Interval mapping
indel	insertion/deletion
INRA	Institut National de la Recherche Agronomique (France)
IPGRI	International Plant Genetic Resources Institute
ISO	Isozyme marker
ISOL@	Italian SOLAnaceae genomics resource
ISSR	Inter-simple sequence repeat
ITAG	International Tomato Annotation Group
ITS	Internal transcribed spacer
JCVI	J Craig Venter Institute
JRC	Joint Research Centre
KafTom	Kazusa full-length tomato cDNA database
Kb	Kilobase
KNOX	*Knotted1-like homeobox* gene
LA	Leaf area
LC-PDA-FD	Liquid chromatography with photo-diode array and fluorescence detection
LC-QTOF-MS	Liquid chromatography with quadrupole time-of-flight mass spectrophotometry
LCN	Number of locules
LD	Linkage disequilibrium
LeGI	Tomato gene index
LFR	Leaflet width/length ratio
L-GalLDH	L-galactono-1,4-lactone dehydrogenase
LM	Leaf morphology
LN	Number of leaves to the first cluster
LN	Leaf number
LNC	Leaf sodium concentration
LNO	Leaf number
LRR	Leucine rich repeat
LRT	Lycopene Rich Tomatoes
LS1	Large subunit 1
LTA	Leaf total area
LW	Leaf weight
LYC	Lycopene content
LZ	Leaf size

LZ	Leucine zippers
MA	Malic acid content
MABC	Marker-assisted back cross
MAPK	Mitogen-activated protein kinase
MAS	Marker-assisted selection
MAT	Maturity
Mb	Megabase
MEP	Methylerythritol phosphate
MF	Morphological marker
MIAME	Minimum Information About a Microarray Experiment
MiBase	Kazusa Micro-Tom database
MIPS	Munich Information Center for Protein Sequence
mips-REcat	MIPS repeat element catalog
mips-REdat	MIPS repeat element database
miRNA	microRNA
MNSV	Melon necrotic spot virus
Mo_B	*Beta modifier* gene
MPS	Massively parallel sequencing
MPSS	Massive parallel signature sequencing
MQM	Multiple-QTL Model
mRNA	Messenger RNA
ms	Male sterility
MS	Mass spectrometry
MT/HA	Metric ton/hectares
N	Root knot nematode resistance
NBS	Nucleotide binding site
NCBI	National Center for Biotechnology Information
ncRNA	Non-coding RNA
NDC	Number of nodes to the first cluster
NFL	Number of flowers per inflorescence or cluster
NFLP	Number of flowers per plant
NFP	Number of fruits per plant
NIL	Near-isogenic line
nor	*Non-ripening* gene
NPGS	National Plant Germplasm System
Nr	Never-ripe mutant
NSF	National Science Foundation
OA	Organic acids
ODO	Overdominant
OFP	Ovate family protein
og	*old gold* allele of *B* gene
OL	Ovary length
ORF	Open reading frame

OS	Ovary shape index (OL/OW)
OW	Ovary width
P	Bacterial speck resistance
PAT	Parthenocarpy
PCA	Principal components analysis
PCR	Polymerase chain reaction
PEI	Pericarp elongation index
PepMV	Pepino mosaic virus
PG	Polygalacturonase
PGDB	Pathway/Genome Databases
PGRU	Plant Genetic Resources Unit
pH	Fruit pH
PHE	Total phenolic content or phenolics
PHT	Plant height
PK	Protein kinase
PlantGDB	Plant genome database
PLDW	Total dry weight of aerial portion of the plant
PLFW	Total fresh weight of aerial portion of the plant
PLH	Plant height
PLRV	Potato leaf roll virus
PLW	Plant weight
PLZ	Plant size
PM	Primary metabolites
POC	Plant Ontology Consortium
POX	Gluthatione peroxidase activity
PR	Pathogenesis-related
PSY1	*Phytoene synthase 1* gene
PT	Pericarp thickness
PTM	Petal morphology
PTOX	*Plastid terminal oxidase* gene
Pts	*Petroselinum* gene
PUF	Puffiness
PUT	Plant genome database-generated unique transcript
PVE	Phenotypic variation explained
PVX	Potato virus X
PVY	Potato virus Y
PYG	Percent green yield
QTL/QTLs	Quantitative trait locus/loci
QTOF	Quadruple time-of-flight
qRT	Quantitative reverse transcription
r	yellow flesh gene
R	Resistance gene
RAPD	Random amplified polymorphic DNA

RAU	Root ammonium uptake
RB	Reverse breeding
RFLP	Restriction fragment length polymorphism
RGA	Resistance gene analog
RI	Recombinant inbred
RIL	Recombinant inbred line
rin	ripening inhibitor gene
RL	Inflorescence rachis length
RKN	Root-knot nematode
RL-SAGE	Robust-LongSAGE
RNA	Ribonucleic acid
RNA-seq	RNA sequencing
RS	Reproductive stage
RT	Ripening time
RTQ	Real-time QTL
Rubisco	Ribulose-1,5-bisphosphate carboxylase/oxygenase
SA	Sensory attributes
SAGE	Serial analysis of gene expression
SAM	S-adenosylmethionine
SAM	Significance analysis of microarrays
SC	Self compatibility or self compatible
SCA	Specific combining ability
SCA	Sun scald
SCAR	Sequence characterized amplified region
SDL	Seed length
SDNF	Number of seeds per fruit
SDNP	Number of seeds per plant
SDS	Seed shape index (SDL/SDWI)
SDW	Seed weight
SDWF	Seed weight per fruit
SDWI	Seed width
SDWP	Seed weight per plant
SE	Stigma exsertion
SFP	Single feature polymorphisms
SG	Seed germination
SGe	Selective genotyping
SGN	Solanaceae Genome Network or Sol Genomics Network
Sh	Sherry gene
SI	Self incompatibility or self incompatible
SlGDB	Solanum lycopersicum plant genome database
SIMAP	Similarity matrix of proteins
siRNA	Small interfering RNA
SME	Small to Medium Enterprise

SNI	Number of internodes on the primary stem
SNP	Single nucleotide polymorphism
SOD	Superoxide dismutase
SOL	International Solanaceae Genome Project
SolCAP	Solanaceae Coordinated Agricultural Project
SOLiD	Sequencing by Oligonucleotide Ligation and Detection
SPM	Sepal morphology
SPME-GC-MS	Solid phase microextraction gas chromatography-mass spectrophotometry
S-RNases	Style-specific ribonucleases
SSH	Suppression subtractive hybridization
SSR	Simple sequence repeat
STA	Starch content
STD	Stem diameter at the first internode
STEM	Stem diameter
STIL	Stile length
STIS	Stile shape index (STIL/STIW)
STIW	Stile width
STR	Stem retention or stem release
STS	Stem scar
STS	Sequence tagged site
STVM	Stem vascular morphology
STW	Stem width
SUC	Sucrose content
SUG	Sugar content
SW	Shoot wilting
t	Tangerine gene
T/A	ton/acre
TA	Tomato Analyzer
TA	Total or titratable acidity
TACW	Total antioxidant capacity
TC	Tentative consensus
TCMV	Tomato chlorotic mottle virus
TCS	Tentative consensus sequences
TCSV	Tomato chlorotic spot virus
T-DNA	Transfer DNA
TED	Tomato expression database
TEV	Tobacco etch virus
TFGD	Tomato Functional Genomics Database
TGI	TIGR gene indices
TGICL	TGI clustering
TGRC	CM Rick Tomato Genetic Resource Center
TIGR	The Institute for Genomic Research

TILLING	Targeting induced local lesions in genomes
TLCV	Tomato leaf curl virus
Tm	Transmembrane
TMV	Tobacco mosaic virus
TN	Transported sodium
TNI	Total number of internodes
TOA	Total organic acids
TOM1	Tomato microarray 1
TOM2	Tomato microarray 2
ToMV	Tomato mosaic virus
ToMoV	Tomato mottle virus
TS	Total solids
TSWV	Tomato spotted wilt virus
TW	Total fruit weight
TYLCSV	Tomato yellow leaf curl sardinia virus
TYLCV	Tomato yellow leaf curl virus
TYTV	Tomato yellow top virus
UC	University of California, Davis
UI	Unilateral incongruity or incompatibility
UPOV	International Union for the Protection of New Varieties and Plants
USDA	United States Department of Agriculture
USDA-ARS	United States Department of Agriculture, Agricultural Research Service
UTLIEF	Ultrathin-layer isoelectric focusing
UTR	Untranslated regions
V	*Verticillium* wilt
VG	Vegetative growth
VIGS	Virus-induced gene silencing
VIS	Viscosity
WGS	Whole genome shotgun
WUE	Water use efficiency
Y	Total yield
YAC	Yeast artificial chromosome
YG	Green yield
YR	Red yield
yrGATE	Your gene structure annotation tool for eukaryotes
ZE	*Zeaxanthin epoxidase* gene

1

Basic Information about Tomatoes and the Tomato Group

Ana Caicedo[1],* and Iris Peralta[2]

ABSTRACT

Domesticated tomato, *Solanum lycopersicum*, is both an economically important crop species and a research model organism. Tomatoes are among the most widely consumed vegetables in the world, and many of the compounds found in tomatoes have received much interest in recent years for their potential health benefits. Due to its short generation time and simple growing requirements, cultivated tomato has been used in scientific research for more than a hundred years, serving as a model for fruit development, and for elucidating the genetic basis of quantitative variation. Cultivated tomato is part of a clade of ~13 closely related wild species, which have also been instrumental in the development of the crop as a model system. Wild tomatoes are a rich source of genetic and morphological diversity, and because most species can be crossed with cultivated tomato, they have played a role in the generation of various genetic and breeding resources. Additionally, study of the wild species themselves has contributed to our understanding of mating system evolution, plant defense, and how plant species adapt and diverge. Although questions remain about the exact phylogenetic relationships among wild tomatoes, and about the history of tomato domestication, the concerted efforts carried out by the tomato research community to create access to various genetic, genomic, germplasm, and bioinformatics resources further contributes to the development of this species as a model system.

[1]Biology Department, University of Massachusetts, 221 Morrill Science Center, Amherst, MA 01003 USA.
[2]Department of Agronomy, National University of Cuyo, Almirante Brown 500, 5505, Chacras de Coria, Luján, Mendoza, Argentina; and, IADIZA-CONICET, C.C. 507, 5500, Mendoza, Argentina.
*Corresponding author

Keywords: model system, phylogeny, taxonomy, ecology, biodiversity, domestication

1.1 Introduction

Domesticated tomato, *Solanum lycopersicum*, is both an economically important crop species, and a model organism for genetic, developmental, and physiological research. A long history of scientific interest in tomato has culminated in the development of *S. lycopersicum* as a model genome for the family Solanaceae, which contains several other important crop species. Together with the clade containing its wild relatives, tomato is currently being developed as an unmatched resource-rich system for diverse areas of biological research.

1.2 Economic and Nutritional Importance of Tomato

1.2.1 Economic Importance

Although botanically a fruit, cultivated tomatoes are consumed primarily as a vegetable or used for seasoning and flavoring in various processed forms. Cultivated tomatoes do not form part of the main calorie sources for most cultures worldwide, but their popularity is such that tomatoes are currently grown in almost every country in the world (Robertson and Labate 2007). In 2003, tomato accounted for almost 11% of vegetable calories consumed per capita per day in the world and for almost 21% of calories consumed from vegetables per capita per day in the United States (FAOSTAT: http://faostat.fao.org/site/609/default.aspx#ancor). Consumption of processed tomatoes far outweighs that of fresh: per capita fresh tomato use in the US in 2005 was 20.6 pounds, while per capita processed tomato use was 73.6 pounds (USDA-ARS 2006).

Production of cultivated tomatoes has consistently increased in the last decades, with a 34% increase in worldwide production in the last ten years (FAOSTAT: http://faostat.fao.org/site/567/default.aspx#ancor; Table 1-1). World production of tomatoes in 2006 was 125.5 million tons (FAOSTAT). Increases in tomato production have generally been accompanied by increases in the amount of land devoted to tomato cultivation (Table 1-1). However, comparisons among the seven top tomato-producing nations suggest that there is still much improvement to be made in increasing the crop's yield. Tomato production is currently led by China and US (Table 1-1), but the US is also one of the world's largest importers of tomatoes. The greatest exporters are Spain, followed by Mexico (FAOSTAT).

Table 1-1 Production of tomatoes in 1996 and 2006 worldwide and in the seven largest producing countries. Data adapted from FAOSTAT (http://faostat.fao.org/site/567/default.aspx#ancor).

		1996			2006		
Leading Countries (2006)	Production (Tons)	Area (Ha)	Yield (Kg/Ha)	Production (Tons)	Area (Ha)	Yield (Kg/Ha)	
China	15537463	574382	27051	32540040	1405103	23158	
Egypt	5995411	173152	34625	7600000	195000	38974	
India	6000000	350000	17143	8637700	497600	17359	
Islamic Rep. of Iran	2974598	119415	24910	4781018	138791	34448	
Italy	6527682	123594	52816	6351202	122192	51977	
Turkey	7800000	187000	41711	9854877	260000	37903	
United States	11874000	186060	63818	11250000	170000	66176	
World	93548709	3409872	27435	125543475	4597220	27309	

1.2.2 Nutritional Composition

Tomato fruits are not a particularly nutrient-rich food source; however, tomatoes are consumed in such large quantities, that they make substantial overall health and nutritional contributions to the human diet. In the US, consumption of tomatoes is second only to potatoes and lettuce among fresh vegetables, and tomatoes are the highest consumed canned vegetable (National Agricultural Statistics Service 2005). Though a poor source of macronutrients, tomatoes contain high amounts of many minerals and vitamins (Table 1-2) (Ensminger et al. 1995), which have received much interest in recent years for their potential health benefits.

Roughly half of the dry matter in tomatoes is made up of sugars, mainly fructose and glucose (Davies and Hobson 1981). Tomatoes are also considered a good source of folic acid and potassium, and are noteworthy for high levels of vitamin C, making tomatoes a substantial contributor of this vitamin to the US diet after oranges and potatoes (Table 1-2) (Davies and Hobson 1981; Beecher 1998). The compounds that have received the most attention for their possible health benefits in tomato, however, have been the carotenoids (Beecher 1998).

Various carotenoid pigments possess retinoid activity and are provitamin A compounds. Vitamin A is an essential nutrient in the human diet, and deficiency can lead to blindness, a depressed immune system, and increased infant mortality; this nutritional disorder especially afflicts children in developing countries (Underwood 2004). Provitamin A, in the form of β-carotene, occurs at high levels in tomatoes (Table 1-2). Although levels of this nutrient are lower than in carrots or spinach, tomatoes

Table 1-2 Nutritional composition of raw and processed tomatoes (adapted from Ensminger 1995).

		Raw Tomato (100 g)	Canned Tomato Paste (100 g)
	Calories (kcal)	22	82
	Proteins (g)	1	3.4
	Fats (g)	0.2	0.4
	Carbohydrates (g)	4.7	18.6
	Fiber (g)	0.5	0.9
Minerals	Ca (mg)	13	27
	P (mg)	27	70
	Na (mg)	3	790
	Mg (mg)	17.7	20
	K (mg)	244	888
	Fe (mg)	0.5	3.5
	Zn (mg)	0.2	0.8
	Cu (mg)	0.01	0.59
Vitamins	Vitamin A (IU)	900	3300
	Vitamin D (IU)	0	-
	Vitamin E (mg)	0.4	-
	Vitamin C (mg)	23	49
	Thiamin (mg)	0.06	0.2
	Riboflavin (mg)	0.04	0.12
	Niacin (mg)	0.7	3.1
	Pantothenic Acid (mg)	0.33	0.44
	Vitamin B-6 (mg)	0.1	0.38
	Folic Acid (mg)	39	-
	Biotin (mg)	4	-

constitute one of the most important sources of this vitamin A, due to the amounts consumed (Beecher 1998). Gamma carotene also has vitamin A activity and is found at high levels in tomatoes (Beecher 1998).

The red color of tomatoes is due to lycopene, a carotenoid compound that does not have retinoid activity, but has oxygen-radical quenching capacity. Tomatoes constitute the main source of lycopene to diet in the US (Beecher 1998; Laquatra et al. 2005), and have been subject of much research for health benefits related to cancer, cardiovascular disease, and chronic diseases. Consumption of tomato and tomato products has been shown to make DNA more resistant to oxidative damage that can lead to cancer (Ellinger et al. 2006, 2007); however, further studies suggest that this effect does not come from lycopene alone, but that additive or synergistic effects from other tomato compounds are involved (Giovannucci 1999; Laquatra et al. 2005; Ellinger et al. 2006). Likewise, conclusive evidence that the antioxidant properties of tomato lead to disease prevention has been elusive. Several studies in the 1990s found tomato consumption associated with reduced prostrate cancer and heart disease (Giovannucci 1999; Laquatra et al. 2005), but most studies have been small and inconclusive (Ellinger et al. 2007). While the antioxidant properties of tomato compounds are not in dispute, recent reviews suggest that it is unlikely that eating tomatoes has any effect on the risk for pancreatic, prostate, gastric or ovarian cancer (Kavanaugh et al. 2007). However, circumstantial evidence is strong enough that the putative health benefits of tomato consumption are considered as promising area of research (Giovannucci 2007).

Regardless of the exact health benefits of the various phytonutrients within tomato, fruit composition is known to vary widely depending on the cultivated variety and growth conditions. Growth substrate has been shown to have an effect on nine of 18 major and trace elements measurable in tomato (Gundersen et al. 2001). Light has an effect on sugar and vitamin concentrations (Davies and Hobson 1981), and lycopene synthesis is inhibited at high and low temperatures (Davies and Hobson 1981). Among tomato cultivars, total carotenoids can vary four-fold, establishing a great range in variation in antioxidant ability (Leonardi et al. 2000).

1.3 Academic Importance of Tomato and its Wild Relatives

1.3.1 Tomato as a Model Species

Several properties of the plant and the availability of numerous resources make tomato currently one of the most effective crop model species for basic and applied research. As a diploid with a short generation time and simple growing requirements, the value of tomato in scientific research has been recognized for more than a hundred years. Genetic mapping of morphological traits in tomato began early in the 20th century (Jones 1917). Furthermore, tomato's importance in agriculture has led to breeding efforts that have yielded numerous spontaneous and induced mutations, which provide a valuable resource for genetic research (Giovannoni 2004).

This long historical interest in tomato has made it one of the most resource-rich systems for biological research (see Section 1.6). Among the numerous tools developed for tomato are: transformation techniques (McCormick et al. 1986), various mapping populations, molecular markers, a physical map, microarrays, and expressed sequence tag (EST) collections. Furthermore, concerted efforts have been made to make these resources publicly available via stock centers and the development of databases and websites (e.g., Mueller et al. 2005a). With the advent of the genomic era, tomato has been chosen as the representative genome of the crop-rich Solanaceae family, and international genome sequencing efforts are nearly complete (Mueller et al. 2005b). The high level of synteny between tomato and other solanaceous crops, such as potato, pepper, and eggplant, suggests that many resources developed for tomato will be portable to these other species, increasing the value of the system (e.g., Livingstone et al. 1999). Highlighted below are some areas of research to which tomato has made important contributions.

1.3.1.1 The Genetic Basis of Quantitative Traits

Tomato has been an especially valuable species in understanding the molecular basis of quantitative variation. The initial molecular linkage maps were created for tomato in the 1980s (Tanksley et al. 1981; Paterson et al. 1988), and by the 1990s tomato was one of first plants with high-density molecular map (Bernatzky and Tanksley 1986; Tanksley et al. 1992). This led to some of the first efforts in plants using molecular markers for mapping quantitative variation (Tanksley and Fulton 2007). The early development of mapping resources has led to other "firsts" for tomato. Tomato was one of the first species in which successful map-based positional cloning of quantitative trait loci (QTLs) was carried out (Frary et al. 2000; Fridman et al. 2000). Tomato is also among the first crops for which marker-assisted selection techniques were used to improve cultivars. Marker development has continued for tomato (Frary et al. 2005), facilitating the continued use of available mapping populations.

1.3.1.2 Fruit Development

Tomato has also emerged as one of the most important model systems for fleshy fruit development and morphological variation (Giovannoni 2004). In particular, tomato serves as a model for climacteric fruit (those for which a burst of respiration and ethylene production accompany ripening). Numerous lines with mutations affecting fruit development are publicly available at various institutions (e.g., C.M. Rick Tomato Genetic Resource Center [http://tgrc.ucdavis.edu/]; Hebrew University [http://zamir.sgn.cornell.edu/mutants/]) (Giovannoni 2004), and there has been much research focus on mutations associated with physiological and biochemical changes during ripening (Vrebalov et al. 2002). Other aspects of fruit development and genetics that have continuously been well studied include fruit weight (Grandillo et al. 1999; Frary et al. 2000), and fruit shape (Grandillo et al. 1999; Tanksley 2004), traits that vary within domesticated tomato. The genetic resources available for tomato have also made it an important model system for domestication of fruit-bearing crops. Studies of domestication loci in tomato have been further extended to search for conservation among domestication loci across crop plants in the Solanaceae, such as eggplant and pepper (Thorup et al. 2000; Doganlar et al. 2002; Frary et al. 2003; van der Knaap and Tanksley 2003; Paran and van der Knaap 2007).

1.3.2 The Tomato Clade as a Model System

Cultivated tomato occurs within a clade of ~13 closely related species (see Section 1.3), which have been instrumental in the development of the crop as a model system. Early tomato studies that revealed extremely low levels of genetic diversity within cultivated tomato were accompanied by the realization that this would be a constraint for mapping efforts. Wild tomatoes, on the other hand, are a rich source of genetic and morphological diversity, and most species can be easily crossed with domesticated tomato. This has led to the development of immortal mapping populations such as introgression lines (IL) and recombinant inbred lines (RIL), stemming from interspecific crosses (Eshed and Zamir 1995; Monforte and Tanksley 2000; Villalta et al. 2005). These mapping populations have been instrumental for the map-based cloning approaches described previously. The introgression line approach (ILs are lines carrying a small portion of a foreign genome in an isogenic background), in particular, was first demonstrated in tomato, providing a way to avoid difficulties associated with gene interactions in identifying loci (see Lippman et al. 2007). Given the role of wild tomato species in the generation of tomato mapping resources, and the prevalence of wild tomato species in crop improvement efforts, it is perhaps more accurate to think of the tomato clade as a valuable model system, rather than cultivated tomato per se.

1.3.2.1 Plant Defense

The tomato clade has been instrumental in dissection of molecular mechanisms of plant-pathogen interactions (e.g., Rathjen et al. 1999; Luderer et al. 2001; Kruger et al. 2002; Rooney et al. 2005; Gabriels et al. 2007). One of the first plant disease resistance (R) genes isolated was cloned in tomato, and corresponds to an allele introgressed from a wild species (Martin et al. 1993); many subsequently isolated R-genes correspond to wild alleles (Jones et al. 1994; Dixon et al. 1996; Thomas et al. 1997; Vos et al. 1998). To date, tomato and its relatives represent one of limited number of plant groups in which *R*-gene evolution has been explored at the population and phylogenetic levels (Riely and Martin 2001; Van Der Hoorn et al. 2001; Caicedo and Schaal 2004a; Kruijt et al. 2005; Rose et al. 2005; Rose et al. 2007; Caicedo 2008).

1.3.2.2 Mating System and Speciation

A variety of mating system occurs within the tomato clade, ranging from self-incompatibility, to facultative outcrossers, to high levels of autogamy (inbreeding) (see Sections 1.3 and 1.4), the group has also been instrumental in research of self-incompatibility and mating system evolution. There has been extensive research on the genetic mechanisms of self-incompatibility in the clade (e.g., Kowyama et al. 1994; Bernatzky et al. 1995; Kondo et al. 2002a), which are complemented by molecular evolutionary analyses of the loci involved (e.g., Richman et al. 1996; Igic et al. 2007). Other research efforts have centered on QTL mapping and, in some cases, isolation of loci contributing to floral traits associated with the change from an allogamous to autogamous mating system (e.g., Bernacchi and Tanksley 1997; Georgiady and Lord 2002; Chen and Tanksley 2004; Chen et al. 2007). Lastly, the creation of mapping lines from interspecific crosses in the tomato group has provided ideal material with which to examine the genetic basis of species isolating barriers and the process of speciation (Moyle and Graham 2005; Moyle 2007).

1.3.2.3 Adaptation and Divergence

The extension of tomato resources to the entire clade has made this one of the best model systems for studies of adaptation and divergence. Given the variety of habitats occupied by wild species of the tomato clade (see Section 1.3), it is not surprising that there is evidence for much morphological and physiological variation among species (Taylor 1986). Many of the traits that differ among species may be adaptations to specific environmental conditions, and are important resources for crop improvement. Thus, there has been much focus in dissecting the molecular basis of potentially useful wild traits such as drought tolerance, salt tolerance, chilling tolerance, and disease and herbivore resistance (e.g., Monforte et al. 1997; Foolad and Lin 2001; Frankel et al. 2003). The variety of mating systems among tomato species and differences in levels of genetic diversity, have further made the clade into rich system in for population studies exploring relationships between genetic diversity, mating system, and recombination, and the demographic processes affecting evolution and diversification of populations (e.g., Baudry et al. 2001; Roselius et al. 2005; Arunyawat et al. 2007).

1.4 Taxonomy and Evolution

1.4.1 Taxonomic Position and Phylogenetic Relationships

Tomatoes were introduced to Europe during the early part of 16th century by Spaniards (Peralta and Spooner 2006). However, the first references of tomato cultivation and uses came from the Italian botanist Mattholius (1544), who described that tomatoes were eaten in the same manner as eggplant in Italy. Matthiolus (1554) also cited both yellow and red fruits, and mentioned the Italian name for the tomato "pomi d'oro" and its Latin equivalent "mala aurea" or golden apple. Early European botanists recognized the close relationship of tomatoes with the genus *Solanum*, and commonly referred to them as *S. pomiferum* (Luckwill 1943b). Tournefort (1694) was the first to name cultivated tomatoes as Lycopersicon ("wolf peach" in Greek). Linnaeus proposed a new nomenclature system and in *Species Plantarum* (Linnaeus 1753) he consistently used Latin binomials to name species, included tomatoes in the genus *Solanum* and described *S. lycopersicum* (the cultivated tomato) and *S. peruvianum*. One year later Miller (1754) followed Tournefort and formally described the genus *Lycopersicon*. Afterward and following Miller's early circumscription, tomatoes have been traditionally recognized under *Lycopersicon* by the majority of taxonomists.

At present, phylogenetic studies have unambiguously shown tomatoes to be deeply nested within *Solanum*, and their inclusion in this genus has gained wide acceptance supported by evidences from phylogenetic studies of the Solanaceae family based on molecular and morphological characters (Spooner et al. 1993; Bohs and Olmstead 1997; Olmstead and Palmer 1997; Olmstead et al. 1999; Peralta and Spooner 2001; Bohs 2005; Peralta and Spooner 2005; Spooner et al. 2005).

Recently, Peralta et al. (2008), in the monograph of wild tomatoes and their relatives, recognized 13 species of wild tomatoes including four species from the previously polymorphic *S. peruvianum sensu lato* (Peralta and Spooner 2005) and the cultivated tomato (*Solanum lycopersicum*) as part of the tomato clade (*Solanum* section *Lycopersicon*) (Tables 1-3, 1-4). This taxonomic treatment also includes four species of the closely related Sections *Lycopersicoides* and *Juglandifolia*. Two species of *Solanum* section *Juglandifolia*, *S. juglandifolia* and *S. ochranthum*, are distributed in Colombia, Ecuador, and Peru; and two species of *Solanum* section *Lycopersicoides*, *S. lycopersicoides* and *S. sitiens*, are distributed in southern Peru and northern Chile (Smith and Peralta 2002; Tables 1-3, 1-4).

From an evolutionary point of view the tomato clade is a young group that diversified recently in Western South America. The age of the genus *Solanum* is estimated at 12 million years based on nuclear and chloroplast markers (Wikström et al. 2001), while the radiation of the tomato clade has been estimated as 7 million years based on four nuclear genes (Nesbitt and Tanksley 2002). Inferring evolutionary relationships within the tomato clade has been difficult because of the group's recent diversification, the presence of shared polymorphism, and the low levels of variation in the self-compatible species. Inference of wild tomato phylogeny have been attempted using morphological (Peralta and Spooner 2005) and molecular characters such as chloroplast and mitochondrial restriction sites, nuclear restriction fragment length polymorphisms (RFLPs), isozymes, internal transcribed spacer (ITS), and granule bound starch synthase (GBSSI or waxy) sequences, simple sequence repeats (SSRs), and amplified fragment length polymorphisms (AFLPs) (Spooner et al. 2005). Although evolutionary relationships are not completely resolved in wild tomatoes, results based on sequences of the structural gene GBSSI (Peralta and Spooner 2001), AFLPs

Table 1-3 Species list for *Solanum* sect. *Lycopersicoides*, sect. *Juglandifolia* and sect. *Lycopersicon*, with equivalent names in *Lycopersicon*.

Names in *Solanum* (Peralta et al. 2008)	*Lycopersicon* equivalent
Solanum lycopersicoides Dunal	*Lycopersicon lycopersicoides* (Dunal) A. Child ex J.M.H. Shaw
Solanum sitiens I.M. Johnst.	*Lycopersicon sitiens* (I.M. Johnst.) J.M.H. Shaw
Solanum juglandifolium Dunal	*Lycopersicon juglandifolium* (Dunal) J.M.H. Shaw
Solanum ochranthum Dunal	*Lycopersicon ochranthum* (Dunal) J.M.H. Shaw
Solanum pennellii Correll	*Lycopersicon pennellii* (Correll) D'Arcy
Solanum habrochaites S. Knapp and D.M. Spooner	*Lycopersicon hirsutum* Dunal
Solanum chilense (Dunal) Reiche	*Lycopersicon chilense* Dunal
Solanum huaylasense Peralta	Part of *Lycopersicon peruvianum* (L.) Miller
Solanum peruvianum L.	*Lycopersicon peruvianum* (L.) Miller
Solanum corneliomuelleri J.F. Macbr. (1 geographic race: Misti near Arequipa)	Part of *Lycopersicon peruvianum* (L.) Miller; also known as *L. glandulosum* C.F. Müll.
Solanum arcanum Peralta (4 geographic races: 'humifusum', lomas, Marañon, Chotano-Yamaluc)	Part of *Lycopersicon peruvianum* (L.) Miller
Solanum chmielewskii (C.M. Rick, Kesicki, Fobes and M. Holle) D.M. Spooner, G.J. Anderson and R.K. Jansen	*Lycopersicon chmeilewskii* C.M. Rick, Kesicki, Fobes and M. Holle
Solanum neorickii D.M. Spooner, G.J. Anderson and R.K. Jansen	*Lycopersicon parviflorum* C.M. Rick, Kesicki, Fobes and M. Holle
Solanum pimpinellifolium L.	*Lycopersicon pimpinellifolium* (L.) Miller
Solanum lycopersicum L.	*Lycopersicon esculentum* Miller
Solanum cheesmaniae (L. Riley) Fosberg	*Lycopersicon cheesmaniae* L. Riley
Solanum galapagense S.C. Darwin and Peralta	Part of *Lycopersicon cheesmaniae* L. Riley

Table 1-4 Taxonomic treatment of wild tomatoes and their relatives.

Taxonomic treatment (Peralta et al. 2008)		
Section *Lycopersicon*		
	Lycopersicon group	
		S. lycopersicum
		S. pimpinellifollium
		S. cheesmaniae
		S. galapagense
	Neolycopersicon group	
		S. pennellii
	Eriopersicon group	
		S. habrochaites
		S. huaylasense
		S. corneliomulleri
		S. peruvianum
		S. chilense
	Arcanum group	
		S. arcanum
		S. chmielewskii
		S. neorickii
Section *Lycopersicoides*		
		S. lycopersicoides
		S. sitiens
Section *Juglandifolia*		
		S. juglandifolium
		S. ochranthum

(Spooner et al. 2005), morphology (Peralta and Spooner 2005), and more recently with multiple single-copy genes (Rodriguez et al. 2005) support the interpretation of our current understanding of relationships within the group (Fig. 1-1).

Wild tomatoes (section *Lycopersicon*) and their closest relatives (sections *Lycopersicoides* and *Juglandifolia*) are clearly monophyletic and sister to the potatoes, section *Etuberosum* (non tuberous species) is clearly monophyletic and sister to potatoes + tomatoes. Section *Lycopersicoides* and section *Juglandifolia* are clearly monophyletic and sister to section *Lycopersicon*. Within this section, *S. pennellii* in most of the results appears at the base of the trees as a polytomy with *S. habrochaites*, or sometimes forms a clade with this species. Although this relationship is unresolved, the lack of the sterile anther appendage character shared with the outgroups suggest that *S. pennellii* is sister to the rest of the tomatoes (sect. *Lycopersicon*). *Solanum chilense, S. corneliomulleri, S. huaylasense, S. peruvianum, S. habrochaites* and

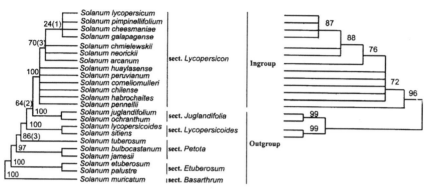

Figure 1-1 Comparison of phylogenetic results obtained with molecular and morphological characters. On the left the abstracted cladistic tree from the analysis of GBSSI gene sequences of the 65 accessions of the 13 tomato species (ingroup sect. Lycopersicon) and ten outgroup taxa (outgroups sect. Juglandifolia, sect. Lycopersicoides, sect. Petota, sect. Etuberosumand sect. Basarthrum) (Peralta and Spooner 2001). Numbers indicate bootstrap values, and decay values are indicated between parentheses. On the right, the strict consensus tree of the 28 most parsimonious trees based on 26 morphological characters (Peralta and Spooner 2005). Numbers indicate bootstrap values.

S. pennellii appear as a polytomy at the base of the GBSSI nuclear gene tree, and the first four species clustered with morphological characters, but the relationships among them are still unresolved. *Solanum chmielewskii* and *S. neorickii* are closely related to *S. arcanum*, comprising a monophyletic group supported in almost all data sets, and sister to the brightly carotenoid colored fruit tomatoes. *Solanum cheesmaniae, S. galapagense, S. lycopersicum, S. pimpinellifolium*, species with typical red, orange or yellow fruits, unambiguously constitute a closely related and recently evolved monophyletic group. Based on an integrative approach that considers morphological and molecular studies, species distribution, mating systems and phylogeny, a taxonomic treatment and a new classification of tomatoes and their relatives have been recently published (Peralta et al. 2008) and are summarized in Table 1-4.

The evolution of wild tomatoes can be has been likely influenced by the need for adaptation to changing environments, as well as the interaction with biotic factors such as pollinator availability (Peralta and Spooner 2005). A hypothetical scenario could interpret that putative self-incompatible ancestral populations probably occupied a wide area of distribution in central Peru. The environmental pressure due to drastic climatic changes during the Holocene and the formation of the Peruvian–Chilean deserts could trigger the process of adaptation to the new habitats leading to differentiation and speciation. Some populations, adapted to more mesic and humid conditions, colonized the northern areas, while populations more adapted to arid habitats may have differentiated and migrated along the Pacific coast and colonized the adjacent western slopes, but limited in their altitudinal distribution by frost. Putative ancestors from the Rio Marañón area may have originated self-compatible taxa (Rick 1986), while ancestral populations along the equatorial coast might have migrated and diversified more recently to the Galápagos Islands (Darwin et al. 2003). Analogously, putative ancestors of *S. juglandifolium* and *S. ochranthum* became adapted to more wet and warm environments, while ancestors of *S. sitiens* and *S. lycopersicoides* evolved in the deserts. Reproductive isolating barriers, such as self-incompatibility arose to prevent gene flow between sympatric populations. The shift from self-incompatibility toward self-compatibility and reduction of floral structures also played an important role in the speciation process; further accentuating isolation and probably genetic drift (see Section 1.4).

1.5 Botanical Descriptions

1.5.1 Morphology

Wild tomatoes are herbaceous plants that can develop secondary growth at the base of the stems and the principal root. Wild tomatoes most probably behave as annuals in their natural environment in the Andean mountains and deserts, because frost or drought kills the plants after the first growing season (Müller 1940a). When the climatic conditions are favorable, wild tomatoes can behave as biennials and perennials depending on the plant capacity for developing secondary growth in basal stems and roots. In the next season several new shoots arise from buds on the woody base or crown. The plant's life span is related to the environmental conditions of every season and its ability to accumulate reserves in the main root and crown. The shoots are initially erect, but later, due to the weight of the branches, the plants become decumbent or prostrate and in some cases can develop adventitious roots from basal nodes. In the cultivated species it is a regular practice to mount up soil at the base of the plant to facilitate the formation of adventitious roots. A well-developed radical system will help the plant anchorage and assure the vegetative and reproductive growth.

Wild tomatoes have an indeterminate growth and the main axis of the plant is a sympodium, typical of the genus *Solanum*, formed by a succession of lateral axes with alternate leaves arranged in a 1/3 phyllotaxic spiral in some species or in 1/2 leaf phyllotaxis in others, and inflorescences are terminal at the end of each sympodial unit (Luckwill 1943b; Danert 1958). Phyllotaxis patterns have taxonomic value to differentiate wild tomatoes (Peralta et al. 2008).

All species are spreading; some are more robust and can produce long branches, to 3–4 m in *S. lycopersicum*, *S. pimpinellifolium*, *S. cheesmaniae*, and *S. galapagense*, and to 6 m in vine forms of *S. habrochaites* (Peralta et al. 2008). In the cultivated species, *S. lycopersicum*, there are cultivars with semi-determinate or determinate growth habit, with short branches and more compact development.

The leaves of tomatoes have usually been considered as pinnate, although the presence of a small leaf tissue connecting the blade dissections along the principal rachis could be interpreted as the leaf being simple and highly dissected or deeply pinnatifid. The degree of leaf dissection is greatly variable among wild tomatoes and is sometimes useful to differentiate species (Peralta et al. 2008). The first leaf of a plantlet is often simple, entire to lobed, or compound with only 1 or 2 leaflets; consecutive leaves are more dissected and present a gradual increase in complexity. Normally the leaves of the 10 to 12 nodes of the main stem axis develop a mature size and a typical dissection pattern of the blade (Luckwill 1943b). Current research in leaf development has characterized the leaf of tomatoes as pinnate and has tried to elucidate the molecular genetic aspects of leaf dissection (Kessler et al. 2001; Bharathan et al. 2002; Gleissberg 2002; Tsiantis et al. 2002; Piazza et al. 2005). Recently, natural variation in patterns of leaf dissection in two endemic species of the Galápagos Islands, *Solanum cheesmaniae* and *S. galapagense*, have been attributed to a single-nucleotide deletion in the promoter of the Petroselinum (*Pts*) gene, which encodes a novel knotted1-like homeobox (*Knox*) protein (Kimura et al. 2008).

Leaves of most tomato species are imparipinnate (with 2–6 [7] opposite, subopposite, or alternate primary lateral leaflets pairs and a terminal leaflet), or, in some species, leaves are interrupted imparipinnate with pairs of primary leaflets alternatively smaller and larger along the main rachis. The terminal leaflet can be of equal size or larger than primary lateral leaflets. The leaves can also develop small leaflets or interjected leaflets usually under half the length of the primary leaflets (Peralta et al. 2008). In the botanical literature the interjected leaflets are also called interstitial or intermediate leaflets. Smaller and sessile interjected leaflets can also develop in the main leaf rachis, between the primary leaflets and interjected leaflets. In some species the primary lateral and terminal leaflets are further divided, to produce secondary to tertiary lateral leaflets. In the deeply dissected leaf of *S. galapagense*, the tertiary leaflets are further lobed to form quaternary lobes (Darwin et al. 2003). Leaflets are quite variable in size and shape from narrowly elliptic, elliptic to broadly elliptic, ovate or orbicular. Leaflet base is usually asymmetric, truncate or rounded to cordate; and sometimes decurrent basiscopically. Leaflet apex is rounded, acute, or acuminate. The margins are straight or more commonly undulate, entire to regularly or irregularly crenate, serrate to dentate, to deeply lobed, the lobes are usually deeper at the leaflet base. The primary leaflets can be sessile to petiolulate; the interjected, secondary and tertiary leaflets are usually sessile to subsessile.

Tomatoes have stipule-like structures that are not attached to the stem or petiole and are interpret as the first leaf pair or prophylls of axillary shoots. These structures are traditionally described as pseudostipules (Child and Lester 1991) and appear to be morphologically identical to those in potato (Spooner et al. 2004). Pseudostipules are entirely absent in four tomato species: *S. cheesmaniae, S. galapagense, S. lycopersicum, S. pimpinellifolium.*

Wild tomatoes present a variety of trichome types and density patterns useful taxonomically. Trichome length range between 10–2,500 µm and four types of glandular trichomes and also four types of non-glandular trichomes have been described in wild tomatoes (Peralta et al. 2008). A dense distribution of short and bent tapering trichomes produce the typical canescent surface of *S. chilense* and *S. peruvianum*, while the large trichomes produce the villous pubescence of different organs (stem, leaves, peduncles, pedicels, sepals, petals, and fruits) and are often combined with other glandular and non-glandular types (*S. lycopersicum, S. habrochaites, S. corneliomulleri*). The color of the plant is the result of the type, combination, and density of trichomes, and varies from bright green in subglabrous plants (*S. arcanum* and *S. huaylasense*) to grayish in canescent plants. Glandular trichomes accumulate essential oils that produce the characteristic smell of tomato leaves that varies considerably among species (Darwin et al. 2003).

The basic inflorescence in wild tomatoes is a cyme with different branching patterns, unbranched monochasia, dichasia or 2–3 times dichotomously branched axis. The number of flower per inflorescence axis varies from 4 to 14. Inflorescences are usually bracteate, some times poorly developed and absent in *S. cheesmaniae, S. galapagense, S. lycopersicum, S. pimpinellifolium*. Wild tomatoes have a typical flower pedicel articulation above the middle or in the distal half, except in *S. pennellii*, where the pedicel is articulate at the base. Flowers are yellow, symmetrical or slightly zygomorphic, stellate to rotate, gamosepalus at the base, gamopetalus. The self-compatible species typically have smaller flowers than the self-incompatible species. Tomato anthers are laterally connivent and form a tube, they have a sterile apical appendage and dehisce by introrse longitudinal slits that first appear as small pores and then develop basipetally, except in *S. pennellii* where the anthers do not have a sterile apical appendages and dehisce by apical pores and later by introrse slits opening just part way down the anther. The anthers are usually of equal length and straight, but in some species (*S. corneliomulleri, S. huaylasense, S. peruvianum*, and *S. pennellii*) they are slightly to strongly curved, because the three upper anthers develop more than the two lower anthers.

The gynoecium is typically bicarpellate (multicarpellate in many cultivars of *S. lycopersicum*), the ovary is superior and globose. Fruits of all wild species are mainly globose to globose-depressed bilocular berries, in the cultivated species fruits are bilocular or multilocular and a wide diversity of shapes and sizes has been developed in different cultivars. The fruit color is produced by a combination of pigments in the epicarp and subepidermical tissues. Four species, *S. cheesmaniae*, *S. galapagense*, *S. lycopersicum*, *S. pimpinellifolium*, members of the "Lycopersicon" species group have carotenoid pigments (red, orange, yellow) uniformly distributed throughout the berry. *Solanum lycopersicum* and *S. pimpinellifolium* have glabrous and typical bright red fruit color produced by the accumulation of lycopene at maturity, while *S. cheesmaniae*, *S. galapagense* have yellow to orange fruits. All other species have green fruits due to the presence of chlorophyll in the pericarp, sometimes with purplish coloration by accumulation of anthocyanin pigments and a green, dark green or purple stripe along the union of the carpels, and are usually pubescent at maturity. Tomato seeds are oval, obovate, or orbicular and flattened laterally with "pseudohairs" due to the development of radial wall thickenings of the cells of the epidermis in mature seeds (Lester 1991; Lester and Durrands 1984) that produce a hairy and silky appearance to the seed surface.

1.5.2 *Ploidy Levels, Genome Size and Karyotype*

All wild tomato species are diploid ($2n = 2x = 24$) and can be crossed to the cultivated tomato, but sometimes with difficulty and using special techniques (Rick 1979). Great advances in tomato genetics have been achievable because of the understanding of mating systems and the possibility of controlled hybridization within and among species, the naturally occurring variability in the species, the occurrence of self-pollination that leads to the expression of recessive mutations, the lack of gene duplication, and the possibility to easily identify the 12 chromosomes (Rick 1978). New methodological approaches like molecular mapping of important agronomical characters and the development of advanced-backcross and introgression lines have provided powerful tools for the improvement of the tomato crop and to understand the processes of domestication (Labate et al. 2007).

The cultivated tomato genome has been selected as a model for the Solanaceae family due to its simple diploid genetics, small genome size, short reproduction period, adjusted transformation methodologies, and the availability of a great diversity of genetic resources within the cultivated species and in wild related species (http://www.sgn.cornell.edu/solanaceae-project/). The tomato genome size (1C amount) is generally considered as approximately 95 pg of DNA (Michaelson et al. 1991). Cultivated tomato has been chosen for an international sequencing project due to its small genome size compared with other species of the Solanaceae, and in addition with the availability of high-density molecular map (Tanksley et al. 1992; Broun and Tanksley 1996) and an advanced BAC (bacterial artificial chromosome) based physical map (http://www. sgn.cornell.edu/solanaceae-project/). It has been estimated that one-quarter of the tomato genome consists of euchromatin, which contains the majority of protein-coding genes (ca. 220 Mb), and three-quarters consist of heterochromatin, which is rich in repetitive sequences and poor in genes. Tomato sequencing efforts have been focused on the euchromatin portion of the genome and are nearly complete (http://www.sgn.cornell.edu/ solanaceae-project/). It has been predicted that tomato has ~35,000 genes occurring largely in euchromatic regions, with an average gene density of 9.8 kb/gene (Van Der Hoeven et al. 2002).

1.6 Ecology and Biodiversity of the Tomato Clade

1.6.1 Distribution and Habitats

Wild tomato species are entirely American in distribution, growing in western South America from Ecuador to northern Bolivia and Chile and with two endemic species in the Galápagos Islands (Darwin et al. 2003; Peralta et al. 2008). Weedy escaped forms related to the cultivated species (*Solanum lycopersicum*) are widespread throughout warm regions of the world.

Wild tomatoes grow in a variety of habitats, from near sea level to over 3,600 m in elevation (Peralta et al. 2008). In western South America some species are distributed along the arid Pacific coastal lowlands and adjacent hills or "lomas formations" where highly endemic plant communities grow as a consequence of the "camanchaca" or fog arising from the ocean that provides enough moisture for the development of vegetation. The Pacific coast of South America is also affected by periodic and recurring climatic phenomena El Niño Southern Oscillation Events (ENSO) that produce short periods of heavy rainfall and high temperatures, and as a consequence mass flowering of plant communities. During El Niño events Sifres et al. (2006) documented that floods facilitated seed dispersal of *S. pimpinellifolium* and reported the increase of population extent in northern Peru. El Niño events occur every 30–50 years and minor events approximately 3.5–7 years (Quinn and Neal 1992; Allan et al. 1996).

Six wild tomato species, *S. chilense, S. habrochaites, S. pennellii, S. peruvianum*, some populations of *S. arcanum*, and *S. pimpinellifolium* grow in the lomas habitats along the Pacific coastal range (Dillon 2005; Peralta et al. 2008). Other species developed in valleys, formed by rivers draining into the Pacific, and uplands in the high Andes. *Solanum arcanum* is mainly found in the Valley of the Río Marañón, *S. huaylasense* in the Valley of the

Río Santa, *S. corneliomulleri* in valleys from central (near Lima) to southern Peru, *S. chmielewskii* in the upper Apurímac Valley of Peru and the Sorata Valley of Bolivia, and *S. neorickii* in dry valleys distributed from Ecuador to southern Peru. *S. chilense, S. corneliomulleri, S. pennellii* and *S. habrochaites* are also found in high altitudes in Andean environments, and the last one in open areas of cloud forests (Peralta et al. 2008). In the Galápagos Islands, *S. cheesmaniae* grows from sea level to rocky places in the volcanoes, while *S. galapagense* occurs most commonly in lower elevation habitats close to the ocean spray, though some populations are also found on rocky volcano slopes (Darwin et al. 2003; Peralta et al. 2008).

Our current knowledge of wild tomatoes distribution patterns and awareness of habitat destruction resulting from human activities point out the urgent necessity of assure the conservation of this valuable genetic resources in their natural habitats, especially endemic species or species with small population or narrow geographic distribution. It is critical to determine the current vulnerability of the species (Peralta et al. 2008).

1.6.2 Mating System Evolution in the Tomato Clade

1.6.2.1 Mating System Diversity

The occurrence of several closely related species with a wide range of mating systems has made wild tomatoes a model system for the study of mating system effects on genetic diversity (Robertson and Labate 2007). Wild tomatoes vary both between and within species in the presence or absence of self-incompatibility mechanisms, and in the degree of allogamy (outcrossing) or autogamy (inbreeding). Although the specific selective pressures giving rise to diverse mating systems throughout the tomato clade are unknown, shifts in mating system are common in plant evolution (Stebbins 1950) and are believed to be driven by mate availability. These theories have been formalized as the reproductive assurance model, which states that it is advantageous to self-fertilize when mates are scarce, even in the face of inbreeding depression (Darwin 1876), and Baker's rule, which observes that colonizing species that disperse over long distances are generally self-compatible (Baker 1955).

Self-incompatibility (SI) is ancestral in the tomato clade (Miller and Tanksley 1990), but has been lost multiple times during tomato evolution. SI and was lost once in the clade containing colored-fruited tomatoes and *S. neorickii* and *S. chmielewskii* (Rick et al. 1976; Taylor 1986; Miller and Tanksley 1990; Marshall et al. 2001), and independently in some populations of *S. pennellii* (Rick and Tanksley 1981) and *S. habrochaites* (Rick et al. 1979). Self-compatible (SC) tomato species vary in their degree of allogamy, ranging from almost complete selfing (e.g., *S. cheesmaniae* and *S. lycopersicum*), to facultative outcrossing (e.g., *S. chmielewskii* and *S. pimpinellifolium)* (Taylor 1986). Variation in outcrossing levels can occur between populations within species (e.g., *S. pimpinellifolium*) (Rick et al. 1977), or between closely related species. *S. chmielewskii* and *S. neorickii*, for example, grow sympatrically, but the first is a facultative outcrosser, while the second is autogamous (Rick et al. 1976), leading to speculations that *S. neorickii* may have evolved from *S. chmielewskii* ancestors by acquiring autogamous reproduction (Rick et al. 1976; Rick 1984). In SC species, levels of outcrossing are reportedly correlated with various morphological features, including larger flowers and stigma exsertion, which aid in cross-pollination (Rick et al. 1976; Rick et al. 1977; Rick 1984).

1.6.2.2 The Genetic Basis of Mating System Diversity

In recent years, some headway has been made in dissecting of the genetic mechanisms underlying mating system evolution in the tomato clade. The genus *Solanum* has a gametophytic type of self-incompatibility, with rejection of self-pollen occurring in styles once pollen tube elongation has initiated (Franklin-Tong and Franklin 2003). The SI mechanism is controlled by a single locus (*S*-locus), which produces style-specific ribonucleases (S-RNases) (McClure et al. 1989) and is localized on chromosome 1 in *S. peruvianum* (Tanksley and Loaiza-Figueroa 1985). A second gene, *HT*, which is unlinked to the *S*-locus and produces an Asn-rich protein of unknown function, is also necessary for the SI response (McClure et al. 1999). SC accessions from various tomato species have been shown to have lower levels of stylar RNase activity and the gene seems absent in SC colored-fruited species (Kondo et al. 2002a; Kondo et al. 2002b). In green-fruited SC species low S-RNase activity is apparently due to various frame-shift mutations or modifications in the promoter region of the gene (Kondo et al. 2002a; Kondo et al. 2002b). However, it seems that mutations affecting the expression of *HT* may have been responsible for loss of SI in the clade, rather than mutations at the *S*-locus. *HT* expression has not been detected in styles of any SC species examined to date (Kondo et al. 2002a).

Dissection of the genetic basis of SC evolution has focused on morphological traits that facilitate cross-pollination. Among autogamous and allogamous *S. pimpinellifolium* populations, stigma exsertion can range from 0 to 3 mm (Rick et al. 1977), and the degree of stigma exsertion differentiates allogamous *S. chmielewskii* from autogamous *S. neorickii*. Similarly, stigma exsertion, corolla size, and anther tube length differentiate the SI from the SC tomato species (Rick and Lamm 1955; Rick 1982a). Recent fine mapping studies using *S. pennellii* introgression lines have identified mutations in the promoter region of *Style2.1*, a gene encoding a transcription factor regulating cell elongation, that contribute to interspecific variation in style length; the gene is known to be linked to other loci influencing stamen length and anther dehiscence (Bernacchi and Tanksley 1997; Fulton et al 1997a; Chen and Tanksley 2004; Chen et al. 2007). A separate study involving a cross between allogamous and autogamous *S. pimpinellifolium* accessions has found evidence for a QTL of large effect on a separate chromosome, as well three QTLs affecting aspects of anther length on different chromosomes (Georgiady et al. 2002). Because floral characters are under likely multigenic control, and autogamy and SC have evolved several times within the tomato clade, it is likely that different loci underlie changes in floral traits across groups.

1.6.3 Genetic Diversity in Wild Tomatoes

Because of the potential of wild tomato species for tomato breeding, there has been some effort to characterize levels of genetic diversity occurring throughout the tomato clade. Levels of genetic variation can vary among species and within species among populations. Variation in levels of genetic diversity can be due to mating system, historical events, or selection for local environmental conditions, and there is interest in determining which of these factors have played a greater role in tomato evolution (e.g., Arunyawat et al. 2007; Stadler et al. 2008). Further interest in tomato genetic diversity involves the utility of species for classical and association mapping studies (e.g., Arunyawat et al. 2007), and identification of species and populations likely to harbor traits for crop improvement.

1.6.3.1 Genetic Diversity and Mating System

Both self-incompatibility and the degree of allogamy can have an impact on levels of polymorphism within species due to their effect on species' colonization potential, population structure, and effective population size (Charlesworth 2003). Various molecular markers have documented lower levels of genetic variation in SC wild tomato species (e.g., Miller and Tanksley 1990; Breto et al. 1993; Egashira et al. 2000; Alvarez et al. 2001; Baudry et al. 2001; Roselius et al. 2005). However, the magnitude of difference between levels of polymorphism among species varies according to the marker and study sample size.

In an RFLP study, diversity, measured as average genetic distance among accessions, was ten-fold higher in *S. peruvianum* and five-fold greater in *S. habrochaites* and *S. pennellii* compared to all of the SC species examined (Miller and Tanksley 1990). However, SSR and allozyme studies have found lower levels of variation in *S. habrochaites* than in *S. pimpinellifolium* (Rick 1984; Alvarez et al. 2001); interpretation of these results is complicated by varying sample sizes and the existence of SC *S. habrochaites* accessions. DNA sequence diversity has been measured within single populations of three SI (*S. habrochaites*, *S. chilense*, *S. peruvianum*) and two SC (*S. pimpinellifolium*, *S. chmielewskii*) species (Baudry et al. 2001; Roselius et al. 2005) (Table 1-5). Average levels of silent nucleotide diversity across 14 loci, measured as θ (Watterson 1975), ranged from 0.58% (*S. habrochaites*) to 2.23% (*S. peruvianum*) in SI species and 0.03% (*S. chmielewskii*) to 0.16% (*S. pimpinellifolium*) in SC species; this translates into average levels of variation that are 3.6 to 83-fold higher in SI than SC (Roselius et al. 2005). It is important to note that these are population-specific rather than species-wide estimates, but they do give an idea of the variation in polymorphism levels across the group. It has been suggested that these differences in polymorphism levels are too great to be explained by an increase in inbreeding alone, and other demographic forces, such as population substructure, differences in soil bank dynamics, or greater extinction rates in SC taxa, must be invoked to account for their reduction in diversity (Baudry et al. 2001; Roselius et al. 2005).

Table 1-5 SNP variation in tomato, wild relatives, and other model species.

species	nucleotide diversity (π or θ)[a]	# of samples	# of loci		Reference
S. lycopersicum (cultivated tomato)	0–0.0019[b]	4	4	SC	Nesbitt and Tanksley 2002
S. lycopersicum var. *cerasiforme*	0.0018–0.0054	4	10–39	SC	Nesbitt and Tanksley 2002
S. pimpinellifolium	0.0016	5	14	SC	Roselius et al. 2005
	0.0064	128	1		Caicedo and Schaal 2004b
S. chmielewskii	0.0003	5	14	SC	Roselius et al. 2005
S. habrochaites	0.0058	3–4	14	SI	Roselius et al. 2005
S. chilense	0.0064	5	14	SI	Roselius et al. 2005
S. peruvianum	0.0223	5	14	SI	Roselius et al. 2005

[a]single values are averages of surveyed loci
[b]only a single surveyed locus contained variation

1.6.3.2 Genetic Diversity and Geography in SI Species

Besides varying across species, levels of genetic diversity can also vary greatly among populations within species. Geographical patterns of intraspecific variation have been explored for three tomato SI species: *S. pennelli*, *S. habrochaites*, and *S. peruvianum* (Rick 1963; Rick et al. 1979; Rick and Tanksley 1981). Interpretations of previous studies of variation in *S. peruvianum* are complicated by the recent splitting of the group into four species (Peralta et al. 2005). Morphologically and genetically, the northernmost accessions (now considered part of the species, *S. arcanum*) are the least variable within populations, but the greatest between populations (Rick 1963; Egashira et al. 2000).

Solanum pennelli and *S. habrochaites* are notable among SI taxa in containing both SI and SC populations (Rick et al. 1979; Rick and Tanksley 1981). *Solanum habrochaites* occurs in central Ecuador to central Peru, with Peruvian populations likely to be SI, and Ecuadorean and southernmost accessions tending to be SC (Rick et al. 1979). Isozyme studies document low levels, and often complete lack of, polymorphism in northern and southern populations, while the percentage of polymorphic loci can reach 50 and 70% in populations from northern Peru (Rick et al. 1979). Levels of population diversity are thus associated with mating system, and also with morphological trends, as SC *S. habrochaites* populations tend to have smaller corollas, thinner stems, and more basal branching (Rick et al. 1979). The geographic pattern of variation suggests that SC populations arose as the species colonized areas north and south of its center of origin.

Solanum pennellii has been studied in the Peruvian portion of its range, and the mean number of allozymes per locus tends to be higher in the north central region of its distribution and lowest in the northern and southern edges (Rick and Tanksley 1981). In this case, SC accessions tend to occur in the south, are rarer than in *S. habrochaites,* and can still contain more than 50% of the variation observed in central populations (Rick and Tanksley 1981). Because genetic diversity and morphological trends in *S. pennellii* tend to be less pronounced than in *S. habrochaites,* it has been suggested that *S. pennelli* colonized its geographic range more slowly (Rick and Tanksley 1981; Rick 1984), perhaps allowing for more gene flow between older and newly established populations.

1.6.3.3 Genetic Diversity and Geography in SC Species

Geographic patterns of variation have been explored in detail in two SC tomato groups: *S. pimpinellifolium* and the tomato species inhabiting the Galapagos Islands. In *S. pimpinellifolium,* early allozyme studies established that levels of diversity within the species were highest in the center of its distribution in northern Peru, and declined north towards Ecuador and south towards Southern Peru (Rick et al. 1977; Rick 1984). Levels of heterozygosity were found to range between 0 and 20% and were correlated with levels of outcrossing, flower size, and stigma exsertion (Rick et al. 1977; Rick 1984). The diversity cline in *S. pimpinellifolium* has been confirmed with DNA sequence data. While overall levels of nucleotide diversity in the species were found to be about 0.64% for a nuclear gene, levels within Peruvian populations ranged from 0 to 0.68% (Caicedo and Schaal 2004b). As in the two variable SI species described above, the *S. pimpinellifolium* diversity cline is believed to be a consequence of the spread of the species along the Pacific coast from its center of origin in Northern Peru, with increased levels of inbreeding favored as the species colonized new habitats (Rick 1984; Caicedo and Schaal 2004b).

Wild tomatoes are widespread in the Galapagos, occurring in 19 islands and islets (Darwin et al. 2003). As self-compatible and highly autogamous taxa, they display very little allozyme variation within populations and no heterozygosity, although some variability occurs between populations (Rick and Fobes 1975; Rick 1983). Diversity estimates in Galapagos taxa based on AFLPs are 70 to 80% of those found in *S. pimpinellifolium* (Nuez et al. 2004). At least two distinct morphological forms have long been recognized to inhabit the Galapagos (Rick and Fobes 1975), leading to the recent reclassification into two species: *S. galapagense* and *S. cheesmaniae* (Darwin et al. 2003). *S. galapagense*, often referred to as *L. cheesmaniae* f. *minor*, is characterized by highly subdivided leaflets, orange-red fruit, short internodes, dense pubescence and is often found in coastal zones (Rick and Fobes 1975; Rick 1983; Darwin et al. 2003). AFLP and allozyme studies have confirmed the genetic distinctness of this group as well as its lower level of variation compared to *S. cheesmaniae* (Rick and Fobes 1975; Darwin et al. 2003; Nuez et al. 2004). *S. cheesmaniae* also displays higher levels of morphological diversity, including variation in degree of fruit coloration, leaf structure, and internode length (Rick and Fobes 1975; Rick 1983; Darwin et al. 2003; Nuez et al. 2004). Geographically, there is no support for island endemism in the distribution of variation (Rick 1983), suggesting a great colonizing ability for both Galapagos tomato species.

1.7 Tomato Domestication and Dissemination

Two competing hypotheses about the place of domestication of tomato have been proposed, one supporting Peru, another in Mexico. The Mexican hypothesis states that feral populations of tomatoes migrated from Peru into Central America and were domesticated in Mexico (Jenkins 1948; Rick et al. 1974). This hypothesis is reasonable and supported by some linguistic, cultural and historical evidences (Rick 1995; Cox 2000). Arguments supporting a Peruvian domestication hypothesis also relied on linguistic evidence and were first presented by De Candolle (1886) and latter maintained by Moore (1935), Müller (1940a,b) and Luckwill (1943a, b). However, none of the evidence is conclusive regarding an initial site of domestication, and tomatoes may have been domesticated independently in both areas (Peralta and Spooner 2007).

Tomatoes were first recorded outside the Americas in Italy in 1544 by the botanist Matthiolus, but at present there are no documents that clearly identified the first European place of introduction. Introduced tomatoes were cultivated first as an ornamental or curiosity plants and thought by many to be poisonous. Tomato was first accepted as a vegetable in southern Europe during the late 16th century and afterwards the crop expanded to northern countries and other parts of the world (McCue 1952). The first European cultivars had yellow to red flattened fruits with deep furrows, and flowers with stigmas exserted from the anther tube. Derived cultivars had a wider range of fruit colors and shapes, smoother fruits, and stigmas included in the anther tube that led to increased fruit set in modern cultivars but reduced the genetic variation of the crop.

The uncertainty about tomato domestication has complicated assessment of the domestication syndrome (i.e., suites of traits differentiating domesticated tomato from it wild progenitor) and dissection of the genetic basis of domestication traits. Some features that are believed to have been under selection during tomato domestication and later improvement efforts are: greater fruit size, fruit quality, increased propensity for self-fertilization, greater apical dominance and erect growth, and various traits enhancing mechanical harvest (Paran and van der Knaap 2007). Curiously, as in some other crops, despite low levels of genetic diversity, domestication and subsequent selection has led to a greater variety of morphological shapes and colors in domesticated tomato fruits than in wild species.

Most progress in determining the molecular basis of domestication and improvement traits has been made for phenotypes involving fruit morphology. Wild tomatoes produce small round fruits, which are believed to be suited to seed dispersal by small vertebrates. Mapping studies suggest that the increase in size and shape variability observed in domesticated tomato can be accounted for by as few as nine loci (reviewed in Tanksley 2004). Mutations in the promoter region of *fw2.2*, a regulator of cell division, have been shown to account for about 30% of fruit size difference between cultivated tomato and wild *S. pennellii* (Frary et al. 2000; Tanksley 2004). Headway has also been made in identification of genes controlling fruit shape like the gene *ovate* (Liu et al. 2002; van der Knaap and Tanksley 2003). Loci involved in cultivar improvement traits affecting plant architecture (Pnueli et al. 1998), fruit sugar content (Fridman et al. 2000; Fridman et al. 2004) have also been recently identified, as well as loci underlying fruit cuticle changes that preadapted tomato to domestication (Hovav et al. 2007). Many of the loci involved in varietal improvement represent introgressions from wild relatives during breeding efforts (e.g., Budiman et al. 2003).

1.8 Tomato Breeding and Research Resources

A long interest in tomato as an important agricultural commodity and genetic model species has led to the development of multiple resources available for both research and crop improvement purposes. These various tools and resources have positioned domesticated tomato and its relatives as an ideal system for characterizing the genetic basis of phenotypic variation. Many of these resources are publicly available, contributing greatly to a thriving cooperative worldwide tomato research community.

1.8.1 Genetic Resources

Wild and cultivated germplasm of tomato, as well as various genetic stocks are maintained and made available through various gene banks within the US and around the world. Ross (1998) considered that the diversity of tomato is likely to be well conserved *ex situ*, and cited 62,832 accessions maintained in gene banks around the world, although most of these accessions correspond to *S. lycopersicum*. This germplasm has been continuously used for a variety of basic and applied research. For recent reviews and listings of available stocks in tomato genebanks see Labate et al. (2007) and Robertson and Labate (2007).

1.8.1.1 Mutant Stocks

The C. M. Rick Tomato Genetic Resource Center (TGRC) at the University of California, Davis curates 1,017 monogenic mutant stocks, accumulated throughout many years from various sources (http://tgrc.ucdavis.edu). Seeds are also available for ~3,500 monogenic mutants in an isogenic background (Menda et al. 2004), along with data and images of the mutant phenotypes observed (http://zamir.sgn.cornell.edu/mutants/index.html). Nearly 100 cytogenetic mutant stocks, comprising translocations, trisomics, and autotetraploids are also available through the TGRC.

1.8.1.2 Mapping Lines

Permanent mapping populations have been the cornerstone of ongoing efforts to determine the molecular basis of quantitative and qualitative phenotypic variation in tomatoes. All publicly available mapping lines correspond to crosses between wild and cultivated germplasm. Introgression lines (ILs), in particular, have contributed to the identification of thousands of QTLs in tomato, avoiding difficulties due to epistasis inherent in other mapping populations (Lippman et al. 2007). Publicly available ILs include: 76 lines of *S. pennelli* in *S. lycopersicum* background (100% genome coverage (Eshed and Zamir 1994)), 98 ILs of *S. habrochaites* in *S. lycopersicum* (85% genome coverage (Monforte and Tanksley 2000)), and 99 ILs of *S. lycopersicoides* in *S. lycopersicum* (96% genome coverage (Chetelat and Meglic 2000; Canady et al. 2005)) (http://tgrc.ucdavis.edu). A set of 99 lines from a recombinant inbred line (RIL) population of an *S. pimpinellifolium* x *S. lycopersicum* cross is also available from the TGRC. Other public mapping lines include alien substitution lines and monosomic addition lines sources (http://tgrc.ucdavis.edu). Though not publicly available as permanent populations, additional efforts in individual labs have led to the creation of various interspecific crosses including *S. cheesmaniae* x *S. lycopersicum* (Paran et al. 1995) and *S. peruvianum* x *S. lycopersicum* (Fulton et al. 1997b).

1.8.1.3 Wild and Cultivated Germplasm

Beginning with a 1942 published survey of disease resistance within the tomato clade (Alexander et al. 1942), wild germplasm has been key to tomato improvement in the US (Stevens and Rick 1986). Examples of genes and traits transferred from wild species into modern cultivars are: resistance to fungal and bacterial diseases (e.g., Bohn and Tucker 1940; Langford 1937; Pitbaldo and Kerr 1980; Laterrot and Rat 1981), fruit color and fruit set (Rick 1982b; Stevens and Rick 1986), and traits facilitating machine harvesting (e.g., Reynard 1961). Today, commercial tomato varieties can contain up to 15 wild disease resistance alleles (Pan et al. 2000). Agronomically useful traits continue to be identified in exotic germplasm, suggesting that wild species will continue to play an important role in tomato breeding (Fulton et al. 2002; Frary et al. 2004; Labate et al. 2007).

Several genebanks around the world maintain wild species and tomato cultivar collections for breeding or research purposes. In the US, the United States Department of Agriculture, Agricultural Research Service (USDA-ARS) Plant Genetic Resources Unit (PGRU) in Geneva, NY has the largest collection of tomato cultivars, and the Seed Savers Exchange in Decorah, Iowa (http://www.seedsavers.org/) maintains heirloom varieties. Internationally, cultivated tomato germplasm can be obtained from The World Vegetable Center in Taiwan (http://www.avrdc.org/). The major genebank for wild tomato species is the TGRC, at the University of California, Davis, which currently contains over 1,000 wild accessions. For detailed reviews on publicly available germplasm within the tomato group see Robertson and Labate (2007).

1.8.2 Genomic Resources

The euchromatic regions of the tomato (*S. lycopersicum*) genome are currently being sequenced through an international effort, using a BAC approach (Mueller et al. 2005b) (see Chapter 10). Additional genomic resources that have been generated recently, either to aid the genome sequencing effort or complement it include: BAC libraries, BAC end sequences (http://www.sgn.cornell.edu). Over 150,000 ESTs from *S. lycopersicum* have been analyzed and clustered into ~30,000 unigenes, from 27 cDNA libraries (Van Der Hoeven et al. 2002; Fei et al. 2004). More limited EST datasets have been generated and analyzed for two key wild species: *S. habrochaites* and *S. pennellii* (http://www.ncbi.nlm.nih.gov/dbEST/). Three microarrays are currently available for expression studies in tomato: an Affymetrix genome array with 10,000 probes corresponding to 9,200 transcripts; a cDNA array with 12,000 elements corresponding to 8,500 genes (http://bti.cornell.edu/CGEP/CGEP.html); and an 11,000 element oligo array (http://bti.cornell.edu/CGEP/CGEP.html).

1.8.3 Bioinformatic Resources

There have been concerted efforts by the tomato research community to develop informational networks that will facilitate research within the group. The SOL Genomic Network (SGN; http://www.sgn.cornell.edu/) currently serves as a portal to databases containing comparative genomic, genetic and taxonomic information for various species within the Solanaceae (Mueller et al. 2005a). Among the data maintained by the SGN is the updated dataflow from the updated International Tomato Genome Sequencing project (Mueller et al. 2005b). Related sites include a tomato expression database (http://ted.bti.cornell.edu/), a tomato metabolite database (http://tomet.bti.cornell.edu/), and searchable mutant databases (http://zamir.sgn.cornell.edu/mutants/; (Menda et al. 2004)). Efforts are also underway to produce a taxonomic monograph of all ~1,500 species in the genus *Solanum*, and progress is being made publicly available on the PBI *Solanum* Project webpage (http://www.nhm.ac.uk/research-curation/projects/solanaceaesource/). The increased availability of species descriptions, ecological information, and taxonomic resolution for close relatives of tomato, will contribute the development of this species as a model system for research in evolution, physiology, genetics, genomics, and other areas of biology.

References

Alexander LJ, Lincoln RE, Wright VA (1942) A survey of the genus *Lycopersicon* for resistance to the important tomato disease occurring in Ohio and Indiana. Plant Dis Rep (Suppl 136): 51–85.

Allan R, Lindesay J, Parker D (1996) El Niño southern oscillation and climatic variability. CSIRO, Australia.

Alvarez AE, van de Wiel CCM, Smulders MJM, Vosman B (2001) Use of microsatellites to evaluate genetic diversity and species relationships in the genus *Lycopersicon*. Theor Appl Genet 103: 1283–1292.

Arunyawat U, Stephan W, Stadler T (2007) Using multilocus sequence data to assess population structure, natural selection, and linkage disequilibrium in wild tomatoes. Mol Biol Evol 24: 2310–2322.

Baker HG (1955) Self-compatibility and establishment after "long-distance" dispersal. Evolution 9: 347–349.

Baudry E, Kerdelhue C, Innan H, Stephan W (2001) Species and recombination effects on DNA variability in the tomato genus. Genetics 158: 1725–1735.

Beecher GR (1998) Nutrient content of tomatoes and tomato products. Proc Soc Exp Biol Med 218: 98–100.

Bernacchi D, Tanksley SD (1997) An interspecific backcross of *Lycopersicon esculentum* x *L. hirsutum* linkage analysis and a QTL study of sexual compatibility factors and floral traits. Genetics 147: 861–877.

Bernatzky R, Glaven RH, Rivers BA (1995) S-related ptotein can be recombined with self-compatibility in interspecific derivatives of *Lycopersicon*. Biochem Genet 33: 215–225.

Bernatzky R, Tanksley SD (1986) Toward a saturated linkage map in tomato based on isozymes and random cDNA sequences. Genetics 112: 887–898.

Bharathan G, Goliber TE, Moore C, Kessler S, Pham T, Sinha NR (2002) Homologies in leaf form inferred from KNOX1 gene expression during development. Science 296: 1858–1860.

Bohn GW, Tucker CM (1940) Studies on fusarium wilt of the tomato. I. Immunity in *Lycopersicon pimpinellifolium* Mill. and its inheritance in hybrids. MO Agri Exp Sta Res Bull 311: 82.

Bohs L (2005) Major clades in *Solanum* based on *ndh*F sequences. In: Keating RC, Hollowell VC, Croat TB (eds) A Festschrift for William G. D'Arcy: The Legacy of a Taxonomist. MBG Press, St. Louis pp 27–49.

Bohs L, Olmstead RG (1997) Phylogenetic relationships in *Solanum* (Solanaceae) based on *ndh*F sequences. Syst Bot 22: 5–17.

Breto MP, Asins MJ, Carbonell EA (1993) Genetic variability in *Lycopersicon* species and their genetic relationships. Theor Appl Genet 86: 113–120.

Broun P, Tanksley SD (1996) Characterization and genetic mapping of simple repeat sequences in the tomato genome. Mol Gen Genet 250: 39–49.

Budiman MA, Chang S-B, Lee S, Yang TJ, Zhang H-B, et al. (2003) Localization of *jointless-2* gene in the centromeric region of tomato chromosome 12 based on high resolution genetic and physical mapping. Theor Appl Genet 108: 190–196.

Caicedo AL (2008) Geographic diversity cline of R gene homologs in wild populations of *Solanum pimpinellifolium* (Solanaceae). Am J Bot 95: 393–398.

Caicedo AL, Schaal BA (2004a) Heterogeneous evolutionary processes affect R gene diversity in natural populations of *Solanum pimpinellifolium*. Proc Natl Acad Sci USA 101: 17444–17449.

Caicedo AL, Schaal BA (2004b) Population structure and phylogeography of *Solanum pimpinellifolium* inferred from a nuclear gene. Mol Ecol 13: 1871–1882.

Canady MA, Meglic V, Chetelat RT (2005) A library of *Solanum lycopersicoides* introgression lines in cultivated tomato. Genome 48: 685–697.

Candolle A (1886) Origin of cultivated plants. D Appleton, New York, USA.

Charlesworth D (2003) Effects of inbreeding on the genetic diversity of populations. Phil Trans Roy Soc Lond Sr B-Biol Sci 358: 1051–1070.

Chen KY, Cong B, Wing R, Vrebalov J, Tanksley SD (2007) Changes in regulation of a transcription factor lead to autogamy in cultivated tomatoes. Science 318: 643–645.

Chen KY, Tanksley SD (2004) High-resolution mapping and functional analysis of se2.1: A major stigma exsertion quantitative trait locus associated with the evolution from allogamy to autogamy in the genus *Lycopersicon*. Genetics 168: 1563–1573.

Chetelat RT, Meglic V (2000) Molecular mapping of chromosome segments introgressed from *Solanum lycopersicoides* into cultivated tomato (*Lycopersicon esculentum*). Theor Appl Genet 100: 232–241.

Child A, Lester RN (1991) Life form and branching within the Solanaceae, pp 151–159. In: JG. Hawkes, RN Lester, M Nee and N Estrada [eds] Solanaceae III: taxonomy, chemistry, evolution. Kew: Royal Botanic Gardens.

Cox S (2000) From discovery to modern commercialism: the complete story behind *Lycopersicon esculentum*: http://www.landscapeimagery.com/articles.html.

Danert S (1958) Die Verzweigung der Solanaceen im reproduktiven Bereich. Abh Deutsch Akad Wiss Berlin, Kl Chem 6: 1–183.

Darwin C (1876) The Effects of Cross and Self Fertilization in the Vegetable Kingdom. John Murray, London, UK.

Darwin SC, Knapp S, Peralta IE (2003) Taxonomy of tomatoes in the Galapagos Islands: native and introduced species of *Solanum* section *Lycopersicon* (Solanaceae). Syst Biodiver 1: 29–53.

Davies JN, Hobson GE (1981) The constituents of tomato fruit—the influence of environment, nutrition, and genotype. CRC Critical Reviews in Food Science and Nutrition 15: 205–280.

Dillon MO (2005) The Solanaceae of the lomas formations of coastal Peru and Chile, pp 27–49. In: RC Keating, VC Hollowell and TB Croat [eds] A Festschrift for William G D'Arcy: the legacy of a taxonomist. Monogr Syst Bot Missouri Bot Gard 104: 27–49.

Dixon MS, Jones DA, Keddie JS, Thomas CM, Harrison K, Jones JD (1996) The tomato *Cf-2* disease resistance locus comprises two functional genes encoding leucine-rich repeat proteins. Cell 84: 451–459.

Doganlar S, Frary A, Daunay MC, Lester RN, Tanksley SD (2002) Conservation of gene function in the Solanaceae as revealed by comparative mapping of domestication traits in eggplant. Genetics 161: 1713–1726.

Egashira H, Ishihara H, Takashina T, Imanishi S (2000) Genetic diversity of the 'peruvianum-complex' (*Lycopersicon peruvianum* (L.) Mill. and *L. chilense* Dun.) revealed by RAPD analysis. Euphytica 116: 23–31.

Ellinger S, Ellinger J, Stehle P (2006) Tomatoes, tomato products and lycopene in the prevention and treatment of prostate cancer: do we have the evidence from intervention studies? Curr Opin Clin Nutr Metabol Care 9: 722–727.

Ellinger S, Ellinger J, Stehle P (2007) Tomatoes, tomato products and lycopene in prevention and therapy of carcinoma of the prostate—what has been verified? Ernahrungs-Umschau 54: 318–323.

Ensminger AH, Ensminger ME, Konlande JE, Robson JRK (1995) The Concise Encyclopedia of Foods and Nutrition. CRC Press, Boca Raton, FL, USA.

Eshed Y, Zamir D (1994) A genomic library of *Lycopersicon pennellii* in *Lycopersicon esculentum* —a tool for fine mapping of genes. Euphytica 79: 175–179.

Eshed Y, Zamir D (1995) An introgression line population of *Lycopersicon pennellii* in the cultivated tomato enables the identification and fine mapping of yield-associated QTL. Genetics 141: 1147–1162.

Fei Z, Tang X, Alba RM, White JA, Ronning CM, Martin GB, Tanksley SD, Giovannoni JJ (2004) Comprehensive EST analysis of tomato and comparative genomics of fruit ripening. Plant J 40: 47–59.

Foolad MR, Lin GY (2001) Genetic analysis of cold tolerance during vegetative growth in tomato, *Lycopersicon esculentum* Mill. Euphytica 122: 105–111.

Frankel N, Hasson E, Iusem ND, Rossi MS (2003) Adaptive evolution of the water stress-induced gene Asr2 in *Lycopersicon* species dwelling in arid habitats. Mol Biol Evol 20: 1955–1962.

Franklin-Tong VE, Franklin FCH (2003) The different mechanisms of gametophytic self-incompatibility. Phil Trans Ro Soc Lond Sr B-Biol Sci 358: 1025–1032.

Frary A, Doganlar S, Daunay MC, Tanksley SD (2003) QTL analysis of morphological traits in eggplant and implications for conservation of gene function during evolution of solanaceous species. Theor Appl Genet 107: 359–370.

Frary A, Fulton TM, Zamir D, Tanksley SD (2004) Advance backcross QTL analysis of a *Lycopersicon esculentum* x *L. pennellii* cross and identification of possible orthologs in the Solanaceae. Theor Appl Genet 108: 485–496.

Frary A, Nesbitt TC, Grandillo S, van der Knaap E, Cong B, Liu J, Meller J, Elber R, Alpert KB, Tanksley SD (2000) fw2.2: A quantitative trait locus key to the evolution of tomato fruit size. Science 289: 85–88.

Frary A, Xu Y, Liu J, Mitchell S, Tedeschi E, Tanksley S (2005) Development of a set of PCR-based anchor markers encompassing the tomato genome and evaluation of their usefulness for genetics and breeding experiments. Theor Appl Genet 111: 291–312.

Fridman E, Carrari F, Liu Y-S, Fernie AR, Zamir D (2004) Zooming in on a quantitative trait for tomato yield using interspecific introgressions. Science 305: 1786–1789.

Fridman E, Pleban T, Zamir D (2000) A recombination hotspot delimits a wild-species quantitative trait locus for tomato sugar content to 484 bp within an invertase gene. Proc Natl Acad Sci USA 97: 4718–4723.

Fulton TM, Beck-Bunn T, Emmatty D, Eshed Y, Lopez J, Petiard V, Uhlig J, Zamir D, Tanksley SD (1997a) QTL analysis of an advanced backcross of *Lycopersicon peruvianum* to the cultivated tomato and comparisons with QTLs found in other wild species. Theor Appl Genet 95: 881–894.

Fulton TM, Bucheli P, Voirol E, Lopez J, Petiard V, Tanksley SD (2002) Quantitative trait loci (QTL) affecting sugars, organic acids and other biochemical properties possibly contributing to flavor, identified in four advanced backcross populations of tomato. Euphytica 127: 163–177.

Fulton TM, Nelson JC, Tanksley SD (1997b) Introgression and DNA marker analysis of *Lycopersicon peruvianum*, a wild relative of the cultivated tomato, into *Lycopersicon esculentum*, followed through three successive backcross generations. Theor Appl Genet 95: 895–902.

Gabriels SHEJ, Vossen JH, Ekengren SK, van Ooijen G, Abd-El-Haliem AM, van den Berg GCM, Rainey DY, Martin GB, Takken FLW, de Wit PJGM, Joosten MHAJ (2007) An NB-LRR protein required for HR signalling mediated by both extra- and intracellular resistance proteins. Plant Journal 50: 14–28.

Georgiady MS, Lord EM (2002) Evolution of the inbred flower form in the currant tomato, Lycopersicon pimpinellifolium. Int J Plant Sci 163: 531–541.

Georgiady MS, Whitkus RW, Lord EM (2002) Genetic analysis of traits distinguishing outcrossing and self-pollinating forms of currant tomato, *Lycopersicon pimpinellifolium* (Jusl.) Mill Genetics 161: 333–344.

Giovannoni JJ (2004) Genetic Regulation of Fruit Development and Ripening. Plant Cell 16: S170–180.

Giovannucci E (1999) Tomatoes, Tomato-Based Products, Lycopene, and Cancer: Review of the Epidemiologic Literature. J Natl Cancer Inst 91: 317–331.

Giovannucci E (2007) Does prostate-specific antigen screening influence the results of studies of tomatoes, lycopene, and prostate cancer risk? J Natl Cancer Inst 99: 1060–1062.

Gleissberg S (2002) Comparative developmental and molecular genetic aspects of leaf dissection, pp 404–417. In: QCB Cronk, RM Bateman and JA Hawkins [eds] Developmental Genetics and Plant Evolution. Taylor & Francis, London, UK.

Grandillo S, Ku HM, Tanksley SD (1999) Identifying the loci responsible for natural variation in fruit size and shape in tomato. Theor Appl Genet 99: 978–987.

Gundersen V, McCall D, Bechmann IE (2001) Comparison of major and trace element concentrations in Danish greenhouse tomatoes (*Lycopersion esculentum* Cv. Aromata F_1) cultivated in different substrates. J Agri Food Chem 49: 3808–3815.

Hovav R, Chehanovsky N, Moy M, Jetter R, Schaffer AA (2007) The identification of a gene (*Cwp1*), silenced during *Solanum* evolution, which causes cuticle microfissuring and dehydration when expressed in tomato fruit. Plant J 52: 627–639.

Igic B, Smith WA, Robertson KA, Schaal BA, Kohn JR (2007) Studies of self-incompatibility in wild tomatoes: I. S-allele diversity in *Solanum chilense* Dun. (Solanaceae). Heredity 99: 553–561.

Jenkins JA (1948) The origin of the cultivated tomato. Econ Bot 2: 379–392.

Jones DA, Thomas CM, Hammond-Kosack KE, Balint-Kurti PJ, Jones JDG (1994) Isolation of the tomato *Cf-9* gene for resistance to *Cladosporium fulvum* by transposon tagging. Science 266: 789–793.

Jones DF (1917) Linkage in *Lycopersicum*. Am Nat 51: 608.

Kavanaugh CJ, Trumbo PR, Ellwood KC (2007) The US Food and Drug Administration's evidence-based review for qualified health claims: tomatoes, lycopene, and cancer. J Natl Cancer Inst 99: 1074–1085.

Kessler S, Kim M, Pham T, Weber N, Sinha NR (2001) Mutations altering leaf morphology in tomato. Int J Plant Sci 162: 475–492.

Kimura S, Koenig D, Kang J, Yoong FY, Sinha N (2008) Natural variation in leaf morphology results from mutation of a novel KNOX gene. Curr Biol 18: 672–677.

Kondo K, Yamamoto M, Itahashi R, Sato T, Egashira H, Hattori T, Kowyama Y (2002a) Insights into the evolution of self-compatibility in Lycopersicon from a study of stylar factors. Plant Journal 30: 143–153.

Kondo K, Yamamoto M, Matton DP, Sato T, Hirai M, Norioka S, Hattori T, Kowyama Y (2002b) Cultivated tomato has defects in both *S-RNase* and *HT* genes required for stylar function of self-incompatibility. Plant Journal 29.

Kowyama Y, Kunz C, Lewis I, Newbigin E, Clarke AE, Anderson MA (1994) Self compatibility in a *Lycopersicon peruvianum* variant (LA2157) is associated with a lack of style s-RNase activity. Theor Appl Genet 88: 859–864.

Kruger J, Thomas CM, Golstein C, Dixon MS, Smoker M, Tang S, Mulder L, Jones JDG (2002) A tomato cysteine protease required for Cf-2-dependent disease resistance and suppression of autonecrosis. Science 296: 744–747.

Kruijt M, Kip DJ, Joosten M, Brandwagt BF, de Wit P (2005) The *Cf-4* and *Cf-9* resistance genes against *Cladosporium fulvum* are conserved in wild tomato species. Mol Plant-Microbe Interact 18: 1011–1021.

Labate JA, Grandillo S, Fulton T, Muños S, Caicedo AL, Peralta I, Ji Y, Chetelat RT, et al. (2007) Tomato, p 1–125. In: C Kole (ed) Genome mapping and molecular breeding in plants: Vol 5 Vegetables. Springer Publishing Co, NY.

Langford AN (1937) The parasitism of *Cladosporium fulvum* Cooke and the genetics of resistance to it. Can J Res 15: 108–128.

Laquatra I, Yeung DL, Storey M, Forshee R (2005) Health benefits of lycopene in tomatoes—conference summary. Nutr Today 40: 29–36.

Laterrot H, Rat B (1981) Efficacite de l'origine canadienne de resistance a *Pseudomonas* tomato. In: Genetics and breeding of tomato: Proceedings of the Meeting of the Eucarpia Tomato Working Group. Institut National de la Recherche Agronomique, Avignon, France, pp 257–266.

Leonardi C, Ambrosino P, Esposito F, Fogliano V (2000) Antioxidative activity and carotenoid and tomatine contents in different typologies of fresh consumption tomatoes. J Agri Food Chem 48: 4723–4727.

Lester RN (1991) Evolutionary relationships of tomato, potato and pepino, and wild species of *Lycopersicon* and *Solanum*, pp 283–301. In: JG Hawkes, RN Lester, M Nee and N Estrada [eds] Solanaceae III: taxonomy, chemistry, evolution. Royal Botanic Gardens, Kew.

Lester RN, Durrands P (1984) Enzyme treatment as an aid in the study of seed surface structures of Solanum species. Ann Bot 53: 129–131.

Linnaeus C (1753) Species Plantarum. Holmiae, Stockholm, Sweden.

Lippman ZB, Semel Y, Zamir D (2007) An integrated view of quantitative trait variation using tomato interspecific introgression lines. Curr Opin Genet Dev 17: 545–552.

Liu J, Van Eck J, Cong B, Tanksley SD (2002) A new class of regulatory genes underlying the cause of pear-shaped tomato fruit. Proc Natl Acad Sci USA 99: 13302–13306.

Livingstone KD, Lackney VK, Blauth JR, van Wijk R, Jahn MK (1999) Genome mapping in *Capsicum* and the evolution of genome structure in the *Solanaceae*. Genetics 152: 1183–1202.

Luckwill LC (1943a) The evolution of the cultivated tomato. J Royal Hortic Soc 68: 19–25.

Luckwill LC (1943b) The Genus *Lycopersicon*: an historical, biological, and taxonomic survey of the wild and cultivated tomatoes. Aberdeen University Press, Aberdeen, UK.

Luderer R, Rivas S, Nurnberger T, Mattei B, Van den Hooven HW, Van der Hoorn RAL, Romeis T, Wehrfritz J-M, Blume B, Nennstiel D, Zuidema D, Vervoort J, De Lorenzo G, Jones

JDG, De Wit PJGM, Joosten MHAJ (2001) No evidence for binding between resistance gene product Cf-9 of tomato and avirulence gene product AVR9 of *Cladosporium fulvum*. Mol Plant-Microbe Interact 14: 867–876.

Marshall JA, Knapp S, Davey MR, Power JB, Cocking EC, Bennett MD, Cox AV (2001) Molecular systematics of *Solanum* section *Lycopersicum* (*Lycopersicon*) using the nuclear ITS rDNA region. Theor Appl Genet 103: 1216–1222.

Martin GB, Brommonschenkel SH, Chunwongse J, Frary A, Ganal MW, Spivey R, Wu T, Earle ED, Tanksley SD (1993) Map-based cloning of a protein-kinase gene conferring disease resistance in tomato. Science 262: 1432–1436.

Matthioli PA (1544) Di Pedacio Dioscoride Anazarbeo libri cinque della historia, et materia medicinale trodotti in lingua uolgare Italiana. N de Bascarini, Venice.

Matthioli PA (1554) Commentarii, in libros sex Pedacci Dioscoricis Anazarbei, de medica materia. V Valgrisi, Venice.

McClure BA, Haring V, Ebert PR, Anderson MA, Simpson RJ, Sakiyama F, Clarke AE (1989) Style self-incompatibility gene products of *Nicotiana alata* are ribonucleases. Nature 342: 955–957.

McClure BA, Mou B, Canevascini S, Bernatzky R (1999) A small asparagine-rich protein required for S-allele-specific pollen rejection in *Nicotiana*. Proc Natl Acad Sci USA 96: 13548–13553.

McCormick S, Niedermeyer J, Fry J, Barnason A, Horsch R, Fraley R (1986) Leaf disc transformation of cultivated tomato (*L. esculentum*) using *Agrobacterium tumefaciens*. Plant Cell 5: 81–84.

McCue GA (1952) The history and use of the tomato: an annotated bibliography. Ann. MO Bot Gard 39: 289–348.

Menda N, Semel Y, Peled D, Eshed Y, Zamir D (2004) In silico screening of a saturated mutation library of tomato. Plant J 38: 861–872.

Michaelson M, Price H, Ellison J, Johnston J (1991) Comparison of plant DNA contents determined by feulgen microspectrophotometry and laser flow cytometry. Am J Bot 78: 183–188.

Miller JC, Tanksley SD (1990) RFLP analysis of phylogenetic relationships and genetic variation in the genus *Lycopersicon*. Theor Appl Genet 80: 437–448.

Miller P (1754) The Gardeners Dictionary, London, UK.

Monforte AJ, Asins MJ, Carbonell EA (1997) Salt tolerance in Lycopersicon species .6. Genotype-by-salinity interaction in quantitative trait loci detection: constitutive and response QTLs. Theor Appl Genet 95: 706–713.

Monforte AJ, Tanksley SD (2000) Development of a set of near isogenic and backcross recombinant inbred lines containing most of the *Lycopersicon hirsutum* genome in a *L. esculentum* genetic background: A tool for gene mapping and gene discovery. Genome 43: 803–813.

Moore JA (1935) The early history of the tomato or loveapple. MO Bot Gard Bull 23: 134–138.

Moyle LC (2007) Comparative genetics of potential prezygotic and postzygotic isolating barriers in a *Lycopersicon* species cross. J Hered 98: 123–135.

Moyle LC, Graham EB (2005) Genetics of hybrid incompatibility between *Lycopersicon esculentum* and *L hirsutum*. Genetics 169: 355–373.

Mueller LA, Solow TH, Taylor N, Skwarecki B, Buels R, Binns J, Lin C, Wright MH, Ahrens R, Wang Y, Herbst EV, Keyder ER, Menda N, Zamir D, Tanksley SD (2005a) The SOL Genomics Network. A comparative resource for Solanaceae biology and beyond. Plant Physiol 138: 1310–1317.

Mueller LA, Tanksley SD, Giovannoni JJ, et al. (2005b) The Tomato Sequencing Project, the first cornerstone of the International Solanaceae Project (SOL). Compar Funct Genom 6: 153–158.

Müller CH (1940a) A revision of the genus *Lycopersicon*. United States Department of Agriculture, Washington DC, USA.

Müller CH (1940b) The taxonomy and distribution of the genus *Lycopersicon*. Nat Hort Magaz 19: 157–160.

National Agricultural Statistics Service and USDA (2005) Agricultural Statistics 2005. Unites States Government Printing Office, Washington DC, USA, 527 pp.

Nesbitt TC, Tanksley SD (2002) Comparative sequencing in the genus *Lycopersicon*: Implications for the evolution of fruit size in the domestication of cultivated tomatoes. Genetics 162: 365–379.

Nuez F, Prohens J, Blanca JM (2004) Relationships origin, and diversity of Galapagos tomatoes: Implications for the conservation of natural populations. Am J Bot 91: 86–99.

Olmstead RG, Palmer JD (1997) Implications for the phylogeny, classification, and biogeography of *Solanum* from cpDNA restriction site variation. Syst Bot 22: 19–29.

Olmstead RG, Sweere JA, Spangler RE, Bohs L, Palmer JD (1999) Phylogeny and provisional classification of the Solanaceae based on chloroplast DNA. In: Nee M, Symon DE, Lester RN, Jessop JP (eds). Solanaceae IV, advances in biology and utilization. Royal Botanic Gardens, Kew, UK, pp 111–137.

Pan Q, Liu Y-S, Budai-Hadrian O, et al. (2000) Comparative genetics of nucleotide binding site-leucine rich repeat resistance gene homologues in the genomes of two dicotyledons: tomato and *Arabidopsis*. Genetics 309–322.

Paran I, Goldman I, Tanksley SD, Zamir D (1995) Recombinant inbred lines for genetic mapping in tomato. Theor Appl Genet 90: 542–548.

Paran I, van der Knaap E (2007) Genetic and molecular regulation of fruit and plant domestication traits in tomato and pepper. J Exp Bot 58: 3841–3852.

Paterson AH, Lander ES, Hewitt JD, et al. (1988) Resolution of quantitative traits into Mendelian factors by using a complete linkage map of restriction fragment length polymorphisms. Nature 335: 721–726.

Peralta IE, Knapp SK, Spooner DM (2005) New species of wild tomatoes (*Solanum* section Lycopersicon: Solanaceae) from Northern Peru. Syst Bot 30: 424–434.

Peralta IE, Spooner DM (2001) Granule-bound starch synthase (GBSSI) gene phylogeny of wild tomatoes (*Solanum* L. section *Lycopersicon* [Mill.] Wettst. subsection *Lycopersicon*). Am J Bot 88: 1888–1902.

Peralta IE, Spooner DM (2005) Morphological characterization and relationships of wild tomatoes (*Solanum* L. Section *Lycopersicon*), pp 227–257. In: RC Keating, VC Hollowell and TB Croat [eds] A Festschrift for William G D'Arcy: The Legacy of a Taxonomist. Monogr Syst Bot Missouri Bot Gard 104: 227–257.

Peralta IE, Spooner DM (2007) History, origin and early cultivation of tomato (Solanaceae), pp 1–27. In: MK Razdan and AK Mattoo [eds]. Genetic improvement of Solanaceous crops. Science Publishers, Enfield, NH, USA.

Peralta IE, Spooner DM, Knapp S (2008) Taxonomy of wild tomatoes and their relatives (*Solanum* Sect. *Lycopersicoides*, Sect *Juglandifolia*, Sect *Lycopersicon; Solanaceae*). Syst Bot Monogr 84: 183.

Piazza P, Sjinski S, Tsiantis M (2005) Evolution of leaf developmental mechanisms. New Phytol 167: 693–710.

Pitbaldo RE, Kerr EA (1980) Resistance to baterial speck (*Pseudomonas syringae* pv. *tomato*) in tomato. Acta Hort 100: 379–382.

Pnueli L, Carmel-Goren L, Hareven D, et al. (1998) The SELF-PRUNING gene of tomato regulates vegetative to reproductive switching of sympodial meristems and is the ortholog of CEN and TFL1. Development 125: 1979–1989.

Quinn WH, Neal VT (1992) The historial record of El Niño events, pp 623–648. In: P Bradley and P Jones [eds]. Climate since 1500. Routledge, London, UK.

Rathjen JP, Chang JH, Staskawicz BJ, Michelmore RW (1999) Constitutively active Pto induces a Prf-dependent hypersensitive response in the absence of avrPto. EMBO J 18: 3232–3240.

Reynard GB (1961) New Source of the j2 Gene Governing Jointless Pedicel in Tomato. Science 134: 2102.

Richman AD, Uyenoyama MK, Kohn JR (1996) Allelic diversity and gene genealogy at the self-incompatibility locus in the Solanaceae. Science 273: 1212–1216.

Rick CM (1963) Barriers to Interbreeding in *Lycopersicon peruvianum*. Evolution 17: 216–217.

Rick CM (1978) The tomato. Sci Am 239: 76–87.

Rick CM (1979) Biosystematic studies in *Lycopersicon* and closely related species of *Solanum*, pp 667–677. In: JG Hawkes, RN Lester and AD Skelding [eds]. The biology and taxonomy of Solanaceae. Academic Press, New York.

Rick CM (1982a) Genetic-relationships between self-incompatibility and floral traits in the tomato species. Biologisches Zentralblatt 101: 185–198.

Rick CM (1982b) The potential of exotic germplasm for tomato improvement, pp 1–28. In: IK Vasil, WR Scowcroft and KJ Frey [eds]. Plant improvement and somatic cell genetics. Academic Press, New York.

Rick CM (1983) Genetic variation and evolution in Galapagos tomatoes, pp 97–106. In: RI Bowman, M Berson and AE Leviton [eds]. Patterns of evolution in Galapagos organisms. American Association for the Advancement of Science.

Rick CM (1984) Evolution of mating systems: Evidence from allozyme variation. In: Chopra VL, Joshi BC, Sharma RP, Bansal HC (eds) V X International Congress of Genetics. Bowker, Epping, Essex, UK, pp 215–221.

Rick CM (1986) Reproductive isolation in the *Lycopersicon peruvianum* complex, pp 477–495. In: WD. D'arcy [eds]. Solanaceae, Biology and Systematics. Columbia Univ Press, New York, USA.

Rick CM (1995) Tomato: *Lycopersicon esculentum* (Solanaceae), pp 452–457. In: SJ and S Nw [eds] 1995. Evolution of crop plants. Longman, Harlow, Essex, London.

Rick CM, Fobes JF (1975) Allozymes of Galapagos tomatoes—Polymorphism, geographic distribution, and affinities. Evolution 29: 443–457.

Rick CM, Fobes JF, Holle M (1977) Genetic variation in *Lycopersicon pimpinellifolium*: evidence of evolutionary change in mating systems. Plant Syst Evol 127: 139–170.

Rick CM, Fobes JF, Tanksley SD (1979) Evolution of mating systems in *Lycopersicon hirsutum* as deduced from genetic—Variation in electrophoretic and morphological characters. Plant Syst Evol 132: 279–298.

Rick CM, Kesicki E, Fobes JF, Holle M (1976) Genetic and biosystematic studies on two new sibling speces of *Lycopersicon* from interandean Peru. Theor Appl Genet 47: 55–68.

Rick CM, Lamm R (1955) Biosystematic Studies on the Status of *Lycopersicon chilense*. Amer J Bot 42: 663–675.

Rick CM, Tanksley SD (1981) Genetic variation in *Solanum pennellii*—Comparisons with 2 other sympatric tomato species. Plant Syst Evol 139: 11–45.

Rick CM, Zobel RW, Fobes JF (1974) Four peroxidase loci in red-fruited tomato species: Genetics and geographic distribution. Proc Natl Acad Sci USA 71: 835–839.

Riely BK, Martin GB (2001) Ancient origin of pathogen recognition specificity conferred by the tomato disease resistance gene *Pto*. Proc Natl Acad Sci USA 98: 2059–2064.

Robertson LD, Labate JA (2007) Genetic resources of tomato (*Lycopersicon esculentum* var. *esculentum*) and wild relatives, pp 25–75. In: MK Razdan and AK Mattoo [eds]. Genetic improvement of Solanaceous crops vol I: Tomato. Science Publishers Inc, Enfield, NH.

Rodriguez G, Pratta G, Zorzoli R, Picardi LA (2005) Characterization of the segregating generation of a tomato hybrid carrying nor and exotic genes. Pesquisa Agropecuaria Brasileira 40: 41–46.

Rooney HCE, van 't Klooster JW, van der Hoorn RAL, et al. (2005) *Cladosporium* Avr2 inhibits tomato Rcr3 protease required for Cf-2-dependent disease resistance. Science 308: 1783–1786.

Rose LE, Langley CH, Bernal AJ, Michelmore RW (2005) Natural variation m the *Pto* pathogen resistance gene within species of wild tomato (*Lycopersicon*). I. Functional analysis of *Pto* alleles. Genetics 171: 345–357.

Rose LE, Michelmore RW, Langley CH (2007) Natural variation in the Pto disease resistance gene within species of wild tomato (*Lycopersicon*). II. Population genetics of *Pto*. Genetics 175: 1307–1319.

Roselius K, Stephan W, Stadler T (2005) The relationship of nucleotide polymorphism, recombination rate and selection in wild tomato species. Genetics 171: 753–763.

Ross RJ (1998) Review paper: global genetic resources of vegetables. Plant Var Seeds 11: 39–60.

Sifres A, JBP, Blanca M, Frutos RD, Nuez F (2006) Genetic structure of *Lycopersicon pimpinellifolium* (Solanaceae) populations collected after the ENSO event of 1997–1998. Genet. Resour. Crop Evol 54: 359–377.

Smith S, Peralta IE (2002) Ecogeographic surveys as tools for analyzing potential reproductive isolating mechanisms: an example using *Solanum juglandifolium* Dunal, *S. ochranthum* Dunal, *S. lycopersicoides* Dunal, and *S. sitiens* IM Johnston. Taxon 51: 341–349.

Spooner DM, Anderson GJ, Jansen RK (1993) Chloroplast DNA evidence for the interrelationships of tomatoes, potatoes, and pepinos (Solanaceae). Amer J Bot 80: 676–688.

Spooner DM, Berg RGvd, Rodríguez A, et al. (2004) Wild potatoes (*Solanum* section Petota; Solanaceae) of North and Central America. Syst Bot Monogr 68: 1–209.

Spooner DM, Peralta IE, Knapp S (2005) Comparison of AFLPs with other markers for phylogenetic inference in wild tomatoes [*Solanum* L. section *Lycopersicon* (Mill.) Wettst.]. Taxon 54: 43–61.

Stadler T, Arunyawat U, Stephan W (2008) Population genetics of speciation in two closely related wild tomatoes (*Solanum* section *Lycopersicon*). Genetics 178: 339–350.

Stebbins GL (1950) Variation and Evolution in Plants. Columbia Unversity Press, New York.

Stevens MA, Rick CM (1986) Genetics and breeding, pp 35–109. In: JG Atherton and J Rudich [eds]. The tomato crop. Chapman and Hall, London.

Tanksley SD (2004) The genetic, developmental, and molecular bases of fruit size and shape variation in tomato. Plant Cell 16: S181–189.

Tanksley SD, Fulton TM (2007) Dissecting quantitative trait variation—examples from the tomato. Euphytica 154: 365–370.

Tanksley SD, Ganal MW, Prince JP, et al. (1992) High density molecular linkage maps of the tomato and potato genomes. Genetics 132: 1141–1160.

Tanksley SD, Loaiza-Figueroa F (1985) Gametophytic self incompatibility is controlled by a single locus on chromosome 1 in *Lycopersicon peruvianum*. Proc Am Soc Hort Sci 82: 5093–5096.

Tanksley SD, Medina-Filho H, Rick H (1981) The effect of isozyme selection on metric characters in a interspecific backcross of tomato—basis of an early screening procedure. Theor Appl Genet 60: 291–296.

Taylor IB (1986) Biosystematics of the tomato, pp 1–34. In: JG Atherton and J Rudich [eds]. The tomato crop: A scientific basis for improvement. Chapman and Hall, London.

Thomas CM, Jones DA, Parniske M, et al. (1997) Characterization of the tomato *Cf-4* gene for resistance to *Cladosporium fulvum* identifies sequences that determine recognitional specificity in *Cf-4* and *Cf-9*. Plant Cell 9: 2209–2224.

Thorup TA, Tanyolac B, Livingstone KD, et al. (2000) Candidate gene analysis of organ pigmentation loci in the Solanaceae. Proc Natl Acad Sci USA 97: 11192–11197.

Tsiantis M, Hay A, Ori N, et al. (2002) Developmental signals regulating leaf form, pp 418–430. In: QCB Cronk, RM Bateman and JA Hawkins [eds]. Developmental genetics and plant evolution. Taylor & Francis, London.

Underwood BA (2004) Vitamin A deficiency disorders: International efforts to control a preventable "pox". J Nutr 134: 231S–236.

USDA-ARS (2006) Tomatoes at a glance. In Tomato briefing room. Edited by the Economic Research Service, USDA (http://www.ers.usda.gov/Briefing/Archive/Tomatoes/index.htm).

Van der Hoeven R, Ronning C, Giovannoni J, Martin G, Tanksley S (2002) Deductions about the number, organization, and evolution of genes in the tomato genome based on analysis of a large expressed sequence tag collection and selective genomic sequencing. Plant Cell 14: 1441–1456.

Van der Hoorn RAL, Kruijt M, Roth R, et al. (2001) Intragenic recombination generated two distinct *Cf* genes that mediate AVR9 recognition in the natural population of *Lycopersicon pimpinellifolium*. Proc Natl Acad Sci USA 98: 10493–10498.

van der Knaap E, Tanksley SD (2003) The making of a bell pepper-shaped tomato fruit: identification of loci controlling fruit morphology in Yellow Stuffer tomato. Theor Appl Genet 107: 139–147.

Villalta I, Reina-Sanchez A, Cuartero J, Carbonell EA, Asins MJ (2005) Comparative microsatellite linkage analysis and genetic structure of two populations of F-6 lines derived from *Lycopersicon pimpinellifolium* and *L. cheesmanii*. Theoretical and Applied Genetics 110: 881–894.

Vos P, Simons G, Jesse T, et al. (1998) The tomato Mi-1 gene confers resistance to both root-knot nematodes and potato aphids. Nat Biotechnol 16: 1365–1369.

Vrebalov J, Ruezinsky D, Padmanabhan V, White R, Medrano D, et al. (2002) A MADS-box gene necessary for fruit ripening at the tomato ripening-inhibitor (Rin) locus. Science 296: 343–346.

Wikström N, Savolainen V, Chase MW (2001) Evolution of the angiosperms: calibrating the family tree. Proc Roy Soc Lond Sr B-Biol Sci 268: 2211–2220.

2

Classical Genetics and Traditional Breeding

John W. Scott,[1], James R. Myers,[2] Peter S. Boches,[3]
Courtland G. Nichols[4] and Frederic F. Angell[5]*

ABSTRACT

The linkage work done on tomato during the latter half of the 20th century laid the foundation for it to become a genetic model species with a recently sequenced genome. Over the last 20 years huge advances have been made in developing molecular linkage maps of tomato based on crosses with several wild relatives. Although many traits of interest have been mapped, there has been only modest work with many morphological markers. Herein a morphological map is presented for the first time in over 19 years that incorporates the changes that have been made in that time frame. Advancements made by traditional tomato breeders have made tomato the second most valuable vegetable crop in the world. Major varieties of the California processing industry that accounts for about 40% of the world's production are outlined as are some of the major fresh market field varieties in North America. Some of the major qualitative and quantitative traits incorporated into varieties by tomato breeders are outlined including disease resistances and horticultural traits such as blossom scar smoothness, crack resistance, earliness, extended shelf life, firmness, heat tolerance, machine harvest, and viscosity. A historical record of important breeding achievements is included. Several limitations faced by conventional tomato breeders

[1]University of Florida, Gulf Coast Research & Education Center, 14625 CR 672, Wimauma, FL 33598, USA.
[2]Oregon State University, Dept. of Horticulture, ALS 4017, Corvallis, OR 97331-7304, USA.
[3]Fall Creek Farm and Nursery, 39318 Jasper-Lowell Rd, Lowell, OR 97452, USA.
[4]1241 Sunset Dr, Hollister, CA 95023, USA.
[5]7281 Miller Ave, Gilroy CA 95020, USA.
*Corresponding author

are also pointed out. Properly done experiments with the coordinated efforts of breeders and molecular geneticists may provide information that will foster improved varieties that can be developed more efficiently in the future. Such varieties may be more broadly adapted, have good combining ability, and have improved stress tolerance, among other attributes.

Keywords: Morphological linkage map, processing tomato varieties, fresh-market tomato varieties, qualitative traits, quantitative traits, disease resistance, stress tolerance

2.1 Introduction

One could write an entire book on classical breeding and genetics of tomato, thus writing only a chapter will not allow us to cover all the topics adequately. Our goal is to cover some of the highlights on breeding accomplishments and to bring out some of the research areas that have proven difficult with conventional breeding approaches. We hope this problem identification will be useful for those working in genomics or genetic transformation and perhaps better solutions will be found in the future by employing modern technologies. Tomato is a model system for plant genetic studies, whose genome has been sequenced, largely because of the pioneering plant exploration and collection trips of Charlie Rick (http://tgrc.ucdavis.edu) along with linkage work that was carried out by Rick and a number of other dedicated scientists who had the vision to found the Tomato Genetics Cooperative in 1951 (http://tgc.ifas.ufl.edu/). The germplasm resources and genetic information was put to good use by a talented group of tomato breeders on whose shoulders we now stand. Some of the important early publications are cited herein but many valuable publications have not been cited due to space limitations. In delving back into the older literature we were reminded of just how much was learned by our predecessors. One useful publication that outlines many tomato horticultural phenotypes and what was known about their inheritance was that of Young and MacArthur (1947). New information is available for some of the traits but not for others so it is a good resource for modern day tomato scientists to be aware. Also a recommended publication for disease and disorder resistance is the review of Walter (1967).

In this chapter the early linkage work based on morphological traits and then isozymes is summarized (Section 2.1). Also in this section is an updated classical linkage map, the first published in over 19 years. Section 2.2 follows with some limitations of classical linkage studies. Many of the major breeding achievements are outlined in Section 2.3. Section 2.4 covers some breeding objectives that have met with various levels of success emphasizing areas that have been difficult to solve by conventional approaches. Section 2.5 briefly summarizes some of the reasons why molecular markers are useful for breeding programs.

2.2 Classical Mapping Efforts with Morphological and Isozyme Markers

The first summary of genetic linkage for tomato was presented by Hedrick and Booth (1907) for what is now known to be chromosome 2. In a series of early studies, MacArthur (1926, 1928, 1931) summarized linkage for all known linkage groups. In 1952, Butler described the seven linkage groups that were known for tomato, including a graphical map (Butler 1952). By 1958, the Tomato Genetics Cooperative Linkage Committee (with L. Butler as the Chairperson) was publishing a regularly updated map and summary of all known linked genes (Linkage Committee 1958). The first linkage report of the Tomato Genetics Cooperative placed many of the known tomato linkage groups on the physical map of tomato via deletion mapping (Linkage Committee 1958). The first genetic map to include all 12 chromosomes was published in 1968 (Butler 1968). By 1975, the published map included more than 258 morphological and physiological markers (Rick 1975). Tanksley (1993) and Tanksley et al. (1992) describe approximately 300 mapped morphological mutants in tomato and provide the most recent, comprehensive treatment of linkage mapping in tomato. Tanksley (1993) also summarizes all prior descriptions of gene linkage in tomato. In addition, these papers summarize the correspondence between molecular and classical genetic linkage maps. The molecular and classical maps were aligned by mapping morphological and isozyme markers with respect to restriction fragment length polymorphism (RFLP) markers (Tanksley et al. 1992; Tanksley 1993). Since some classical markers have also been located via deletion mapping on the cytological map (Linkage Committee 1958), the classical and molecular maps were aligned to the cytological map as well (Tanksley et al. 1992; Tanksley 1993).

The aligned classical, molecular, and cytological maps presented in Tanksley et al. (1992) and Tanksley (1993) provide some of the most comprehensive genome maps available in any plant species. There are, however, some problems associated with the classical linkage map as presented. These maps were assembled from linkage data collected from the literature (sometimes from small populations) and analyzed without the application of statistical procedures that take into account the statistical accuracy of the data or the frequency of double crossovers (Weide et al. 1993). Subsequent to the publication of the classical linkage maps in Tanksley et al. (1992) and Tanksley (1993), four major classical linkage mapping studies have been conducted and all have required the re-ordering of several to many loci. Revised maps are available for numerous chromosomes including the short arm of chromosome 1 (Balint-Kurti et al. 1995), chromosome 3 (van der Biezen et al. 1994), chromosome 6 (Weide et al. 1993; van Wordragen et al. 1996), the long arm of chromosome 7 (Burbidge et al. 2001) chromosome 10 (van Tuinen et al. 1997), chromosome 11 (van Tuinen et al. 1998), and chromosome 12 (van Tuinen et al. 1997) (Table 2-1).

Figure 2-1 is an updated classical linkage map reflecting some of these revisions. The updated map includes the addition of the new corolla color gene *Bco* to chromosome 7 (Chetelat 1998), the addition of *Pt-4, v, pet, uni* and *ds* to chromosome 11 (van Tuinen et al. 1998), and the removal of several loci from chromosome 11, which were not placed with regard to more than one locus on the classical map (van Tuinen et al. 1998). The map in Figure 2-1 includes updated maps of the short arm of chromosome 1 and chromosomes 3, 6, and 11. These were created from unified data for a majority of loci (Table 2-1), and take into account the frequency of double crossovers. The updated maps of these chromosomes include changes to the order and relative distances of several markers relative to the classical map of Tanksley (1993) or Tanksley et al. (1992). Readers should therefore, interpret other portions of the classical map with some caution. Limitations of classical genetic markers in research include dominance, epistasis, pleiotropy, and incomplete penetrance or expressivity.

Many 'classical loci' that have been characterized using Mendelian genetics but not placed on the classical linkage map have been mapped using molecular markers. Examples include numerous disease resistance loci (reviewed in Foolad 2007). In other cases, homologs of classical loci have been discovered using molecular techniques, for example the Milky Way and related loci (MW, Au, and OR) and *Cf,* resistance to *Cladosporium fulvum* (Soumpourou et al. 2007).

The most recent summary of isozyme mapping in tomato (Tanksley 1993) describes 41 isozyme loci from 15 unique enzymic reactions, 36 of which were placed on the linkage map. Since then four isozyme loci have been described, of which three have been mapped, including *Dia-3, Fdh-1, Mdh-1* (Table 2-1). *Mdh-1* was placed on 3L between *sy* and *sf*, but could not be positioned because recombination was highly suppressed in the *S. lycopersicoides* population relative to the distances expected from the classical map (Chetelat and DeVerna 1993).

2.3 Limitations of Classical Endeavors and Utility of Molecular Mapping

This topic has been recently reviewed by Chetelat and Ji (2007). They report the number of morphological markers that can be simultaneously and independently genotyped in this fashion is in many cases severely

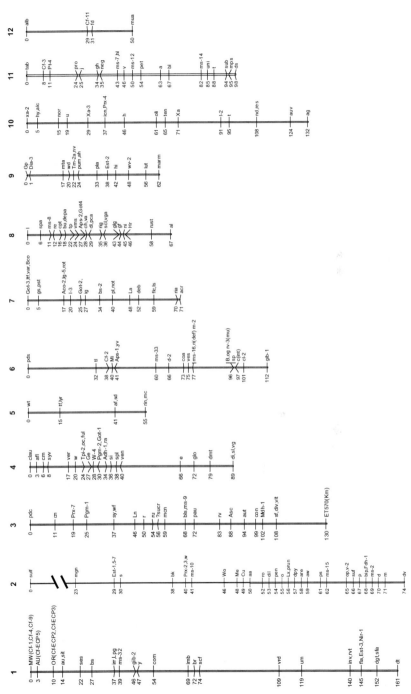

Figure 2-1 Revised classical tomato linkage map.

Table 2-1 Revisions to classical map listed by chromosome number.

Gene Name(s)	Chromosome	Reference
Loci with revised locations		
Cf-1, Cf-4, Cf-9, sit, au, ses, bs, irr, Lpg, ms-32, y, com, imb, br, scf	1	Balint-Kurti et al. 1995; van Tuinen et al. 1997
pdc, cn, Prx-7, Pgm-1, sy, wf, Ln, r, ru, mcn, bls, ms-9, pau, rv, sf	3	van der Biezen et al. 1994
Cf-2	6	Jones et al. 1992
Cf-2, Aps-1, yv, ms-33, d-2, coa, ms-16, ri(def), B, og, rv-3 (mu),sp,c(int)	6	Weide et al. 1993
alc, hy, nor, u, oli, Fs, Prx-4, Xa, Xa-3,icn, h, nd, l-2, t, res, auv, ag, alb, fd, mua	10, 12	van Tuinen et al. 1997
tab, Cf-3, j, gh, neg, ms-7, hl, ms-12, a, bl, ms-14, t, sub, mps, ds	11	van Tuinen et al. 1998
Loci removed due to lack of accurate placement		
co, Idh-1, Prx-1, mts	1	Balint-Kurti et al. 1995
Xa-2, Cf-11	10, 12	van Tuinen et al. 1997
wo-3, ele, apn, pco, mnt, ini, bi, I, I-2, x, mon, j-2, up	11	van Tuinen et al. 1998
Newly placed and newly described loci		
Cf-9	1	Jones et al. 1992
rot	1	Jones & Jones 1996
gib-2	1	Balint-Kurti et al. 1995
tri, hp-2	1	van Tuinen et al. 1997
Fdh-1 (isozyme)	2	Chetelat et al. 1993b
Asc, ET570 (Km)	3	van der Biezen et al. 1994
sucr	3	Chetelat et al. 1993a
Mdh-1 (isozyme)	3	Chetelat & DeVerna 1993

Cf-5	6	Jones et al. 1992
gib-1, ves	6	Weide et al. 1993
Bco	7	Chetelat 1998
Dia-3 (isozyme)	9	Chetelat et al. 1993
fri, yg-2	10,12	van Tuinen et al. 1997
pro, pt-4, v, pet, uni, ds	11	van Tuinen et al. 1998
Classical loci placed with the use of molecular markers		
AU (Cf-ECP5), OR (Cf-ECP2, Cf-ECP3)	1	Soumpourou et al. 2007
S	1	Liedl et al. 1993
Sod-2 (isozyme)	1	Tanksley et al. 1992
Ph-3	9	Chunwongse et al. 1998
Ve	9	Zamir et al. 1993
pct	9	Madishetty et al. 2006
Lv	12	Blaker et al. 1995
Loci found allelic to previously mapped loci		
int(com), dp(dgt), pu-2(al)	1	Jones & Jones 1996a
def(ri), mu(ro-3), int(c)	6	Weide et al. 1993
poc(pct)	9	Madishetty et al. 2006
ics(clau)	4	Rick 1998
Loci assigned to chromosomes but not mapped		
cb, cjf, era, gas, Jau, lu, ms-6, Nr-2, pe, pen, pr, tmf, Vi	1	Mutschler et al. 1987
Dia-2	1	Chetelat et al. 1993c

Table 2-1 contd....

Table 2-1 contd....

Gene Name(s)	Chromosome	Reference
gra?, ms-5, ne, vig	2	Mutschler et al. 1987
dis, ful-3, Od, pli, Prx-6, sts, sft	3	Mutschler et al. 1987
cox, 6pdgh-1, Rs	4	Mutschler et al. 1987
at, Fw, inf, n, 6Pgdh-3,	5	Mutschler et al. 1987
Dia-1	5	Chetelat et al. 1993c
Mae-1 (isozyme)	5	Chetelat et al. 1999
anr, inc?, prc, x-2	6	Mutschler et al. 1987
Anr, int, mu, prc, ri, ves, x-2	7	Mutschler et al. 1987
ltm	7	Moreau et al. 1997
ae, afr, deli, glau, spl-4, trs, yg-8	8	Mutschler et al. 1987
Crk, fsc, Nr, rela, Sn	9	Mutschler et al. 1987
ten	10	van Tuinen et al. 1997
ms-31, sh, xa-2	10	Mutschler et al. 1987
Abg	10	Rick et al. 1994
wv-3, ele, apn, pcv, mnt, ini, bi, I, 1-2, x	11	van Tuinen et al. 1998
ap, fgv, I, mn, ms-3, ne-2, Ora, Sod-1, up, v	11	Mutschler et al. 1987
Sm	11	Laterrot & Moretti 1995
Ora	11	Avdeyev & Scherbinin 1997
Aco-1, brt, Est-4, hp, 6Pgdh-2, Pgi-1, yg-2, yg-3, yg-4	12	Mutschler et al. 1987
Cf-11	12	van Tuinen et al. 1997

limited by their phenotypic effects. Problems frequently encountered are epistasis among genes controlling related traits, sterility or inviability of multiple marker combinations, and lack of markers with seedling stage expression for some genomic regions. Additionally, linkage tests of new morphological markers typically segregate in repulsion phase with respect to the tester combination, which limits precision of recombination fraction estimates for recessive genes. Coupling phase linkage tests provide greater precision, but require the prior synthesis of new marker combinations, which becomes limiting with more than just a few markers per chromosome. Isozymes overcame many of these limitations, but available enzyme staining technology restricted the number of protein markers. The advance of molecular markers based largely on interspecific populations to attain much greater marker saturation with neutral phenotypic effects has overcome the above limitations. More recently, marker saturation made possible by genomic technology is allowing for more precise identification and mapping of genes of interest.

2.4 Classical Tomato Breeding Achievements

2.4.1 Disease Resistance

Adding disease resistance to tomatoes was a major accomplishment during the 20th century. Field screening is often supplemented with seedling inoculation, which saves time and space. Fortunately, resistance to many important diseases is simply inherited and dominant. Most sources of resistance have come from the wild species and genes have been introgressed into tomato inbreds with good horticultural type. Wild species collected in Ecuador, Peru, and Chile by Charles Rick of UC Davis have contributed many important traits.

2.4.2 Quantitatively Inherited Traits and Cultivars—Processing Tomatoes

California typically produces about 40% of the world's processing tomatoes and breeding for this region will be emphasized herein. Many traits required by the California food processors for the best tomato cultivars are quantitatively inherited. These include yield, concentrated fruit setting, earliness, fruit firmness, crack resistance, and heat-tolerant fruit setting. In addition, successful cultivars require soluble solids of >5.4 °Brix, a pH of <4.4, viscosity of <4.0 Paste Bostwick, puree color of < 24 Agtron scale, fruit size of 75–80 g, and blocky fruit shape. Combining these traits plus disease resistance into one cultivar for the processors has been a huge challenge for the breeders, who have had great success. This success required many years of field selection and replicated trials along with quality evaluation of the fruit in the laboratory. This could not have been accomplished without the help of plant pathologists, geneticists, UC farm advisors, and tomato growers who provided field space for replicated trials.

In the 1950s the tomato varieties for California food processors had large (250 g), soft fruit, hand-harvested several times, yielding 36 MT/HA (16 T/A) over 40,500 hectares (100,000 acres). This has changed dramatically. Today an 80 g, firm, durable, elongated fruit is harvested by machine with state average yields of 78–90 MT/HA (35–40 T/A) over 101,000 hectares (250,000 acres). The visionary breeding of processing tomato cultivars for mechanical harvest was done by G.C. Hanna at the University of California, Davis (UC). For more detail on this achievement see Stevens and Rick (1986). Cultivars now have excellent processing quality with multiple disease resistances and stress tolerance. The most important processing cultivars in California over time along with some of their characteristics are listed in Table 2-2. The eight cultivars with the most usage in the California processing industry are listed in Table 2-3 along with their time of use and their tonnage.

2.4.3 *Quantitatively Inherited Traits and Cultivars—Fresh Market Tomatoes*

Great improvements in the 20th Century were achieved with many quantitatively inherited traits of fresh market tomatoes. Fruit firmness, earliness, large fruit size, crack resistance, symmetrical fruit shape, vine cover, and high yields all involve quantitative inheritance that has often not been well characterized genetically. Increased fruit firmness (Scott 1984) was very important for shipping to distant markets. One of the first important shipping cultivars was Fla. MH-1 a cultivar developed for possible machine harvest (Crill et al. 1971). More recently, firm-fruited breeding lines are being developed that combine the brychytic (*br*) gene to increase side shoot production without apical dominance, prostrate growth habit, and jointless stems to develop a new generation of machine harvest tomatoes in case labor is not available to harvest fresh market tomato crops (Ozminkowski et al. 1990). Very smooth, large, round fruit with tiny blossom end scars added to market appeal. Nipple blossom scar genes help to provide blossom scar smoothness (Barten et al. 1994) are now widely used by tomato breeders. However, other quantitative genes also affect the percentage of smooth fruit in genotypes with nipple genes under stressful growing conditions. Multiple disease resistance allowed tomatoes to be grown successfully in many more areas. Introduction of F_1 hybrids in the 1970s greatly increased early yields and larger fruit size with the highest monetary value (e.g., 'Royal Flush' and 'Jackpot'). The long shelf-life genes *rin* and *alc*, when used in hybrids, reduced fruit softening due to reduced enzyme activity

Table 2-2 The most important California processing tomato cultivars over the last 100 years.

Variety[1]	Source	Best Years[2]	Avg. Fruit Wt. (g)	Quality			Yield (MT/HA)	Disease Resistance[3]	Notes
				Soluble Solids[2]	Viscosity/ Consistency	Firmness			
Santa Clara Canner	Italy Trophy	1895–1936	250	high	low	soft	11	none	indeterminate vine with rough fruit
Pearson	Eastern States	1937–1955	200	high	low	soft	22	none	determinate vine with smooth fruit
VF6 & VFI1	UC Davis	1955–1962	200	high	low	soft	35	VF	
Ace	Campbell Soup	1955–1963	250	high	low	med/soft	45	none	
VF145B7879	UC & Castle	1962–1974	75	5.5	low	med/soft	56	VF	1st mech harvest
VF13-L	UC Davis	1970–1977	70	low	med	good	45	VF	v durable
VF315	Niagara/HM[4]	1970–1976	75	med	low	med	56	VF	
VF134	UC & Peto	1973–1977	70	5.3	med/high	good	45	VF, ASC	
Peto 81	Peto	1974–1982	75	5.2	med/high	good	56	VFF, ASC	
UC82	UC Davis	1977–1980	65	4.8	high	good	56	VF, ASC	
*GS 12	Goldsmith	1978–1982	75	5.3	low/med	med	56	VF	early
*Peto 31	Peto	1980–1985	70	5.0	low	med	56	VF	v early
Murrieta	Harris Moran	1980–1987	60	5.3	low	med	56	VF	peeler
*Peto Pride II	Peto	1980–1985	75	5.3	med	good	67	VFF, ASC	
*Castle 1017	Castle	1981–1985	75	5.5	low	good	67	VFF, ASC	
FM6203	Ferry Morse	1981–1993	75	5.1	med/high	good	67	VF, ASC	peeler

Table 2-2 contd....

Table 2-2 contd....

Variety[1]	Source	Best Years[2]	Avg. Fruit Wt. (g)	Soluble Solids[2]	Quality Viscosity/ Consistency	Firmness	Yield (MT/HA)	Disease Resistance[3]	Notes
*FM785	Ferry Morse	1985–1991	80	5.6	low/med	good	78	VFF, ASC	flavor
Diego (UC204)	UC & HM[4]	1987–1991	75	5.4	med	good	67	VFF, ASC	
*Alta	Campbell Soup	1987–1992	70	5.4	med	good	67	VF, ASC	
Apex 1000	Ferry Morse	1989–1994	70	5.3	low	v good	67	VFF, ASC	peeler
*Nema 1401	Peto	1989–1993	75	5.3	high	good	67	VFFN, ASC	
*Brigade	Asgrow	1990–1995	70	5.2	med	good	67	VFF, ASC	
*Nema 512	Peto	1991–1994	75	5.0	med/high	good	78	VFFN	early
*BOS 3155	Orsetti	1992–2005	80	5.4	med/high	good	67	VFF	v durable
*H3044	Heinz	1993–1996	75	4.8	med/high	good	78	VFFN	early, hi color
*H8892	Heinz	1994–2001	60	5.1	high	good	78	VFFN	v hi color
*Sun 6117	Sun/Nunhems	1998–2007	75	5.2	med	good	78	VFFN	early
*H9665	Heinz	1999–2006	75	5.0	med	good	78	VFFNP	v durable
*APT410	Asgrow/ Seminis	2000–2006	70	5.2	med	good	78	VFFNP	early
*Hypeel 303	Peto/Seminis	2000–2007	75	5.2	high	good	78	VFFNP	peeler
*CDX179	Campbell Soup	2001–2005	75	5.5	med	good	78	VFFNP	
*H9557	Heinz	2001–2008	65	5.3	high	good	78	VFFNP	
*H9663	Heinz	2002–2009	75	5.1	med/high	good	90	VFFNP	v hi color

*H9780	Heinz	2003-2011+	75	5.4	v high	v good	90	VFFNP	v durable
*AB2	DeRuiter	2004-2011+	80	5.6	med	good	90	VFFNP	
*H2401	Heinz	2006-2011+	75	5.3	high	good	90	VFFNP	v durable
*Sun 6366	Sun/Nunhems	2007-2011 +	75	5.5	med	good	110	VFFNP	early
*H8504	Heinz	2008-2011+	65	5.3	v high	good	110	VFFNP	quality

[1]George 2008a,b; *indicates F$_1$ hybrid
[2]California Processing Tomato Advisory Board 1986–2011
[3]V=verticillium wilt, F=fusarium wilt race 1, FF=fusarium wilt races 1&2, ASC=alternaria stem canker, N=root knot nematode, and P=bacterial speck
[4]Harris Moran

Table 2-3 The eight most widely used varieties in California history.

Primary decade	Cultivar	Total tons (MT) used (Millions)
1960s	VF145B-7879	7.3–9
1980s	FM6203	10.8
1990s	Brigade	6.4
1990s + 2000s	BOS 3155	22.7
1990s	Heinz 8892	11.8
2000s + 2010s	AB2	9.8
2000s + 2010s	Heinz 9780	6.6
2000s + 2010s	SUN6366	8.3

as described by Tigchelaar et al. (1978) and Mutschler et al. (1992). Since the early 1990s, cultivars heterozygous for the *rin* gene with firm genetic backgrounds have been widely grown in Mexico, Spain, North Africa, Australia, and elsewhere to facilitate shipment of vine-ripe fruit to distant markets (Rabinowitch and Kedar 1995). The old gold or crimson (og^c) gene boosts lycopene levels and gives excellent interior fruit color grown in warm weather: 15–30°C (Tomes1963; Thompson et al. 1967). Old gold refers to the gold colored flower petals that are linked to this gene. The gold color expresses under cool conditions but disappears when temperature are high. Over the years, a number of crimson cultivars have been released including several in Canada by Ernie Kerr. More recently, interest in this gene has increased due to the antioxidant properties of lycopene (Vrieling et al. 2007). An example of a recent release is the superior flavor hybrid Fla. 8153 being sold under the name Tasti-Lee™ (Scott et al. 2008). The crimson gene, although originally discovered in an indeterminate background and tightly linked to the *sp+* gene, has been bred primarily into determinate backgrounds and only recently has been transferred back into modern, indeterminate types.

Information on major fresh-market cultivars has not been well documented. Florida has been a major producer of fresh-market tomatoes and leads the USA in production and crop value. 'Homestead 24' was widely grown in the 1950s and 1960s. When Fusarium wilt race 2 became problematic 'Walter' was widely grown in the late 1960s and 1970s. 'Flora-Dade' and 'Florida MH-1' were popular in the 1970s. Hybrids 'Duke' and 'FTE 12' were predominant in South Florida in the 1980s. Other major cultivars in the hybrid era were 'Sunny' in the 1980s, 'Agriset 761' in the early 1990s, and 'Florida 47' for the last decade. 'Solar Set' was an important heat-tolerant cultivar during the 1990s while 'Florida 91' has taken over much of that hectarage since then. 'Sebring' has resistance to all three races of Fusarium wilt and Fusarium crown rot and in the last nine years been widely grown. There has been an increase in plum tomatoes for fresh market and 'Peto 882', 'Mariana', and 'BHN 410', 'BHN 411', and more recently 'BHN 685' have had considerable successes. In the southeast, some important cultivars other than those mentioned in Florida have been 'Sunguard' and a number of the Mountain series developed at North Carolina State University. 'Mountain Fresh' and 'Mountain Spring' have been the most widely grown of these. California is another large USA producer of fresh tomatoes and some of the major cultivars are listed in Table 2-4.

Table 2-4 Major California cultivars for the last 50 years.

Decade(s)	Cultivar
1960s/1970s	Pearson, Ace, Early Pak
1970s/1980s	Royal Flush, Valerie, Jackpot, Casino Royale
1980s	Castle 17, Count Fleet, Sweepstakes
1990s	Shady Lady, Olympic, Mt. Spring, Sun Bright, SVR 2935
2000s	Quality 21, Quality 23, Bobcat

2.4.4 Qualitative Inheritance

Geneticists of the 20th century identified hundreds of simply inherited genetic traits and mapped many of them on the chromosomes (Fig. 2-1). Some of these genes were mentioned in the previous section. Seedling markers closely linked to important traits promoted efficient selection, e.g., $Tm2^2/ah$ (green stem) developed at Cornell in the 1970s and $ms\text{-}10^{35}/aa$. Several of the other genes used in tomato improvement for plant or fruit traits are listed in Table 2-5.

2.4.5 Allozymes

Some of the isozymic traits are closely linked to disease resistant genes, such as *Aps*-1 with the *Mi* gene for nematodes (Rick and Fobes 1974) and *Got*--2 with *I-3* gene for Fusarium wilt race 3 (Bournival et al. 1989). Electrophoresis of these isozymes allowed breeders to save only resistant plants. Unfortunately, crossovers were possible so periodic testing with live organisms was important. Isozymic traits, including *Adh-1*, are also very useful for quick and inexpensive testing of hybridity prior to hybrid seed sales (Tanksley and Jones 1981). Nowadays, PCR markers are often used instead of isozyme markers.

2.4.6 Stress Tolerance

To comprehensively breed for good levels of stress tolerance is an in-depth process and no attempt will be made here to cover these topics in detail.

Table 2-5 Examples of single genes that have been useful for horticultural improvement in tomato (adapted from Tigchelaar 1986).

Gene designation	Gene symbol	Cultivar
	Growth habit	
Brachytic	*br*	Redbush, Smarty
Dwarf	*d*	Epoch, Tiny Tim*
Jointless pedicel	*j-1*	Penn Red
	j-2	Many
Potato leaf	*c*	Geneva 11
Self-pruning	*sp*	Many
	Fruit characters	
Colorless peel	*y*	Traveller, Ohio MR-13
Green stripe	*gs*	Tigerella, Green Zebra
High beta	*B*	Caro-Rich, Jaune Flamme
High pigment	*hp*	Redbush
Low total carotene	*r*	Snowball, Livingston's Gold Ball
Male-sterility	many genes	Some F_1 hybrids
Nipple blossom scar	*n-2*	NC140
	n-4	Mountain Spring, Valerie
Nonripening	*alc (nor ^{A})*	Lenor
Old gold crimson	*og^c*	Vermillion, Plum Crimson*
Parthenocarpic fruit	*pat-2*	Severianin*
Ripening inhibitor	*rin/+*	Daniela*
Tangerine	*t*	Sunray, Jubilee, Carolina Gold, Mountain Gold
Uniform ripening	*u*	Heinz 1350*

*plus many others

Much more detail on this topic can be found in a recent review (Foolad 2007). Heat-tolerant fruit setting ability was originally achieved by planting in hot climates where night temperatures remained above 20°C (Schaible 1962). The optimum night temperature range for fruit set is 15–20°C. Selection for fruit set at elevated temperatures also can improve fruit set at low night temperatures (Young 1963; J. Scott, unpublished observations). The resulting wide adaptability has allowed tomatoes to be grown in almost every country of the world when the necessary disease resistances and other stress tolerances have been added. Many heat-tolerant cultivars have been developed at the Asian Vegetable Research and Development Center (AVRDC) in Taiwan now AVRDC—The World Vegetable Center, at the University of Florida, USA and at several private companies. Further horticultural improvements may well expand the use of heat-tolerant cultivars in seasons where there is not excessive heat. Possibly their overall improved yielding ability could one day render most other cultivars obsolete.

Improved nutrient utilization also provides wider adaptability. Many developing countries have poor soil conditions for growing crops, e.g., low nitrogen, phosphorous, and potassium. Most tomatoes require high levels of N, P, and K. Tomato lines that were very efficient at utilizing low levels of N, P, and K have been identified at the University of Wisconsin (O'Sullivan et al. 1974; Makmur et al. 1978; Coltman et al. 1982).

Salt tolerant inbreds have been identified that provide normal fruit and vine size under saline soil conditions. In a field of over 200 breeding lines, FME1205 was the only cultivar with a normal vine and crop. All others had very small vines with very poor fruit set growing in a field with very saline soil conditions (Rush and Epstein 1976; CG Nichols, personal observations). Stock seed of E1205 was produced in 1972 but discarded before salt tolerance was verified, since it was too soft for machine harvest.

Drought tolerance due to large, deep, root systems or low transpiration has been identified. In a one acre field of four 0.25 acre varieties, that missed a critical irrigation while fruit was sizing up, FM 785 exhibited high yield while the other three good varieties had very small fruit and poor yield (CG Nichols, personal observations).

Tropical rain or flood tolerance has been found by the University of West Indies in Jamaica in the 1970s. Unfortunately, this selection had an enormous vine with poor fruit set and was very difficult to work with. The variety Nagcarlan, in the Philippines, has shown very good rain/flood tolerance (Rebigan et al. 1977; Villareal et al. 1977).

Defoliation tolerance due to air pollution has been observed in New Jersey by researchers of the USDA, Beltsville in the 1970s (Webb et al. 1973).

2.4.7 Achievement Efforts

Plant breeders, generally with rather low budgets, have made great improvements in the tomato throughout the 20th century. Early on much of the breeding was done at public institutions with seed companies taking the cultivars to produce and sell the seed. Once hybrids became popular, there was a dramatic shift of the breeding effort where private companies took a much more active role in tomato breeding. This was primarily due to the ability of companies to have reasonable protection for their hybrid cultivars vs. open-pollinated cultivars. More discussion on breeding tomato hybrids is available (Scott and Angell 1998). However, there was still considerable breeding done by private companies in the days of open-pollinated cultivars and in fact progress in the 1950s and 1960s relied on cooperation between not only private and public breeding programs, but also between private companies themselves. This is in stark contrast to today's world of plant protection and rampant patenting of plants, genes, and even breeding concepts. Tomato improvement in the USA resulted from a cooperative effort among universities, federal agencies, food processors, and seed companies. Some of the major tomato breeding programs at USA universities were University of Arkansas, University of California (Davis), Cornell University, University of Florida, University of Hawaii, University of Missouri, Michigan State University, University of New Hampshire, North Carolina State University, The Ohio State University, Oregon State University, Purdue University, Rutgers University and Texas A&M University. Genetic studies on tomato were done at other universities as well. There was also important tomato breeding work done at the USDA especially at Beltsville, MD and Charleston, SC. Important tomato processing companies included Campbell Soup, Heinz, Hunt Wesson, and Libby's. Some key seed companies were Asgrow/Seminis/Monsanto, Castle, Ferry-Morse, Goldsmith, Harris-Moran (and previously the Joseph Harris Company), Northrup King/Rogers/ Syngenta, Orsetti, Peto/Seminis/Monsanto, and Sun Seeds/Nunhems. Of course, much tomato improvement work took place outside of the US. A number of Dutch seed companies made important contributions, often for greenhouse cultivar development: Bruinsma, DeRuiter/Monsanto, Enza-Zaden, Royal Sluis, Seminis/Monsanto and Sluis and Groot. Historically, several countries have done considerable work on tomato breeding: Australia, Brazil, Bulgaria, Canada, France, Israel, India, Japan, and the Netherlands. In addition, several national research centers have made important contributions: Asian Vegetable Research and Development Center (AVRDC, Taiwan) now AVRDC—The World Vegetable Center, Canadian Department of Agriculture (CDA), and Institut National de la Recherche Agronomique (INRA, France). The insightful publication of Walter (1967) listed a number of the pioneering institutions and scientists who were doing multiple disease resistance work at that time. These included "Glasshouse Crops Research Institute, Littlehampton, Sussex, England, led by Darby; Horticultural Research Institute, Ontario Department of Agriculture and Food, Canada, led by Kerr; Redlands Horticultural Research Station, Queensland, Australia, led by Bonner; Horticultural Research Station, Shimokuriyagawa, Morioka, Japan, led by Kamimura; Horticultural Division of National Tokai-Kinki Agricultural Experiment Station, Japan, led by Suzuki; Escola Superior de Agricultura, Vicosa, Minas Gerais, Brazil, led by Couto; Station d'Amelioration des Plantes, Montflavet, France, led by Laterrot; University of Puerto Rico A.E.S., Rio Piedras, P.R., led by Azzam; Pretoria Horticultural Research Station, Union of South Africa, led by Joubert; Agricultural Experimental Institute of Duna-Tiszakoz, Kecskemét, People's Republic of Hungray, led by Meszolv; and Department of Field and Vegetable Crops, Hebrew University, Rehovot, Israel, led by Retig." Several of the cultivars first developed for various traits and their developers are listed in Table 2-6.

Table 2-6 Some important tomato breeding achievements.

Genetic Trait	Cultivar	Where Developed	Breeder	Year	Reference
I. Horticultural					
Blossom-scar smoothness	Valerie Jackpot Casino Royale Mt. Spring & Mt. Fresh	Goldsmith (CA) Ferry Morse (CA) Ferry Morse (CA) NC State Univ.	V. Morris C. Nichols C. Nichols R. Gardner	1970's 1970's 1980's 1990's	Barten et al. 1994 FM Veg Varieties 1981 FM Veg Varieties 1981 Gardner 1992, 1999
Crack resistance	Burgess Crack Proof used to develop Campbell 135 & 146	Campbell Soup, Riverton, NJ	G. Reynard	1950's	Reynard 1951 Reynard 1960
Determinate vine	Pritchard Pearson	U. of Calif., Davis	O. Pearson	1935	Porter & MacGilliveray 1937
Earliness	Farthest North, Fargo Fireball Campbell 28	North Dakota Joseph Harris Campbell Soup	A. Yaeger W. Scott J. Moore	1933 1960's 1970's	Yaeger 1933
Extended shelf life *rin/+* gene	Devine Ripe™	Rehovot, Israel	N. Kedar	1992	Rabinowitch & Kedar 1995
Firmness	ES24	Eastern States, MA	O. Pearson & C. John	1954	Scott 1984
	Heinz 3 & 4	H.J. Heinz, Ohio	C. John	1960's	Scott 1984
	Fla. 1346	Univ. of Florida	J. Strobel	1960's	Scott 1984
Heat-tolerant fruit set	Narcarlang/Nagcarlan used to develop Campbell Exp1-20	Philippines Campbell Soup, Paris, TX	L. Schaible & C. Nichols	1950's 1962–1965	Schaible 1962, Reynard personal comm.
	Saladette, FM-9, Processor 40	Texas A&M	P. Leeper	1970's	Leeper personal comm.

Table 2-6 contd....

Table 2-6 contd....

Genetic Trait	Cultivar	Where Developed	Breeder	Year	Reference
Machine harvest	VF145 A&B	U. of Calif., Davis	G. Hanna	1961	Stevens & Rick 1986
	Fla. MH-1	Univ. of Florida	P. Crill	1971	Crill et al. 1971
	Jackpot	Ferry Morse CA	C. Nichols	1976	FM Veg. Varieties 1981
Viscosity (consistency)	UC82	U. of Calif. Davis	A. Stevens	1976	Stevens et al. 1977
II. Disease Resistance					
Alternaria stem canker resistance	Homestead	USDA and Univ. of Florida	C. Andrus	1953	Scott 2002b
Alternaria alternata f. *lycopersici*	Fla. 1346	Univ. of Florida	J. Strobel	1960's	Grogan et al. 1975
Anthracnose fruit rot resistance (*Colletotrichum* sp.)	Campbell 135 & 146	Campbell Soup Riverton, NJ	G. Reynard	1950's	Hoadley 1960
	several inbreds	USDA, Beltsville	T. Barksdale	1960's	Barksdale 1972
Bacterial speck resistance *Pseudomonas syringae* po tomato	Ontario 7710	Ontario, Canada	E. Kerr	1983	Kerr & Cook 1983
Bacterial wilt resistance *Ralstonia solanacearum*	Hawaii 7526 Louisiana Pink Rodade	Hawaii Louisiana South Africa	J. Gilbert S. Bosch	1968 1982? 1985	Gilbert et al. 1969 Sheela & Peter 1987 Bosch et al. 1985
	Hawaii 7996	Hawaii	J. Gilbert	1970's	Thoquet et al. 1996
	several inbreds including CL5915 & CL657	AVRDC	Opeña	1992	C Nichols personal comm.
	several inbreds	Univ. of Philippines		1980's	C Nichols personal comm.

Trait	Variety/Line	Location	Breeder	Year	Reference
	Fla. 8109	Univ. of Florida	J. Scott	2003	Scott et al. 2003
Black mold tolerance *Alternaria alternata*	Apex 1000	Ferry Morse, CA	C. Nichols	1982	FM Veg Varieties 1984[1]
	Cannery Row	Ferry Morse, CA	C. Nichols	1983	FM Veg Varieties 1984[1]
Curly top virus resistance	CVF4	Utah AES & USDA	O. Cannon	1966	Martin 1969
	Columbia, Roza & RowPak	Washington State & USDA	M. Martin	1976	Martin 1985
Early blight tolerance *Alternaria solani:* Collar Rot	Southland	USDA Charleston Univ. of Florida	C. Andrus	1945	Reynard & Andrus 1945
	Manalucie	USDA Beltsville	J. Walter	1950's	Walter & Kelbert 1953
	68B134		T. Barksdale	1969	Barksdale 1969
Foliage	NC EBR 3,4,5 and 6	N.C. State Univ.	R. Gardner	1990's	Gardner & Shoemaker 1999
Fusarium crown and root rot *Fusarium oxysporum f. sp. radicis-lycopersici*	Ohio CR-6	Ohio State Univ.	J. Scott & J. Farley	1983	Scott & Farley 1983
Fusarium wilt resistance *Fusarium oxysporum f. sp. lycopersici* Race 1	Pan America	USDA Beltsville	W. Porte	1941	Porte & Walker 1941
Race 2	Walter	Univ. of Florida	J. Strobel et al.	1970's	Strobel et al. 1969
Race 3	*S. pennellii* BC into New Yorker = NY87-1095-6	Cornell, NY	M. Mutschler	1988	Goffreda & Mutschler 1988
	BHRS-2-3	Australia	D. McGrath	1988	McGrath 1988
Races 1, 2, 3	Fla. 7481 & Fla. 7547	Univ. of Florida	J. Scott	1995	Scott & Jones 1995

Table 2-6 contd....

Table 2-6 contd....

Genetic Trait	Cultivar	Where Developed	Breeder	Year	Reference
Gray leafspot resistance *Stemphylium solani*	seven varieties Manalucie Homestead Fla. 1346	Univ. of Hawaii Univ. of Florida Univ. of Florida Univ. of Florida	W. Frazier J. Walter C. Andrus et al. J. Strobel	1949 1953 1953 1960's	Frazier et al. 1950 Walter & Kelbert 1953 Scott 2002b
Late blight tolerance *Phytophthora infestans*	W. Virginia 63 NC 1CELBR, NC 2CELBR (*Ph-2* + *Ph-3* combined)	W. Virginia North Carolina	M. Gallegly R. Gardner	1960's 2008	Gallegly 1960 Gardner & Panthee 2010
Nematode resistance *Meloidogyne* spp.	Anahu VFN 8 Calmart	Univ. of Hawaii U. of Calif. Davis U. of Calif. Davis	J. Gilbert P. Smith P. Smith	1955 1960's 1969	Gilbert et al. 1969 Stevens & Rick 1986 C Nichols personal com.
Southern blight *Sclerotium rolfsii*	several breeding lines	Texas A&M Univ. and Univ. of Georgia	P. Leeper et al.		Leeper et al. 1992
Tobacco mosaic virus resistance $Tm2^2$ gene	Ohio M-R 9 several inbreds including 72T24/25 & 74T17/19	Ohio State Univ. U. of Calif. Davis	L. Alexander P. Smith	1966 1970's	Pelham 1966 C Nichols personal com.
Tomato mottle virus resistance	several inbreds including Fla. 719	Univ. of Florida	J. Scott	1990's	Griffiths & Scott 2001
Tomato spotted wilt virus resistance *Sw-1* gene *Sw-5* gene *Sw-7* gene	Pearl Harbor seven varieties Stevens Fla. 8516	Hawaii Hawaii South Africa Univ. of Florida	W. Frazier W. Frazier J. Van Zijl J. Scott	1945 1949 1988 2007	Kikuta et al. 1945 Frazier et al. 1950 van Zijl et al. 1986 Stevens et al. 1992 Price et al. 2007

Tomato yellow leaf curl virus					
5 recessive genes					
Ty-1 gene	TY20 F$_1$	Israel	M. Pilowsky	1988	Pilowsky & Cohen 1990
Ty-2 gene	Ty52	Israel	D. Zamir	1994	Zamir et al. 1994
	H24	India	G. Kalloo	1990's	Kalloo & Bannerjee 1990
	Several inbreds	AVRDC	P. Hanson	2000's	Hanson et al. 2000
					Hanson et al. 2006
Ty-3 gene	Gc9, Gc173	Univ. of Florida	J. Scott	2007	Ji et al. 2007
Ty-4 gene	Gc171, Fla. 726	Univ. of Florida	J. Scott	2009	Ji et al. 2009
Ty-5 gene	TY172	Volcani Center, Israel	I. Levin	1995	Anbinder et al. 2009
ty-5 gene	Tyking	Royal Sluis, The Netherlands	J. Hoogstraten	1990	Hutton et al. 2012
Verticillium Wilt resistance	VR Moscow	Utah	O. Cannon	1952	Cannon & Waddoups 1952
	Loren Blood	U. of Calif. Davis	G. Hanna	1950's	
Verticillium albo-atrium	VF6 & VF11				
III. Insect resistance					
Colorado potato beetle tomato horn worm	*L. hirsutum f. glabratum* BC to FM 6203	Cornell	M. Mutschler	1980's	Williams et al. 1980
Spider mite	Hawaii 7526	Hawaii	J. Gilbert	1968	Gilbert et al. 1969
Spider mite	68B134	USDA Beltsville	A. Stoner	1968	Stoner 1970

2.5 Breeding Objectives Positive and Negative

2.5.1 Historical Background

Some of the historical aspects of tomato breeding were presented over 25 years ago (Stevens and Rick 1986; Tigchelaar 1986). Indigenous people did the initial selection for productivity, which involved selection for fruit size and non-exserted styles, the latter likely being selected indirectly as seed was saved from productive plants that bred true. Italians made improvements in fruit characteristics over 200 years ago (Stevens and Rick 1986). Breeding of cultivars in the modern era began in the late 1800s as seed companies began to sell tomato seed to the public. Many of the accomplishments up to 1937 were reported by Boswell (1937). The twentieth century was a golden age for classical breeding of tomato and other crops. Profound improvements were made in plant type, fruit yield and quality, adaptability to diverse growing conditions, and disease resistance. In the field, the use of the self pruning (*sp*) gene allowed for determinate plants to be developed, which allowed for more fruit to develop on smaller vines and thus plants to be grown with closer spacing and harvested over shorter time periods with little or no pruning. Improvements for earliness, fruit firmness, and defect-free fruit were made. Breeders then developed cultivars with more concentrated fruit set that allowed for once over mechanical harvest of primarily processing tomatoes but also fresh market tomatoes to a lesser extent. Incorporation of the jointless pedicel (*j2*) gene facilitated harvest of fruit without stems, which has been of critical importance to mechanical harvesting. Breeding difficulties have been encountered with traits controlled both qualitatively (single genes) and quantitatively (multiple genes). Several breeding issues are discussed below.

2.5.2 Jointless Pedicels

The jointless pedicel trait (*j, j-2*) is desirable for ease of harvest because the fruit stem remains on the plant during fruit picking. This trait has been bred intensively by numerous breeders and successful cultivars, mostly small-fruited processing and fresh-market types, have been developed. However, cultivars with jointed pedicels are often superior in horticultural attributes to those with jointless pedicels, especially where large-fruited cultivars are being developed. It has been particularly difficult to develop large-fruited, jointless, heat-tolerant tomatoes (Scott et al. 1998). Many of the other difficulties encountered in the breeding of jointless cultivars have not been documented in the literature. Possibly without the smaller cells of a jointed pedicel, the flow of water into the fruit is not as well regulated. This could cause more catfacing during early development and greater cracking as fruit mature. Flavor of many jointless lines tested has been bland (JW Scott, unpublished data), and this could also relate to more water entering the fruit and diluting the flavor. Under stress conditions large fruited, jointless tomatoes tend to have rough blossom-end scars and poor pollination, which results in off-shapes. The former can also happen with adequate pollination.

2.5.3 Parthenocarpy

Tomato breeders have long been attracted to the facultative parthenocarpic trait (conditioned by several loci—*pat-1* through *pat-4*) because it allows fruit set under conditions of cold and heat stress. Tomato varieties with the trait will set fruit under the cool maritime climatic conditions of the Pacific Northwest US as well as setting fruit through an Arizona summer. A group of heat-tolerant lines were sent to Saudi Arabia for testing under extremely hot temperatures and the only one which set fruit was a parthenocarpic line as pollination and fertilization failed in the others (JW Scott, unpublished data). Despite this result, after considerable breeding efforts in Florida it was seen that parthenocarpic expression was better under cool than warm, humid conditions (JW Scott, unpublished data). With the exception of certain early season heirloom types and most cultivars developed by the Oregon State University breeding program (Baggett 1995), parthenocarpy has not been incorporated into commercial backgrounds. One reason is that cultivars with parthenocarpy have a significantly higher frequency of rough fruit (catfacing and extruded secondary ovaries). A second reason is that parthenocarpic fruit often lack the firmness required by the fresh market and processing industries. Parthenocarpic cultivars will set seed when growing temperatures are sufficient for normal fertilization, but amounts of seed are often less than non-parthenocarpic types, making seed production more expensive and less efficient. If *pat-2* was cloned, it would be interesting to see if the defects mentioned would still be encountered.

2.5.4 High Pigment

Tomato fruit with the high pigment trait (conditioned by *hp-1* through *hp-3*) have higher levels of several nutrients including carotenoids (lycopene and any other carotenoids present in the fruit), ascorbic acid, phenolics and flavonoids (Levin et al. 2003). Unripe fruit are dark green and ripen to an intense deep red appearance when mature. Considerable effort has gone into developing high pigment cultivars (Levin et al. 2003; Lenucci et al. 2006) but they have not been widely accepted. While high pigment conditions desirable fruit traits, deleterious pleiotropic effects at the whole plant level cause brittle stems and reduced productivity and it is this latter characteristic that has prevented wide-spread adoption. It may be possible to overcome the deleterious plant characteristics while retaining the desirable fruit attributes by selection for increased productivity in a high-pigment background or by developing fruit specific promoters for *hp*. However, even with fruit specific promoters another problem has been that the fruit often have blotchy ripening or yellow top and this seems to relate to the high chlorophyll concentrations in the dark green fruit (JW Scott, unpublished data).

2.5.5 Fruit Disorders

Whereas great progress has been made in developing cultivars with minimal disorders, progress is often slowed by the variable expressivity of many disorders due to the interactions between genetic and environmental factors. Many times selections are made in segregating populations where a disorder such as graywall is not present, and it is only later on that sensitivity is discovered. This can "kill" or limit production of an otherwise promising cultivar. Certainly having molecular markers linked to tolerance to such disorders would be a great boon to combining tolerance with other traits in new cultivars.

One change that occurred in the last decades of the 20th century was the large-scale replacement of cultivars with green-shouldered fruit, with uniform or light-green shoulders by use of the *u* or *ug* genes. This change has resulted in the reduction of the disorder yellow top also called persistent green shoulder. Whereas some green-shouldered cultivars are tolerant to yellow top, many were not. Likewise, some yellow top can occur with uniform green shoulder fruit, but it is not common. Ripening disorders such as yellow top and blotchy ripening are less common when chlorophyll concentrations are not so high. Fruit with glossy, light green shoulders reflect more sun light than dark green-shouldered fruit. These results in less heat buildup in the fruit, thus reducing sunburn and allowing the fruit to develop red color rather than the yellow color brought on by high temperatures.

2.5.6 Flavor

Tomato breeders are often blamed for breeding poor flavor into tomatoes due to their concentration on other traits such as firmness for shipping, high yields, and disease resistance. In fact, it is challenging to develop successful cultivars with necessary production and shipping traits without even considering flavor, and this has not helped the flavor situation. Development of determinate tomato cultivars with concentrated fruit set gives a lower foliage/fruit ratio than in indeterminate cultivars, resulting in limiting carbohydrates that could go into fruit flavor components. Indeterminate lines had higher soluble solids than their isogenic determinate counterparts (Emery and Munger 1970) and high soluble solids selections with large fruit had larger vines with less concentrated fruit set (Rick 1974). Recently, the failed attempts to develop determinate grape tomatoes with the quality of the indeterminates supports this concept. Moreover, it should be pointed out that there is no easy way for breeders to select for flavor, a trait subject to large environmental influences (Scott 2002a), and beyond this there are differences in individual consumer preference. Furthermore, harvesting of less mature fruit for shipping has added to the problem of providing consumers with tomatoes as flavorful as those grown in home gardens.

Flavor results primarily from a combination of sugars, acids, and about 16 aromatic volatiles, and genetic control is not simple, to say the least. In some stocks alkaloid compounds further affect flavor causing bitterness. Conditions that optimize fruit flavor such as water stress, high sunlight, a large differential between day and night temperatures, high salts, and soils with a high nutrient holding capacity are often not present in major tomato production areas. Once a genetic background is attained with good flavor and other desirable attributes, it should be exploited to the fullest with new traits backcrossed into it as the favorable flavor background may be difficult to regain once the background is broken up.

2.5.7 Large Fruit Size

For fresh tomatoes, many markets, especially in the US, require very large fruit (>200 g). A disturbing trend at present is produce buyers who want to market even larger fruit than what is presently considered extra-large, e.g., 5 x 5 and 5 x 4 tomatoes over 5 x 6 tomatoes. In general, large fruits are more prone to defects such as cracking as fruit undergo rapid expansion. Increased fruit size is generally achieved through increasing locule number and arrangement, which can lead to greater incidence of rough blossom scar and off-shaped fruit. Smaller fruited tomatoes with two to four locules of equal size and consistently arranged in a symmetrical pattern in the fruit result in very smooth fruits, which are highly resistant to rough blossom scar and off-shape. Although breeders have done an admirable job in developing many acceptable cultivars with higher percentages of large, marketable fruit, this requirement has undoubtedly slowed progress in combining desirable traits into single cultivars. Parental options for tomato breeders are limited since many desirable parents with "pretty good" fruit size have to be discarded because they are not quite big enough. Adding new traits is more difficult. For instance, many disease resistance genes are introgressed from wild species, and the fruit in the donor parents are always very small. Sometimes there is a linkage between the resistance and small fruit (Acosta 1964) and other times it just takes longer to transfer the resistance to such large fruited backgrounds. Introgression of traits under complex genetic control takes longer when the goal is very large fruit. Scott et al. (1986) indicated that heat-tolerant inbreds tend to be reduced in fruit size, and a hybrid approach was used to help overcome this problem where by heat-tolerant parents were crossed to very large heat-sensitive parents to attain hybrids with commercially acceptable fruit size. It has also been difficult to move high sugars from small-fruited progenitors into large-fruited recurrent parents. DeVerna and Paterson (1991) reported on several studies, which showed soluble solids (and indirectly sugars) were greater in small fruit than in large fruit, and that this was true in segregating populations. Although Rick (1974) was able to recover plants with large fruit and high soluble solids from a *S. chmielewskii* source, as mentioned previously, these plants did not have concentrated fruit set. Presently, high-sugar cultivars are indeterminate grape, cherry and cocktail type tomatoes with fruit size less than 50 g.

2.5.8 Multiple Disease Resistance

As indicated previously conventional tomato breeders have done a tremendous job in developing multiple disease resistant cultivars. However, there are some problems that are difficult to overcome without molecular markers. One is the linkage drag so often associated with introgressions from wild species that harbor the resistance genes. Improved plants have been selected phenotypically that were later found to have shorter introgressions. Molecular markers, saturated in the introgressed region allow for systematic approaches to reduce the linkage drag as genes are fine-mapped. It is also difficult to combine certain resistance genes that are linked in repulsion. This phenomenon is somewhat common because resistance genes are often clustered in the genome. Molecular markers will help to combine genes linked in repulsion via fine-mapping each resistance gene and in selection of recombinant plants with respective markers for each of the resistance genes.

Overall, breeders have not had great success with diseases that are not controlled by single dominant genes. Quantitatively controlled traits, although less prone to being overcome by genetic changes in the disease organisms, generally confer only a moderate level of disease resistance, which is more difficult to distinguish in breeding because of genotype-environment interactions. If disease pressure is not high, escapes and partial escapes confound selection efficiency. Also, developing disease resistances based on multiple genes has been very difficult and slow because of the inferior horticultural traits often associated with multiple gene resistances. Some of this may be due to linkage drag at each QTL or to undesirable pleiotropic effects of genes. If molecular markers for quantitative trait loci (QTLs) can be developed under high disease pressure, then perhaps future breeding for quantitative resistances will be enhanced. However, success in this area will remain challenging.

2.5.9 Genetic Backgrounds

Successful cultivars have good genetic backgrounds that allow for favorable performance over varied environmental conditions. Discovery of this requires extensive trials over several years and locations. This is mentioned because the importance of the genetic background is often underappreciated by people who do not really understand what constitutes successful plant breeding. For instance, the biotechnology firm Calgene originally thought they could transform the fresh market tomato industry with an antisense polygalactuonase gene that provided an increase in shelf-life in Flavr-Savr™ tomatoes. Yet the genetic background was not really considered until it was too late and too much money was lost to save Flavr-Savr. The intent here is not to disparage Calgene. In the 1980s many other scientists with molecular genetics training did not understand the importance of genetic backgrounds, and there were predictions that their science would replace conventional plant breeding. Politically, this thinking had much to do with the decline in public sector plant breeding as funding shifted toward molecular genetics research and thus faculty positions. However, nowadays almost all plant molecular geneticists and several major funding agencies realize that for plant improvement the newer technologies are useful only when combined with conventional breeding programs. Everyone in the field, including plant breeding students, needs to appreciate the importance of genetic backgrounds and how they can affect the functioning of new genes of interest.

2.5.10 Environmental Variation/Strong G x E Interactions

Generally, G x E interactions precludes cultivars from being grown in diverse growing regions and necessitates selection within regions to find the optimal cultivar. The most successful cultivars have been those developed under the environmental conditions most similar to the areas where they are grown. This has necessitated numerous breeding programs around the world committed to breeding for open field and protected culture to come up with highly successful cultivars. Despite this, breeders have done a commendable job in developing cultivars with good adaptability. Two prime examples for fresh market varieties are 'Flora-Dade' and 'Daniela' that were widely grown on several continents for several years. The processing variety E6203 also showed wide adaptation. There are likely genes, which control such adaptation but they are as yet unknown. If such genes could be identified, they would be of great value for future cultivar improvement. Likewise, great progress has been made in developing heat-tolerant cultivars, but locating molecular markers linked to these genes would be a boon to future cultivar development, especially in the face of global warming. Furthermore, heat-tolerant cultivars often set fruit better under cool temperatures, so this would be an added bonus. Of course, marking other stress related genes such as salt or drought tolerance would also help with developing resilient cultivars for future tomato growers. To date molecular markers have made their profound commercial advances with single genes and have generally not been identified until genes have been crossed from wild tomatoes into cultivated backgrounds. If genomic technologies can improve selection for quantitatively controlled traits, it would add great versatility to tomato improvement in the future.

2.5.11 Combining Ability

As mentioned, hybrids have become prevalent in many production regions around the world for both fresh market and processing cultivars. Tomato breeders spend considerable time developing inbreds that are generally selected for superior horticultural characteristics, disease resistance, and sometimes for other characteristics. The better looking inbreds are test-crossed with known parents to determine how the new "improved" inbreds perform as parents. Those that combine well have good general combining ability (GCA) while those that do not combine well have poor GCA and are generally discarded or otherwise receive only limited crossing. Many times inbreds that look really nice make less desirable hybrids than poorer looking inbreds. Thus, the discovery of good GCA is a trial and error process that takes considerable time and effort. Moreover, it can be frustrating if the phenotypes selected do not necessarily produce the best parents. Perhaps there are genes that are common to parent lines with superior GCA. If molecular markers linked to such genes could be identified, such genes could be more easily incorporated into parental material and breeding efforts could become more focused. Highly successful hybrids likely also have a significant specific combining ability (SCA), but finding these superior combinations would seem to be more likely if crosses are made between good GCA types. It is beyond the scope of this chapter to adequately discuss high-throughput breeding and the theories of genetic distance regarding hybrid vigor. "Good GCA" genes may relate to specific traits such as fruit size, fruit shape, vine vigor, and other traits. One does not want to get parents that are too "homogenized", but certain genes may be universally beneficial. Discovery of valuable markers for GCA and/or wide adaptation will likely take some elegant experimentation involving the cooperation of breeders making careful observations and astute molecular geneticists committed to this type of research.

2.6 Limitations of Traditional Breeding and Rationale for Molecular Breeding

Marker-assisted breeding is the best approach to speed up development of new varieties. In populations segregating for desirable traits, it allows the breeder to eliminate plants and lines not possessing the desirable traits. Where molecular markers are available for several disease resistance genes in a segregating population, the breeder can simultaneously select for multiple disease resistances and identify segregates homozygous for multiple resistance genes in a single plant. Use of molecular markers allows the breeder to quickly become familiar with phenotypic traits associated with particular genes. A good example of this is the smaller fruit size and increased blossom-end rot susceptibility associated with the *I-3* gene in homozygous condition (Scott 1999) and the taller, more open, dark green plant type associated with *I-3* in both homozygous and heterozygous condition. The astute breeder thus learns to combine knowledge of molecular markers with associated phenotypic traits to further assist selection. Marker-assisted breeding requires large numbers of DNA tests at considerable cost but saves time and field space. Markers can also allow for rapid backcrossing where plants heterozygous for the marker(s) are selected for the next backcross without the necessity to resort to modified backcrossing where the trait of interest is selected in F_2 or more advanced generations between crosses. Molecular markers are especially helpful to the breeder when selecting for quantitatively inherited traits but often such markers are not available since locating QTLs definitively can be difficult. Markers are also useful for traits where the biological assay is difficult. Some examples of cumbersome biological tests are for root-knot nematode resistance, tomato spotted wilt resistance, and Fusarium crown rot resistance. Other situations where molecular markers are useful are situations where one is unable to work with a pathogen due to contamination concerns. Subsequent to marker assisted selection, the breeder can evaluate the best selections in many field trials to determine wide adaptability.

Molecular advancements are being made at an accelerated pace and it is an exciting time for plant improvement. However, threats to successful agriculture are also omnipresent in a shrinking world and diligent, cooperative work by breeders and molecular geneticists will be required to keep pace. Political support will also be needed but that's definitely beyond the scope of this chapter and book.

Acknowledgements

The authors thank Randy Gardner for review of the chapter. Thanks also to Linda Nolan and Christine Cooley for help in preparation of this chapter. Court Nichols and Fred Angell, the primary writers for Classical Breeding Achievements section, were impressed by the outstanding tomato research work of M. Allen Stevens and John W. Scott, which inspired us to stress crucial variety development work by tomato breeders of the 20th Century. We also wish to thank Ben George, Gene Miyao, and Charles Rivera for providing valuable information.

References

Acosta JC, Gilbert JC, Quinon VL (1964) Heritability of bacterial wilt resistance in tomato. Proc Am Soc Hort Sci 84: 455–462.

Anbinder I, Reuveni M, Azari R, Paran I, Nahon S, Shlomo H, Chen L, Lapidot M, Levin I (2009) Molecular dissection of *Tomato leaf curl virus* resistance in tomato line TY172 derived from *Solanum peruvianum*. Theor Appl Genet 119: 519–530.

Avdeyev YI, Scherbinin BM (1997) Additional analyses of the localization of the gene *Ora*. Rep Tomato Genet Coop 47: 7.

Balint-Kurti PJ, Jones DA, Jones JDG (1995) Integration of the classical and RFLP linkage maps of the short arm of tomato chromosome 1. Theor Appl Genet 90: 17–26.

Baggett JR (1995) A historical summary of vegetable breeding at Oregon State University. Proc Oregon Hort Soc 86: 89–99.

Barksdale TH (1969) Resistance of tomato seedlings to early blight. Phytopathology 59: 443–446.

Barksdale TH (1972) Resistance in tomato to six anthracnose fungi. Phytopathology 62: 660–3.

Barten JHM, Scott JW, Gardner RG (1994) Characterization of blossom-end morphology genes in tomato and their usefulness in breeding for smooth blossom-end scars. J Am Soc Hort Sci 119(4): 798–803.

Blaker N, Reiwich S, Yoder JI (1995) Mapping the *Lv* gene. Rep Tomato Genet Coop 45: 14.

Bosch SE, Louw AJ, Aucamp E (1985) Rodade, bacterial wilt resistant tomato. Hort Science 20(3): 458–459.

Boswell VR (1937) Improvement and Genetics of Tomatoes, Peppers, and Eggplant. Yearbook of Agriculture, USDA, Washington DC, USA, pp 176–206.

Bournival BL, Scott JW, Vallejos CE (1989) An isozyme marker for resistance to race 3 of *Fusarium oxysporum* f. sp. *Lycopersici* in tomato. Theor Appl Genet 78: 489–494.

Burbidge A, Lindhout P, Grieve TM, Schumacher K, Theres K, van Heusden AW, Bonnema AB, Woodman KJ, Taylor IB (2001) Re-orientation and integration of the classical and interspecific linkage maps of the long arm of tomato chromosome 7. Theor Appl Genet 103: 443–454.

Butler L (1952) The linkage map of the tomato. J Hered 43: 25–35.

Butler L (1968) Linkage Summary. Rep Tomato Genet Coop 18: 4–6.

Cannon OS, Waddoups V (1952) Loran Blood and VR Moscow, two new Verticillium wilt resistance tomatoes for Utah. Utah Farm Home Sci 13: 74.

Chetelat, RT (1998) *Bco*, a corolla intensifier on chromosome 7. Rep Tomato Genet Coop 48: 10–12.

Chetelat RT, DeVerna JW (1993) Map location of Malate dehydrogenase-1 (Mdh-1), a new isozyme marker for 3L. Rep Tomato Genet Coop 43: 13–14.

Chetelat, RT, DeVerna JW, Klann E, Bennett AB (1993a) Sucrose accumulator *(sucr)* a gene controlling sugar composition in the fruit of *L. chmielewskii* and *L. hirsutum*. Rep Tomato Genet Coop 43: 14–16.

Chetelat RT, Pudlo W, DeVerna JW (1993b) Map location of Formate dehydrogenase-1 *(Fdh-1)*, a new isozyme marker on 2L. Rep Tomato Genet Coop 43: 11–13.

Chetelat RT, Takahashi M, DeVerna JW(1993c) Map location of Diaphorase-3 (Dia-3), a new isozyme marker on 9S. Rep Tomato Genet Coop 43: 10.

Chetelat RT, Adams DF, Adams DO (1999) Mae-1, a malic enzyme encoding gene on chromosome 5. Rep Tomato Genet Coop 49: 12–13.

Chetelat RT, Ji Y (2007) Cytogenetics and evolution. In: Razdan MK, Mattoo AK (eds) Genetic Improvement of Solanaceous Crops Vol 2: Tomato. Science Publishers, Enfield, NH, USA, pp 77–112.

Chunwongse J, Chunwongse C, Black L, Hanson P (1998) Mapping of the Ph-3 gene for late blight from L. pimpinellifolium L. 3708. Rep Tomato Genet Coop 48: 13–14.

Coltman R, Gerloff G, Gabelman WH (1982) Intraspecific variation in growth phosphorus acquisition and phosphorus utilization in tomatoes under phosphorus deficiency stress. In: Scaife A (ed) Proc 9th Int Plant Nutrition Colloquium Vol 1. Warwick University, Coventry, United Kingdom, August 22–27, 1982, pp 117–22.

Crill JP, Strobel JW, Burgis DS, Bryan HH, John CA, Everett PH, Bartz JA, Hayslip NC, Dean WW (1971) Florida MH-1, Florida's first machine harvest fresh market tomato. Fla Agri Exp Sta Circ S-212. 12 p.

DeVerna JW, Paterson AH (1991) Genetics of *Lycopersicon*. In: Kalloo G (ed) Genetic Improvement of Tomato. Springer, Berlin, Germany, pp 21–38.

Emery GC, Munger HM (1970) Effects of inherited differences in growth habit on fruit size and soluble solids in tomato. J Am Soc Hort Sci 95: 410–412.

Foolad MR (2007) Tolerance to abiotic stresses. In: Razden MK, Mattoo AK (eds) Genetic Improvement of Solanaceous crops. Vol 2: Tomatoes. Science Publishers Enfield, NH, USA, pp 521–591.

FM Vegetable Varieties (1981) Ferry-Morse Seed Company, Mountain View, CA, USA.

FM Vegetable Varieties (1984) Ferry-Morse Seed Company, Mountain View, CA, USA.

Frazier WA, Dennett RK, Hendrix JW, Poole CF, Gilbert JC (1950) Seven new tomatoes: varieties resistant to spotted wilt, Fusarium wilt, and grey leaf spot. Hawaii Agri Exp Sta Bull 103.

Gallegly ME (1960) Resistance to the late blight fungus in tomato. In: Proc Plant Sci Seminar, Campbell Soup Co. Camden, NJ, USA, pp 113–135.

Gardner RG (1992) 'Mountain Spring' tomato; NC 8276 and NC 84173 tomato breeding lines. Hort Science 27: 1233–1234.

Gardner RG (1999) NC109 tomato breeding line. 'Mountain Fresh' F1 hybrid. Hort Science 34: 941–942.

Gardner, RG, Panthee, DP (2010) NC 1 CELBR and NC 2 CELBR: Early. blight and late blight-resistant fresh market tomato breeding lines. Hort Science 45: 975–976.

Gardner RG, Shoemaker PB (1999) Mountain Supreme: early blight resistant hybrid tomato and its parents, NC EBR-3 and NC EBR-4 tomato breeding lines. Hort Science 34: 745–746.

George B (2008a) Development of the California Processing Tomato—1859–2004. Tomato Magaz 12(3): 19–22.

George B (2008b) A Brief History of the California Processing Tomato. Presentation at the Annual Meeting of the DJ Thompson Group (California commercial seedsmen), 10 p.

Gilbert JC, Brewbaker JL, Tanaka JS, Chinn JT, Hartmann RW, Crozier JA Jr, Ito PJ (1969) Vegetable Improvement at the Hawaii Agr Exp Sta Res Report 175.

Goffreda JC, Mutschler MA (1988) Origin and release of breeding line NY 871095-6 resistant to Fusarium race 3. Rep Tomato Genet Coop 38: 50.

Griffiths PD, Scott JW (2001) Inheritance and linkage of Tomato mottle virus resistance genes derived from *Lycopersicon chilense* accession LA 1932. J Am Soc Hort Sci 126: 462–467.

Grogan RG, Kimble KA, Misaghi I (1975) A stem canker disease of tomato caused by *Alternaria alternata f.* sp. *lycopersici*. Phytopathology 65: 880–886.

Hanson PM, Bernacchi D, Green S, Tanksley SD, Muniyappa V, Padmaja AS, Chen H, Kuo G, Fang D, Chen J (2000) Mapping a wild tomato introgression associated with tomato yellow leaf curl virus resistance in a cultivated tomato line. J Am Soc Hort Sci 125(1): 15–20.

Hanson P, Green SK, Kuo G (2006) *Ty-2*, a gene on chromosome 11 conditioning geminvirus resistance in tomato. Rep Tomato Genet Coop 56: 17–18.

Hedrick UP, Booth, NO (1907). Mendelian characters in tomatoes. Proc Soc Hort Sci 5: 19–24.

Hoadley AD (1960) The development of anthracnose resistant tomatoes. In: Proc Plant Sci Seminar, Campbell Soup Co. Camden, NJ, USA, pp 19–36.

Hutton SF, Scott JW, Schuster DJ (2012) Recessive resistance to *Tomato yellow leaf curl virus* from the tomato cultivar Tyking is located in the same region as *Ty-5* on chromosome 4. HortScience 47: 324–327.

Ji Y, Schuster DJ, Scott JW (2007) *Ty-3*, a begomovirus resistance locus near the tomato yellow leaf curl virus resistance locus *Ty-1* on chromosome 6 of tomato. Mol Breed 20: 271–284.

Ji Y, Scott JW, Schuster DJ, Maxwell DP (2009) Molecular mapping of *Ty-4*, a new tomato yellow leaf curl virus resistance locus on chromosome 3 of tomato. J Am Soc Hort Sci 134: 281–288.

Jones DA, Balint-Kurti PJ, Dickinson MJ, Dixon MS, Jones JDG (1992) Locations of genes for resistance to *Cladosporium fulvum* on the classical and RFLP maps of tomato. Rep Tomato Genet Coop 42: 19–22.

Jones DA, Jones JDG (1996) The *rvt* gene maps close to *inv* on the long arm of chromosome 1. Rep Tomato Genet Coop 46: 16.

Kalloo G, Banerjee MK (1990) Transfer of tomato leaf curl virus resistance from *Lycopersicon hirsutum f.* glabratum to *L. esculentum*. Plant Breed 105: 156–159.

Kerr EA, Cook FI (1983) Ontario 7710—a tomato breeding line with resistance to bacterial speck, *Pseudomonas syringae pv. tomato* (Okabe). Can J Plant Sci 63: 1107–1109.

Kikuta K, Hendrix JW, Frazier WA (1945) Pearl Harbor—A tomato variety resistant to spotted wilt in Hawaii. Hawaii Agri Exp Sta Circ 24.

Laterrot H, Moretti A (1995) *Sm* without vascular fusariosis resistance. Rep Tomato Genet Coop 45: 29.

Leeper PW, Phatak SC, Bell DK, George BF, Cox EL, Oerther GE, Scully BT (1992) Southern blight-resistant tomato breeding line: 5635M, 5707M, 5719M, 5737M, 5876M, and 5913M. Hort Science 27(5): 475–478.

Lenucci MS, Cadinu D, Taurino M, Piro G, Dalessandro G (2006) Antioxidant composition in cherry and high-pigment tomato cultivars. *J Agri Food Chem* 54: 2606–2613.

Levin I, de Vos CHR, Tadmor Y, Bovy A, Lieberman M, Oren-Shamir M, Segev O, Kolotilin I, Keller M, Ovadia R, Meir A, Bino RJ (2003) High pigment tomato mutants—more than just lycopene. Isr J Plant Sci 54: 179–190.

Liedl BE, Liu SC, Esposito D, Mutschler MA (1993) Identification and mapping of S in a self-compatible F_2 population of *L. esculentum* x *L. pennellii*. Rep Tomato Genet Coop 43: 33–34.

Linkage Committee (1958) Linkage Report. Rep Tomato Genet Coop 8: 7–8.

MacArthur JW (1926) Linkage studies with the tomato. Genetics 11: 387–405.

MacArthur JW (1928) Linkage studies with the tomato. II. Three linkage groups. Genetics 13: 410–420.

MacArthur JW (1931) Linkage studies with the tomato. III. 15 factors in six groups. Trans Roy Can Inst 18: 1–20.

Madishetty K, Bauer P, Sharada MS, Al-Hammadi ASA, Sharma R (2006) Genetic characterization of the *polycotyledon* locus in tomato. Theor Appl Genet 113: 673–683.

Makmur A, Gerloff GC, Gabelman WH (1978) Physiology and inheritance of efficiency in potassium utilization in tomatoes grown under potassium stress. J Am Soc Hort Sci 103: 545–9.

Martin MW (1969) Inheritance of resistance to curly top virus in the tomato breeding line CVF4. Phytopathology 59: 1040 (Abstr).

Martin MW (1985) Naming and release of four tomato varieties: Roza, Columbia, Rowpac, and Saladmaster. Release notice dated 4/2/76 by ARS, USDA, Washington Agric Expt Sta and Idaho Agric Expt Sta in Rep Tomato Genet Coop 35: 55.

McGrath DJ (1988) BHRS 2-3 Fusarium wilt resistant tomato. Hort Science 23(6): 1093–1094.

Moreau P, Thoquet P, Laterrot H, Moretti A, Olivier J, Grimsley NH (1997) A locus, *ltm*, controlling the development of intumescences, is present on chromosome 7. Rep Tomato Genet Coop 47: 15.

Mutschler MA, Wolfe DW, Cobb ED, Yourstone KS (1992) Tomato fruit quality and shelf life in hybrids heterozygous for the *alc* ripening mutant. Hort Science 27(4): 352–355.

Mutschler A, Rick C (1987) 1987 Linkage maps of tomato. Rep Tomato Genet Coop 37: 5–34.

O'Sullivan J, Gabelman WH, Gerloff GC (1974) Variations in efficiency of nitrogen utilization in tomatoes (*Lycopersicon esculentum* Mill.) grown under nitrogen stress. J Am Soc Hort Sci 99: 543–7.

Ozminkowski RH Jr, Gardner RG, Henderson WP, Moll RH (1990) Prostrate growth habit enhances fresh-market tomato fruit yield and quality. Hort Science 25: 914–915.

Pelham J (1966) Resistance in tomato to tobacco mosaic virus. Euphytica 15: 258–267.

Pilowsky M, Cohen S (1990) Tolerance to tomato yellow leaf curl virus. Rep Tomato Genet Coop 40: 29–30.

Porte WS, Walker HB (1941) The Pan America tomato, a new red variety highly resistant to Fusarium wilt. USDA Circ 611 6 p.

Porter DR, MacGillivray JH (1937) The production of tomatoes in California. Calif Agri Ext Serv Circ 104.

Price DL, Memmott FD, Scott JW, Olson SM, Stevens MR (2007) Identification of molecular markers linked to a new *tomato spotted wilt virus* resistance source in tomato. Rept Tomato Genet Coop 57: 35–36.

Processing Tomato Advisory Board (of CA), Reports 1986–2008: www.ptab.org.

Rabinowitch H, Kedar N (1995) Development of long shelf-life greenhouse tomatoes. Proc Tomato Quality Workshop and Tomato Breeders Roundtable, Davis, California, USA, pp 22–23 (Abstr).

Rebigan JB, Villareal RL, Lai S-H (1977) Reactions of three tomato cultivars to heavy rainfall and excessive soil moisture. Phil J Crop Sci 2(4): 221–226.

Reynard GB (1951) Inherited resistance to radial cracks in tomato fruits. Proc Am Soc Hort Sci 58: 231–44.

Reynard GB (1960) Breeding tomatoes for resistance to fruit cracking. In: Proc Plant Sci Seminar, Campbell Soup Co. Camden, NJ, USA, pp 19–36.

Reynard GB, Andrus CF (1945) Inheritance of resistance to the collar-rot phase of *Alternaria solani* on tomato. Phytopathology 35: 25–36.

Rick CM (1974) High soluble-solids content in large-fruited tomato lines derived from a wild green fruited species. Hilgardia 42: 493–510.

Rick CM (1975) The tomato. In: King RC (ed) Handbook of Genetics, vol 2. Plenum Press, New York, NY, USA, pp 247–280.

Rick CM (1998) New alleles at old loci. Rep Tomato Genet Coop 48: 43.

Rick CM, Fobes J (1974) Association of an allozyme with nematode resistance. Rep Tomato Genet Coop 24: 25.

Rick CM, Cisneros P, Chetelat RT, DeVerna JW (1994) Abg—a gene on chromosome 10 for purple fruit derived from S. *lycopersicoides*. Rep Tomato Genet Coop 44: 29–30.

Rush DW Epstein E (1976) Genotypic responses to salinity differences between salt sensitive and salt tolerant genotypes of tomato. Plant Physiol 57: 162–66.

Schaible LW (1962) Fruit setting response of tomatoes to high temperatures. Plant Sci Symp Campbell Soup Co. p 89–98.

Scott JW (1984) Genetic sources of tomato firmness. Proc. Fourth Tomato Quality Workshop. Univ Fla Res Rep BEC 83-1: 60–67.

Scott JW (1999) Tomato plants heterozygous for fusarium wilt race 3 resistance develop larger fruit than homozygous resistant plants. Proc Fla State Hort Soc 112: 305–307.

Scott JW (2002a) A breeder's perspective on the use of molecular techniques for improving fruit quality. Hort Science 37: 464–467.

Scott JW (2002b) Tomato. In: Wehner T (ed) Vegetable Varieties from North America. ASHS Press, Alexandria, VA, USA: http: //cuke.hort.ncsu.edu/cucurbit/wehner/vegcult/tomatoai.html.

Scott JW, Angell FF (1998) Tomato. In: Banga SS, Banga SK (eds) Hybrid Cultivar Development. Narosa Publishing House, New Delhi, India, pp 451–474.

Scott JW, Baldwin EA, Klee HA, Brecht JA, Olson SM, Bartz JA, Sims CA (2008) Fla. 8153 hybrid tomato; Fla. 8059 and Fla. 7907 breeding lines. Hort Science 43(7): 2228–2230.

Scott JW, Bryan HH, Ramos LJ (1998) High temperature fruit setting ability of large-fruited, jointless pedicel tomato hybrids with various combinations of heat-tolerance. Proc Fla State Hort Soc 110: 281–284.

Scott JW, Farley JD (1983) 'Ohio CR-6' tomato. Hort Science 18(1): 114–115.

Scott JW, Jones JB, Somodi GC (2003) Development of a large fruited tomato with a high level of resistance to bacterial wilt (*Ralstonia solanacearum*). Rep Tomato Genet Coop 53: 36–37.

Scott JW, Jones JP (1995) Fla. 7547 and Fla. 7481 tomato breeding lines resistant to *Fusarium oxysporum f.* sp. *Lycopersici* races 1, 2, and 3. Hort Science 30: 645–646.

Scott JW, Volin RB, Bryan HH, Olson SM (1986) Use of hybrids to develop heat tolerant tomato cultivars. Proc Fla State Hort Soc 99: 311–314.

Sheela AG, Peter KV (1987) Additional sources of resistance to bacterial wilt in tomato. Rep Tomato Genet Coop 37: 63–64.

Soumpourou E, Iakovidis M, Chartrain L, Lyall L, Thomas CW (2007) The *Solanum pimpinellifolium Cf-ECp1* and *Af-ECP4* genes for resistance to *Cladosporium fulvum* are located at the *Milky Way* locus on the short arm of chromosome 1. Theor Appl Genet 115: 1127–1136.

Stevens MA, Dickinson GL, Aquirre MS (1977) UC82, a high yielding processing tomato. U Calif Davis Veg Crops Series 183, 8 pp.

Stevens MA, Rick CM (1986) Genetics and Breeding. In: Atherton JG, Rudich J (eds) The Tomato Crop, a scientific basis for improvement. Chapman and Hall, London, England, pp 35–110.

Stevens MR, Scott JW, Gergerich RC (1992) Inheritance of a gene for resistance to tomato spotted wilt virus (TSWV) from *Lycopersicon peruvianum* Mill. Euphytica 59: 9–17.

Stoner AK (1970) Plant breeding for resistance to insects in vegetables. Hort Science 5: 72.

Strobel JW, Hayslip NC, Burgis DS, Everett PH (1969) Walter; A determinate tomato resistant to races 1 and 2 of the Fusarium wilt pathogen. Fla Agri Expt Sta Circ S-202 9 p.

Tanksley SD, Ganal MW, Prince JP, de Vicente MC, Bonierbale MW, Broun P, Fulton TM, Giovannoni JJ, Grandillo S, Martin GB, Messeguer R, Miller JC, Miller L, Paterson AH, Pineda O, Röder MS, Wing RA, Wu W, Young ND (1992) High density molecular linkage maps of the tomato and potato genomes. Genetics 132: 1141–1160.

Tanksley SD (1993) Linkage map of tomato (*Lycopersicum esculentum*) (2N=24). In: SJ O'Brien (ed) Genetic Maps: Locus Maps of Complex Genomes, Ed. 6. Cold Spring Harbor Press, New York, USA, pp 6.39–6.60.

Tanksley SD, Jones RA (1981) Application of alcohol dehydrogenase allozymes in testing the genetic purity of F_1 hybrids of tomato. Hort Science 16(2): 179–181.

Thompson AE, Tomes ML, Erickson HT, Wann EV, Armstrong RJ (1967) Inheritance of crimson fruit color in tomatoes. Proc Am Soc Hort Sci 91: 495–504.

Thoquet P, Olivier J, Sperisen C, Rogowsky P, Laterrot H, Nigel G (1996) Quantitative trait loci determining resistance to bacterial wilt in tomato cultivar Hawaii 7996. Mol Plant-Micr Intract 9: 826–836.

Tigchelaar EC (1986) Tomato Breeding In: Bassett JJ (ed) Breeding Vegetable Crops. AVI Publishing, Westport, CT, USA, pp 135–171.

Tigchelaar EC, McGlasson WB, and Buescher RW (1978) Genetic regulation of tomato fruit ripening. Hort Science 13: 508–13.

Tomes ML (1963) Temperature inhibition of carotene synthesis in tomato. Bot Gaz 124: 180–185.

van der Biezen EA, Overduin B, Nijkamp HJJ, Hille J (1994) Integrated genetic map of tomato chromosome 3. Rep Tomato Genet Coop 44: 8–11.

van Tuinen A, Peters AHLJ, Koorneef M (1998) Mapping of the pro gene and revision of the classical map of chromosome 11. Rep Tomato Genet Coop 48: 62–70.

van Tuinen A, Cordonnier-Pratt MM, Pratt LH, Verkerk R, Zabel P, Koorneef M (1997) The mapping of phytochrome genes and photomorphogenic mutants in tomato. Theor Appl Genet 94: 115–122.

van Wordragen MF, Weide RI, Coppoolse E, Koorneef M, Zabel P (1996) Tomato chromosome 6: a high resolution map of the long arm and construction of a composite integrated marker-order map. Theor Appl Genet 92: 1065–1072.

van Zijl JJB, Bosch SE, Coetzee CPJ (1986) Breeding tomatoes for processing in South Africa. Acta Hort 194: 69–75.

Villareal RL, Rebigan JB, Lai SH (1977) Fruit setting ability of heat-tolerant, moisture-tolerant and traditional cultivars grown under field and greenhouse conditions. Phil J Crop Sci 2: 55–61.

Vrieling A, Voskuil DW, Bonfrer JM, Korse CM, van Doorn J, Cats A, Depla AC, Timmer R, Witteman BJ, van Leeuwen FE, van Veer LJ, Rookus MA, Kampman E (2007) Lycopene supplementation elevates circulating insulin-like growth factor-binding protein-1 and -2 concentrations in persons at greater risk of colorectal cancer. Am J Clin Nutr 86: 1456–1462.

Walter JM (1967) Hereditary resistance to disease in tomato. Annu Rev Phytopathol 5: 131–162.

Walter J M, Kelbert DGA (1953) Manalucie, a tomato with distinctive new features. Fla Agri Exp Sta Circ S-59: 10 p.

Webb RE, Barkesdale TH Stoner AK (1973) Chapter 21: Tomatoes. In: Nelson RR (ed) Breeding Plants for Disease Resistance; Concepts and Applications. Pennsylvania University Press, University Park, PA, USA, pp 344–361.

Weide R, van Wordragen MF, Lankhorst RK, Verkerk R, Hanhart C, Liharska T, Pap P, Stam P, Zabel P, Koorneef M (1993) Integration of the classical and molecular linkage maps of tomato chromosome 6. Genetics 135: 1175–1186.

Williams WG, Kennedy GG, Yamamoo RT, Thacker JD, Bordner J (1980) 2-tridecanone: A naturally occurring insecticide from the wild tomato *Lycopersicon hirusutum f. glabratum.* Science 207: 888–889.

Yeager AF (1933) Tomato breeding. North Dakota Agri Exp Sta Bull 276.

Young PA (1963) Two-way varieties for hot or cold climates Amer. Veg. Grower 11: 73.

Young PA, MacArthur JW (1947) Horticultural characters of tomatoes. Texas Agri Exp Sta Bull 698: 61.

Zamir D, Bolkan H, Juvik JA, Watterson JC, Tanksley SD (1993) New evidence for placement of Ve—the gene for resistance to *Verticillium* race 1. Rep Tomato Genet Coop 43: 51–52.

Zamir D, Michelson IE, Zakay Y, Navot N, Zeidan M, Sarfatti M, Eshed Y, Harel E, Pleban T, van-Oss H, Kedar N, Rabinowitch HD, Czosnek H (1994) Mapping and introgression of a tomato yellow leaf curl virus tolerance gene, *Ty-1*. Theor Appl Genet 88: 141–146.

Diversity within Cultivated Tomato

Esther van der Knaap, * *Claire Anderson* and
Gustavo Rodriguez

ABSTRACT

Cultivated tomato is morphologically more diverse than the wild relatives from which the crop was domesticated. Quantification and employment of the phenotypic diversity in breeding programs require an objective evaluation of morphology. For fruit, the morphological characterization is facilitated by the publicly available software program Tomato Analyzer. Attributes of shape are used to detect quantitative trait loci (QTL) in segregating populations. Tomato Analyzer is also effectively used to classify tomato varieties based on fruit shape. Despite the extensive morphological diversity, genetic diversity among the cultivated types is quite low. Recent efforts to identify nucleotide polymorphisms that differentiate cultivated accessions and the cloning of key fruit shape genes has yielded important insights into the selection process of tomato from a small round berry to a large and diversely shaped fruit.

Keywords: morphology, phenotype, fruit shape, classification, Tomato Analyzer and genetics

3.1 Introduction

Tomato (*Solanum* section *lycopersicon*), which includes wild and cultivated types, represents a diverse group of plants. This chapter focuses on the phenotypic and genetic diversity within the cultivated germplasm. As

Department of Horticulture and Crop Science, The Ohio State University/Ohio Agricultural Research and Development Center, Wooster, OH 44691, USA.
*Corresponding author

discussed in Chapter 1, tomato species are indigenous to the Andes region in South America. The domestication of tomato traces to a Mexican origin of cultivation although a Peruvian origin has also been proposed (Jenkins 1948). Through Spanish contact, tomato was brought to Europe and other parts of the world starting in the early part of the 16th century. One of the first written record of an early cultivated type appeared in Italy in 1544 as a short paragraph in a manuscript describing a poisonous distant relative of tomato, the mandrake. This record details human consumption of the fruit with oil and salt, thus tomato was already being established in the Italian cuisine at that time (Matthiolus 1544). Additionally, the fruit was described as flattened, segmented, and of yellow color, hence its Italian name pomo d'oro (golden apple; Matthiolus 1544). Other fruit types were documented following Matthiolus's original publication, most notably describing variation in color as well as size and shape. More in-depth discussion about the early period of tomato improvement can be found in the tomato monograph by Peralta et al. (2008). Although details about tomato domestication and initial improvements are unclear, it is certain that the last 500 years resulted in a significant increase in morphological diversity among domesticated tomato accessions throughout the world.

For cultural and historical reasons, varieties of tomato are classified as landraces (also called regional), heirlooms, vintage, modern (also called contemporary and includes hybrids), breeding or elite breeding lines. However, definitions for these categories are neither accepted by all nor clearly delineated. Tomatoes classified as landraces are farmer or gardener-selected and are adapted to the local environment, typically in areas of local subsistence. Heirloom tomatoes were first defined as varieties that were handed down from generation to generation within a family and were known to exist prior to about 1940 when commercial hybrids first started becoming available. Heirlooms are not necessarily adapted to local environments. All heirlooms and landraces are open-pollinated but not all open-pollinated varieties are heirlooms or landraces. Other definitions of heirloom define them as varieties that have been in existence for about 50 years and yet another less recognized definition states that any variety, which is "treasured", is an heirloom (Male 1999; C Male pers comm). The term "vintage" and "modern" tomato as defined by Williams and St. Clair (1993) and Park et al. (2004), refers to tomato accessions released before or after a certain year, respectively. Breeding and elite breeding lines are the materials used in modern breeding programs, most of which are conducted by private companies.

Diversity within cultivated tomato is particularly evident in the fruit. Fruit size variation ranges from small, weighing only a few grams, to very large and weighing more than 500 grams. Fruit shape ranges from round to pear-shaped, torpedo, oval and even bell pepper shaped, whereas fruit color includes various shades of red, pink, yellow, gold, green when ripe, black, white, gold/red bicolor, and striped (Male 1999; Paran and van der Knaap 2007). Wild tomato relatives on the other hand, produce small round fruits that are often green in color. Leaf forms of the cultivated accessions include regular leaf, potato leaf, angora, rugose and variations on all of these. There is even one variety with variegated foliage (C Male pers comm).

Plant form in the cultivated tomato varies in degree of erectness and apical dominance, as well as in the degree of axillary shoot formation and growth. One of the most dramatic differences in plant form is related to shoot determinacy. Tomato features sympodial growth, and coincident with the transition from vegetative to reproductive growth, the primary shoot meristem terminates in an inflorescence. Growth of the main shoot continues from the flanks of the inflorescence meristem, which after formation of two to three leaves, terminates again in an inflorescence meristem. In most tomatoes, including all wild relatives, growth is indeterminate, which means that shoot termination and continued growth from the flanks goes on indefinitely. Determinate tomatoes, however, feature accelerated termination of the sympodial shoot, where growth terminates after two successive inflorescences. These types are represented by the processing varieties that are grown in the field and mechanically harvested (Pnueli et al. 1998; Paran and van der Knaap 2007). The degree of determinacy varies among cultivated types from accessions exhibiting very strong determinacy, resulting in extremely compact plants, to more loosely determinate accessions.

Inflorescence architecture varies from wild type, which features a single peduncle bearing few or many flowers, to highly branched inflorescence types (Lippman et al. 2008). Typically, smaller fruited tomatoes such as the cherry types produce many flowers and fruits per inflorescence, whereas larger fruited types produce fewer fruits. The very large fruited types, such as the beefsteaks, produce fruits with many locules and often display inflorescences with only 2–3 flowers (Welty et al. 2007). Interestingly, loci controlling fruit weight often overlap with loci controlling flowers per inflorescence, suggesting a pleiotropic effect of one gene controlling both traits (van der Knaap and Tanksley 2003).

The flavor and aroma of tomato varieties differ also and loci linked to the traits have been found to segregate in cultivated tomato (Carli et al. 2011; Zanor et al. 2009). Although it is difficult to find consensus, the perception of those who have grown thousands of heirloom varieties as well as modern accession is that the best of the heirlooms usually taste much better than the majority of the modern accessions (C Male pers comm.). There is agreement, however, that vine-ripened tomatoes have superior taste in comparison to the majority of varieties that are harvested at green or breaker stages of fruit development and shipped for sale in retail markets.

Other quality traits such as firmness, fruit texture and crack resistance are also variable among the varieties and QTLs controlling this trait have been identified (Chaib et al. 2007).

The diversity within cultivated tomato germplasm provides a great resource for variety improvement. Understanding the molecular mechanisms underlying this variation will help us to understand the evolution of traits that have been selected during domestication. This, in turn, will allow implementation of the knowledge in the development of novel crops or crops with improved characteristics while also providing a basic understanding of the mechanisms of plant diversity. In order to harness this phenotypic variation, and to employ this diversity in basic and applied research projects, it is important to measure and quantify these traits accurately and objectively.

3.2 Phenotypic Analysis of Tomato Fruit

Evaluating phenotypic traits such as fruit morphology, color intensity, nutritional quality, firmness, flavor and aroma are complicated because of the quantitative nature of these traits. Moreover, the objective and accurate quantification of these traits can be difficult. In this chapter, we discuss the phenotypic analysis of tomato fruit shape as an example of how one particular quantitative trait can be reproducibly analyzed. Utilization of the software program "Tomato Analyzer" (TA) to quantify fruit attributes is presented (Brewer et al. 2006; Gonzalo et al. 2009). Using scanned images of longitudinally and transversely cut fruit, TA automatically identifies the objects and their boundaries in the digital image. Utilizing the boundary information, as many as 39 fruit attributes are measured. Fruit size measurements include fruit perimeter, area, width, and height. Fruit shape is assessed by measurement of fruit shape index, homogeneity, asymmetry, and proximal and distal fruit end shapes (i.e., angles, blockiness, protrusion and indentation). The software also offers morphometric output of fruit shape (Gonzalo et al. 2009). In addition, internal features of the fruit such as internal eccentricity (to measure the eccentric position of the seed in the fruit) and pericarp thickness and area are calculated by TA after a few manual adjustments (Gonzalo et al. 2009). The software program and a detailed manual are available at http://www.oardc.ohio-state.edu/vanderknaap/.

To ensure the utility of this tool beyond tomato and other Solanaceous vegetable crops, terms for the attributes described by TA were developed that are applicable across the plant kingdom. For example, the distal fruit end is the part of the fruit farthest away from the stem end and is also called the blossom end in tomato. The proximal end of the fruit is a more universal term and is preferred over the often used stem end expression.

The terms are arranged in a controlled trait vocabulary and are found at Sol Genomics Network (SGN: http://solgenomics.net/). For each attribute term, a unique mathematical descriptor was developed. For example, the distal end angle measures the slope of the fruit boundary at a chosen position from the tip of the fruit. The blockiness attribute calculates the ratio of fruit width at a chosen position above the distal or proximal end over the mid-width. Although not discussed further, Tomato Analyzer also measures hue, luminosity and Chroma, and the L, a and b values for color evaluation (Rodriguez et al. 2011b).

Using the morphological output of TA, quantitative trait loci (QTL) were mapped in five tomato populations that segregated for fruit shape. The populations were derived from a cross between different tomato varieties with elongated fruit shape and tomato's closest wild relative, *S. pimpinellifolium* accession LA1589 that produces small round fruit (Brewer et al. 2007; Gonzalo and van der Knaap 2008). These studies identified loci that had the largest effect on elongated fruit shape. The loci and effect that they have on fruit shape in descending order are: *sun, ovate, fs8.1* and *fs2.1* (Gonzalo and van der Knaap 2008). Additional loci were also identified that affected at least two attributes of shape. These were found on nearly every chromosome: middle of chromosome (chr) 1, top of chr 2, bottom of chr 3, bottom of chr 5, middle of chr 6, bottom of chr 7, bottom of chr 11 and top of chr 12 (Brewer et al. 2007; Gonzalo and van der Knaap 2008). Locule number is also an important feature that dramatically affects tomato morphology by resulting in a flattened fruit shape. The locule number trait is controlled by two major loci, *lc* and *fas* (Lippman and Tanksley 2001; van der Knaap and Tanksley 2003; Barrero and Tanksley 2004).

The genes underlying *sun, ovate, fas* and *lc* were identified by positional cloning. *SUN* encodes a positive regulator of growth and is hypothesized to alter hormone or secondary metabolite homeostasis (Wu et al. 2011; Xiao et al. 2008). *OVATE* encodes a negative regulator of growth by acting as a repressor of transcription (Liu et al. 2002; Hackbusch et al. 2005; Wang et al. 2007). *FAS* encodes a YABBY protein involved in organ polarity and carpel number (Cong et al. 2008). *LC* is most likely encoded by the ortholog of *WUSCHEL*, controlling stem cell fate in plants (Munos et al. 2011). The identity of genes underlying the other shape loci has not been determined.

TA is also used effectively in the area of fruit classification and standardization. The International Union for the Protection of New Varieties and Plants (UPOV) and International Plant Genetic Resources Institute (IPGRI) scales classify tomato based on visual observation of the fruit, ranging from flattened to strongly pear shaped (International Union for the Protection of New Varieties and Plants 2001; International Plant Genetic Resources Institute 1996). Using a collection of over 300 accessions consisting of heirloom, Spanish, Italian, Latin American, modern varieties and wild accessions, we modified the morphological scales: flattened and slightly flattened categories became flat; rectangular and cylindrical became rectangular; ovoid, pear-shaped and strongly pear shaped became obovoid; heart-shaped was split in heart and oxheart; and two new categories were introduced: long and ellipsoid. The round category remained the same (Fig. 3-1). Visual observations suggested that these eight groups were consistently distinguishable and that the merging of similar categories facilitated fruit classification. Using a blind shape categorization test, seven participants (six of whom were not previously familiar with tomato

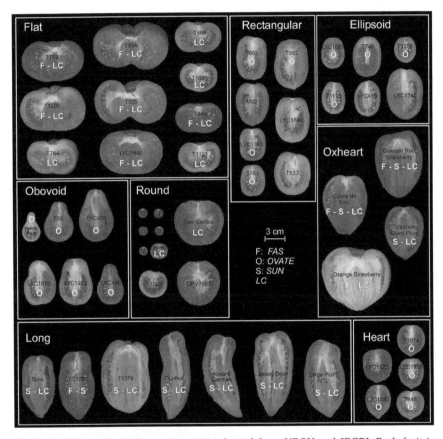

Figure 3-1 Tomato fruit shape categories adapted from UPOV and IPGRI. Each fruit is identified by variety name (information available at http://solgenomics.net/ and Rodriguez et al. 2011a) and presence of the variant allele of *SUN, OVATE, LC* and/or *FAS* (abbreviated as S, O and F, respectively). Figure reprinted from Rodriguez et al. 2011, Plant Physiol 156: 275-285 (www.plantphysiol.org). Copyright American Society of Plant Biologists.

Color image of this figure appears in the color plate section at the end of the book.

shape categories) classified 70 images of fruit into eight categories. The results supported the basis for the establishment of these new categories. Discrepancies occurred only when a fruit was of an intermediate shape. In addition, these eight categories can be distinguished based on TA measurements for seven attributes that are most diverse among the categories (Table 3-1). Fruit shape index is an attribute that clearly separates the flat and round fruit from the elongated tomatoes. Long tomatoes show the highest fruit shape index followed by obovoid whereas rectangular and ellipsoid fruit display similar fruit shape indices. Long tomatoes also exhibit a pronounced tip measured by distal end protrusion and lower values for distal end angles. Obovoid fruits are distinguished by their pronounced neck constriction at the proximal end, which results in lower values for the proximal end blockiness and proximal eccentricity attributes. The flat category fruit are differentiated from round tomatoes since they exhibit fruit shape indices much smaller than 1 (Fig. 3-1). Oxheart and heart categories are distinguished by the values for distal end protrusion and proximal eccentricity.

3.3 The Effect of the Tomato Shape Genes on Fruit Morphology

SUN and *OVATE* control fruit elongation whereas *FAS* and *LC* control locule number and fruit size (Liu et al. 2002; Cong et al. 2008; Xiao et al. 2008; Munos et al. 2011). To determine which fruit presented in Fig. 3-1 carry the mutant allele of *SUN*, *OVATE*, *FAS* and/or *LC*, the accessions were genotyped for these genes. Some flat tomatoes carry *FAS* but nearly all carry *LC* (Fig. 3-1; Rodriguez et al. 2011a) causing high locule number in addition to the flat shape. Three of the four oxheart fruits shown in Fig. 3-1 also carry the *FAS* allele whereas all carry the *LC* allele. All of the varieties shown in the long category bear the *SUN* allele, which causes very elongated and tapered fruits. These long accessions also carry *LC* resulting in fruits with more than three locules, yet fruit shape is elongated and not flat. This implies that *LC* does not affect fruit elongation perse. *OVATE* is present in every obovoid category shown in Fig. 3-1, and this gene also underlies fruit shape in other categories including ellipsoid, rectangular, and heart shaped fruit. Interestingly, some accessions with elongated fruit shape carry neither the *SUN* nor *OVATE* alleles. Also, some flat and oxheart accessions with high locule number do neither carry *FAS* nor *LC* (Fig. 3-1; Rodriguez et al. 2011a). It seems plausible that shape in these varieties is controlled by other loci for which the underlying genes have not yet been identified. A few predictions can be made regarding loci that control fruit shape in these varieties. For example, *fs8.1* and *fs2.1* controlling fruit shape index segregated in several studies (Grandillo and Tanksley 1996; Gonzalo and van der Knaap 2008) and therefore are likely to control fruit shape in

Table 3-1 Morphological characteristics of eight tomato fruit shape categories.

	Fruit shape Index	Width widest Position	Distal end protrusion	Proximal Fruit Blockiness at 20%	Distal and angle at 20%	Rectangular	Proximal eccentricity
Long[a]	> 1.70	< 0.56	> 0.15	0.75–0.99	37.9–63.5	0.46–0.53	0.73–0.89
Obovoid	> 1.20	> 0.53	0.00	0.56–0.81	63.1–103.4	0.38–0.54	0.50–0.91
Ellipsoid	1.10–1.80	< 0.52	< 0.08	0.76–0.91	59.8–98.8	0.46–0.55	0.73–0.89
Rectangular	1.10–1.90	< 0.55	< 0.03	0.85–0.92	52.5–99.0	0.53–0.58	0.68–0.82
Oxheart	0.80–1.30	< 0.47	> 0.10	0.89–0.94	77.9–114.0	0.48–0.51	0.90–1.18
Heart	0.95–1.40	< 0.48	> 0.05	0.83–0.95	74.6–108.4	0.45–0.51	0.80–0.90
Round	0.90–1.00	< 0.48	< 0.02	0.84–0.92	98.1–114.1	0.51–0.55	0.84–1.0
Flat	< 0.85	< 0.53	< 0.06	0.74–0.96	110.7–150.0	0.52–0.60	0.85–1.27

[a]Visually determined fruit shape categories (first column). The range of values for the attributes in each shape category is listed. Table reprinted from Rodriguez et al. 2011, Plant Physiol 156: 275–285 (www.plantphysiol.org). Copyright American Society of Plant Biologists.

some of these lines. Nevertheless, the diversity in tomato fruit form within the cultivated tomato germplasm is to a large extent explained by the alleles *SUN, OVATE, LC* and *FAS*, despite a few exceptions. Additionally, all four mutant alleles were found in commercial varieties, highlighting the importance of the shape genes to modern breeding (Rodriguez et al. 2011a).

3.4 Genotype-Based Diversity Analysis

Identifying the genes that control morphological traits by positional cloning and a general understanding of genome diversity is efficiently investigated by the application of molecular markers. Multiple platforms for detecting polymorphisms exist, and the appropriate system will depend on the goals of the individual researcher and the type of markers to be mapped. DNA-based markers such as restriction fragment length polymorphisms (RFLP) were used to generate the first reference linkage maps in tomato (Tanksley et al. 1992). RFLP markers are cumbersome as they require Southern blotting and probe hybridizations to detect the polymorphisms. Higher throughput genome diversity and genetic studies are obtained with PCR-based markers. One of the first PCR-based markers was random amplified polymorphic DNA (RAPD), which uses random decamer sequences as primers to amplify products from genomic DNA (Williams et al. 1990). Another PCR-based marker which, like RAPD, does not require a priori genome sequence information is amplified fragment length polymorphism (AFLP; Haanstra et al. 1999). Transposon-based markers are also PCR-based and largely anonymous with the exception that the transposon sequences of the species of interest are known. These markers are based on insertion polymorphisms of the same transposon in related accessions and have been used successfully in tomato (Tam et al. 2005, 2007). Unfortunately, these sequence-independent markers are population-specific and dominant, and therefore not easily transferred between laboratories. However, both AFLP and transposon-based markers have been quite useful in tomato diversity analyses within cultivated accessions because of the high level of polymorphism and the automation of marker generation and analysis (Park et al. 2004; Tam et al. 2005; van Berloo et al. 2008).

Sequence-dependent PCR based markers display useful characteristics including co-dominance of alleles and robust assays and are now commonly used in most plant species. The markers are easily shared with other researchers, which facilitates comparisons among studies. Simple sequence repeat (SSR) or microsatellites, insertion and deletion polymorphisms (indel), and nucleotide substitutions provide the basis for sequence-dependent PCR-based markers. Mutations located in coding and regulatory regions may be the direct cause of phenotypic variation.

Therefore, sequence-dependent PCR-based markers are very promising in association mapping studies which aim to link phenotype to genotype without the development of structured mapping populations.

Tomato sequence-dependent PCR based markers have been developed from a variety of sources including genomic and expressed sequence tag (EST) sequences as well as intron-based markers and oligonucleotide array hybridizations (Smulders et al. 1997; Areshchenkova and Ganal 1999; Suliman-Pollatschek et al. 2002; He et al. 2003; Yang et al. 2004; Frary et al. 2005; Labate and Baldo 2005; Van Deynze et al. 2007; Sim et al. 2009). SSRs and single nucleotide polymorphisms (SNPs) can be difficult to identify in transcribed sequences due to their low occurrence between closely related tomato species and within the cultivated germplasm. Instead, SSR and SNP polymorphisms are more readily identified by mining intron sequences. This approach is feasible for crops where genome sequence is not yet available since intron position is commonly conserved between diverged species, especially for highly conserved genic sequences. Using conserved orthologous sequence (COS) derived from tomato, the position of introns was determined by comparison with the *Arabidopsis* genome (Timms et al. 2006; Wu et al. 2006; Van Deynze et al. 2007). In tomato, the intron mining approach led to the identification of 937 COS from which 243 were polymorphic within cultivated tomato (Van Deynze et al. 2007). In addition, an oligonucleotide array corresponding to 22,714 tomato unigenes and hybridization of genomic DNA to the array led to the identification of many potential single feature polymorphisms within cultivated tomato. Of these, 306 were found in 111 genes and validated by sequencing (Sim et al. 2009). Additional information about tomato markers is provided in chapter 4 of this volume.

3.5 Genetic Diversity within Cultivated Tomato Germplasm

Compared to most crop plants, the level of genetic diversity within cultivated tomato germplasm is very low (Miller and Tanksley 1990; Williams and St. Clair 1993). This is likely the result of bottlenecks that took place during initial domestication and the transfer of a narrow set of germplasm to Europe and other parts of the world (Rick 1976; Nesbitt and Tanksley 2002). Recent breeding efforts have improved genetic diversity within cultivated germplasm, primarily via introgression of genomic regions derived from wild relatives that confer increased disease resistance and other quality traits (Williams and St. Clair 1993; Labate and Baldo 2005; Sim et al. 2009). Despite the low level of polymorphism within cultivated tomato, sequence-independent (AFLP, RAPD, transposon-based) and sequence-dependent (SSR, indel, SNP) markers are useful for distinguishing tomato cultivars (Table 3-2). These markers are also useful for grouping accessions

Table 3-2 Polymorphism levels among cultivated tomato.

Study	Number of varieties	Type of varieties	Class of markers	Main findings in distinguishing accessions
(Miller and Tanksley 1990)	23	Vintage, few landraces, cherry tomato	RFLP	Vintage varieties show the lowest genetic diversity but exhibit unique polymorphisms
(Williams and St. Clair 1993)	46	Modern, vintage, South American, cherry tomato	RFLP, RAPD	Vintage varieties show the lowest polymorphism level, followed by Modern, South American, and cherry tomato
(Smulders et al. 1997)	7	European breeding lines	SSR	Polymorphism among cultivars increase with the total length of the repeat
(Villand et al. 1998)	96	South and Central American, other areas, cherry tomato	RAPD	Higher level of polymorphism among South and Central American accessions in comparison to other accessions
(Bredemeijer et al. 2002)	521	European breeding lines	SSR	Most varieties could be identified with the 20 SSR markers used
(Archak et al. 2002)	27	Indian, American vintage	RAPD	Newer Indian varieties show reduced diversity compared to older varieties
(He et al. 2003)	19	Mixed areas	SSR	The 65 polymorphic SSR distinguished each variety
(Park et al. 2004)	74	Mainly California vintage, modern hybrids	AFLP	Most vintage and modern hybrids clustered separately
(Carelli et al. 2006)	35	Brazilian landraces, few vintage	RAPD	Most Brazilian landraces clustered in one group
(Garcia-Martinez et al. 2006)	48	Traditional Spanish cultivars, few hybrids and cherry tomato	SSR, AFLP	Clustering was based in part on the traditional classes
(Tam et al. 2007)	58	Vintage, French breeding lines, cherry tomato	Transposon-based	Some clusterings based on fruit types and weights

(Mazzucato et al. 2008)	61	Italian landraces, few vintage and modern	SSR	Some Italian landraces clustered according to vegetative and reproductive traits
(Yi et al. 2008)	87	Myanmar landraces, other areas	SSR	Most Myanmar landraces cluster separately from other varieties
(Benor et al. 2008)	39	Chinese, Japanese, S. Korean, US breeding and commercial lines	SSR	Clustering consistent with geographical location
(Van Berloo et al. 2008)	94	Historical and breeding materials from seed companies	AFLP	Clustering based on fruit size
(Sim et al. 2009, 2011)	88	Processing and fresh market (US and Canada), landraces (S and Central America), vintage	SNP, InDel	Fresh market, processing and landrace groups show higher genetic diversity than vintage. Significant genetic differences exist between market classes
(Rodriguez et al. 2011)	116	Heirloom, regional (mainly US and Italy)	SNP, SSR	Subpopulation structure driven by fruit shape and fruit shape genes

based upon genetic diversity and determining diversity within and among individual tomato accessions.

A number of methods are available for estimating the level of polymorphism or genetic diversity in a collection of germplasm. For example, the mean polymorphic information content (PIC) for SSR markers sampled in a selection of cultivated tomato accessions ranged from 0.31 to 0.4 (Bredemeijer et al. 2002; He et al. 2003; Benor et al. 2008). The PIC value reflects the probability that the sampled alleles would be different between two randomly chosen varieties. Another measure of diversity is the Jaccard index or Jaccard similarity coefficient, which is typically used with molecular data obtained from dominant markers. Depending on marker and germplasm type, Jaccard similarity coefficient values between 0.05 and 0.98 have been reported (Park et al. 2004; Carelli et al. 2006; Yi et al. 2008). A low Jaccard coefficient denotes low genetic similarity between two accessions. Another common measure of diversity is Nei's genetic distance, whose value is based on allele frequencies between or within populations. A Nei's genetic distance value of "0" indicates that the two varieties carry identical alleles for all evaluated loci. Using RFLP and RAPD markers, Williams and St. Clair (1993) estimated Nei's distance values ranging from 0 to 0.33 in a broad collection of tomato germplasm. Bredemeijer et al. (2002) applied SSR markers to a large collection of tomato breeding lines and estimated Nei's distance values ranging from 0 to 0.70. Clustering of the molecular marker data and using the genetic distances demonstrated groupings within the cultivated tomato germplasm based on either geography or tomato types (Table 3-2).

More recent analyses of genetic diversity within cultivated tomato were obtained using Bayesian approaches (Mazzucato et al. 2008; van Berloo et al. 2008; Rodriguez et al. 2011a; Sim et al. 2011). Implemented in the software STRUCTURE (Pritchard et al. 2000), these analyses also allow the estimation of the proportion of an accession that belongs to a particular subpopulation and thus the identification of admixture in the entire population under study. The STRUCTURE analyses demonstrated that cultivated tomato can be clustered based on market type such as processing and fresh market (Sim et al. 2011) and fruit size (van Berloo et al. 2008), and that certain fruit shape categories appear to be overrepresented in one compared to another subpopulation (Rodriguez et al. 2011a). The STRUCTURE analyses showed that the selection for adaptation to the environment (Sim et al. 2011) and fruit shape (Rodriguez et al. 2011a) played an important role in population structure within the cultivated germplasm. It should be noted that results from the clustering analysis are dependent on the selection of the accessions for the study as well as the selection of the markers. For example, the number of subpopulations in cultivated tomato may range from three (van Berloo et al. 2008) to six (Sim et al. 2011). When including the wild relative *S. pimpinellifolium* and the presumed ancestor of cultivated tomato *S. lycopersicum* cherry tomato, the cultivated tomato clusters in just two subpopulations (Ranc et al. 2008). The latter is easily explained by the fact that diversity in the wild relative is much higher than in the cultivated tomato and therefore the clustering is biased towards separating wild from cultivated. The selection of markers may also affect clustering. For example, using 25 neutral markers and four fruit shape gene markers (*SUN*, *OVATE*, *FAS* and *LC*) led to the identification of five clusters among the group of 116 cultivated accessions that varied for shape. These five clusters were significantly different from one another based on Nei's standard genetic distances and pairwise F_{st} values (Rodriguez et al. 2011a). When removing the four fruit shape gene markers from the dataset, the optimal number of clusters could not be determined implying that selection of markers could have a significant impact on the identification of population structure. Yet, the data also showed that those accessions exhibiting different fruit shapes have a different genetic background. This finding was supported by the notion that the mutant allele of *OVATE* was rarely found in the same accession that harbor one of the other three mutant shape gene alleles. On the other hand, mutation in *LC* was often found together with either *SUN* or *FAS*. Since nearly all of the 116 accessions used in the STRUCTURE analysis were heirloom or regional types, this finding suggests the existence of at least two genetically distinct lineages among the older cultivated tomato accessions (Rodriguez et al. 2011a). Another important finding from the study was that *LC*, *OVATE* and *FAS* were found in *S. lycopersicum* cherry tomato as well as regional accessions from South and Central America while *SUN* was only found in regional and heirloom accessions originating from Europe. This finding suggested that the *SUN* mutation was likely a post-domestication event while the other mutants arose during or before domestication (Rodriguez et al. 2011a).

3.6 Future Prospects

With respect to fruit morphology, the genes *SUN*, *OVATE*, *LC* and *FAS* explain to a large extend the variation in the tomato germplasm. Accurate measurements of shape will facilitate the identification of additional fruit shape and size genes including enhancers and suppressors of the known genes. With the availability of new molecular markers from the Sol CAP project (http://solcap.msu.edu/tomato.shtml) and the ease of sequencing entire tomato genomes using second and third generation technologies, the discovery of additional genes controlling tomato fruit morphology is within reach. Therefore, novel insights into the molecular basis of the morphological diversity of the tomato fruit and the driving forces of its domestication will be further elucidated in the near future.

Acknowledgements

Tomato fruit shape research in the Van der Knaap laboratory is supported by the National Science Foundation grant numbers DBI0227541 and IOS0922661. We thank Dr. Carolyn Male for information on heirloom tomatoes and inspiring discussions.

References

Archak S, Karihaloo JL, Jain A (2002) RAPD markers reveal narrowing genetic base of Indian tomato cultivars. Curr Sci 82: 1139–1143.

Areshchenkova T, Ganal MW (1999) Long tomato microsatellites are predominantly associated with centromeric regions. Genome 42: 536–544.

Barrero LS, Tanksley SD (2004) Evaluating the genetic basis of multiple-locule fruit in a broad cross section of tomato cultivars. Theor Appl Genet 109: 669–679.

Benor S, Zhang M, Wang Z, Zhang H (2008) Assessment of genetic variation in tomato (*Solanum lycopersicum* L.) inbred lines using SSR molecular markers. J Genet Genom 35: 373–379.

Bredemeijer GMM, Cooke RJ, Ganal MW, Peeters R, Isaac P, Noordijk Y, Rendell S, Jackson J, Roder MS, Wendehake K, Dijcks M, Amelaine M, Wickaert V, Bertrand L, Vosman B (2002) Construction and testing of a microsatellite database containing more than 500 tomato varieties. Theor Appl Genet 105: 1019–1026.

Brewer MT, Lang L, Fujimura K, Dujmovic N, Gray S, van der Knaap E (2006) Development of a controlled vocabulary and software application to analyze fruit shape variation in tomato and other plant species. Plant Physiol 141: 15–25.

Brewer MT, Moyseenko JB, Monforte AJ, van der Knaap E (2007) Morphological variation in tomato: A comprehensive study of quantitative trait loci controlling fruit shape and development. J Exp Bot 58: 1339–1349.

Carelli BP, Gerald LTS, Grazziotin FG, Echeverrigaray S (2006) Genetic diversity among Brazilian cultivars and landraces of tomato *Lycopersicon esculentum* Mill. revealed by RAPD markers. Genet Resour Crop Evol 53: 395–400.

Carli P, Barone A, Fogliano V, Frusciante L, Ercolano MR (2011) Dissection of genetic and environmental factors involved in tomato organoleptic quality. BMC Plant Biol 11: 58.

Chaib J, Devaux MF, Grotte MG, Robini K, Causse M, Lahaye M, Marti I (2007) Physiological relationships among physical, sensory, and morphological attributes of texture in tomato fruit. J Exp Bot 58: 1915–1925.

Cong B, Barrero LS, Tanksley SD (2008) Regulatory change in yabby-like transcription factor led to evolution of extreme fruit size during tomato domestication. Nat Genet 40: 800–804.

Frary A, Xu Y, Liu J, Mitchell S, Tedeschi E, Tanksley S (2005) Development of a set of PCR-based anchor markers encompassing the tomato genome and evaluation of their usefulness for genetics and breeding experiments. Theor Appl Genet 111: 291–312.

Garcia-Martinez S, Andreani L, Garcia-Gusano M, Geuna F, Ruiz JJ (2006) Evaluation of amplified fragment length polymorphism and simple sequence repeats for tomato germplasm fingerprinting: Utility for grouping closely related traditional cultivars. Genome 49: 648–656.

Gonzalo MJ, Brewer MT, Anderson C, Sullivan D, Gray S, van der Knaap E (2009) Tomato fruit shape analysis using morphometric and morphology attributes implemented in tomato analyzer software program. J Am Soc Hort Sci 134: 77–87.

Gonzalo MJ, van der Knaap E (2008) A comparative analysis into the genetic bases of morphology in tomato varieties exhibiting elongated fruit shape. Theor Appl Genet 116: 647–656.

Grandillo S, Tanksley SD (1996) QTL analysis of horticultural traits differentiating the cultivated tomato from the closely related species *Lycopersicon pimpinellifolium*. Theor Appl Genet 92: 935–951.

Haanstra JPW, Wye C, Verbakel H, Meijer-Dekens F, van den Berg P, Odinot P, van Heusden AW, Tanksley S, Lindhout P, Peleman J (1999) An integrated high-density RFLP-AFLP map of tomato based on two *Lycopersicon esculentum* x *L. pennellii* F_2 populations. Theor Appl Genet 99: 254–271.

Hackbusch J, Richter K, Muller J, Salamini F, Uhrig JF (2005) A central role of *Arabidopsis thaliana* ovate family proteins in networking and subcellular localization of 3-aa loop extension homeodomain proteins. Proc Natl Acad Sci USA 102: 4908–4912.

He C, Poysa V, Yu K (2003) Development and characterization of simple sequence repeat (SSR) markers and their use in determining relationships among *Lycopersicon esculentum* cultivars. Theor Appl Genet 106: 363–373.

International Plant Genetic Resources Institute (1996) Descriptors for tomato (*Lycopersicon* spp). International Plant Genetic Resources Institute, Rome, Italy.

International Union for the Protection of New Varieties and Plants (2001) Guidelines for the conduct of tests for distinctness, homogeneity and stability (tomato). International Union for the Protection of New Varieties and Plants, Geneva, Switzerland.

Jenkins JA (1948) The origin of cultivated tomato. Econ Bot 2: 379–392.

Labate JA, Baldo AM (2005) Tomato SNP discovery by EST mining and resequencing. Mol Breed 16: 343–349.

Lippman Z, Cohen O, Alvarez JP, Abu-Abied M, Pekker I, Paran I, Eshed Y, Zamir D (2008) The making of a compound inflorescence in tomato and related nightshades. PLoS Bio 18: e288.

Lippman Z, Tanksley SD (2001) Dissecting the genetic pathway to extreme fruit size in tomato using a cross between the small-fruited wild species *Lycopersicon pimpinellifolium* and *L. esculentum* var. Giant heirloom. Genetics 158: 413–422.

Liu J, Van Eck J, Cong B, Tanksley SD (2002) A new class of regulatory genes underlying the cause of pear-shaped tomato fruit. Proc Natl Acad Sci USA 99: 13302–13306.

Male CJ (1999) 100 heirloom tomatoes for the American garden. Workman Publishing, New York, USA.

Matthiolus PA (1544) Di pedacio dioscoride anazerbeo libri cinque della historia, et material medicinale trodotti in lingua vulgare Italiana. In: Anazarbos PDo (ed) De materia medica, Venetia de Bascaraini.

Mazzucato A, Papa R, Bitocchi E, Mosconi P, Nanni L, Negri V, Picarella ME, Siligato F, Soressi GP, Tiranti B, Veronesi F (2008) Genetic diversity, structure and marker-trait associations in a collection of Italian tomato (*Solanum lycopersicum* L.) landraces. Theor Appl Genet 116: 657–669.

Miller JC, Tanksley SD (1990) Effect of different restriction enzymes, probe source, and probe length on detecting restriction fragment length polymorphism in tomato. Theor Appl Genet 80: 385–389.

Munos S, Ranc N, Botton E, Berard A, Rolland S, Duffe P, Carretero Y, Le Paslier MC, Delalande C, Bouzayen M, Brunel D, Causse M (2011) Increase in tomato locule number is controlled by two single-nucleotide polymorphisms located near WUSCHEL. Plant Physiol 156: 2244–2254.

Nesbitt TC, Tanksley SD (2002) Comparative sequencing in the genus *Lycopersicon*: Implications for the evolution of fruit size in the domestication of cultivated tomatoes. Genetics 162: 365–379.

Paran I, van der Knaap E (2007) Genetic and molecular regulation of fruit and plant domestication traits in tomato and pepper. J Exp Bot 58: 3841–3852.

Park YH, West MA, St Clair DA (2004) Evaluation of AFLPs for germplasm fingerprinting and assessment of genetic diversity in cultivars of tomato (*Lycopersicon esculentum* L.). Genome 47: 510–518.

Peralta IE, Spooner DM, Knapp S (2008) Taxonomy of wild tomatoes and their relatives (*Solanum* sect. *Lycopersicoides*, sect. *Juglandifolia*, sect. *Lycopersicon*; Solanaceae), vol 84. American Society of Plant Taxonomists, Ann Arbor, MI, USA.

Pnueli L, Carmel-Goren L, Hareven D, Gutfinger T, Alvarez J, Ganal M, Zamir D, Lifschitz E (1998) The self-pruning gene of tomato regulates vegetative to reproductive switching of sympodial meristems and is the ortholog of cen and tfl1. Development 125: 1979–1989.

Pritchard JK, Stephens M, Donnelly P (2000) Inference of population structure using multilocus genotype data. Genetics 155: 945–959.

Ranc N, Munos S, Santoni S, Causse M (2008) A clarified position for Solanum lycopersicon var. cerasiforme in the evolutionary history of tomatoes (Solanaceae). BMC Plant Biol 8: 130.

Rick CM (1976) Tomato. In: Simmonds NW (ed) Evolution of Crop Plants. Longman Group, London, UK, pp 268–273.

Rodriguez GR, Munos S, Anderson C, Sim SC, Michel A, Causse M, McSpadden-Gardener BB, Francis D, van der Knaap E (2011a) Distribution of *SUN, OVATE, LC* and *FAS* in the tomato germplasm and the relationship to fruit shape diversity. Plant Physiol 156: 275–285.

Rodriguez GR, Francis D, van der Knaap E, Strecker J, Njanji I, Thomas J, Jack A (2011b) New features and many improvements to analyze morphology and color of digitalized plant organs are available in Tomato Analyzer 3.0. Proceedings of The 22nd Midwest Artificial Intelligence and Cognitive Science Conference Cincinnati, USA, April 16–17, 2011.

Sim SC, Robbins MD, Chilcott C, Zhu T, Francis DM (2009) Oligonucleotide array discovery of polymorphisms in cultivated tomato (*Solanum lycopersicum* L.) reveals patterns of SNP variation associated with breeding. BMC Genomics 10: 466.

Sim SC, Robbins MD, Van Deynze A, Michel AP, Francis DM (2011) Population structure and genetic differentiation associated with breeding history and selection in tomato (*Solanum lycopersicum* L.). Heredity 106: 927–935.

Smulders MJM, Bredemeijer G, Rus-Kortekaas W, Arens P, Vosman B (1997) Use of short microsatellites from database sequences to generate polymorphisms among *Lycopersicon esculentum* cultivars and accessions of other *Lycopersicon* species. Theor Appl Genet 94: 264–272.

Suliman-Pollatschek S, Kashkush K, Shats H, Hillel J, Lavi U (2002) Generation and mapping of AFLP, SSRs and SNPs in *Lycopersicon esculentum*. Cell Mol Biol Let 7: 583–597.

Tam SM, Causse M, Garchery C, Burck H, Mhiri C, Grandbastien MA (2007) The distribution of copia-type retrotransposons and the evolutionary history of tomato and related wild species. J Evol Biol 20: 1056–1072.

Tam SM, Mhiri C, Vogelaar A, Kerkveld M, Pearce SR, Grandbastien MA (2005) Comparative analyses of genetic diversities within tomato and pepper collections detected by retrotransposon-based SSAP, AFLP and SSR. Theor Appl Genet 110: 819–831.

Tanksley SD, Ganal MW, Prince JP, de Vicente MC, Bonierbale MW, Broun P, Fulton TM, Giovannoni JJ, Grandillo S, Martin GB (1992) High density molecular linkage maps of the tomato and potato genomes. Genetics 132: 1141–1160.

Timms L, Jimenez R, Chase M, Lavelle D, McHale L, Kozik A, Lai Z, Heesacker A, Knapp S, Rieseberg L, Michelmore R, Kesseli R (2006) Analyses of synteny between *Arabidopsis thaliana* and species in the Asteraceae reveal a complex network of small syntenic segments and major chromosomal rearrangements. Genetics 173: 2227–2235.

van Berloo R, Zhu A, Ursem R, Verbakel H, Gort G, van Eeuwijk FA (2008) Diversity and linkage disequilibrium analysis within a selected set of cultivated tomatoes. Theor Appl Genet 117: 89–101.

van der Knaap E, Tanksley SD (2003) The making of a bell pepper-shaped tomato fruit: Identification of loci controlling fruit morphology in yellow stuffer tomato. Theor Appl Genet 107: 139–147.

Van Deynze AE, Stoffel K, Buell RC, Kozik A, Liu J, van der Knaap E, Francis D (2007) Diversity in conserved genes in tomato. BMC Genom 8: 465.

Villand J, Skroch PW, Lai T, Hanson P, Kuo CG, Nienhuis J (1998) Genetic variation among tomato accessions from primary and secondary centers of diversity. Crop Sci 38: 1339–1347.

Wang S, Chang Y, Guo J, Chen JG (2007) *Arabidopsis* ovate family protein 1 is a transcriptional repressor that suppresses cell elongation. Plant J 50: 858–872.

Welty N, Radovich C, Meulia T, van der Knaap E (2007) Inflorescence development in two tomato species. Can J Bot 85: 111–118.

Williams CE, St Clair DA (1993) Phenetic relationships and levels of variability detected by restriction fragment length polymorphism and random amplified polymorphic DNA analysis of cultivated and wild accessions of *Lycopersicon esculentum*. Genome 36: 619–630.

Williams JG, Kubelik AR, Livak KJ, Rafalski JA, Tingey SV (1990) DNA polymorphisms amplified by arbitrary primers are useful as genetic markers. Nucl Acids Res 18: 6531–6535.

Wu F, Mueller LA, Crouzillat D, Petiard V, Tanksley SD (2006) Combining bioinformatics and phylogenetics to identify large sets of single-copy orthologous genes (cosii) for comparative, evolutionary and systematic studies: A test case in the euasterid plant clade. Genetics 174: 1407–1420.

Wu S, Xiao H, Cabrera A, Meulia T, van der Knaap E (2011) SUN regulates vegetative and reproductive organ shape by changing cell division patterns. Plant Physiol doi: 10.1104/pp 111.181065.

Xiao H, Jiang N, Schaffner EK, Stockinger EJ, van der Knaap E (2008) A retrotransposon-mediated gene duplication underlies morphological variation of tomato fruit. Science 319: 1527–1530.

Yang W, Bai X, Kabelka E, Eaton C, Kamoun S, van der Knaap E, Francis D (2004) Discovery of single nucleotide polymorphisms in *Lycopersicon esculentum* by computer aided analysis of expressed sequence tags. Mol Breed 14: 21–34.

Yi SS, Jatoi SA, Fujimura T, Yamanaka S, Watanabe J, Watanabe KN (2008) Potential loss of unique genetic diversity in tomato landraces by genetic colonization of modern cultivars at a non-center of origin. Plant Breed 127: 189–196.

Zanor MI, Rambla JL, Chaib J, Steppa A, Medina A, Granell A, Fernie AR, Causse M (2009) Metabolic characterization of loci affecting sensory attributes in tomato allows an assessment of the influence of the levels of primary metabolites and volatile organic contents. J Exp Bot 60: 2139–2154.

Molecular Markers, Genetic Maps and Association Studies in Tomato

Martin W. Ganal

ABSTRACT

This chapter provides an overview about the history of tomato molecular marker analyses with a focus on the different marker systems and on future developments regarding the use of molecular markers in tomato with the special emphasis on single nucleotide polymorphism (SNP) markers. Furthermore, the status of the genetic reference map of tomato is summarized. Based on molecular marker data, our current knowledge concerning population structure of the cultivated tomato and the extent of linkage disequilibrium in various tomato germplasm pools is discussed with respect to applications such as association studies and the analysis of candidate genes for specific traits. Finally, future developments in the area of molecular marker analysis, population analysis and the association of molecular markers with specific traits are outlined considering the soon available full genome sequence of tomato.

Keywords: Single nucleotide polymorphism, population structure, linkage disequilibrium, genome sequencing, phenotyping

4.1 Introduction

With the advent of large scale molecular marker development and availability, it is possible to not only map individual genes but also quantitatively inherited traits in segregating populations and to identify molecular markers that are tightly linked to such traits. While such an

Trait Genetics GmbH, Am Schwabeplan 1b, 06466 Gatersleben, Germany.

endeavour is relatively simple for traits that are controlled by individual genes, this is more complicated for loci controlling quantitative trait loci (QTL) since it requires sophisticated population development and genetic marker analyses over many generations until such QTLs are identified, validated and used for plant breeding.

In order to shorten the process of QTL identification, over the last years and especially in the area of human genetics a technology termed association genetics has been developed that is based on the use of unrelated individuals with defined phenotypes, which are analyzed with many markers. Based on the genetic variation observed with each of these markers, it is attempted to identify markers directly that show significant associations with a trait of interest (reviewed in McCarthy et al. 2008). The prerequisite for such an analysis is, beside excellent phenotyping, that ideally markers associated with each individual gene of an organism can be tested. This requires the analysis of millions of sequence variations in a given organism. Fortunately, it has been observed that not all single nucleotide polymorphisms (SNPs) in a given gene are inherited as individual segregating units but that only a small proportion of the theoretically possible combinations of linked SNPs are reflected in the actual sequence context of a genomic region. Such combinations of adjacent SNPs are termed haplotypes. Haplotypes can thus be considered as alleles at a given locus (Cardon and Abecasis 2003). Detailed analyses in many organisms have demonstrated that haplotypes can extend over smaller or larger genomic regions. For example in organisms such as human (Conrad et al. 2006) and maize (Rafalski and Morgante 2004), the extend of conserved haplotypes (also termed linkage disequilibrium) varies between less than 1 kilobase and several hundred kilobases and is highly linked to the analyzed population and the frequency of genetic recombination in the respective region.

For association analyses in humans meanwhile several millions of SNPs have been identified and the size range of haplotypes and linkage disequilibrium has been identified in many individuals (Serre and Hudson 2006). For large scale marker analyses highly multiplexed arrays containing more than 1 million SNPs based on haplotype data generated within the HapMap activity have been generated, which are now routinely used in association studies (Maresso and Broeckel 2008). This process has been termed genome-wide association studies (GWAS) and it has successfully been used for the identification of genes influencing complex traits such as disease factors (Evans and Cardon 2006; Gibson and Goldstein 2007; Kruglyak 2008).

Association genetics holds also great potential in the analysis of quantitatively inherited traits in plants. So far the association of specific markers in the form of defined sequence variation has been successfully demonstrated mainly in maize where large numbers of markers are available

(Yu and Buckler 2006). However, even in maize, such analyses have mainly been confined to the area of candidate gene analysis (Thornsberry et al. 2001) with few examples of successful genome scanning (Belo et al. 2008). In contrast to whole genome scanning, the candidate gene approach makes more or less educated guesses about the involved genes based on their possible function, chromosomal location or other available information and in this way reduces the number of individual sequence variants that are being analyzed. The problem with this approach is that such guesses towards candidate genes can be wrong and thus such an approach will frequently be unsuccessful (Tabor et al. 2002).

This summary will provide an overview about the current status of molecular marker development and its potential use for the association of markers with interesting traits either on a candidate gene level or in the future on a full genome scale in tomato. It will also summarize our currently knowledge about haplotype structure and linkage disequilibrium in tomato and will outline strategies for generating essential tools for large scale marker analyses.

4.2 Molecular Markers for Tomato Genome Mapping and Marker/Trait Association Studies

In general, tomato genetic marker analyses have largely benefited from the high level of polymorphism that is being observed between cultivated tomato (*Solanum lycopersicum*) and its wild relatives. Many molecular markers in tomato have been mapped in crosses between *S. lycopersicum* and *S. pennellii* (Tanksley et al. 1992). Since in this cross nearly all markers are polymorphic due to the high level of sequence diversity, it has led to the fast development of high density genetic maps that were useful for many purposes such as the map-based cloning of individual disease resistance genes (Martin et al. 1993) or whole genome analysis of QTLs in segregating populations (Paterson et al. 1988).

This approach has however disregarded that sequence variation in the cultivated tomato germplasm is very low compared to the sequence variation in the wild species of the tomato section within the genus *Solanum* (Miller and Tanksley 1990). Such studies have demonstrated that only approximately 10% of the total genetic variation in tomato is found in the cultivated tomato with the largest level of genetic variation being in the small-fruited (cherry) material that is genetically related to the wild species *Solanum pimpinellifolium* and *Solanum lycopersicum* ssp. *cerassiforme* considered to be possible ancestors of the cultivated red-fruited tomato. The genetic diversity of the fresh-market tomato and processing tomato varieties representing other major commercially used germplasm is even lower than that of the small-fruited varieties (van Berloo et al. 2008b).

Taking this into consideration, a discussion of the currently employed marker systems that can be used for association studies in tomato is being presented in the following section.

4.2.1 RFLPs and RFLP-based PCR Markers

The first molecular markers that were used for tomato genetic mapping and genome analysis were based on restriction fragment length polymorphisms (RFLPs). Such markers were derived either from tomato genes or single copy sequences derived from clones of non-methylated DNA and were analyzed through the hybridization of such probes onto filters containing DNA digested with restriction enzymes. A large wealth of RFLP markers (between 1,000 and 2,000) have been generated and mapped in the years from 1985 to 2000 (Tanksley et al. 1992). Their value was predominant in the generation of genetic maps, the study of synteny with other Solanaceae resulting in the fact that hundreds of potato markers can also be used for tomato mapping and the association of markers and traits through QTL analyses in segregating populations. Many QTLs have been localized with RFLP markers onto the tomato genome including QTLs for important traits ranging from disease resistance to quality and morphological traits. For general association mapping, RFLP markers have only been used in special cases due to the fact that they display low levels of polymorphism in breeding material and the analysis of many RFLP markers (hundreds) requires major efforts although some multiplexing (use of several markers simultaneously) has been achieved (Young et al. 1988). In the last years, RFLP marker were mostly replaced by cleaved amplified polymorphic sequences (CAPS) markers derived from sequence polymorphisms in the respective DNA sequence for the routine analysis of large numbers of plant samples based on the polymerase chain reaction (PCR) principle. A number of CAPS markers have been generated so that they detect genetic variation in *Solanum lycopersicum* (Bai et al. 2004; Frary et al. 2005). Such markers can be used in association studies although too few of these markers are currently available for such experiments.

4.2.2 PCR-based Anonymous Markers

Starting from approximately 1990, PCR-based analyses were used in tomato. At the beginning, random amplified polymorphic DNA (RAPD) markers were used for the location of interesting traits in the tomato genome using nearly isogenic lines (Martin et al. 1991) and bulked segregant analysis (Michelmore et al. 1991), which is a special case of association mapping used in segregating populations. The RAPD technique was soon replaced by the more robust amplified fragment length polymorphism (AFLP) technique that permits the simultaneous analysis of more than 50 individual fragments in a single reaction (Vos et al. 1995). The advantage of PCR-based anonymous markers are that thousands of loci can be analyzed in the tomato genome in a multiplexed fashion so that a dense coverage of the genome can be achieved with molecular markers (Haanstra et al. 1999). The benefit of these technologies is most pronounced in the area of generating high density genetic maps and specifically for saturating chromosomal regions containing interesting genes for breeding and map-based cloning using bulked segregant analysis where in some cases marker densities of more than one marker per 50 kilobases have been achieved (Vos et al. 1998). Some of the weaknesses of PCR-based anonymous markers are that such markers are frequently derived from repeatitive DNA sequences, that the precise chromosomal position of such markers in various crosses or lines is not guaranteed so that quite frequently the same size fragment is not derived from the same locus and that such markers are not evenly distributed in the genome (Bonnema et al. 2002). Nevertheless, since such markers can be generated in very large numbers, marker densities can be reached that make them very useful for association studies.

4.2.3 Microsatellite or SSR Markers

Microsatellite or simple sequence repeat (SSR) markers based on short tandem repeats are of great interest in tomato genetic analyses since such markers detect a high level of polymorphism even in closely related material such as the cultivated tomato. They are multiallelic (i.e., one marker can identify a number of different alleles), which is very useful for association studies (Bredemeijer et al. 2002). Microsatellites can be readily isolated from genomic libraries or the screening of existing sequence databases such as expressed sequence tags (ESTs). Approximately 2000 well characterized and mapped microsatellite markers have been described and characterized in tomato lines (e.g., Smulders et al. 1997; Areshchenkova and Ganal 1999; Shirasawa et al. 2010a). One problem with the development of SSR markers in tomato is the fact that genomic tomato microsatellites are, in contrast to many other plants, frequently localized in the centromeric heterochromatin and less frequent in the euchromatin that contains most genes (Areshchenkova and Ganal 1999). Furthermore, microsatellites in coding regions are relatively short and significantly less polymorphic (Smulders et al. 1997). However, with the advent of more sequencing data from the tomato sequencing project, it can be expected that microsatellite markers might gain some renewed interest since then, they can be generated in a targeted approach based on the genomic sequence and microsatellite markers could provide highly polymorphic markers that have the ability to detect also relatively rare alleles due to their multiallelic nature.

4.2.4 *Single Nucleotide Polymorphism (SNP) Markers*

As in the genetic analysis of the human genome, single nucleotide polymorphism (SNP) markers have recently attracted major interest in tomato. A useful SNP is in principle every single base sequence variation that occurs between individuals in single copy DNA. These SNPs can be used for molecular marker analysis and technologies have been developed that permit the analysis of many (thousands) SNPs simultaneously through the analysis of highly multiplexed arrays (Fan et al. 2006; Maresso and Broeckel 2008). The main disadvantage of SNP markers is the amount of work that is required for their identification. Specifically, until the more recent development of the Next Generation Sequencing technologies, the identification and validation of an SNPs required the screening of DNA sequences from different lines (Yang et al. 2004; Labate and Baldo 2005; Labate et al. 2009) for such sequence polymorphisms through the amplification of a given single copy locus (e.g., a gene), the screening of arrays (Sim et al. 2009) or database analyses (Jimenez-Gomez and Maloof 2009). For example, published data have demonstrated that it is possible to identify SNPs in tomato genes and non-coding sequences through the comparative sequencing of individual lines with the result of identifying genes that contain SNPs in germplasm that is either representing cultivated tomato lines or individual unadapted lines, which are very close relatives thereof (van Deynze et al. 2007). Larger numbers of SNPs have been recently identified in tomato through comparative transcriptome sequencing using conventional and Next Generation Sequencing technologies (Jimenez and Maloof 2009; Shirasawa et al. 2010b; Blanca et al. 2011). It has also demonstrated that sequence variation in genes is low in accessions of cultivated tomato with only a small proportion of the genes being polymorphic at all. This level of polymorphism can be slightly increased by the use of non-coding sequences such as intron sequences and other genomic single copy sequences.

4.3 Genetic Maps and Available Markers

The mapping of the markers on a genetic map and if available also on a physical map are prerequisites for high quality association studies. The first landmark of tomato genetic mapping was achieved in 1992 with the publication of a saturated genetic map with more than 1,000 mapped RFLP markers. These markers were either based on genes (cDNAs) or single copy genomic genomic probes. At that time with an average marker distance of less than one million base pairs, this was one of the most densely populated genetic map in crop plants. In reality the marker density on a per kilobase basis was higher in euchromatin which contains most genes since many mapped markers were derived from genes (Tanksley et al. 1992).

Over the years, the number of mapped markers in a comparable population (Tomato-EXPEN 2000) was increased with the mapping of additional markers such as the COS markers (Fulton et al. 2002), SSR and sequence tagged site (STS) markers (Shirasawa et al. 2010a,b) and it now serves as the backbone of the current tomato genetic map. At present, this map contains several thousand markers. Furthermore, the set of the mapped tomato markers can be complemented by a large number of potato markers since the potato genome is, except for a few inversions, largely syntenic with the tomato genome (Mueller et al. 2005a,b; www.sgn.cornell.edu). The mapping of RFLP, SSR and STS markers has also been complemented through the mapping of more than 1,000 AFLP markers (Haanstra et al. 1999).

Through the hybridization of oligonucleotide probes (overgo-probes) onto bacterial artificial chromosome (BAC) libraries of tomato, the fingerprinting of these BAC clones and BAC end sequencing, a wealth of information has been generated that permits the location of many mapped probes onto a physical contig map of tomato. Together with the progress on sequencing the tomato euchromatin on a BAC by BAC basis and the whole genome sequencing approach using Next Generation sequencing the tomato genome becomes accessible for saturating it at a very high density with additional markers for specific association studies (Mueller et al. 2005a, b; www.sgn.cornell.edu).

4.4 Population Structure, Linkage Disequilibrium and Association of Markers and Traits

Despite this high marker density, so far very few studies towards the association of markers with specific traits have been attempted and published. The main reason for this is probably the low level of polymorphism in cultivated tomato resulting in few actually useful markers and the lack of detailed studies on the extent of linkage disequilibrium until very recently. In the following part, the limited data will be summarized that have been published in that area.

4.4.1 Population Structure in Tomato

The tomato section of the *Solanum* genus contains a wide variety of different species. Some of the species are highly outbreeding with very strong self-incompatibility systems that display a very high level of sequence diversity between individuals (Arunyawat et al. 2007). However, beside their importance as source for disease resistance genes and other traits, the information derived from these species cannot be transferred to the cultivated tomato (*Solanum lycopersicum*).

The cultivated tomato is highly inbreeding and has passed through a number of genetic bottlenecks during domestication and commercial use so that only a small amount of the original variation is still present in current tomato varieties (Tanksley and McCouch 1997). Already with RFLP markers, it has been discovered that only approximately 10% of the entire genetic variation observed within the crossing range of tomato species is present in the cultivated tomato germplasm (Miller and Tanksley 1990).

Few comprehensive studies exist concerning the population structure of current tomato germplasm. These studies are based on the analysis of large samples of tomato varieties with microsatellite, AFLP or SNP markers. For example within a variety identification project, more than 500 current tomato varieties have been investigated with a set of microsatellite markers. Although this marker set was quite small (20 markers), due to their multiallelic nature, it was possible to obtain information on the population structure of individual cultivated tomato germplasm pools (Breedemeijer et al. 2002). A more comprehensive study of 94 cultivated tomato lines with 882 AFLP markers revealed a clear population structure in tomato varieties that is based on their fruit size (van Berloo et al. 2008b). Specifically, the highest level of polymorphism is found in the small fruited (cherry) type tomato varieties. This is most likely due to the fact that cherry tomatoes are most closely related to the wild red fruited ancestors of the cultivated tomato. A lower level of polymorphism is observed in the regular sized fresh-market tomato varieties and an even lower level is present in the large fruited beef-type tomato varieties probably because each of these types has mainly been a selection from the previous germplasm pool.

4.4.2 Linkage Disequilibrium

Very little data exist on the extend of linkage disequilibrium (LD) in tomato. As indicated in the previous section, the data are completely different comparing wild tomato species and cultivated tomato. Data from wild tomato species show that LD in some wild outbreeding tomato species does not extend much further than the size of a gene and thus is comparable to some other wild germplasm such as in maize (Städler et al. 2008).

For cultivated tomato only few comprehensive studies have been published recently. In one study, the extent of LD has been investigated in 94 tomato lines with 304 mapped markers. These data demonstrate that the extent of LD is very high and extends over a considerable range (up to 10–20 centiMorgan). Such data indicate that since the development of the modern cultivated tomato varieties, only a limited amount of genetic recombination or relatively few effective generations with recombination events have occurred (van Berloo et al. 2008b).

Two other more recent reports confirm the extent of LD in cultivated material (Robbins et al. 2011; Sim et al. 2011). In one of these studies, the population structure and extent of LD has been investigated with a number of PCR-based markers. This study reveals that tomato germplasm is structured according to usage groups such as fresh-market and processing material. In the other study, through the analysis of more than 400 markers in 102 tomato lines, the extent of LD was determined to 6–8 cM over all tomato types and up to 14–16 cM in specific usage types.

4.4.3 First Efforts towards Association Mapping in Tomato

Tomato is in fact one of the first crop plants for which association mapping has been attempted and successfully employed to identify markers that are associated with known traits. The approach has been exploited to localize markers that were associated with genes that confer disease resistance against important tomato pathogens. In tomato, many monogenic disease resistance genes have been introgressed from highly polymorphic wild relatives such as *Solanum habrochaites* and *Solanum peruvianum*. Examples for such genes are the resistance genes against tobacco mosaic virus or root-knot nematodes from *Solanum peruvianum*. Furthermore, introgressed regions from wild tomato relatives usually suppress genetic recombination in subsequent generations so that large linkage blocks with complete LD are conserved even after many generations. In some cases such linkage blocks derived from introgressed regions comprise parts or full chromosome arms. In the early phases of tomato genome mapping, whole genome scans with RFLP markers have been used to successfully identify markers associated with important disease resistance genes in tomato either by using individual markers or pools of five markers evenly spread over the tomato genome (Young et al. 1988). Later the same approach has been used with RAPD and AFLP markers that permit the analysis of higher multiplexes and/or more markers in order to achieve higher marker densities and find very tightly linked markers.

4.4.4 Candidate Gene Analysis

Over the last years, a considerable number of candidate genes for many traits have been identified (Causse et al. 2004; Bermudez et al. 2008). Such genes were either selected based on the function of the respective genes determined by biochemical characterization or sequence homology to known genes. In tomato, this approach has been successfully been used to identify genes that are associated with specific mutants (Thorup et al.

2000; De Jong et al. 2004) and in the closely related potato, this approach has been also used for the association of specific candidate genes through the analysis of the genetic variation in a large panel of potato lines (Li et al. 2005, 2008).

True association studies where a panel of unrelated tomato lines is being characterized for phenotypic traits and the genetic variation in candidate genes are still in its infancy (Mazzucato et al. 2008). Only a limited number of studies have been published where associations of markers or candidate genes have been investigated. For example, for the fruit size QTL *fw2.2* a study of primitive tomato germplasm has been performed with respect to that QTL and it was found that the *fw2.2* QTL was not mainly associated with varying fruit size there as it has been demonstrated for that gene in cultivated tomato varieties (Cong and Tanksley 2002; Nesbitt and Tanksley 2002). In another study it could be shown that allelic diversity in the *SUN*, *OVATE*, *LC* and *FAS* genes were associated with fruit shape in tomato (Rodriguez et al. 2011).

Candidate gene analysis in tomato will gain increased importance since in the recent years major efforts have been undertaken to identify QTLs and underlying candidate genes in segregating material and introgression lines. During these analyses large numbers of potential candidate genes have been identified (Bermudez et al. 2008) and one of the next steps in validation of such candidate genes needs to be the association of specific sequence variants in a panel of unrelated lines as it has been successfully demonstrated for maize where a specific candidate gene has been clearly linked to increase carotenoid content in kernels (Harjes et al. 2008). Once such associations have been demonstrated and validated, the specific allele can be used via marker assisted selection in the improvement of the crop.

4.5 Future Developments

4.5.1 Requirements on Markers for Marker/Trait Associations

For marker/trait association studies, a very high number of molecular markers, which are uniformly distributed throughout the genome or at least the gene-containing regions, is necessary (Peleman and van der Voort 2003). The average distance of polymorphic markers used for association studies in a given organism or population should be smaller than the average size of the chromosomal segments conserved by LD and different alleles or haplotypes present in the respective germplasm should be identifiable.

The marker system that satisfies the above described criteria is based on SNPs since very large numbers of markers need to be identified and analyzed. Since individual SNP markers are biallelic and do not easily permit the identification of more than two haplotypes either very large numbers of SNPs need to be identified or specific efforts towards the identification of haplotypes need to be undertaken.

Given that LD extends in cultivated tomato over several centiMorgan and the total genome has a genetic length of approximately 1,400 to 1,500 centiMorgan, a minimal marker density requires at least several hundred highly polymorphic and evenly distributed markers for a reasonable genome coverage. Assuming that not every SNP marker is informative in every line and that for each locus SNPs should be available that can discriminate more than two different haplotypes, the necessary numbers increase further. Usually in association studies, the interesting alleles that have a significant (and positive) effect on a trait of interest at a specific chromosomal region are not among the most prevalent alleles but are relatively rare and present in only a limited number of lines in an association panel. A final point that increases the number of SNPs is the fact that not every single SNP is accessible to high-throughput SNP analysis on arrays or other high throughput SNP analysis technologies. Given these points, it can be concluded that for reasonable genome coverage with SNPs a marker density of at least one and better several polymorphic markers per centiMorgan will be essential for whole genome association studies. This corresponds to a minimum of thousands of polymorphic SNPs that need to be identified in cultivated tomato material.

4.5.2 *Large Scale SNP Identification for Marker/Trait Associations*

Currently with a full genomic tomato sequence just becoming available, one of the most efficient ways of identifying such large SNP numbers is the comparative sequencing of tomato genes or other single copy sequences in a panel of tomato breeding germplasm. In the public, such efforts have recently started and reports on the identification of SNPs in genes have been published (Van Deynze et al. 2007; Jimenez and Maloof 2009; Shirasawa et al. 2010b; Blanca et al. 2011). The advantage of the described approaches is that large numbers of genes in tomato have been identified during large scale EST identification and that in many (but not all) cases variation in or adjacent (within conserved haplotypes) to genes is the reason for genetic variation on traits.

With the requirements described in the previous section, it is clear that for the identification of thousands of SNPs in cultivated tomato germplasm thousands of genes need to be analyzed through comparative sequencing. Ideally, the genetic variation of every single copy gene in tomato needs to be probed since this would ensure that most of the genetic variation in the gene containing regions of the tomato genome is sampled. It is estimated that the tomato genome contains approximately 35,000 genes (van der Hoeven et al. 2002) of which a considerable proportion will be present in closely related copies (e.g., closely related dispersed or tandemly repeated gene families, recently duplicated genes) so that it can be estimated that 25,000 to 30,000 genes will be available for large scale SNP identification projects within genes.

We have approached the analysis of as many single copy tomato genes as possible for SNPs through a comparative sequencing approach. In this approach, we have specifically focussed our efforts towards the identification of SNPs in tomato breeding germplasm through the sequencing of genes in elite tomato germplasm and varieties in order to target such SNPs that are useful for genetic studies such as mapping and trait associations in breeding material. Specifically, the efforts are targeted towards the identification in cherry, fresh-market, beef-type and processing varieties.

In general, tomato is one of the least polymorphic crop plants within breeding material (one SNP every 250 base pairs in a panel of 15 lines). For comparison, a sample from the wild tomato species *Solanum pennellii*, which is commonly used as a mapping parent and which was included in the panel, displayed nearly five times that many SNPs for the analyzed genes. Similar values have been reported for smaller numbers of genes (Bermudez et al. 2008).

As expected and recently described by others (Robbins et al. 2011; Sim et al. 2011), the highest level of polymorphism was observed within the cherry tomato types. A significant decrease (approximately 50%) of polymorphism compared to the cherry material is observed in the analyzed fresh-market tomato material with further smaller decreases in the number of useful SNPs in beef-type tomato varieties and processing type tomato varieties compared to the fresh-market material. An evaluation of the identified SNPs concerning their haplotype structure revealed that the vast majority of tomato genes displayed only two to three different haplotypes with very few cases where more than three haplotypes are observed confirming the assumption that the cultivated tomato has passed through severe bottlenecks during evolution (Tanksley and McCouch 1997).

4.5.3 Genome-wide Association Studies

So far no genome-wide association studies have been published using a very large number of SNP or other markers. However, with the recent identification of large numbers of SNP markers in several thousand tomato genes, the development of arrays with large numbers of SNPs will become feasible in the very near future. Together with information about identified haplotypes, SNP arrays with thousands of SNPs that are polymorphic in tomato varieties are currently being developed (http://solcap.msu.edu). Such arrays will get close to the necessary marker density (at least one polymorphic SNP per centiMorgan) that is necessary to identify associations of markers with traits of interest in elite tomato germplasm. These arrays will open the door to large scale association studies for traits of economic importance such as the content of important metabolites (e.g., lycopine, vitamins) and other quality traits. Furthermore, such markers can be used for the targeted and precise introgression of important genes and QTLs from wild species into the cultivated tomato germplasm (Fernie et al. 2006).

4.5.4 The Future—The Tomato Genome Sequence, Large Scale Phenotyping and Association Studies

Although the entire tomato genome sequence has not yet been published, it is expected that the full sequence will be available in 2012. With the full tomato genome sequence available, new avenues for the identification of molecular markers (e.g., SNPs) will be possible through the comparative sequencing of the genome of many different tomato lines using the novel high-throughput DNA sequencing methods. In this way, it will be possible to identify basically all interesting SNPs within single copy tomato sequences and make them available for large scale SNP arrays and association studies on a truly genome-wide level (Mardis 2008).

With that in mind, it is now absolutely necessary to generate large panels consisting of hundreds of phenotypically well characterized tomato lines that can be employed in such association studies. Associations of markers to specific traits can only be reliably detected with precisely genotyped and phenotyped material (Schauer et al. 2006). Such a phenotypic characterization should cover phenotypic traits, traits of economic importance in the field (e.g., yield) and important metabolites (e.g., soluble solids, flavonoids, carotenoids, vitamins and flavor compounds) that are stored in databases (reviewed by Yano et al. 2007). As in associations studies in other organisms such as maize and human, with large scale marker analyses and precisely phenotyped material, it will be possible to associate markers with specific traits of scientific or economic importance (Van Berloo et al. 2008a) and employ them in the improvement of tomato towards such traits as taste, soluble solid content or metabolites that are beneficial for human health.

Furthermore, technologies should be developed that permit the fast validation of identified candidates in elite germplasm through the generation of specific introgression lines and other advanced genetic material that can be directly used in the breeding process. This challenge is even more pronounced in tomato, since the vast majority of the genetic variation in the tomato germplasm is found in the wild species and few concepts (Fernie et al. 2006; Lippman et al. 2007) have been presented to tap into this resource for tomato improvement.

Acknowledgements

The input and efforts of Hartmut Luerssen, Andreas Polley, Joerg Plieske, Eva-Maria Graner, Markus Wolf and Gregor Durstewitz toward the realization of the TraitGenetics SNP identification project in tomato is gratefully acknowledged.

References

Areshchenkova T, Ganal M (1999) Long tomato microsatellites are predominantly associated with centromeric regions. Genome 42: 536–544.

Arunyawat U, Stephan W, Städler T (2007) Using multilocus sequence data to assess population structure, natural selection, and linkage disequilibrium in wild tomatoes. Mol Biol Evol 24: 2310–2322.

Bai Y, Feng X, van der Hulst R, Lindhout P (2004) A set of simple PCR markers converted from sequence specific RFLP markers on tomato chromosome 9 to 12. Mol Breed 13: 281–287.

Belo A, Zheng P, Luck S, Shen B, Meyer DJ, Li B, Tingey S, Rafalski A (2008) Whole genome scan detects an allelic variant of *fad2* associated with increased oleic acid levels in maize. Mol Genet Genom 279: 1–10.

Bermúdez L, Urias U, Milstein D, Kamenetzky L, Asis R, Fernie AR, van Sluys MA, Carrari F, Rossi M (2008) A candidate gene survey of quantitative trait loci affecting chemical composition in tomato fruit. J Exp Bot 59: 2875–2890.

Blanca JM, Pascual L, Ziarsolo P, Nuez F, Canizares J (2011) ngs-backbone: a pipeline for read cleaning, mapping and SNP calling using next generation sequence. BMC Genomics 12: 285.

Bonnema G, van den Berg P, Lindhout P (2002) AFLPs mark different genomic regions compared with RFLPs: a case study in tomato. Genome 45: 217–221.

Bredemeijer GMM, Cooke RJ, Ganal MW, Peeters R, Isaac P, Noordijk Y, Rendell S, Jackson J, Röder M, Wendehake K, Dijcks M, Amelaine M, Wickart V, Bertrand L, Vosman B (2002) Construction and testing a microsatellite database containing more than 500 tomato varieties. Theor Appl Genet 105: 1019–1026.

Cardon LR, Abecasis GR (2003) Using haplotype blocks to map human complex trait loci. Trends Genet 19: 135–140.

Causse M, Duffe P, Gomez MC, Buret M, Damidaux R, Zamir D, Gur A, Chevalier C, Lemaire-Chamley M, Rothan C (2004) A genetic map of candidate genes and QTLs involved in tomato fruit size and composition. J Exp Bot 55: 1671–1685.

Cong B, Liu J, Tanksley SD (2002) Natural alleles at a tomato fruit size quantitative trait locus differ by heterochronic regulatory mutations. Proc Natl Acad Sci USA 99: 13606–13611.

Conrad DF, Jakobsson M, Coop G, Wen X, Wall JD, Rosenberg NA, Pritchard JK (2006) A worldwide survey of haplotype variation and linkage disequilibrium in the human genome. Nat Genet 38: 1251–1260.

De Jong WS, Eannetta NT, de Jong DM, Bodis M (2004) Candidate gene analysis of anthocyanin pigmentation loci in the Solanaceae. Theor Appl Genet 108: 423–432.

Evans DM, Cardon LR (2006) Genome-wide association: a promising start to a long race. Trends Genet 22: 350–354.

Fan J-B, Chee MS, Gunderson KL (2006) Highly parallel genomic assays. Nat Rev Genet 7: 632–644.

Fernie AR, Tadmor Y, Zamir D (2006) Natural genetic variation for improving crop quality. Curr Opin Plant Biol 9: 196–202.

Frary A, Xu Y, Mitchell S, Tedeschi E, Tanksley S (2005) Development of a set of PCR-based anchor markers encompassing the tomato genome and evaluation of their usefulness for genetics and breeding. Theor Appl Genet 111: 291–312.

Fulton TM, vam der Hoeven R, Eannetta NT, Tanksley SD (2002) Identification, analysis and utilization of conserved ortholog set markers for comparative genomics in higher plants. Plant Cell 14: 1457–1467.

Gibson G, Goldstein DB (2007) Human genetics: the hidden text of genome wide associations. Curr Biol 17: 929–932.

Haanstra JPW, Wye C, Verbakel H, Meijer-Dekens F, van den Berg P, Odinot P, van Heusden AW, Tanksley S, Lindhout P, Peleman J (1999) An integrated high-density RFLP-AFLP map of tomato based on two *Lycopersicon esculentum* x *L. pennellii* F$_2$ populations. Theor Appl Genet 99: 254–271.

Harjes CE, Rocheford TR, Bai L, Brutnell TP, Kandianis CB, Sowinski SG, Stapleton AE, Vallabhaneni R, Williams M, Wirtel ET, Yan J, Buckler ES (2008) Natural genetic variation on lycopene epsilon cyclase tapped for maize biofortification. Science 319: 330–333.

Jimenez-Gomez JM, Maloof JN (2009) Sequence diversity in three tomato species: SNPs, markers, and molecular evolution. BMC Plant Biol 9: 85.

Kruglyak L (2008) The road to genome-wide association studies. Nat Rev Genet 9: 314–318.

Labate JA, Baldo AM (2005) Tomato SNP discovery by EST mining and resequencing. Mol Breed 16: 343–349.

Labate JA, Robertson LD, Baldo AM (2009) Multilocus sequence data reveal extensive departure from equilibrium in domesticated tomato (*Solanum lycopersicum* L.). Heredity 103: 257–267.

Li L, Strahwald J, Hofferbert HR, Lübeck J, Tacke E, Junghans H, Wunder J, Gebhardt C (2005) Dna variation in the invertase locus *invGE/GF* is associated with tuber quality traits in populations of potato breeding clones. Genetics 170: 813–821.

Li L, Paulo M-J, Strahwald, Lübeck J, Hofferbert H-R, Tacke E, Junghans H, Wunder J, Draffehn A, van Eeuwijk F, Gebhardt C (2008) Natural DNA variation at candidate loci is associated with potato chip color, tuber starch content, yield and starch yield. Theor Appl Genet 116: 1167–1181.

Lippman ZB, Semel Y, Zamir D (2007) An integrated view of quantitative trait variation using tomato interspecific introgression lines. Curr Opin Genet Dev 17: 1–8.

Mardis ER (2008) The impact of next-generation sequencing technology on genetics. Trends Genet 24: 133–141.

Maresso K, Broeckel U (2008) Genotyping platforms for mass-throughput genotyping with SNPs, including human genome wide scans. Adv Genet 60: 107–139.

Martin GB, Brommonschenkel SH, Chunwangse J, Frary A, Ganal MW, Spivey R, Wu T, Earle ED, Tanksley SD (1993) Map-based cloning of a protein-kinase gene conferring disease resistance in tomato. Science 262: 1432–1436.

Martin GB, Williams JGK, Tanksley SD (1991) Rapid identfication of markers linked to a *Pseudomonas* resistance gene in tomato by using random primers and near-isogenic lines. Proc Natl Acad Sci USA 88: 2336–2340.

Mazzucato A, Papa R, Bitocchi E, Mosconi P, Nanni L, Negri V, Picarella ME, Siligato F, Soressi GP, Tiranti B, Veronesi F (2008) Genetic diversity, structure and marker-trait associations in a collection of Italian tomato (*Solanum lycopersicum* L.) landraces. Theor Appl Genet 116: 657–669.

McCarthy MI, Abecasis GR, Cardon LR, Goldstein DB, Little J, Ionnidis JPA, Hirschhorn JN (2008) Genome-wide association studies for complex traits: consensus, uncertainty and challenges. Nat Rev Genet 9: 356–369.

Michelmore RW, Paran I, Kesseli RV (1991) Identification of markers linked to disease-resistance genes by bulked segregant analysis: a rapid method to detect markers in specific genomic regions by using segregating populations. Proc Natl Acad Sci USA 88: 9828–9832.

Miller JC, Tanksley SD (1990) RFLP analysis of phylogenetic relationships and genetic variation in the genus *Lycopersicon*. Theor Appl Genet 80: 437–448.

Mueller LA, Solow T, Taylor N, Skwarecki B, Buels R, Binns J, Lin C, Wright M Ahrens R, Wang Y, Herbst E, Keyder E, Menda N, Zamir D, Tanksley SD (2005a) The SOL Genomics Network. A comparative resource for Solanaceae biology and beyond. Plant Physiol 138: 1310–1317.

Mueller LA, Tanksley SD, Giovannoni JJ, van Eck J, Stack S, Choi D, Kim BD, Chen M, Cheng Z, Li C, Ling H, Xue Y, Seymour G, Bishop G, Bryan G, Sharma R, Khurana J, Tyagi A, Chattopadhyay D, Singh NK, Stiekema W, Lindhout P, Jesse T, Klein-Lankhorst R, Bouzayen M, Shibata D, Tabata S, Granell A, Botella MA, Giuliano G, Frusicante L, Causse M, Zamir D (2005b) The Tomato Sequencing Project, the first cornerstone of the International Solanaceae Project (SOL). Comp Funct Genom 6: 153–158.

Nesbitt TC, Tanksley SD (2002) Comparative sequencing in the genus *Lycopersicon*: implications for the evolution of fruit size in the domestication of cultivated tomatoes. Genetics 162: 365–379.

Paterson AH, Lander ES, Hewitt JD, Peterson S, Lincoln SE, Tanksley SD (1988) Resolution of quantitative traits into Mendelian factors by using a complete linkage map of restriction fragment length polymorphisms. Nature 335: 721–726.

Peleman JD, van der Voort JR (2003) Breeding by design. Trends Plant Sci 8: 330–334.

Rafalski A, Morgante M (2004) Corn and humans: recombination and linkage disequilibrium in two genomes of similar size. Trends Genet 20: 103–111.

Robbins MD, Sim SC, Yang W, van Deynze A, van der Knaap E, Joobeur T, Francis DM (2011) Mapping and linkage disequilibrium analysis with a genome-wide collection of SNPs that detect polymorphism in cultivated tomato. J Exp Bot 62: 1831–1845.

Rodriguez GR, Munos S, Anderson C, Sim S-C, Michel A, Causse M, McSpadden Gardener BB, Francis D, van der Knaap E (2011) Distribution of *SUN, OVATE, LC*, and *FAS* in the tomato germplasm and the relationship to fruit shape diversity. Plant Physiol 156: 275–285.

Schauer N, Semel Y, Roessner U, Gur A, Balbo J, Carrari F, Pleban T, Perez-Melis A, Bruedigam C, Kopka J, Willmitzer L, Zamir D, Fernie AR (2006) Comprehensive metabolic profiling and phenotyping of interspecific introgression lines for tomato improvement. Nat Biotechnol 24: 447–454.

Serre D, Hudson TJ (2006) Resources for genetic variation studies. Annu Rev Genom Hum Genet 7: 443–457.

Shirasawa K, Asamizu E, Fukuoka H, Ohyama A, Sato S, Nakamura Y, Tabata S, Sasamoto S, Wada T, Kishida Y, Tsuruoka H, Fujishiro T, Yamada M, Isobe S (2010a) An interspecific linkage map of SSR and intronic polymorphism markers in tomato. Theor Appl Genet 121: 731–739.

Shirasawa K, Isobe S, Hirakawa H, Asamizu E, Fukuoka H, Just D, Rothan C, Sasamoto S, Fujishiro T, Kishida Y, Kohara M, Tsuruoka H, Wada T, Nakamura Y, Sato S, Tabata S (2010b) SNP discovery and linkage map construction in cultivated tomato. DNA Res 17: 381–391.

Sim SC, Robbins MD, van Deynze A, Michel AP, Francis DM (2011) Population structure and genetic differentiation associated with breeding history and selection in tomato (*Solanum lycopersicum* L.). Heredity 106: 927–935.

Smulders MJM, Bredemeijer G, Rus-Kortekaas W, Arens P, Vosman B (1997) Use of short microsatellites from database sequences to generate polymorphisms among *Lycopersicon* species. Theor Appl Genet 97: 264–272.

Städler T, Arunyawat U, Stephan W (2008) Population genetics and speciation in two closely related wild tomatoes (*Solanum* section *Lycopersicon*). Genetics 178: 339–350.

Tabor HK, Risch NJ, Myers RM (2002) Candidate-gene approaches for studying complex genetic traits: practical considerations. Nat Rev Genet 3: 1–7.

Tanksley SD, Ganal MW, Prince JP, de Vicente MC, Bonierbale MW, Broun P, Fulton TM, Giovannoni JJ, Grandillo S, Martin GB, Messeguer R, Miller JC, Miller L, Paterson AH, Pineda O, Röder MS, Wing RA, Wu W, Young ND (1992) High-density molecular linkage maps of the tomato and potato genomes. Genetics 132: 1141–1160.

Tanksley SD, McCouch SR (1997) Seed banks and molecular maps: unlocking genetic potential from the wild. Science 277: 1063–1066.

Thornsberry JM, Goodman MM, Doebley J, Kresovich S, Nielsen D, Buckler ES (2001) *Dwarf8* polymorphisms associate with variation in flowering time. Nat Genet 28: 286–289.

Thorup TA, Tanyolac B, Livingstone KD, Popovsky S, Paran I, Jahn M (2000) Candidate gene analysis of organ pigmentation loci in the Solanaceae. Proc Natl Acad Sci USA 97: 11192–11197.

Van Berloo R, van Heusden S, Bovy A, Meijer-Dekens F, Lindhout P, van Eeuwijk F (2008a) Genetic research in a public private research consortium: prospects for indirect use of elite breeding germplasm in academic research. Euphytica 161: 293–300.

Van Berloo R, Zhu A, Ursem R, Verbakel H, Gort G, van Eeuwijk FA (2008b) Diversity and linkage disequilibrium analysis within a selected set of cultivated tomatoes. Theor Appl Genet 117: 89–101.

Van Deynze A, Stoffel K, Buell RC, Kozik A, Liu J, van der Knaap E, Francis D (2007) Diversity in conserved genes in tomato. BMC Genom 8: 465.

Van der Hoeven R, Ronning C, Giovannoni J, Martin G, Tanksley S (2002) Deductions about the number, organization, and evolution of genes in the tomato genome based on analysis of a large expressed sequence tag collection and selective genome sequencing. Plant Cell 14: 1441–1456.

Vos P, Hogers R, Bleeker M, Reijans M, van der Lee TH, Hornes M, Frijters A, Pot J, Peleman J, Kuiper M, Zabeau M (1995) AFLP: a new technique for DNA fingerprinting. Nucl Acids Res 23: 4407–4414.

Vos P, Simons G, Jesse T, Wijbrandi J, Heinen L, Hogers R, Frijters A, Groenendijk J, Diergaarde P, Reijans M, Fierens-Onstenk J, de Both M, Peleman J, Liharska T, Hontelez J, Zabeau M (1998) The tomato *Mi-1* gene confers resistance to both root-knot nematodes and potato aphids. Nat Biotechnol 16: 1365–1369.

Yang W, Bai X, Kabelka E, Eaton C, Kamoun S, van der Knaap EDF (2004) Discovery of single nucleotide polymorphism in *Lycopersicon esculentum* by computer aided analysis of expressed sequence tags. Mol Breed 14: 21–24.

Yano K, Aoki K, Shibata D (2007) Genomic databases for tomato. Plant Biotechnol 24: 17–25.

Young ND, Zamir D, Ganal MW, Tanksley SD (1988) Use of isogenic lines and simultaneous probing to identify DNA markers tightly linked to the *Tm-2a* gene in tomato. Genetics 120: 579–585.

Yu J, Buckler ES (2006) Genetic association mapping and genome organization of maize. Curr Opin Biotechnol 17: 155–160.

Mapping and Tagging of Simply Inherited Traits

*Ilan Levin** and *Arthur A. Schaffer*

ABSTRACT

Simply inherited traits are qualitative traits controlled by one or two genes, none of which is significantly influenced by the environment. Methods are currently available that can dissect the genetic basis of complex quantitative traits into distinct genes, which can be eventually identified and cloned, thus bridging the gap between qualitative and quantitative traits.

This book chapter focuses on mapping and tagging of simply inherited traits in the tomato plant. As we shall be demonstrating, many of these traits share profound importance from both practical as well as scientific perspectives. The molecular dissection of several of these traits has now set the scene for tagging genes controlling highly complex quantitative traits and established the tools for exploiting gene interactions affecting traits in a more than additive or synergistic manner.

Previous manuscripts on this subject matter have summarized the mapping and tagging of simply inherited traits in several plant species. In addition, a comprehensive list of disease and insect resistance genes, as well as a summary of flower, fruit and yield related traits for which genes or quantitative trait loci (QTL) have been identified and mapped in tomato was lately presented. We have, therefore, chosen to focus in this book chapter on simply inherited traits related to tomato fruit quality, with special emphasis on taste and color.

Keywords: fruit, quality, taste, color, qalitative, quantitative, traits

Department of Vegetable Research, Institute of Plant Sciences, Agricultural Research Organization, the Volcani Center, P.O. Box 6, Bet Dagan 50250, Israel.
*Corresponding author

5.1 Introduction

Traditionally, traits are divided into two categories: 1) qualitative or Mendelian traits, and 2) quantitative traits. Qualitative traits are those traits controlled by one or two (and rarely a few) genes, none of which is significantly influenced by the environment. The effect of each gene is typically "large" and discernable in nature and overall, will result in discrete, observable, phenotypic classes. For qualitative traits, the individual's phenotype is usually a clear representation of its genotype. These traits are also referred to as simply inherited traits and the genetic discipline that focuses on these traits is known as Mendelian genetics. Traits such as seed color, seed shape, pod shape, pod color and petal color observed by Gregor Mendel among pea plants are good examples of such traits. These traits enabled Mendel to formulate the two basic laws of genetics: the law of segregation and the law of independent assortment. These two laws, formulated between 1856 and 1863, are still fundamental in predicting the mode of genetic inheritance of traits.

Unlike qualitative traits, quantitative traits are influenced by multiple genes (~three genes or more) as well as the effects of the environment. These traits display continuous phenotypic classes and are also known as polygenic or complex traits. Examples of such traits are: yield, fruit size, and fiber quality in plants; growth rate, meat quality, and milk production in animals; height, weight and blood pressure in humans.

Many desired traits are not readily noticeable and their coherent identification is sometimes considered to be an art. In many instances such identification also requires special instrumentation and dedicated technologies. Good examples of the latter traits are sugar and carotenoid content of vegetables and fruits, metabolites often associated with taste and functional qualities of these agricultural products, respectively. Once scored, quantified or measured, geneticists and breeders can rank individuals within a certain population according to levels of expression of a desired phenotype and select for individuals with outstanding performances or traits as parents of the subsequent generation. Further on, breeders seek to transmit, or introgress, these valuable traits into the different populations or lines by breeding, i.e., a time consuming process of crosses and selection (Fig. 5-1). The time and cost requirements of a breeding process are highly dependent upon the nature of the target traits and accordingly upon the means needed for their quantitaion or scoring, the number of genes controlling their expression, whether they can be measured directly or indirectly at an early developmental stage, and whether they are associated with other undesired characteristics.

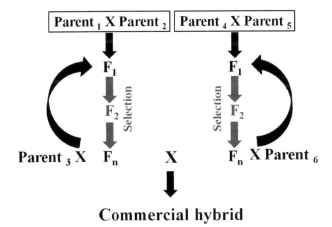

Commercial hybrid

Figure 5-1 Breeding is a time consuming repeated process which involves both crosses and selection.

Methods are currently available that can dissect the genetic basis of quantitative traits into distinct genes, which can be eventually identified and cloned, thus bridging the gap between qualitative and quantitative traits. Good examples of such gene identification are *FW2.2* controlling fruit size (Frary et al. 2000), *SUN* controlling fruit shape (Xiao et al. 2008) and *LIN5* as well as *AGPL1* controlling fruit sugar content in tomato (Schaffer et al. 2000; Fridman et al. 2004). Once identified, these gene sequences can serve as DNA markers, sharing characteristics of simply inherited traits, to breed valuable qualities into tomato germplasm by marker-assisted selection. This book chapter focuses on mapping and tagging of simply inherited traits. As we shall be demonstrating, many of these traits share profound importance from both practical as well as scientific perspectives. The molecular dissection of several of these traits has now set the scene for tagging genes controlling highly complex quantitative traits and established the tools for exploiting gene interactions affecting traits in a more than additive or synergistic manner.

A book chapter summarizing the mapping and tagging of simply inherited traits in plant species, focusing primarily on tomato and pepper, was recently published (Paran and Levin 2008). In addition, a comprehensive list of disease (fungal, bacterial and nematode) and insect resistance genes, as well as a summary of flower, fruit and yield related traits for which genes or quantitative trait loci (QTL) have been identified and mapped in tomato was lately presented (Foolad 2007; Labate et al. 2007). We have, therefore, chosen to focus in this book chapter on simply inherited traits related to tomato fruit quality, with special emphasis on taste and color.

5.2 Importance of the Simply Inherited Traits

Thousands of simply inherited traits have been identified among tomato accessions (Menda et al. 2004), which are extensively listed in Stevens and Rick (1986), http://tgrc.ucdavis.edu/ and in http://zamir.sgn.cornell.edu/mutants/. Initially, such traits were identified as abnormal or unusual phenotypes spontaneously appearing among pure-bred lines or populations. Later on, mutant phenotypes were induced, either by radiation, including fast neutron and gamma radiation, or by chemical treatments, including ethyleneamine and ethylenemethane sulfonate (EMS). Today, transgenic plants generated by overexpression or silencing of specific genes are also considered as mutant accessions when they show a particular phenotype, which is different from their azygous control counterparts and cosegregates with the transgene.

Simply inherited traits share both academic and economic importance. In tomato, simply inherited traits were exclusively used for the production of the first generation genetic linkage maps. Five distinctive morphological traits: dwarf plant habit, potato leaf, peach (fuzzy) fruits, yellow fruit flesh and colorless fruit epidermis, identified among tomato accessions, and cited as early as 1905, are good examples of such traits (Halsted et al. 1905). Monogenic inheritance for these traits was demonstrated later on, in addition to two additional traits: lutescent foliage and pyriform fruit shape (Price and Drinkard 1908). The first linkage study in tomato, which ranks as one of the classical organism for such studies, can be traced back to Jones (1917), who reinterpreted data of Hedrick and Booth (1907) on the cosegregation of dwarfness (d) and elongate (ovate) fruit shape (o) as the consequence of linkage between them. E.W. Lindstrom followed with an intensive study of linkage on chromosome 2, utilizing Jones' markers in addition to peach fruit (p) and compound inflorescence (s; Rick 1991). Throughout the years about 1,200 such traits have been mapped or assigned to the 12 chromosomes (Stevens and Rick 1986). However, since then another 3,417 induced mutations have been cataloged (Menda et al. 2004), and a collection of those have been curated (http://zamir.sgn.cornell.edu/mutants/). Among them are most of the previously described phenotypes from the monogenic mutant collection of the Tomato Genetics Resource Center (http://tgrc.ucdavis.edu/), and over a thousand new mutants, with multiple alleles per locus. This rather new, ambitious and successful attempt to saturate the tomato genome with induced mutations indicated that some organs such as leaves are more prone to alterations in comparison to others (Menda et al. 2004). This study is also a comprehensive resource describing how saturated mutational studies can be carried out and how phenotypes could be categorized. In this study a total of 13,000 M_2 families, derived from EMS and fast-neutron mutagenesis, were visually phenotyped in the field and categorized into a morphological catalog that includes 15 primary and 48 secondary categories. These categories, presented in Table 5-1, demonstrate the type and nature of mutant simply inherited traits that can be discovered in tomato and other plant species.

Many simply inherited traits, such as those listed in Tables 5-2 and 5-3, share economic importance and have been introduced by traditional breeding into leading tomato cultivars. For instance, most of the 50 top ranking processing tomato varieties grown today in the USA are characterized by resistance to Verticillium wilt race 1, Fusarium wilt race 1, Fusarium wilt race 2 and root-knot nematode (http://www.ipm.ucdavis. edu/PMG/r783900511.html). Simply inherited resistance traits were also introgressed into fresh-market tomato cultivars grown either in the open field or under protected environmental conditions (greenhouses and screenhouses). Today, it seems highly unlikely that a tomato variety can be acceptable for cultivation without harboring a cassette of 3–7 disease resistance traits, and their number is constantly increasing.

Two of the simply inherited traits presented in Table 5-2, *ripening inhibitor* (*rin*; Fig. 5-2) and *non-ripening* (*nor*), have been and are still being introgressed into fresh-market tomato cultivars in order to extend the shelf-life of their fruits' post-harvest quality. Although *rin* and *nor* are considered recessive, and tomato plants harboring these traits in a homozygous state yield fruits that fail to develop their characteristic red color, it was found that in a heterozygous state, they behave as semi-dominant yielding red tomato fruits with a significantly extended post-harvest shelf-life (Fig. 5-2). The genes that encode the *rin* and *nor* traits have subsequently been cloned and characterized. Positional map-based cloning of the *rin* locus revealed two tandem MADS-box genes (LeMADS-RIN and LeMADS-MC),

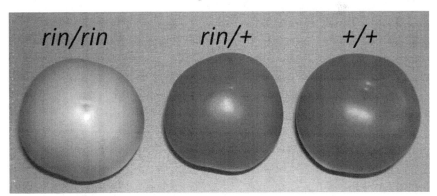

Figure 5-2 Mature *rin/rin* tomato fruit in comparison to its nearly isogenic normal-ripening counterpart (+/+) and their F$_1$ hybrid (*rin/+*).

Color image of this figure appears in the color plate section at the end of the book.

Table 5-1 Tomato mutants phenotypic catalog (Menda et al. 2004, http://zamir.sgn.cornell. edu/mutants/).

Major category	Sub category
Seed	Germination Seedling lethality Slow germination
Plant size	Extremely small Small plant Large plant
Plant habit	Internode length Branching Aborted growth Other plant habit
Leaf morphology	Leaf width Leaf size Leaf complexity Leaf texture Other leaf development
Leaf color	Purple leaf White leaf Yellow leaf Yellow-green leaf Dull green/gray leaf Dark green leaf Variegation
Flowering	Flowering timing
Inflorescence	Inflorescence structure
Flower morphology	Flower homeotic mutation Flower organ size Flower organ width Other flower morphology
Flower color	White flower Pale yellow flower Strong yellow flower
Fruit size	Small fruit Large fruit
Fruit morphology	Long fruit Rounded fruit Other fruit morphology
Fruit color	Yellow fruit Orange fruit Dark red fruit Epidermis Green fruit
Fruit ripening	Early ripening Late ripening
Sterility	Partial sterility Full sterility
Disease and stress response	Necrosis Wilting Other disease response

Table 5-2 Examples of simply inherited traits of commercial importance according to category. Gene symbol, effect and chromosome location are shown.

Category	Symbol	Effect	Chromosome location
Disease resistance	*Ve*	Resistance to Verticillium wilt	9
	I	Immunity to Fusarium wilt race 0	7
	Mi	Resistance to root-knot nematode species	6
	Tm	Resistance to tobacco mosaic virus	2
	Sm	Resistance to *Stemphylium solani*	11
	Ph	Resistant to *Phytophthora infestans*	7
	Pto	Resistance to *Pseudomonas syringae*	5
	Cf-1	*Cladosporium fulvum* resistance-1	1
	Ty-1	Tolerance to tomato yellow leaf curl virus	6
	Sw-5	Resistance to tomato spotted wilt virus	9
Plant habit	*sp*	Plant habit determinate	6
Fruit attachment	*j-2*	Jointless pedicel	12
Fruit ripening	*rin*	Retarded ripening—positively affecting shelf life in heterozygous state	5
	nor	Retarded ripening—positively affecting shelf life in heterozygous state	10
Fruit color	*Aft*	Fruit anthocyanins and flavonols increased	10
	y	Colorless fruit epidermis resulting in pink fruits	1
Fruit sugar content or profile	*Lin5*	Higher fruit sugar content	9
	AgpL1	Higher fruit sugar content	1
	Fgr	Fructose to glucose ratio increased	4
Sterility	*ms-17*	One of about 34 loci imparting male sterility. Useful in crosses to eliminate emasculation	2
Fruit weight and size	*FW2.2*	Fruit weight and size	2
Fruit shape	*o*	Ovate or pear shaped fruits	2
	SUN	Elongated fruit shape	7
Parthenocarpy	*pat*	Fruit mostly parthenocarpic resulting in seedless fruits	3

whose expression patterns suggested roles in fruit ripening (Vrebalov et al. 2002). Cloning of the *NOR* gene, revealed that it encodes a transcription factor related to a family of plant transcription factors associated with multiple aspects of plant development including meristem and cotyledon development and leaf senescence (Giovannoni et al. 2004).

Significant efforts have also been invested in characterizing and utilizing simply inherited traits associated with tomato fruit quality and in particular, color and taste. These traits will be dealt with extensively below.

Table 5-3 Gene identification and map location for selected mutants that increase or modulate carotenoid content in ripe tomato fruits.

Mutation	Description	Gene[a]	Map location	Reference
r	Yellow color of ripe fruit flesh	*PSY1*	Chromosome 3	Fray and Grierson 1993
B	Orange color of fruits, fruit β-carotene highly increased	*βLCY*	Chromosome 6	Ronen et al. 2000
og, og^c	Corolla tawny orange, fruit β-carotene highly reduced, ~15–20% increase in fruit lycopene content	*βLCY*	Chromosome 6	Ronen et al. 2000
DEL	Fruit color orange due to the accumulation of δ-carotene at the expense of lycopene	*εLCY*	Chromosome 12	Ronen et al. 1999
t	Fruit flesh and stamens orange colored. Fruits accumulate prolycopene (7Z,9Z,7'Z,9'Z-tetra-*cis*-lycopene) instead of the all-*trans*-lycopene	*CRTISO*	Chromosome 10	Isaacson et al. 2002
hp-1, hp-1^w	Fruit carotenoids and other phytonutrients highly increased	*DDB1*	Chromosome 2	Levin et al. 2006; Lieberman et al. 2004; Liu et al. 2004
hp-2, hp-2^j, hp-2^dg	Fruit carotenoids and other phytonutrients highly increased	*DET1*	Chromosome 1	Bino et al. 2005; Levin et al. 2003; Mustilli et al. 1999
hp-3	Fruits accumulate 30% more carotenoids	*ZE*	Chromosome 2	Galpaz et al. 2008
gh	Fruits milky white, fruit phytoene increased	*PTOX*	Chromosome 11	Josse et al. 2000; Mackinney et al. 1956

[a]Gene abbreviations: *CRTISO = CAROTENOID ISOMERASE, DET1 = DEETIOLATED 1, DDB1 = UV-DAMAGED DNA BINDING PROTEIN 1, βLCY = β-LYCOPENE CYCLASE, εLCY = ε-LYCOPENE CYCLASE, PSY1 = PHYTOENE SYNTHASE 1, PTOX = PLASTID TERMINAL OXIDASE, ZE = ZEAXANTHIN EPOXIDASE*

5.3 Genetic and Molecular Dissection of Simply Inherited Traits

Genetic and molecular dissection of simply inherited traits share both commercial and academic importance. Detection of DNA makers, which are highly associated with commercial traits can reduce time, space and labor needed for the breeding process. The process of detection of DNA markers linked to a trait of interest and further identification of the gene or genes that control these traits rely heavily on genetic maps that have become more sophisticated in the recent years.

About 30 genetic linkage maps have been developed for the tomato based on different intra- and interspecific crosses, as recently summarized (Foolad 2007). Several of these maps can be accessed on the worldwide web at the Solanaceae Genomics Network (SGN; http://www.sgn.cornell. edu/).

Several basic features are required to place a gene, a marker or a trait onto a genetic linkage map: two different alleles of the gene (marker or trait), two different alleles of a second gene (marker or trait) to map relative to, and an opportunity for meiotic recombination. The latter is achieved by constructing segregating populations, usually second filial (F_2) or backcross (BC) population, originating between two phenotypically or genetically divergent parental lines. Genetic maps report the linear order of genes and the amount of recombination between linked genes (Fig. 5-3). They do not contain information on physical distance, neither cytological distance nor number of DNA base pairs between markers. The tomato sequencing project, launched in November 2003, is now generating easily accessible tools that bridge the gap between recombination distances and the particular physical distances, in base pairs, among DNA markers (Mueller et al. 2005; http://www.sgn.cornell.edu/; Fig. 5-3).

Earlier maps were based on simply inherited traits, showing visible and discrete variation in a population. Color and morphological traits that show clear segregation patterns in populations generated by cross-breeding were among these traits. These first generation maps were limited in terms of size, density, and further progress.

The advent of molecular markers, based on protein and DNA molecules, advanced substantially the size and density of molecular maps. This was primarily due to fact that two genetic stocks may harbor many DNA and protein sequence polymorphisms, which are phenotypically neutral. The first isozyme gene linkage map of the tomato and its application in genetics and breeding was published by Tanksley and Rick (1980).

An example of the use of a combination of morphological and isozyme markers is the double tagging of male sterility gene in tomato (Tanksley and Zamir 1988). Male sterility genes are useful for hybrid seed production, eliminating the need to emasculate the female parent. Several nuclear male sterility genes have been identified in tomato but their use suffers from two drawbacks. First, because male sterility is recessive, its transfer in backcross programs requires progeny testing. Second, once the gene has been transferred into the parent line, it will continue to segregate, which requires roguing fertile plants before pollination. To overcome these problems, a male sterility gene *ms-10* was placed in *cis* linkage configuration with the codominant isozyme marker *Prx-2* (at 0.5 cM distance) and the recessive morphological marker *aa* (at 5 cM distance), which results in the absence of anthocyanin in the seedling hypocotyls. The use of the

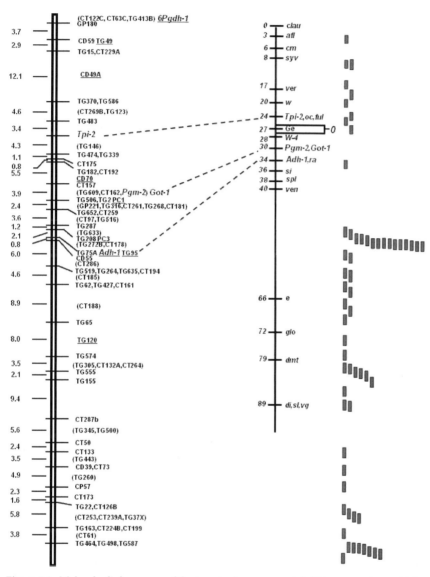

Figure 5-3 Molecular linkage map of the tomato chromosome 4 (left) in comparison with its classical map (center) and red bars (right) showing the approximate location of sequenced contigs (positions of markers from the classical map on the molecular map are shown by dashed lines. Classical and molecular maps were taken from Tanksley et al. (1992), and presented with the kind permission of SD Tanksley. Location of sequenced contigs can be accessed through http://www.sgn.cornell.edu/).

Color image of this figure appears in the color plate section at the end of the book.

codominant *Prx-2* marker allows selection of heterozygous plants during backcrossing, eliminating the need for progeny testing, while the use of the recessive morphological marker allows roguing of fertile plants at the seedling stage.

Several drawbacks should be noted with regard to isozyme markers. First, the number of isozyme loci that can be scored is limited. To date, only 40–50 reagent systems have been developed that permit the staining of a particular protein in a gel. Furthermore, not all of these reagent systems work efficiently with all plant species. Therefore, for many species only 15–20 loci can be mapped. A second drawback is tissue variability. Some isozymes are better expressed in certain tissues such as roots, whereas others are best sampled in leaf tissue. Therefore, several samplings of individual plants of a segregating population are necessary to score all the available isozymes for mapping purposes. Due to these limitations, isozymes could not offer much progress in the development of linkage maps in most species. Because neither of these drawbacks affects DNA markers, isozyme loci are rarely scored today.

Morphological and isozyme markers have given way to molecular markers that reveal neutral sites of variation or polymorphism at the DNA sequence level. These DNA markers have been adopted almost universally by map-makers for at least three major reasons: 1) such markers do not usually compromise the viability or fitness of an organism; 2) they are not tissue specific and can therefore be scored from the very initial phase of development; and 3) they are much more numerous compared to isozyme and/or morphological markers. These advantages afforded the creation of denser maps with greater accuracy and utility. To date, as was recently summarized (Peters et al. 2003), more than 20 DNA marker systems are routinely used.

Second generation linkage maps rely heavily on DNA markers (Fig. 5-3). The main and early marker system used to generate these maps was restriction fragment length polymorphism (RFLP; Botstein et al. 1980). Many other DNA marker systems have been developed throughout the years and most of them were reviewed by Peters et al. (2003). These include microsatellites or simple sequence repeats (SSR), random amplified polymorphic DNA (RAPD), amplified fragment length polymorphism (AFLP) and single nucleotide polymorphisms (SNP).

5.4 Identification of Markers Linked to Traits of Interest

In order to identify molecular markers linked to simply inherited traits, segregation analysis should be performed in F_2 or other types of segregating populations constructed from parents that differ at alternate alleles of the trait and are polymorphic at the DNA level. However, when *a priori* information about the location of genes controlling the trait does not exist the task of finding linkage could be formidable, especially for species with large number of chromosomes. To alleviate the problem, methods that utilize special genetic stocks have been developed that include the use of near-isogenic lines (NILs) and bulked segregant analysis (BSA). These methods allow tagging molecular markers to many traits in different plant species; some examples were listed by Mohan et al. (1997).

5.4.1 Use of NILs to Identify Linkage with Molecular Markers

NILs are produced by repeated backcrossing of a donor parent, characterized by a unique or mutant phenotype with a recurrent parent characterized by a common phenotype often referred to as wild-type. An initial F_1 is crossed to the recurrent parent to produce BC_1 progeny. BC_1 progeny (in case of dominant mutation) or BC_1S_1 progeny (in case of recessive mutation) carrying the mutant phenotype are crossed again to the recurrent parent. This process usually continues for at least 5–6 backcross generations, consequently reducing the sequence divergence between the donor and recurrent parents and limiting it to the region spanning the gene of interest (Fehr 1987; Muehlbauer et al. 1988). The result of such backcrossing program is NILs that share common genetic background throughout the genome except for a region containing the locus that controls the selected trait (Fig. 5-4). Models were developed to describe the proportion of the donor parent genome in the background of the recurrent genome parent as a function of chromosome length (Stam and Zeven 1981). For example, the length of the donor parent segment in BC_6 is estimated to be 32 cM for a chromosome of 100 cM, assuming no other chromosomes are carrying donor-parent segments. In practice, the size of the introgressed segment may

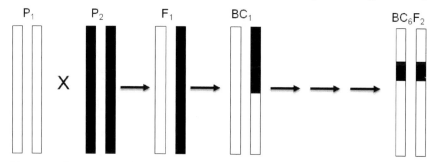

Figure 5-4 Construction of near-isogenic lines (NILs). The F_1 results from crossing the recurrent parent (P_1) and the donor parent (P_2). The F_1 is backcrossed to P_1 to produce BC_1 progenies and the mutant phenotype is selected among the progenies. Backcrossing continues until BC_6. After this generation, selfing (BC_6F_2) results in the production of a homozygous line identical to the recurrent parent except for a segment from the donor parent (represented as black bar) containing the selected locus.

vary in different sets of NILs. For example, the size of the introgressed region containing the *Tm-2a* gene in different NILs of tomato varied between 4 cM to 51 cM even after 11 backcross generations (Young and Tanksley 1989). In addition, unlinked chromosome segments may also be introgressed, causing ambiguities in data interpretation.

NILs for many traits have been produced in breeding programs for various crops, including tomato (Bernard 1976; Maxon Smith and Ritchie 1983). However, the major limitation for using these NILs for identification of linked markers is the low level of DNA polymorphism between the recurrent and donor parents DNA. Therefore, the use of NILs was applied mainly for tagging disease resistance genes because these genes are often introgressed from wild species that exhibit a high level of DNA polymorphism relative to the recurrent parent DNA (Young et al. 1988; Hinze et al. 1991; Klein-Lankhorst et al. 1991; Martin et al. 1991; Paran et al. 1991).

Polymorphism between NILs visualized using DNA markers indicates a putative linkage between the marker and the target gene. However, linkage must be confirmed by analysis of segregating progenies because false positive polymorphisms can arise due to the occurrence of additional unlinked donor segments in the background of the recurrent parent genome. Reduction of false positive markers can be achieved by screening independent sets of NILs because the chances that the same unlinked segments are introgressed to different NILs is small (Paran et al. 1991).

NILs screening is best suited for high throughput PCR markers such as RAPD or AFLP. However, single locus markers, such as RFLP or SSR, can also be used. In order to speed up the process of screening the NILs, pools of 5–8 RFLP clones can be hybridized together, allowing a significant reduction in the number of hybridizations (Young et al. 1988). Similarly, SSR markers can be multiplexed in a single reaction.

A collection of NILs, such as the *S. pennellii* introgression lines (Eshed et al. 1992), can also serve as an excellent tool to easily map a gene or a DNA marker of interest onto the tomato genome. This mapping procedure has been previously described in detail and is demonstrated in Fig. 5-5 (Levin et al. 2000; Lieberman et al. 2004; Sapir et al. 2008).

5.4.2 Use of Bulked Segregant Analysis (BSA) to Identify Linkage with Molecular Markers

The major limitation of using NILs for tagging target genes is the long time required for their production. A method, bulked segregant analysis (BSA) was developed that alleviates this problem by comparing pools of DNA from individuals that segregate for the target trait in a single cross (Giovannoni et al. 1991; Michelmore et al. 1991; Fig. 5-6).

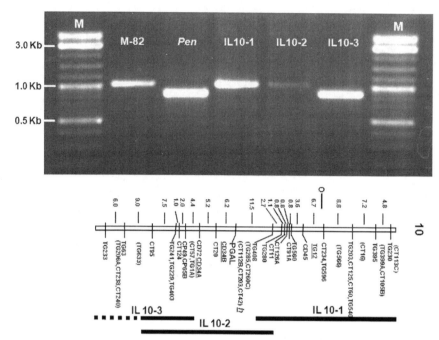

Figure 5-5 Mapping of a gene onto the tomato genome using the *S. pennellii* introgression lines (Eshed and Zamir 1995; Sapir et al. 2008) [*Pen* = DNA extracted from *S. pennellii*, M-82 = DNA extracted from *S. lycopersicum* cv. M-82 (LA3475), IL 10-1 through 10-3 = DNA extracted from the corresponding *S. pennellii* introgression lines, M = DNA size marker, the dashed line on IL 10-3 represents the approximate location of the gene due to the identity in the banding pattern of *S. pennellii* and IL 10-3 and the difference between the banding patterns of *S. pennellii* and IL 10-2, map of the tomato chromosome 10 was adopted from http: //tgrc.ucdavis.edu/ pennellii-ILs.pdf].

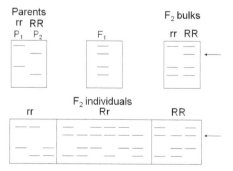

Figure 5-6 Bulked segregant analysis for tagging a disease resistance gene. The four bands of the two parents (two per parent) represent four dominant loci. The bulks are derived from homozygous resistant F_2 (RR) and from homozygous susceptible F_2 (rr) progeny. The locus marked by an arrow is polymorphic between the bulks because of linkage to the R gene as illustrated by its segregation in the F_2. The other three loci are unlinked to the R gene and are therefore monomorphic between the bulks.

BSA has been applied in many plant species using mainly RAPD and AFLP markers. Some examples include identification of markers linked to the gene for resistance to *Cladosporium fulvum* (*Cf-9*) in tomato (Thomas et al. 1995), mapping of nodulation genes in pea (Schneider et al. 2002), identification of AFLP markers linked to resistance gene for parasitism in cowpea (Quedraogo et al. 2001), mapping a stalk rot resistance gene in maize (Yang et al. 2004) and mapping resistance genes to pod weevil resistance in bean (Blair et al. 2006). In the study designed to identify AFLP markers linked to *Cf-9* in tomato (Thomas et al. 1995), 728 primer pairs were used. These primer pairs amplified approximately 42,000 loci, of which three co-segregated with the resistance gene. Because the bulks were constructed from F_2 DNA, only *cis* markers could be identified. Further reduction in the number of potentially informative markers occurs because of lack of DNA polymorphism between the parents. Assuming the frequencies of *cis* and *trans* linked markers are equal, and an average of 50% polymorphism between the parents, about 25% of the screened markers were informative in this study.

5.4.3 Conversion of Dominant to Codominant Markers

A good example for tagging and mapping a target gene using NILs and BSA, as well as the development of locus specific PCR markers was presented by Zhang and Stommel (2000, 2001) on the *Beta* (*B*) gene and its modifier, *beta modifier* (*Mo_B*), in tomato. Red tomatoes accumulate the carotenoid lycopene as the major pigment in the mature fruit. Mutation at the *B* locus results in an orange-colored tomato because of the accumulation of a high level of beta-carotene at the expense of lycopene and is inherited as a single dominant gene (presented visually in Fig. 5-7). The modifier gene, Mo_B, influences the expression of the dominant form of *B* and the relative percentages of lycopene and beta-carotene present in ripened fruit. In the presence of the homozygous recessive form of the allele, $Mo_B Mo_B$, beta-carotene represents more than 90% of the total carotene content, and fruits are orange-pigmented. Expression of the dominant Mo_B+ form of the allele, however, reduces beta-carotene and increases lycopene content resulting in red-orange fruit pigmentation (Zhang and Stommel 2000).

NILs that differ at *B* were developed, in which *B* was introgressed from the wild species *S. habrochaites*, thus providing ample DNA polymorphism between wild type cultivated tomato (*S. lycopersicum* cv. Rutgers) and its isogenic mutant line. For BSA, an interspecific F_2 population was constructed from a cross of red-fruited *S. lycopersicum* and orange-fruited *S. galapagense*. Two bulks were constructed by combining equal amounts of DNA from each of 7 red-fruited and 7 orange-fruited F_2 plants. Using RAPD (1,018 primers) and AFLP (64 primer pairs), polymorphic products were identified, which

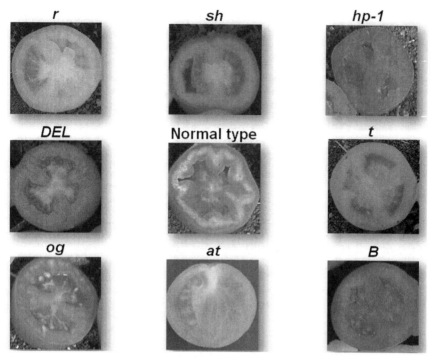

Figure 5-7 Tomato fruit color mutants related to carotenoids biosynthesis (Abbreviations: *B* = *BETA-CAROTENE*, yellow-pink color, high β-carotene, low lycopene in ripe fruit flesh; *DEL* = *DELTA*, reddish-orange mature fruit color, due to inhibition of lycopene and increase of delta-carotene; *hp-1* = *high pigment-1*, chlorophyll, carotenoids, ascorbic acid content of fruit intensified; *og* = *old gold*, increased fruit lycopene content; *r* = *yellow flesh*, yellow color of ripe fruit flesh; *sh* = *sherry*, fruit flesh yellow with reddish tinge; *t* = *tangerine*, fruit flesh and stamens orange colored).

Color image of this figure appears in the color plate section at the end of the book.

distinguished the NILs and the two bulked DNA samples constructed for BSA. A single 100 bp AFLP amplification product, which distinguished the NILs co-segregated with Mo_B and was demonstrated to be tightly linked to the locus and mapped to the long arm of chromosome 6. Two RAPD products, OPAR18$_{1100}$ (Fig. 5-8A) and UBC792$_{830}$, of 1,100 and 830 bp in size, respectively, were polymorphic between orange- and red-fruited bulks constructed from F_2 individuals in the *S. lycopersicum* and *S. galapagense* mating series. OPAR18$_{1100}$ and UBC792$_{830}$ displayed recombination frequencies of 4.2% and 7.6%, respectively, in F_2 progeny. The *B*-linked OPAR18$_{1100}$ marker was also mapped to the long arm of chromosome 6, proximal to Mo_B, and revealed linkage between *B* and Mo_B.

In order to develop specific PCR markers useful for marker-assisted selection, the dominant RAPD markers OPAR18$_{1100}$ and UBC792$_{830}$, found

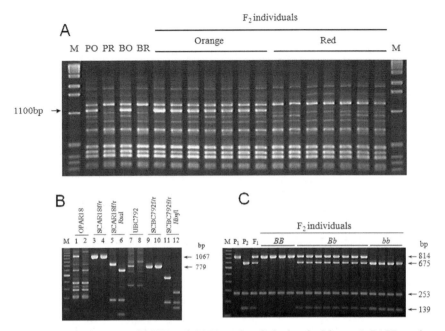

Figure 5-8 Development of RAPD and CAPS markers linked to the *B* locus. **A.** RAPD marker OPAR18$_{1100}$ linked to the *B* locus (PO and PR are orange and red-fruited parents, respectively. BO and BR are bulks derived from orange and red-fruited F$_2$, respectively). **B.** CAPS markers derived from two RAPD markers (Lanes 1 and 2 and lanes 7 and 8 are the RAPD products of OPAR18 and UBC792, respectively amplified from two parents. Lanes 3 and 4 and lanes 9 and 10 are products of the two parents generated after cloning of OPAR18$_{1100}$ and UBC792$_{830}$ and amplification with specific PCR primers. Lanes 5 and 6 and lanes 11 and 12 are CAPS markers generated by cleaving the specific PCR products with restriction enzymes). **C.** Linkage of the codominant CAPS marker derived from the RAPD marker OPAR18$_{1100}$ to *B* (P$_1$, P$_2$ and F$_1$ are the parents of the mapping population and their F$_1$. *BB*, *Bb* and *bb* are F$_2$ individuals segregating for *B*). M—molecular DNA ladder. Fig. 5-8A is taken from Zhang and Stommel (2000). Fig. 5-8B and 5-8C are taken from Zhang and Stommel (2001) and presented with the kind permission of JR Stommel.

associated with *B*, were cloned, characterized and converted to codominant CAPS markers (Zhang and Stommel 2001). For cloning, the two amplified fragments were purified from the gel, cloned and sequenced. On the basis of the DNA sequence, long oligonucleotide primers (22–24 bp) were designed such that each primer contained the original 10 bases of the RAPD primer plus the next 12 or 14 bases. Amplification of the parents used in the cross for BSA analysis resulted in equal sized PCR products. Sequence comparison of the PCR products amplified from the two parents revealed restriction site polymorphisms that enabled the development of CAPS markers (Fig. 5-8B). The CAPS markers were tested and verified for linkage with *B* in the F$_2$ population (Fig. 5-8C).

5.5 Use of Genomics Tools for Mapping

The completion of whole genome sequences of model plants and the ongoing sequencing projects for several additional crop plants, including tomato (Mueller et al. 2005), in conjunction with improving DNA array technologies and bioinformatics tools provide new opportunities to rapidly generate a large number of molecular markers for gene mapping.

Several tomato genome arrays are currently commercially available, such as the Affymetrix GeneChip® array (Affymetrix Inc., Santa Clara, CA; http://www.affymetrix.com/products/arrays/specific/tomato.affx), which consists of over 10,000 *S. lycopersicum* probe sets. A more comprehensive GeneChip® array, which consists of 20,000 tomato probe sets, can be obtained via collaboration with Syngenta Inc. (Research Triangle Park, NC; http://www.syngenta.com/en/index.html).

A combination of array mapping and bulked segregant analysis can provide a robust and fast method to map genes controlling simply inherited traits. This technique was applied to map several developmental EMS mutations in *Arabidopsis* (Hazen et al. 2005). In case the mutation is caused by deletion, direct comparison of the hybridization patterns between mutant and wild type strains can map the mutation to a small interval without the need to construct bulks from a segregating population. Using this approach, the flowering time deletion mutations *fkf1* and *cry2-1* were estimated to be missing 185 features covering 77 kb and 19 features covering 7 kb genomic regions in chromosome 1, respectively (Hazen et al. 2005). With future whole genome tilling arrays, the resolution power of the array mapping approach will be much greater.

5.6 Simply Inherited Traits Associated with Fruit Nutritional Quality

Fruit quality has been a major focus of most classical tomato breeding programs during the past century as was recently reviewed by Foolad (2007). Color and nutritional quality are among the major tomato fruit quality characteristics of interest. The attention to tomato fruit color has recently increased as the health benefits of lycopene, the major carotenoid in tomato that is responsible for its characteristic red fruit color, has become more evident (Di Mascio et al. 1989; Levy et al. 1995; Stahl and Sies 1996; Gerster 1997; Kohlmeier et al. 1997). Several major genes with significant contribution to high concentration of fruit lycopene (e.g., the genes encoding the *hp* and *og^c* mutant phenotypes) and other carotenoids (e.g., *beta-carotene, B*) were previously phenotypically identified and mapped onto the classical linkage map of tomato (Wann et al. 1985; Stevens and Rick 1986). In addition, during the past two decades, numerous QTLs and candidate genes with significant effects on fruit color and/or lycopene content were identified in wild tomato accessions such as *S. pimpinellifolium*, *S. peruvianum*, *S. habrochaites*, *S. chmielewskii* and *S. pennellii* and mapped onto tomato chromosomes along with the previously identified genes (Foolad 2007). While some of the identified QTLs mapped to the chromosomal locations of many of the known genes of the carotenoid biosynthesis pathway, many mapped to other locations. It was, therefore, suggested that there might be more genes affecting fruit color in tomato than those known to affect the carotenoid biosynthesis pathway (Liu et al. 2003).

Tomato mutant accessions with divergent color phenotypes in their fruits were the subject of molecular genetics studies, leading to the identification of the genes responsible for such phenotypes. A selection of such mutants, their gene identification, map location, and characteristic color is presented in Table 5-3 and Fig. 5-7. The sequence of these genes can now serve as recombination-free DNA markers to expedite breeding toward altering pigmentation and enhancing nutritional value of plant foods. Of particular interest are the light-responsive *high pigment* mutations (*hp-1*, *hp-1ʷ*, *hp-2*, and *hp-2ʲ*, *hp-2ᵈᵍ*; Table 5-3). These mutants display shorter hypocotyls with higher anthocyanin levels, darker foliage, and higher fruit pigmentation than their isogenic normal counterparts (Mochizuki and Kamimura 1984; Wann et al. 1985; Peters et al. 1989; Mustilli et al. 1999; Levin et al. 2003). The high pigmentation of fruits of these mutants is due to significantly elevated levels of chlorophylls at most of their immature developmental stages. Mature ripe-red fruits of these mutants are characterized by an intense red color, which is mainly due to increased levels of carotenoids, primarily lycopene. Because of their effect on fruit color, attributed to enhanced lycopene content, *hp* mutations have been introgressed into several commercial processing and fresh-market tomato cultivars that are currently marketed as Lycopene Rich Tomatoes (LRT) (Wann 1997; Levin et al. 2006). The processing tomato varieties are primarily cultivated for the purpose of lycopene extraction, which is further used as food additive, food supplement, and food colorant in many processed products (http://www.lycored.com/). Current processing tomato cultivars harboring such mutations can reach up to a 3.5-fold increase in fruit lycopene content (80–280 ug·g⁻¹ FW). Interestingly, this increase is higher than that reported thus far using genetically modified alternatives (Levin et al. 2006), pointing to the potential of classical breeding in achieving crop improvement.

The origins of *hp-1*, *hp-1ʷ*, *hp-2*, and *hp-2ʲ*, and *hp-2ᵈᵍ* mutations have been extensively summarized (Lieberman et al. 2004; Levin et al. 2006). Further, the *hp-2*, and *hp-2ʲ*, *hp-2ᵈᵍ* mutations have been mapped to the gene encoding the nuclear protein DEETIOLATED1 (DET1), a central negative regulator of photomorphogenesis (Mustilli et al. 1999; Levin et al. 2003). The gene controlling the *hp-1* and *hp-1ʷ* mutant phenotypes has also been recently identified (Lieberman et al. 2004). Results show that *hp-1* and *hp-1*[w] are alternative alleles at the tomato gene encoding UV DAMAGED DNA BINDING protein 1 (DDB1), recently shown to interact both biochemically and genetically with the DET1 protein (Schroeder et al. 2002; Liu et al. 2004). DDB1 is a protein evolutionarily conserved from fission yeast to humans. It was initially identified, together with DDB2, as a subunit of a heterodimeric protein complex that recognizes the UV-induced DNA lesions in the nucleotide excision repair pathway. Mounting evidence has now established a major role of DDB1 as a substrate-recruiting subunit of the Cullin 4 (CUL4)-based E3 ubiquitin ligase complexes that also contain RBX1 (also named ROC1), DET1, and in the case of *Arabidopsis*, COP10 as well (Hu et al. 2004; Wertz et al. 2004; Bernhardt et al. 2006; Chen et al. 2006).

Tomato *hp* mutations are best known for their positive effect on carotenoid (lycopene and other carotenes) levels in ripe red fruits (Mochizuki and Kamimura 1984; Wann et al. 1985; Levin et al. 2003). Interestingly, mature fruits of plants carrying the *hp-1* mutation were also found to exhibit a 13-fold increase of the flavonoid quercetin in tomato fruit pericarp (Yen et al. 1997) and a significant increase in ascorbic acid (vitamin C) content (Mochizuki and Kamimura 1984). In a later study, 8- and 12.8-fold increase was identified in quercetin levels in the fruit peel of the tomato mutants *hp-2* and *hp-2*[j], respectively, compared to their isogenic normal counterparts (Levin et al. 2006). These results suggest that other metabolites may be increased in tomato *hp* mutants. To validate this hypothesis, the overall metabolic modifications between *hp-2*[dg] tomato fruits and their isogenic counterparts were compared (Bino et al. 2005). Targeted metabolite analyses, as well as large-scale non-targeted mass spectrometry (MS)-based metabolite profiling, were used to phenotype the differences in fruit metabolite composition. Metabolite analyses using targeted high-performance liquid chromatography with photodiode array detection (HPLC–PDA) showed higher levels of isoprenoids and phenolic compounds, as well as vitamin C, in *hp-2*[dg] fruits. A selected list of such metabolites, including their average levels in ripe red fruits and their fold increase in *hp-2*[dg] are presented in Levin et al. (2006). Non-targeted GC–MS profiling of red fruits produced 25 volatile compounds that showed a 1.5-fold difference between the genotypes (Bino et al. 2005). Analyses of red fruits using HPLC coupled to high-resolution quadruple time-of-flight mass spectrometry (LC–QTOF–MS) generated 6,168 and 5,401 mass signals in ESI-positive and ESI-negative detection modes, respectively. A total of 142 and 303 mass signals in ESI-positive and ESI-negative modes, respectively, showed a two-fold difference between the genotypes. Of this total of 445 mass signals, 383 (~86%) were up-regulated in the *hp-2*[dg] genotype and 62 (~14%) were down-regulated (Bino et al. 2005). Overall, these results show that the *hp-2*[dg] fruits are more active metabolically, overproducing many metabolites sharing antioxidant or photoprotective activities. Because *hp-2^{dg}* is highly iso-phenotypic to other tomato *hp* mutants, similar metabolic responses are also expected in those mutants.

Using microarray technology, a transcriptional profiling study was also conducted on fruits harvested from *hp-2^{dg}* plants and their isogenic counterparts (Kolotilin et al. 2007). Results show that a large portion of the genes that are affected by the *hp-2^{dg}* mutation display a tendency for up- rather than down-regulation, indicating that this genotype is more active transcriptionally as well. Ontology assignment of these differentially regulated transcripts revealed a consistent up-regulation of transcripts related to chloroplast/chromoplast biogenesis and photosynthesis in *hp-2^{dg}* mutants throughout fruit ripening. A tendency of up-regulation was also observed in structural genes involved in phytonutrient biosynthesis. However, this up-regulation was not as consistent, positioning plastid biogenesis as a more important determinant of phytonutrient overproduction in *hp-2^{dg}* and possibly other *hp* genotypes. These results were linked to microscopic observations that revealed a highly significant increase in chloroplast/chromoplast size and number in pericarp cells of mature-green *hp-2^{dg}/hp-2^{dg}* and *hp-2^{j}/hp-2^{j}* fruits in comparison to their normal counterparts.

As previously noted (Kolotilin et al. 2007; Levin et al. 2003; Liu et al. 2004), the identification of genes controlling the simply inherited *hp* mutant phenotypes have established a link between light signaling and over-production of fruit phytonutrients. It was further shown that in these mutants plastid biogenesis is the major determinant that drives this over-production (Kolotilin et al. 2007). Interestingly, these concepts were also recently documented in an EMS mutant tomato plant termed *hp-3* with an abberant *Zeaxanthin Epoxidase* (ZE) gene (Galpaz et al. 2008). Fruits harvested from the *hp-3* mutant displayed 30% more carotenoids in the mature fruit compared to their isogenic normal counterparts. This increase in fruit carotenoids content was accompanied by at least 2-fold increase in plastid compartment size (Galpaz et al. 2008). Constitutive over-expression of *ZE* in tomato plants characterized in a later study were found to display enhanced sensitivity of the tomato plants to photo-inhibition caused by high light stress (Wang et al. 2008).

Another mutant of special interest is *t* (*tangerine*), which produces orange-colored fruits accumulating prolycopene (7Z,9Z,7'Z,9'Z-tetra-*cis*-lycopene) instead of the all-*trans*-lycopene that accumulates in normal red-fruited tomatoes (Fig. 5-7; Isaacson et al. 2002). *cis*-isomers of lycopene, thought to be powerful antioxidants, have been shown to be more bioavailable than the *trans*-isomer, indicating that they are more efficiently absorbed, and therefore deliver lycopene into the plasma more effectively. This might be interpreted to mean that *cis*-isomers of lycopene are more beneficial, and therefore more valuable to human health than the *trans*-isomer (Ishida et al. 2007). Results recently published support the hypothesis that lycopene *cis*-isomers are highly bioavailable and suggest that special tomato varieties can be utilized to increase both the intake and bioavailability of health-beneficial carotenoids (Unlu et al. 2007). Because light-responsive *hp* mutants are characterized by higher total fruit carotenoids, *hp-1/hp-1 t/t* double mutant fruits share almost double the content of *cis* isomers of lycopene in comparison to non-*hp* (+/+ *t/t*) mutant fruits, demonstrating the power of classical breeding to both modulate the profile and increase the content of selected carotenoids in the tomato fruit (I. Levin, unpublished results).

Efforts to increase the functional quality of tomato fruits have been invested in flavonoid metabolites as well. Despite the relative success obtained in increasing flavonoid content in tomato fruits by transgenic modifications, there is an ongoing interest in breeding a high flavonoid tomato without genetic engineering (Willits et al. 2005). This interest is motivated by consumers' reluctance to accept transgenic fruits and vegetables.

As recently summarized (Jones et al. 2003; Sapir et al. 2008), fruits of several tomato accessions as well as species, which are closely related to the cultivated tomato, contain significantly higher amounts of anthocyanins (Giorgiev 1972; Rick 1964; Rick et al. 1994; Fig. 5-9). The *Anthocyanin fruit* (*AFT*, formerly *AF*) from *S. chilense*, *Aubergine* (*ABG*) from *S. lycopersicoides* (Fig. 5-9), the recessive *atroviolacium* (*atv*) mutation from *S. cheesmaniae*, and the purple-smudge phenotype originating from *S. peruvianum* cause anthocyanin expression in tomato fruit. We have introgressed the *AFT* gene from *S. peruvianum* accessions (Fig. 5-9) into cultivated tomato. Recently, the wild species *S. pennellii v. puberulum* was shown to be a source for enriching tomato fruits with functional flavonoids (Willits et al. 2005).

Another approach to increase fruit flavonoids is through the introgression of the *hp* mutations because, as noted herein, these mutations are also characterized by increased levels of flavonoid metabolites relative to their isogenic counterparts. Results recently presented (van Tuinen et al. 2006) indicated that several phenolic compounds with high antioxidant capacity are new or increased in fruits of double mutant *AFT/AFT hp-1^w/hp-1^w* genotypes in comparison to fruits of single mutant parents. One of these compounds was identified as the flavonoid rutin (van Tuinen et al. 2006). We have more recently shown that: 1) *AFT* fruits are also characterized by significantly higher levels of the flavonols quercetin and kaempferol, thus enhancing their functional value; and 2) double homozygote *AFT/AFT hp-1/hp-1* genotypes displayed a more-than-additive (synergistic) effect on the production of fruit anthocyanidins and flavonols (Sapir et al. 2008). This effect was manifested by ~5-, 19-, and 33-fold increases of petunidin, malvidin and delphinidin, respectively, in the double mutants compared to the cumulative levels of their parental lines (demonstrated visually in Fig. 5-9).

Figure 5-9 Tomato fruit color phenotypes related to flavonoid biosynthesis [A. ANTHOCYANIN *FRUIT* (*AFT*) from *S. chilense*; B. *AUBERGINE* (*ABG*) from *S. lycopersicoides*; C. *S. peruvianum* (PI 128650); D. Purple Smudge introgressed from *S. peruvianum*; E. Fruits of a double homozygous *AFT/AFT hp-1/hp-1* plant, *AFT* introgressed from *S. peruvianum* (PI 128650); F. Fruit skin from a tomato *y* mutant; G. Fruit skin from normal red tomato.

Color image of this figure appears in the color plate section at the end of the book.

Another important simply inherited mutant related to the flavonoid biosynthetic pathway is the *y* mutant (Fig. 5-9). Fruits of this mutant are typified by colorless fruit epidermis, resulting in pinkish fruits that are preferred by consumers in most Asian countries. It is highly likely that the phenotype of the *y* mutant is attributed to major changes in the flavonoid pathway leading to the formation of naringenin chalcone, the yellow pigment accumulating in tomato fruit cuticle.

5.7 Simply Inherited Traits Associated with Fruit Taste: Brix, Sugars, Acids and Aroma Volatiles

The literature relating to the quality traits of tomato fruit taste, including aspects of breeding and selection, was reviewed in the 1980s by Davies and Hobson (1981) and Stevens (1986). The reader is referred to these reviews for coverage of the earlier research. Dorais et al. (2001) reviewed the field of greenhouse tomatoes and most recently two excellent and extensive reviews on breeding for tomato fruit quality traits have been published (Causse et al. 2007; Foolad 2007). The present review will therefore deal with some particular examples of breeding and selection for tomato flavor, with focus on the mapping and tagging of simply inherited traits. The emphasis will be on the traits of soluble solids and sugars, as these have attracted the most research attention and are primary determinants of fruit taste.

At first glance, the parameters of fruit taste do not appear to lend themselves to a discussion of simply inherited traits. It has often been considered that quality traits such as Brix are quantitative traits in the simple sense, i.e., that they are controlled by multiple interacting loci with only small individual effects. Causse et al. (2007) have emphasized that this is one of the major difficulties in selecting for quality traits. While this is likely true, there have recently been some major inroads in breeding for improved fruit taste, particularly sugar content, by selection for simply inherited traits controlled by single genes with major effects. In fact, although quantitative traits are influenced by multiple individual loci, any such loci which bestow a major effect can be treated as a simply inherited trait controlled by a single locus.

The main quality components, which affect taste and flavor of the fruit, are the sugars, acids and volatiles (Davies and Hobson 1981). The relative contribution of each of these components to fruit mass is in that same order: the sugars make up about 50% of the dry weight, the acids about 5% and the volatiles only a fraction of that (Davies and Hobson 1981). Each of these has both quantitative and qualitative components to them. For example, the soluble sugars of all cultivated tomato varieties studied to date are primarily the hexoses, glucose and fructose in approximately equimolar concentrations (Davies 1966). The total soluble sugar content will largely determine Brix, irrespective of the makeup of the sugars. Modifying the qualitative sugar composition, in terms of selecting for novel sugar profiles such as relatively high sucrose or fructose levels, can impact fruit quality and taste. In particular, the accumulation of fructose, which is twice as sweet as glucose (Biester et al. 1925), may lead to a sweeter tasting tomato without a concomitant increase in total sugars. Similarly, organic acid content is a quantitative trait, but the relative ratios of the principal organic acids accumulated, citric acid and malic acid, is determined by simple inheritance (Stevens 1986). The same holds true for the volatile components where qualitative changes in a particular volatile component can have major impact on the aroma and flavor of the fruit. The accumulation of a particular volatile component, generally to minor concentrations, will likely be simply inherited (e.g., Tadmor et al. 2002; Tieman et al. 2006).

Accumulation of all the taste quality components, including sugars, organic acids, amino acids, or volatiles, is determined by complex interacting metabolic pathways. Irrespective of the complexity, which is indicated by recent approaches emphasizing theories of systems biology (e.g., Sweetlove and Fernie 2005; Schauer et al. 2006), metabolic pathways can be broken down to their individual components, or metabolic steps. All pathways have rate limiting steps, which can be particular enzymatic steps, transport steps or any of the more subtle and generally less understood components of the particular metabolic pathway. These limiting steps can be modified to improve horticultural quality attributes, assuming that genetic variability for the component is available and can be identified. Once such genetic variability is identified, breeding and selecting for the modified pathway can become a straightforward program based on simple inheritance.

It has become a truism that the genetic variability within the cultivated tomato species is narrow and limited (Tanksley and McCouch 1997; Gur and Zamir 2004). Early studies, which documented the variation and heritability for components such as Brix within the cultivated tomato, *S. lycopersicum*, have been thoroughly reviewed by Stevens (1986). Suffice to say that regarding soluble solids concentration and Brix, Stevens concluded that: "Most studies of the inheritance of tomato fruit solids are of little practical value". This is largely due to the inverse relationship between fruit yield/ size and soluble solids concentration, which has been well documented and will be referred to in more detail later in this review.

The most detailed analyses of QTLs for fruit quality traits within the cultivated tomato species has been carried out by the group of M. Causse. Since the small fruited cherry tomatoes are generally characterized by rich flavor, their strategy was to attempt to transfer quality traits from the small cherry tomatoes to larger fruited varieties. In an elegant series of research reports (Causse et al. 2001, 2002; Saliba-Colombani et al. 2001; Lecomte et al. 2004; Chaib et al. 2006; reviewed in Causse et al. 2007), QTLs were identified in an RI population developed from an initial cross between a small cherry sized tomato and a medium sized variety. Positive effects of the QTLs identified were attributed to the cherry sized parent. Following the identification of five QTL regions (each in length from 15 to 31 cM), a marker-assisted backcross program was initiated, transferring the five regions to different large-fruited backgrounds. Not surprisingly, increases in soluble solids, acidity and pigment concentration due to the cherry QTLs were accompanied by decreases in fruit weight, suggesting that the effects on sugar and acid were secondary to the effect on fruit size. A similar conclusion was reached by Georgelis et al. (2004) who also searched for QTLs for sugar concentration in a large fruit x small cherry cross. The relationship between fruit size and solids concentration has been explained by Ho (1996) and Stevens (1986), relating the differences in fruit size to the size of the parenchyma cells, with the larger cells having a higher water content, thereby diluting the solids.

The implication from these studies may be that the small cherry sized fruit of the *S. lycopersicum* species may not be good candidates for genes that can increase solids and acids content in large size fruit. This is not to say that genetic variability for certain quality traits such as aroma which may be improved by novel volatiles, or texture, which are not as concentration dependent as Brix, acids or fruit color, will not be found in the small fruited tomatoes. However, the concentration dependent traits may be inherently related to the small fruit size, and not merely a genetically linked trait which can be broken. If so, and if genetic variability for one of the many components for sink solids accumulation can be found within the cultivated tomato species, then the small fruited varieties are likely not an effective genetic resource for the transfer of solids content to large varieties without impacting fruit size and yield.

Most of the QTLs reported for soluble solids content have been derived from interspecific crosses and in many cases an allele of the wild species is responsible for increases in soluble solids and sugar levels. A quick perusal of Table 5-4 (modified from Foolad 2007 and Causse et al. 2007) shows that QTLs for sugar content and Brix are overwhelmingly reported from studies of interspecific crosses with very few derived from intraspecific studies. This is no doubt due to two factors: 1) the low genetic variability for these traits within the species, as described previously, compared to the broader variability retained in the wild species, but also 2) the difficulty of doing QTL studies within the cultivated species due to lack of polymorphisms, in comparison to the relative abundance of interspecific polymorphisms.

However, just as in the case of intraspecific hybridizations, in which fruit size is negatively correlated with solids concentration, so too are most of the QTLs for solids content derived from interspecific crosses. Chen at al. (1999) and Causse (2007) consolidated the QTLs for increased Brix and reduced fruit size from many of the above listed studies and showed that the QTLs generally colocalize, reinforcing the general negative relationship observed between Brix and yield even among the interspecific introgressions.

While most QTL studies from interspecific crosses report data for Brix or soluble sugars as well as fruit size, not all reports present the data for Brix*yield, which is the true indicator of improvement in agricultural yield (Gur and Zamir 2004). One of the reasons is likely due to the type of population studied since total yield is more easily measured in determinate (*sp*) backgrounds. With indeterminate plants, total yield per plant throughout its growth is not always practical to measure, although total yield per fruit cluster may be just as informative. Nevertheless, the comparison of QTLs for Brix*yield, which is the true indication of whether there is indeed genetic potential for increasing the harvest index of the plant is less frequently performed and the number of QTLs for this trait is low. One particular population, the ILs derived from *S. pennellii* (Eshed and

Table 5-4 Summary of QTL mapping populations used for total soluble solids and sugar concentration. (Based on Foolad 2007 and Causse et al. 2007).

Species used	Mapping population	# QTLs for Brix or soluble sugar concentration	Reference
S. lycopersicum	RIL	5	Causse et al. 2001 Saliba-Columbani et al. 2001
	F_2	6	Georgelis et al. 2004
S. pimpinellifolium	BC_2, BC_3	12	Tanksley et al. 1996
	BC_1	3	Grandillo and Tanksley 1996
	BCS_1	13	Chen at al. 1999
	$BC2F_6$	2	Dogenlaar et al. 2002
S. cheesmaniae	F_2, F_3	7	Paterson e al. 1991
	F_8 RIL	12	Goldman et al. 1995
S. chmielewskii	F_2	1	Osborn et al. 1987
	BC_1, BC_2	4	Paterson et al. 1988, 1990
	$BILs/BC_2F_5$	3	Azanza et al. 1994
	BC_2F_5	1	Azanza et al. 1994, 1995
	NILs	1	Frary et al. 2003
S. habrochaites	BC_2, BC_3	5	Bernacchi et al. 1998
	SubNILs	1	Monforte and Tanksley 2000
	NILs	1	Frary et al. 2003
S. pennellii	ILs	3	Eshed and Zamir 1994
	ILs	23	Eshed and Zamir 1995
	ILs	9	Causse et al. 2004
	ILs	44	Schauer et al. 2006
	BC_2/BC_2F_1	3	Frary et al. 2004
S. peruvianum	BC_3, BC_4	9	Fulton et al. 1997
S. parviflorum	BC_3	5	Fulton et al. 2000

Zamir 1994, 1995), has been analyzed repeatedly for QTLs for Brix*yield. In their initial analysis of QTLs from this population, Eshed and Zamir (1995) reported over 20 QTLs for Brix but only 25% of these were also QTLs for Brix*yield. Similarly, the more recent detailed study of Schauer et al. (2006), comparing over 100 metabolite levels in these ILs and determining metabolic QTLs, shows that practically all of the ILs with increased Brix values also had reduced fruit size or harvest index (see especially supplementary manuscript tables at http://www.nature.com/nbt/journal/ v24/n4/suppinfo/nbt1192_S1.html). While approximately 60 of the ILs had increased Brix, only five of these (ILs 1-4, 6-4, 9-2-5, 10-2-2 and 12-1-1) did not show the reduced yield components. Interestingly, two of these QTLs, (1-4 and 9-2-5) are the only Brix QTLs whose genes have been cloned by a map-based strategy, as described below.

The negative correlation between quantitative traits such as Brix and yield has been extensively analyzed by the Zamir and Fernie groups and documented in a series of papers based on analyses using the *S. pennellii* ILs (Schauer et al. 2006, 2008). From their studies, they concluded that a large proportion of the fruit metabolite QTLs (a total of 889) were negatively correlated with plant morphology traits (e.g., growth habit, yield and harvest index). Furthermore, they distinguished between those metabolites that are part of central pathways of primary metabolism, such as the components that make up Brix values (sugars, organic acids, amino acids) and the metabolites that are less central to general metabolism. The latter can be considered secondary metabolites and these generally are less impacted by the source-sink relationships of whole plant morphology and physiology, which determine total photoassimilate partitioning. Accordingly, it will be more straightforward to alter the minor components of secondary metabolism, such as the volatile components of fruit quality, without impacting negatively on yield and harvest index. The impact of introgressions from wild species on secondary metabolism volatiles has been studied by Tadmor et al. (2002) and Tieman et al. (2006).

An indication that the link between Brix and yield is not always unbreakable are the recent data on the heterotic and additive analyses of the ILs in their homozygotic and heterozygotic conditions. Although practically all of the Brix QTLs were associated with reduced yield, some of these negative relationships were broken in the hybrid condition (Gur and Zamir 2004; Schauer et al. 2008). For example, IL 8-3 has increased Brix but reduced yield. However, the hybrid IL8-3 × M-82 retained the increase in Brix but also showed an increase in yield (Gur and Zamir 2004), strongly indicating that the increase in Brix was not yield-related and that the two traits are inherited independently. In fact, the two traits showed opposite modes of inheritance; the *S. pennellii* derived increase in Brix is dominant while the *S. pennellii* derived decrease in yield is recessive. Azanza et al. (1995) reported that introgressions from *S. chmielewskii* chr 7 increased Brix without negatively effecting yield. In a later report (Yousef and Juvik 2001), which focused on one of the chr 7 segments (7M), results indicated that Brix per plant was higher in the heterozygous state than in the homozygous state, again perhaps indicating independently inherited traits of yield and Brix.

Two recent strategies that will no doubt contribute to breeding for fruit quality are the analysis of the fruit metabolome and their respective QTLs (referred to as mQTLs), together with the mapping of the candidate genes that may be involved in determining these metabolite levels. The merging of these two strategies, especially with the increase in number of mapped genes from the Solanaceae Genome Project and the dissection of the *S. pennellii* IL population into 500 sub-ILs (Lippman et al. 2007), increases the chances for identifying the genes responsible for the QTLs. Extensive analyses of metabolites in the tomato germplasm have been carried out (Schauer et al. 2005; Carrari et al. 2006; Moco et al. 2006) and a number of studies have analyzed the metabolites in the mapping populations described above. For example, Fulton et al. (2002) measured individual sugars, organic acids and glutamic acid in four advanced backcross populations based on different wild species and identified over 200 QTLs for 15 traits. Both Causse et al. (2004) and Schauer et al. (2006) analyzed metabolites from the same *S. pennellii* IL population. Causse et al. (2004) targeted their metabolite analyses to the sugars and organic acids while Schauer et al. (2006) carried out large scale metabolomic analyses based on GC-MS following derivatization and identified 74 metabolites and almost 900 mQTLs. More recently, Stevens et al. (2007) assayed these lines, as well as two other populations for ascorbic acid and Tieman et al. (2006) analyzed them for volatile content and citric acid levels.

In addition to the QTL portion of the metabolite studies, there have been attempts to colocalize the QTLs with candidate genes involved in the respective metabolic pathways. The large collection of metabolite data of Schauer et al. (2006) was recently further complemented by the study by Bermudez et al. (2008) who added to the list of mapped candidate genes of Causse et al. (2004) additional gene markers derived from the SGN database. Some of the results of colocalization may be speculative. For example, Schauer et al. (2006) suggested that a QTL for maleic acid on IL7-4 may be related to a vacuolar pyrophosphatase gene, which maps to the same bin and is involved in proton transport into the fruit vacuole. However, the authors quickly point out that the small number of mapped candidate genes, compared to the large number of genes in a particular bin or IL, makes such colocalization studies only the beginning hypothesis for further detailed research. Causse et al. (2004) found only a few colocalized candidate genes, which could be implicated with sugar or acid QTLs, while Stevens et al. (2007) identified two candidates of ascorbic acid metabolism, GDP-mannose epimerase and monodehydroascorbate reductase that colocalized with two QTLs for ascorbic acid levels. Nevertheless, Stevens et al. (2007) reported that the number of QTLs that colocalized with candidate genes would be expected merely by chance. Co-mapping of 23 candidate genes involved in carotenoid metabolism and 16 QTLs for pigmentation levels led to a similar conclusion (Liu et al. 2003). Only five of the QTL cosegregated with the same bins that contained candidate genes — a number that is expected by chance alone. These studies clearly demonstrate that a similar map location of a QTL and a candidate gene is far from a direct causative relationship between a gene and a phenotype, especially at this stage of the sequencing of the tomato genome.

A limited number of major QTLs for quality traits have been identified and entered into tomato breeding schemes. Two such genes increase the total soluble solids content without decreasing the plant yield, fruit size or harvest index. Both originated as QTLs for Brix but their subsequent identification was accomplished by two different strategies: one, the *LIN5* locus, by map-based cloning and the other, the *AGPase-LS1* locus, by candidate gene colocalization and biochemical studies.

LIN5 was initially identified as a Brix QTL in the early *S. pennellii* ILs study (Eshed and Zamir 1995). The introgression was subsequently minimized and found in the 9-2-5 sub-IL. Successful map-based cloning of *LIN5* was aided by the locus being a hot-spot for recombination (Fridman et al. 2000). *LIN5* was determined to encode one of the apoplastic invertases of tomato, Lycopersicon Invertase 5, or LIN5. The metabolic explanation for the effect of *LIN5* in increasing Brix with no concomitant decrease in yield parameters is based on an early hypothesis that increased activity of apoplastic invertase in the young fruit serves to hydrolyze the translocated sucrose from the source leaves upon arrival in the fruit, and thereby increases the sucrose gradient from source to sink, thus increasing net transport of photoassimilate into the fruit (Eschrich 1980). The biochemical explanation proposed is that the *S. pennellii* allele for *LIN5* has a single amino acid change, which makes it a more efficient enzyme kinetically, reducing the Km for the sucrose substrate. QTL studies for Brix based on segregating populations developed from crosses between *S. lycopersicum* and various wild species indicated that only the *S. pennellii* LA716 accession yielded a QTL for Brix at this locus. The conclusion that a particular amino acid was responsible for increased Brix in *LIN5* genotypes was made possible by comparing the amino acid sequence of the LIN5 protein with homologs from all the wild species of tomato. The *S. pennellii* enzyme was unique in having an aspartic acid rather than a glutamic acid at position 239. Biochemical characterization of the different enzymes indicated that the aspartic acid substitution indeed results in an enzyme with an improved Km (Fridman et al. 2004).

The second gene, which has been identified and shown useful for increasing Brix without decreasing yield parameters, is the *AGPase-LS1* locus (Petreikov et al. 2006, 2009). Although there were reports of a Brix-related QTL at the distal portion of chr 1–4 (Eshed and Zamir 1994, 1995; Monforte and Tanksley 2000), the strategy taken in the identification of this gene was one of biochemical analyses and testing of candidate genes. From an initial interspecific cross of *S. lycopersicum* and *S. habrochaites* (LA1777), segregating backcross populations were selected on the basis of sugar level. One of the advanced stable selections for increased Brix was then analyzed for carbohydrate levels and sugar metabolism enzyme activities in the developing fruit in order to identify biochemical parameters that may be associated or causal to the increase in soluble sugars in the mature fruit (Schaffer et al. 2000). We demonstrated that transient starch accumulation of the high sugar lines was significantly increased, thereby increasing the carbohydrate pool in the developing fruit. A comparison of enzyme activities responsible for the sucrose to starch pathway demonstrated that only a single enzyme, ADP-glucose pyrophosphorylase (AGPase), was significantly more active in the high sugar line. Since this enzyme is frequently limiting to starch synthesis (Preiss and Sivak 1996), it stood out as an excellent candidate gene. Mapping of the four genes encoding the two subunits of the heterotetrameric enzyme showed that the high sugar line harbored the *S. habrochaites* allele for *AGPase-LS1* (large subunit 1) on the distal portion of chr 1–4, coinciding with a reported QTL for Brix. Subsequently, the QTL for Brix from chr 1–4 was further divided into multiple QTLs (Frary et al. 2003; Petreikov et al. 2006) and it is not clear that the AGPase locus is indeed the locus responsible for the early QTLs reported.

The physiological basis for the increase in Brix due to *AGPase-LS1* appears to be related to the extended period of gene expression of the wild species allele during fruit development (Petreikov et al. 2006). This lengthier expression, continuing into the major period of fruit expansion, allows for transient starch synthesis to continue in the enlarged immature green fruit, thereby significantly increasing the carbohydrate pool of the developing fruit unit. Calculated increases in starch content in the developing fruit indicate that increasing the transient starch reservoir is an efficient strategy for increasing fruit sink strength.

Together, these two studies show that photoassimilate transported into the fruit sink can be increased by modifying the sink carbohydrate metabolism. In both cases, the increased utilization of photoassimilate by the different enzymatic steps serves to "pull" additional photoassimilate into the developing fruit, thereby increasing the net photoassimilate content, and not merely increasing Brix by increasing the concentration of soluble sugars in a smaller fruit.

An additional gene that modifies fruit sugar content, perhaps the first identified (Klann et al. 1992; Chetelat et al. 1995), is the *sucr* locus. As mentioned above, all *S. lycopersicum* accumulate primarily hexoses, glucose and fructose, in near equimolar amounts. It was long known (Davies 1966) that the green fruited wild species of tomato (the *Eriopersicon* subgenus according to Muller 1940) accumulated sucrose rather than hexose. Biochemical comparisons of sugar metabolism between the tomato and its wild relatives pointed to soluble vacuolar invertase as responsible for the different sugar accumulation patterns (Yelle et al. 1988; Miron and Schaffer 1991; Stommel 1992). Subsequent cloning, colocalization and gene expression analyses showed conclusively that the *TIV* gene encoding the vacuolar invertase was the locus (*sucr*) responsible for sucrose to hexose partitioning (Klann et al. 1993; Chetelat et al. 1995; Hadas et al. 1995). The lack of *TIV* expression and resulting absence of the invertase enzyme in the maturing fruit of the wild species allows for the translocated sucrose to be stored in the vacuole (Miron et al. 2002). Results reported by Chetelat et al. (1995) indicate that the *sucr* locus determines the type of sugar accumulated but does not impact the total amount of sugar in the fruit. Thus, although the wild species are characterized by very high sugar contents it appears that it is not attributed to the accumulation of sucrose.

An additional sugar accumulation trait that has been introgressed from wild species is the ratio of fructose to glucose in the mature fruit. While *S. lycopersicum* fruit have a near equimolar ratio, the green fruited wild species have significantly higher levels of fructose in comparison to glucose (Davies 1966). Since the hexose level is masked by the high sucrose level in the green fruited species this difference in hexoses is not readily apparent. However, when the trait is introgressed into the hexose accumulating tomato, the high fructose level is evident. The trait has been QTL mapped and at least two loci contributing to the trait have been identified. The major locus, *Fgr* (*Fructose to glucose ratio*), mapped to chr 4 (Levin et al. 2000) while a second locus maps to chr 6, near the loci for a fructokinase (*FK2*) and a hexokinase (Levin et al. 2007). The two loci show an epistatic relationship, with *FK2^{hab}* further increasing the fructose to glucose ratio only in the presence of the wild species allele for *Fgr* (Levin et al. 2007). The biochemical mechanism is still not understood and fine- mapping and cloning of the gene is in progress.

5.8 Summary

The classical distinction between simply inherited qualitative traits and complex quantitative traits which are controlled by multiple loci highly influenced by environmental factors, has become less apparent due to advances made in nucleic acid technologies. High-throughput technologies enable rapid discovery of DNA markers or gene sequences that have major effects on quantitative traits. These DNA markers and gene specific sequences can now serve as selection tools and be regarded as a simply inherited trait, since they segregate in a discrete fashion in a segregating population. Because DNA sequences are very rarely influenced by the environment, they can be developed to reflect a unique map position. Furthermore, because DNA sequences exist in virtually every plant cell throughout development, they can be determined and analyzed with great accuracy at the earliest stages of plant development, months before the target trait may otherwise be noticeable or measured.

DNA markers tightly associated with traits or phenotypes of economic importance are known to be an excellent tool to expedite breeding using marker-assisted selection. DNA markers are most informative when they display complete linkage disequilibrium with genes responsible for a desired phenotype. When DNA markers are based on gene sequences that encode proteins that clearly control such phenotypes, these gene-sequences can be further used to study its developmental expression profile under various environmental conditions and to exploit this information to tailor the best environment, which would allow maximal expression of a desired phenotype. These sequences can also be used in transgenic modifications to decipher the role of the nucleotide and/or amino-acid sequence in establishing or maximizing a particular phenotype or directly constitute a phenotype of interest in more than a single species. In this sense, genetic maps proved to be an excellent tool in identifying genes that cause discrete or quantitative phenotypes by map-based cloning (Tanksley et al. 1995; Jander et al. 2002; Alonso-Blanco et al. 2005). Moreover, current technologies allow the possible use of maps of one species to isolate orthologous genes in another, provided that the linear sequence and repertoire of markers is sufficiently well conserved between the two species being compared (see for example, Delseny et al. 2001).

The advances recently made in the field of high-throughput metabolomics are also establishing themselves with great importance in understanding complex quantitative traits by identifying gene products which are otherwise undetected. This knowledge is leading to the discovery of discrete genes which can decrease, increase or divert metabolic fluxes resulting in the accumulation of desired metabolites together or at the expense of other less important constituents.

Genomic, metabolomic and proteomic analyses using sophisticated bioinformatics tools will no doubt significantly contribute to our understanding of "system biology", i.e., the systematic integrated study of complex interactions in biological systems or networks. This understanding is expected to unravel with better accuracy what exactly determines a phenotype and facilitate development of methodologies, which will enable us to produce custom-made agricultural products by modulating simply inherited genes using classical or molecular technologies.

Due to the ever increasing global food shortage and the continuous demand for agricultural products with better quality, our greatest challenge is to discover the gene combinations, which make quality phenotypes without reducing yield. The identification and modulation of discrete simply inherited genes, which control regulatory networks, are expected to become the preferred target to combat this challenge.

Acknowledgements

The author would like to thank Dr. Yaakov Tadmor from the Institute of Plant Sciences, the Volcani Center, Israel for his contribution of most of the tomato fruit photos to this book chapter.

The purple smudge photo was kindly provided by Jim Myers and Peter Boches, Department of Horticulture, Oregon State University, USA.

References

Alonso-Blanco C, Mendez-Vigo B, Koornneef M (2005) From phenotypic to molecular polymorphisms involved in naturally occurring variation of plant development. Int J Dev Biol 49: 717–732.

Azanza F, Kim D, Tanksley SD, Juvik JA (1995) Genes from *Lycopersicon chmielewskii* affecting tomato quality during fruit ripening. Theor Appl Genet 91: 495–505.

Azanza F, Young TE, Kim D, Tanksley SD, Juvik JA (1994) Characterization of the effects of introgressed segments of chromosome 7 and 10 from *Lycopersicon chmielewskii* on tomato soluble solids, pH and yield. Theor Appl Genet 87: 965–972.

Bermudez L, Urias U, Milstein D, Kamenetzky L, Asis R, Fernie AR, Van Sluys MA, Carrari F, Rossi M (2008) A candidate gene survey of quantitative trait loci affecting chemical composition in tomato fruit. J Exp Bot 59: 2975–2990.

Bernacchi D, Beck-Bunn T, Eshed Y, Lopez J, Petiard V, Uhlig J, Zamir D, Tanksley SD (1998) Advanced backcross QTL analysis in tomato. I. Identification of QTL for traits of agronomic importance from *Lycopersicon hirsutum*. Theor Appl Genet 97: 381–397.

Bernard RL (1976) United States national germplasm collections. In: Hill LD (ed) World Soybean Research. Interstate Printers and Publishing, Danville, IL, USA, pp 286–289.

Bernhardt A, Lechner E, Hano P, Schade V, Dieterle M, Anders M, Dubin MD, Benvenuto G, Bowler C, Genschik P, Hellmann H (2006) CUL4 associates with DDB1 and DET1 and its downregulation affects diverse aspects of development in *Arabidopsis thaliana*. Plant J 47: 591–603.

Biester A, Wood MW, Wahlin CS (1925) Carbohydrate studies: I. the relative sweetness of pure sugars. Am J Physiol 73: 387–400.

Bino RJ, de Vos CHR, Lieberman M, Hall RD, Bovy A, Jonker HH, Tikunov Y, Lommen A, Moco S, Levin I (2005) The light-hyperresponsive *high pigment-2^{dg}* mutation of tomato: alterations in the fruit metabolome. New Phytol 166: 427–438.

Blair MW, Munoz C, Garza R, Cardona C (2006) Molecular mapping of genes for resistance to the bean pod weevil (*Apion godmani* Wagner) in common bean. Theor Appl Genet 112: 913–923.

Botstein D, White RL, Skolnick M, Davis RW (1980) Construction of a genetic linkage map in man using restriction fragment length polymorphisms. Am J Hum Genet 32: 314–331.

Carrari F, Baxter C, Usadel B, Urbanczyk-Wochniak E, Zanor MI, Nunes-Nesi A, Nikiforova V, Centero D, Ratzka A, Pauly M, Sweetlove LJ, Fernie AR (2006) Integrated analysis of metabolite and transcript levels reveals the metabolic shifts that underlie tomato fruit development and highlight regulatory aspects of metabolic network behavior. Plant Physiol 142: 1380–1396.

Causse M, Duffe P, Gomez MC, Buret M, Damidaux R, Zamir D, Gur A, Chevalier C, Lemaire-Chamley M, Rothan C (2004) A genetic map of candidate genes and QTLs involved in tomato fruit size and composition. J Exp Bot 55: 1671–1685.

Causse M, Saliba-Colombani V, Buret M, Lesschaeve I, Issanchou S (2001) Genetic analysis of organoleptic quality in fresh market tomato. 2. Mapping QTLs for sensory attributes. Theor Appl Genet 102: 273–283.

Causse M, Saliba-Colombani V, Lecomte L, Duffe P, Rousselle P, Buret M (2002) Genetic analysis of fruit quality attributes in fresh market tomato. J Exp Bot 53: 2090–2098.

Causse M, Damidaux R, Rousselle P (2007) Traditional and enhanced breeding for quality traits in tomato. In: Razdan MK, Matoo AK (eds) Genetic Improvement of Solanaceous Crops. Vol 2: Tomato. Science Publishers, Enfield, NH, USA, pp 153–192.

Chaib J, Lecomte L, Buset M, Causse M (2006) Stability over genetic backgrounds, generations and years of quantitative trait locus (QTLs) for organoleptic quality in tomato. Theor Appl Genet 112: 934–944.

Chen FQ, Foolad MR, Hyman J, St Clair DJ, Beeleman RB (1999) Mapping QTLs for lycopene and other fruit traits in a *Lycopersicon esculentum* X *L. pimpinellifolium* cross and comparison of QTLs across tomato species. Mol Breed 5: 283–299.

Chen H, Shen Y, Tang X, Yu Y, Wang J, Guo L, Zhang Y, Zhang H, Feng S, Strickland E, Zheng N, Deng XW (2006) *Arabidopsis* CULLIN4 forms an E3 ubiquitin ligase with RBX1 and the CDD complex in mediating light control of development. Plant Cell 18: 1991–2004.

Chetelat RT, DeVerna JW, Bennett AB (1995) Introgression into tomato (*Lycopersicon esculentum*) of the *L. chmielewskii* sucrose accumulator gene (*sucr*) controlling fruit sugar composition. Theor Appl Genet 91: 327–333.

Davies JN (1966) Occurrence of sucrose in the fruit of some species of *Lycopersicon*. Nature 209: 640–641.

Davies JN, Hobson GE (1981) The composition of tomato fruit—The influence of environment, nutrition and genotype. Crit Rev Food Sci Nutr 15: 205–280.

Delseny M, Salses J, Cooke R, Sallaud C, Regad F, Lagoda P, Guiderdoni E, Ventelon M, Brugidou C, Ghesquiere A (2001) Rice genomics: present and future. Plant Physiol Biochem 39: 323–334.

Di Mascio P, Kaiser S, Sies H (1989) Lycopene as the most efficient biological carotenoid singlet oxygen quencher. Arch Biochem Biophys 274: 532–538.

Dogenlaar S, Frary A, Ku HM, Tanksley SD (2002) Mapping quantitative trait loci in inbred backcross lines of *Lycopersicon pimpinellifolium* (LA1589). Genome 45: 1189–1202.

Dorais M, Papadopoulos AP, Gosselin A (2001) Greenhouse tomato fruit quality. Hort Rev 26: 239–319.

Eschrich W (1980) Free space invertase, its possible role in phloem loading. Ber Dtsch Bot Ges 93: 363–378.

Eshed Y, Abu-Abied M, Saranga Y, Zamir D (1992) *Lycopersicon esculentum* lines containing small overlapping introgressions from *L. pennellii*. Theor Appl Genet 83: 1027–1034.

Eshed Y, Zamir D (1994) Introgressions from *Lycopersicon pennellii* can improve the soluble solids yield of tomato hybrids. Theor Appl Genet 88: 891–897.

Eshed Y, Zamir D (1995) An introgression line population of *Lycopersicon pennellii* in the cultivated tomato enables the identification and fine mapping of yield-associated QTL. Genetics 141: 1147–1162.

Fehr WR (1987) Principles of Cultivar Development. Vol 1: Theory and Technique. Macmillan Publishing, New York, USA.

Foolad MR (2007) Genome mapping and molecular breeding of tomato. Int J Plant Genom 2007: 1–52.

Frary A, Fulton TM, Zamir D, Tanksley SD (2004) Advanced backcross QTL analysis of a *Lycopersicon esculentum* X *L. pennellii* cross and identification of possible orthologs in the Solanaceae. Theor Appl Genet 108: 485–496.

Frary A, Nesbitt TC, Grandillo S, Knaap E, Cong B, Liu J, Meller J, Elber R, Alpert KB, Tanksley SD (2000) fw2.2: a quantitative trait locus key to the evolution of tomato fruit size. Science 289: 71–72.

Frary A, Doganlar S, Frampton A, Fulton T, Uhlig J, Yates H, Tanksley S (2003) Fine mapping of quantitative trait loci for improved fruit characteristics from *Lycopersicon chmielewskii* chromosome 1. Genome 46: 235–243.

Fray RG, Grierson D (1993) Identification and genetic-analysis of normal and mutant phytoene synthase genes of tomato by sequencing, complementation and co-suppression. Plant Mol Biol 22: 589–602.

Fridman E, Carrari F, Liu YS, Fernie AR, Zamir D (2004) Zooming in on a quantitative trait for tomato yield using interspecific introgressions. Science 305: 1786–1789.

Fridman E, Pleban T, Zamir D (2000) A recombination hotspot delimits a wild-species quantitative trait locus for tomato sugar content to 484 bp within an invertase gene. Proc Natl Acad Sci USA 97: 4718–4723.

Fulton TM, BeckBunn T, Emmatty D, Eshed Y, Lopez J, Petiard V, Uhlig J, Zamir D, Tanksley SD (1997) QTL analysis of an advanced backcross of *Lycopersicon peruvianum* to the cultivated tomato and comparisons with QTLs found in other wild species. Theor Appl Genet 95: 881–894.

Fulton TM, Bucheli P, Voirol E, Lopez J, Petiard V, Tanksley SD (2002) Quantitative trait loci (QTL) affecting sugars, organic acids and other biochemical properties possibly contributing to flavor, identified in four advanced backcross populations of tomato. Euphytica 127: 163–177.

Fulton TM, Grandillo S, Beck-Bunn T, Fridman E, Frampton A, Lopez J, Petiard V, Uhlig J, Zamir D, Tanksley SD (2000) Advanced backcross QTL analysis of a *Lycopersicon esculentum* x *Lycopersicon parviflorum* cross. Theor Appl Genet 100: 1025–1042.

Galpaz N, Wang Q, Menda N, Zamir D, Hirschberg J (2008) Abscisic acid deficiency in the tomato mutant *high-pigment 3* leading to increased plastid number and higher fruit lycopene content. Plant J 53: 717–730.

Georgelis N, Scott JW, Baldwin EA (2004) Relationship of tomato fruit sugar concentration with physical and chemical traits and linkage of RAPD markers. J Am Soc Hort Sci 129: 839–845.

Gerster H (1997) The potential role of lycopene for human health. J Am Coll Nutr 16: 109–126.

Giorgiev C (1972) Anthocyanin fruit tomato. Rep Tomato Genet Coop 22: 10.

Giovannoni J, Tanksley S, Vrebalov J, Noensie E (2004) NOR Gene compositions and methods for use thereof. US Patent No 6,762,347 B1 issued July 13.

Giovannoni JJ, Wing RA, Ganal MA, Tanksley SD (1991) Isolation of molecular markers from specific chromosomal intervals using DNA pools from existing mapping populations. Nucl Acids Res 19: 6553–6558.

Goldman IL, Paran I, Zamir D (1995) Quantitative trait locus analysis of a recombinant inbred line population derived from a *Lycopersicon esculentum* × *Lycopersicon cheesmanii* cross. Theor Appl Genet 90: 925–932.

Grandillo S, Tanksley SD (1996) Genetic analysis of RFLPs, GATA microsatellites and RAPDs in a cross between *L. esculentum* and *L. pimpinellifolium*. Theor Appl Genet 92: 957–965.

Gur A, Zamir D (2004) Unused natural variation can lift yield barriers in plant breeding. PLoS Biol 2: 1610–1615.

Hadas R, Schaffer A, Miron D, Fogelman M, Granot D (1995) PCR-generated molecular markers for the invertase gene and sucrose accumulation in tomato. Theor Appl Genet 90: 1142–1148.

Halsted BD, Owen EJ, Shaw JK (1905) Experiments with tomatoes. NJ Agri Exp Stn Annu Rep 26: 447–477.

Hazen SP, Borevitz JO, Harmon FG, Pruneda-Paz JL, Schultz TF, Yanovsky MJ, Liljegren SJ, Ecker JR, Kay SA (2005) Rapid array mapping of circadian clock and developmental mutations in Arabidopsis. Plant Physiol 138: 990–997.

Hedrick UP, Booth NO (1907) Mendelian characters in tomatoes. Proc Am Soc Hort Sci 5: 19–24.

Hinze K, Thompson RD, Ritter E, salamini F, Schultze-Lefert P (1991) Restriction fragment length polymorphism-mediated targeting of the *ml-o* resistance locus in barley (*Hordeum vulgare*). Proc Natl Acad Sci USA 88: 3691–3695.

Ho LC (1996) The tomato. In: Zamski E, Schaffer AA (eds) Photoassimilate Distribution in Plants and Crops. Dekker, New York, USA, pp 709–728.

Hu J, McCall CM, Ohta T, Xiong Y (2004) Targeted ubiquitination of CDT1 by the DDB1–CUL4A–ROC1 ligase in response to DNA damage. Nat Cell Biol 6: 1003–1009.

Isaacson T, Ronen G, Zamir D, Hirschberg J (2002) Cloning of *tangerine* from tomato reveals a carotenoid isomerase essential for the production of β-Carotene and xanthophylls in plants. Plant Cell 14: 333–342.

Ishida BK, Roberts JS, Chapman MH, Burri BJ (2007) Processing tangerine tomatoes: effects on lycopene-isomer concentrations and profile. J Food Sci 72: C307–C312.

Jander G, Norris SR, Rounsley SD, Bush DF, Levin IM, Last RL (2002) *Arabidopsis* map-based cloning in the post-genome era. Plant Physiol 129: 440–450.

Jones CM, Mes P, Myers JR (2003) Characterization and inheritance of the *Anthocyanin fruit* (*Aft*) tomato. J Hered 94: 449–456.

Jones DF (1917) Linkage in *Lycopersicum*. Am Nat 51: 608–621.

Josse EM, Simkin AJ, Gaffé J, Labouré AM, Kuntz M, Carol P (2000) A plastid terminal oxidase associated with carotenoid desaturation during chromoplast differentiation. Plant Physiol 123: 1427–1436.

Klann E, Chetelat R, Bennett AB (1993) Expression of acid invertase gene controls sugar composition in tomato (*Lycopersicon*) fruit. Plant Physiol 103: 863–870.

Klann E, Yelle S, Bennett AB (1992) Tomato fruit acid invertase complementary DNA. Plant Physiol 99: 351–353.

Klein-Lankhorst R, Rietveld P, Machiels B, Verkerk R, Weide R, Gebhardt C, Koornneef M, Zabel P (1991) RFLP markers linked to the root knot nematode resistance gene *Mi* in tomato. Theor Appl Genet 81: 661–667.

Kohlmeier L, Kark JD, Gomez-Gracia E, Martin BC, Steck SE, Kardinaal AF, Ringstad J, Thamm M, Masaev V, Riemersma R, Martin-Moreno JM, Huttunen JK, Kok FJ (1997) Lycopene and myocardial infarction risk in the EURAMIC Study. Am J Epidemiol 146: 618–626.

Kolotilin I, Koltai H, Tadmor Y, Bar-Or C, Reuveni M, Meir A, Nahon S, Shlomo H, Chen L, Levin I (2007) Transcriptional profiling of *high pigment-2^{dg}* tomato mutant links early fruit plastid biogenesis with its overproduction of phytonutrients. Plant physiol 145: 389–401.

Labate JA, Grandillo S, Fulton T, Muños S, Caicedo AL, Peralta I, Ji Y, Chetelat RT, Scott JW, Gonzalo MJ, Francis D, Yang W, van der Knaap E, Baldo AM, Smith-White B, Mueller LA, Prince JP, Blanchard NE, Storey DB, Stevens MR, Robbins MD, Wang J-F, Liedl BE, O'Connell MA, Stommel JR, Aok K, Iijima Y, Slade AJ, Hurst SR, Loeffler D, Steine MN, Vafeados D, McGuire C, Freeman C, Amen A, Goodstal J, Facciotti D, Van Eck J, Causse M (2007) Tomato. In: Kole C (ed) Genome Mapping and Molecular Breeding in Plants. Vol 5: Vegetables. Springer, Berlin, Heidelberg, New York, pp 1–125.

Lecomte L, Duffé P, Buret M, Servin B, Hospital F, Causse M (2004) Marker-assisted introgression of five QTLs controlling fruit quality traits into three tomato lines revealed interactions between QTLs and genetic backgrounds. Theor Appl Genet 109: 658–668.

Levin I, de Vos CHR, Tadmor Y, Bovy A, Lieberman M, Oren-Shamir M, Segev O, Kolotilin I, Keller M, Ovadia R, Meir A, Bino RJ (2006) *High pigment* tomato mutants—more than just lycopene. Isr J Plant Sci 54: 179–190.

Levin I, Frankel P, Gilboa N, Tanny S, Lalazar A (2003) The tomato *dark green* mutation is a novel allele of the tomato homolog of the *DEETIOLATED1* gene. Theor Appl Genet 106: 454–460.

Levin I, Gilboa N, Cincarevsky F, Oguz I, Petreikov M, Yeselson Y, Shen S, Bar M, Schaffer AA (2007) Epistatic interaction between the Fgr and FK2 genes determines the fructose to glucose ratio in mature tomato fruit. Isr J Plant Sci 54: 215–223.

Levin I, Gilboa N, Yeselson E, Shen S, Schaffer AA (2000) *Frg*, a major locus that modulates fructose to glucose ratio in mature tomato fruit. Theor Appl Genet 100: 256–262.

Levy J, Bosin E, Feldman B, Giat Y, Miinster A, Danilenko M, Sharoni Y (1995) Lycopene is a more potent inhibitor of human cancer cell proliferation than either α-carotene or β-carotene. Nutr Cancer 24: 257–266.

Lieberman M, Segev O, Gilboa N, Lalazar A, Levin I (2004) The tomato homolog of the gene encoding UV-damaged DNA binding protein 1 (DDB1) underlined as the gene that causes the *high pigment-1* mutant phenotype. Theor Appl Genet 108: 1574–1581.

Lippman ZB, Semel Y, Zamir D (2007) An integrated view of quantitative trait variation using tomato interspecific introgression lines. Curr Opin Genet Dev 17: 545–552.

Liu Y, Roof S, Ye Z, Barry C, van Tuinen A, Vrebalov J, Bowler C, Giovannoni J (2004) Manipulation of light signal transduction as a means of modifying fruit nutritional quality in tomato. Proc Natl Acad Sci USA 101: 9897–9902.

Liu YS, Gur A, Ronen G, Causse M, Damidaux R, Buret M, Hirschberg J, Zamir D (2003) There is more to tomato fruit colour than candidate carotenoid genes. Plant Biotechnol J 1: 195–207.

Mackinney G, Rick CM, Jenkins JA (1956) The phytoene content of tomatoes. Proc Natl Acad Sci USA 42: 404–408.

Martin GB, Williams GK, Tanksley SD (1991) Rapid identification of markers linked to a *Pseudomonas* resistance gene in tomato by using random primers and near-isogenic lines. Proc Natl Acad Sci USA 88: 2336–2340.

Maxon Smith JW, Ritchie DB (1983) A collection of near-isogenic lines of tomato: research tool of the future? Plant Mol Biol Rep 1: 41–45.

Menda N, Semel Y, Peled D, Eshed Y, Zamir D (2004) In silico screening of a saturated mutation library of tomato. Plant J 38: 861–872.

Michelmore RW, Paran I, Kesseli RV (1991) Identification of markers linked to disease resistance genes by bulked segregant analysis: A rapid method to detect markers in specific genomic regions by using segregating populations. Proc Natl Acad Sci USA 88: 9828–9832.

Miron D, Petreikov M, Carmi N, Shen S, Levin I, Granot D, Zamski E, Schaffer AA (2002) Sucrose uptake, invertase localization and gene expression in developing fruit of *Lycopersicon esculentum* and the sucrose-accumulating *Lycopersicon hirsutum*. Physiol Plant 115: 35–47.

Miron D, Schaffer AA (1991) Sucrose phosphate synthase, sucrose synthase and acid invertase activities in developing fruit of *Lycopersicon esculentum* Mill and the sucrose accumulating *Lycopersicon hirsutum*. Plant Physiol 95: 623–627.

Mochizuki T, Kamimura S (1984) Inheritance of vitamin C content and its relation to other characters in crosses between *hp* and *og* varieties of tomatoes. In: Synopsis of the 9th Meeting of the Eucarpia Tomato Working Group, 22–24 May 1984. Wageningen, Netherlands, pp 8–13.

Moco S, Bino RJ, Vorst O, Verhoeven HA, de Groot J, van Beek TA, Vervoort J, de Vos CH (2006) A liquid chromatography-mass spectrometry-based metabolome database for tomato. Plant Physiol 141: 1205–1218.

Mohan M, Nair S, Bhagwat A, Krishna TG, Yano M, Bhatia CR, Sasaki T (1997) Genome mapping, molecular markers and marker-assisted selection in crop plants. Mol Breed 3: 87–103.

Monforte AJ, Tanksley SD (2000) Fine mapping of a quantitative trait locus (QTL) from *Lycopersicon hirsutum* chromosome 1 affecting fruit characteristics and agronomic traits: breaking linkage among QTLs affecting different traits and dissection of heterosis for yield. Theor Appl Gen 100: 471–479.

Muehlbauer GJ, Specht JE, Thomas-Compton MA, Staswick PE, Bernard RL (1998) Near-isogenic lines—A potential resource in the integration of conventional and molecular marker linkage maps. Crop Sci 28: 729–735.

Mueller LA, Tanksley SD, Giovannoni JJ, van Eck J, Stack S, Choi D, Kim BD, Chen M, Cheng Z, Li C, Ling H, Xue Y, Seymour G, Bishop G, Bryan G, Sharma R, Khurana J, Tyagi A, Chattopadhyay D, Singh NK, Stiekema W, Lindhout P, Jesse T, Lankhorst RK, Bouzayen M, Shibata D, Tabata S, Granell A, Botella MA, Giuliano G, Frusciante L, Causse M, Zamir D (2005) The tomato sequencing project, the first cornerstone of the international solanaceae project (SOL). Compar Funct Genom 6: 153–158.

Muller CH (1940) A revision of the genus *Lycopersicon*. USDA Misc Publ 328: 29.

Mustilli AC, Fenzi F, Ciliento R, Alfano F, Bowler C (1999) Phenotype of the tomato *high pigment-2* mutant is caused by a mutation in the tomato homolog of *DEETIOLATED1*. Plant Cell 11: 145–157.

Osborn TC, Alexander DC, Fobes JF (1987) Identification of restriction fragment length polymorphisms linked to genes controlling soluble solids content in tomato. Theor Appl Genet 73: 350–356.

Paran I, Kesseli RV, Michelmore RW (1991) Identification of restriction fragment length polymorphism and random amplified polymorphic DNA markers linked to downy mildew resistance genes in lettuce using near-isogenic lines. Genome 34: 1021–1027.

Paran I, Levin I (2008) Mapping and tagging of simply inherited traits. In: Kole C, Abbott AG (eds) Principles and Practices of Plant Genomics. Vol 2: Molecular Breeding. Science Publishers, Enfield, NH, USA, pp 139–174.

Paterson AH, Damon S, Hewitt JD, Zamir D, Rabinowitch HD, Lincoln SE, Lander ES, Tanksley SD (1991) Mendelian factors underlying quantitative traits in tomato: comparison across species, generations and environments. Genetics 127: 181–197.

Paterson AH, DeVerna JW, Lanini B, Tanksley SD (1990) Fine mapping of quantitative trait loci using selected overlapping recombinant chromosomes in an interspecific cross of tomato. Genetics 124: 735–742.

Paterson AH, Lander ES, Hewitt JD, Peterson S, Lincoln SE, Tanksley SD (1988) Resolution of quantitative traits into Mendelian factors by using a complete linkage map of restriction fragment length polymorphisms. Nature 335: 721–726.

Peters JL, Cnudde F, Gerats T (2003) Forward genetics and map-based cloning approaches. Trends Plant Sci 8: 484–491.

Peters JL, van Tuinen A, Adamse P, Kendrick RE, Koornneef M (1989) *High pigment* mutants of tomato exhibit high sensitivity for phytochrome action. J Plant Physiol 134: 661–666.

Petreikov M, Shen S, Yeselson Y, Levin I, Bar M, Schaffer AA (2006) Temporally extended gene expression of the ADP-Glc pyrophosphorylase large subunit (AgpL1) leads to increased enzyme activity in developing tomato fruit. Planta 224: 1465–1479.

Petreikov M, Yeselson L, Shen S, Levin I, Schaffer AA, Efrati A, Bar M (2009) Carbohydrate balance and accumulation during development of near-isogenic tomato lines differing in the AGPase-L1 allele. J Am Soc Hort Sci 134: 134–140.

Preiss J, Sivak MN (1996) Starch synthesis in sink and sources. In: Zamski E, Schaffer AA (eds) Photoassimilate Distribution in Plants and Crops. Dekker, New York, USA, pp 63–96.

Price HL, Drinkard AW Jr (1908) Inheritance in tomato hybrids. Va Agri Exp Stn Bull 177: 1–53.

Quedraogo JT, Maheshwari V, Berner DK, St-Pierre CA, Belzile F, Timko MP (2001) Identification of AFLP markers linked to resistance to cowpea (*Vigna unguiculata* L.) parasitism by *Striga gesnerioides*. Theor Appl Genet 102: 1029–1036.

Rick CM (1964) Biosystematic studies on Galapagos Island tomatoes. Occas Paper Calif Acad Sci 44: 59.

Rick CM (1991) Tomato paste: a concentrated review of genetic highlights from the beginnings to the advent of molecular genetics. Genetics 128: 1–5.

Rick CM, Cisneros P, Chetelat RT, DeVerna JW (1994) *Abg*, a gene on chromosome 10 for purple fruit derived from *S. lycopersicoides*. Rep Tomato Genet Coop 44: 29–30.

Ronen G, Carmel-Goren L, Zamir D, Hirschberg J (2000) An alternative pathway to β-carotene formation in plant chromoplasts discovered by map-based cloning of *Beta* (*B*) and *old-gold* (*og*) colour mutations in tomato. Proc Natl Acad Sci USA 97: 11102–11107.

Ronen G, Cohen M, Zamir D, Hirschberg J (1999) Regulation of carotenoid biosynthesis during tomato fruit development: expression of the gene for lycopene epsilon-cyclase is down-regulated during ripening and is elevated in the mutant *Delta*. Plant J 17: 341–351.

Saliba-Colombani V, Causse M, Langlois D, Philouze J, Buret M (2001) Genetic analysis of organoleptic quality in fresh market tomato. 1. Mapping QTLs for physical and chemical traits. Theor Appl Genet 102: 259–272.

Sapir M, Oren-Shamir M, Ovadia R, Reuveni M, Evenor D, Tadmor Y, Nahon S, Shlomo H, Chen L, Meir A, Levin I (2008) Molecular aspects of *Anthocyanin fruit* tomato in relation to *high pigment-1*. J Hered 99: 292–303.

Schaffer AA, Levin I, Ogus I, Petreikov M, Cincarevsky F, Yeselson E, Shen S, Gilboa N, Bar M (2000) ADP-glucose pyrophosphorylase activity and starch accumulation in immature tomato fruit: the effect of a *Lycopersicon hirsutum*-derived introgression encoding for the large subunit. Plant Sci 152: 135–144.

Schauer N, Semel Y, Balbo I, Steinfath M, Repsilber D, Selbig J, Pleban T, Zamir D, Fernie AR (2008) Mode of inheritance of primary metabolic traits in tomato. Plant Cell 20: 509–523.

Schauer N, Semel Y, Roessner U, Gur A, Balbo I, Carrari F, Pleban T, Perez-Melis A, Bruedigam C, Kopka J, Willmkitzer L, Zamir D, Fernie AR (2006) Comprehensive metabolic profiling and phenotyping of interspecific introgression lines for tomato improvement. Nat Biotechnol 24: 447–454.

Schauer N, Zamir D, Fernie AR (2005) Metabolic profiling of leaves and fruit of wild species tomato: A survey of the *Solanum lycopersicum* complex. J Exp Bot 56: 297–307.

Schneider A, Walker S, Sagan M, Duc G, Ellis T, Downie J (2002) Mapping of the nodulation loci *sym9* and *sym10* of pea (*Pisum sativum* L.). Theor Appl Genet 104: 1312–1316.

Schroeder DF, Gahrtz M, Maxwell BB, Cook RK, Kan JM, Alonso JM, Ecker JR, Chory J (2002) De-etiolated 1 and damaged DNA binding protein 1 interact to regulate *Arabidopsis* photomorphogenesis. Curr Biol 12: 1462–1472.

Stahl W, Sies H (1996) Lycopene: a biologically important carotenoid for humans? Arch Biochem Biophys 336: 1–9.

Stam P, Zeven AC (1981) The theoretical proportion of the donor genome in near-isogenic lines of self-fertilizers bred by backcrossing. Euphytica 30: 227–238.

Stevens MA (1986) Inheritance of tomato fruit quality components. Plant Breed Rev 4: 273–311.

Stevens MA, Rick CM (1986) Genetics and breeding. In: Atherton JG, Rudich J (eds) The Tomato Crop: A Scientific Basis for Improvement. Chapman and Hall, New York, USA, pp 35–109.

Stevens R, Buret M, Duffe P, Garchery C, Baldet P, Rothan C, Causse M (2007) Candidate genes and quantitative trait loci affecting fruit ascorbic acid content in three tomato populations. Plant Physiol 143: 1943–1953.

Stommel JR (1992) Enzymatic components of sucrose accumulation in the wild tomato species *Lycopersicon peruvianum*. Plant Physiol 99: 324–328.

Sweetlove LJ, Fernie AR (2005) Regulation of metabolic networks: understanding metabolic complexity in the systems biology era. New Phytol 168: 9–23.

Tadmor Y, Fridman E, Gur A, Larkov O, Lastochkin E, Ravid U, Zamir D, Lewinsohn E (2002) Identification of *malodorous*, a wild species allele affecting tomato aroma that was selected against during domestication. J Agri Food Chem 50: 2005–2009.

Tanksley S, Ganal M, Prince J, de Vicente M, Bonierbale M, Broun P, Fulton TM, Giovannoni JJ, Grandillo S, Martin GB, Messeguer R, Miller JC, Miller L, Paterson AH, Pineda O, Röder MS, Wing RA, Wu W, Young ND (1992) High density molecular maps of the tomato and potato genomes. Genetics 132: 1141–1160.

Tanksley SD, Ganal MW, Martin GB (1995) Chromosome landing—a paradigm for map-based cloning in plants with large genomes. Trends Genet 11: 63–68.

Tanksley SD, Grandillo S, Fulton TM, Zamir D, Eshed Y, Fulton TM, Zamir D, Eshed Y, Petiard V, Lopez J, Beck-Bunn T (1996) Advanced backcross QTL analysis in a cross between an elite processing line of tomato and its wild relative *L. pimpinellifolium*. Theor Appl Genet 92: 213–224.

Tanksley SD, McCouch SR (1997) Seed banks and molecular maps: unlocking genetic potential from the wild. Science 277: 1063–1066.

Tanksley SD, Rick CM (1980) Isozymic gene linkage map of the tomato: Applications in genetics and breeding. Theor Appl Genet 58: 161–170.

Tanksley SD, Zamir D (1988) Double tagging of a male-sterile gene in tomato using a morphological and enzymatic marker gene. HortScience 23: 387–388.

Thomas CM, Vos P, Zabeau M, Jones DA, Norcott KA, Chadwick BP, Jones DG (1995) Identification of amplified restriction fragment polymorphism (AFLP) markers tightly linked to the tomato Cf-9 gene for resistance to *Cladosporium fulvum*. Plant J 8: 785–794.

Tieman DM, Zeigler M, Schmelz EA, Taylor MG, Bliss P, Kirst M, Klee HJ (2006) Identification of loci affecting flavour volatile emissions in tomato fruits. J Exp Bot 57: 887–896.

Unlu NZ, Bohn T, Francis D, Clinton SK, Schwartz SJ (2007) Carotenoid absorption in humans consuming tomato sauces obtained from tangerine or high-β-carotene varieties of tomatoes. J Agri Food Chem 55: 1597–1603.

van Tuinen A, de Vos CHR, Hall RD, Linus HW, van der Plas LHW, Bowler C, Bino RJ (2006) Use of metabolomics for identification of tomato genotypes with enhanced nutritional value derived from natural light-hypersensitive mutants. In: Jaiwal PK (ed) Plant Genetic Engineering. Vol 7: Metabolic Engineering and Molecular Farming–1. Studium Press, Huston, TX, USA, pp 240–256.

Vrebalov J, Ruezinsky D, Padmanabhan V, White R, Medrano D, Drake R, Schuch W, Giovannoni J (2002) A MADS-box gene necessary for fruit ripening at the tomato ripening-inhibitor (*rin*) locus. Science 296: 275–276.

Wang N, Fang W, Han H, Sui N, Li B, Meng QW (2008) Overexpression of zeaxanthin epoxidase gene enhances the sensitivity of tomato PSII photoinhibition to high light and chilling stress. Physiol Plant 132: 384–396.

Wann EV (1997) Tomato germplasm lines T4065, T4099, T5019, and T5020 with unique genotypes that enhance fruit quality. Hort Sci 32: 747–748.

Wann EV, Jourdain EL, Pressey R, Lyon BG (1985) Effect of mutant genotypes *hp og^c* and *dg og^c* on tomato fruit quality. J Am Soc Hort Sci 110: 212–215.

Wertz IE, O'Rourke KM, Zhang Z, Dornan D, Arnott D, Deshaies RJ, Dixit VM (2004) Human de-etiolated-1 regulates c-Jun by assembling a CUL4A ubiquitin ligase. Science 303: 1371–1374.

Willits MG, Krame CM, Prata RT, De Luca V, Potter BG, Steffens JC, Graser G (2005) Utilization of the genetic resources of wild species to create a nontransgenic high flavonoid tomato. J Agri Food Chem 53: 1231–1236.

Xiao H, Jiang N, Schaffner E, Stockinger EJ, van der Knaap E (2008) A retrotransposon-mediated gene duplication underlies morphological variation of tomato fruit. Science 319: 1527–1530.

Yang DE, Zhang CL, Zhang DS, Jin DM, Weng ML, Chen SJ, Nguyen H, Wang B (2004) Genetic analysis and molecular mapping of maize (*Zea mays* L.) stalk rot resistance gene *Rfg1*. Theor Appl Genet 108: 706–711.

Yelle S, Hewitt JD, Robinson NL, Darnon S, Bennett AB (1988) Sink metabolism in tomato fruit III. Analysis of carbohydrate assimilation in a wild species. Plant Physiol 87: 737–740.

Yen HC, Shelton BA, Howard LR, Vrebalov SLJ, Giovanonni JJ (1997) The tomato *high-pigment* (*hp*) locus maps to chromosome 2 and influences plastome copy number and fruit quality. Theor Appl Genet 95: 1069–1079.

Young ND, Tanksley SD (1989) RFLP analysis of the size of chromosomal segments retained around the *Tm-2* locus of tomato during backcross breeding. Theor Appl Genet 77: 353–359.

Young ND, Zamir D, Ganal MW, Tanksley SD (1988) Use of isogenic lines and simultaneous probing to identify DNA markers tightly linked to the *Tm-2a* gene in tomato. Genetics 120: 579–585.

Yousef GG, Juvik JA (2001) Evaluation of breeding utility of a chromosomal segment from *Lycopersicon chmielewskii* that enhances cultivated tomato soluble solids. Theor Appl Genet 103: 1022–1027.

Zhang Y, Stommel JR (2000) RAPD and AFLP tagging and mapping of *Beta* (*B*) and *Beta* modifier (*Mo_B*), two genes which influence beta-carotene accumulation in fruit of tomato (*Lycopersicon esculentum* Mill.). Theor Appl Genet 100: 368–375.

Zhang Y, Stommel JR (2001) Development of SCAR and CAPS markers linked to the *Beta* gene in tomato. Crop Sci 41: 1602–1608.

6

Molecular Mapping of Complex Traits in Tomato

Silvana Grandillo,[1], Pasquale Termolino[1] and Esther van der Knaap[2]*

ABSTRACT

Many important traits are influenced by multiple loci that each contributes large or small effects to the trait and that interact with each other and the environment. With the advent of molecular markers and genetic linkage maps, quantitative trait loci (QTLs) are mapped in a variety of populations in tomato (*Solanum lycopersicum*). The narrow genetic basis of cultivated tomato has forced tomato geneticists and breeders to also explore the wealth of genetic variation present in the exotic germplasm for the development of mapping populations and for crop improvement. The tomato clade (*Solanum* sect. *Lycopersicon*) and its four allied species (*S. ochrantum*, *S. juglandifolium*, *S. lycopersicoides* and *S. sitiens*) has proven to be a model system for QTL mapping, cloning and for the practical outcomes of QTL breeding based on the combined use of natural biodiversity and marker-assisted selection (MAS) approaches.

In this chapter, we describe the types of populations used for QTL mapping and fine mapping such as primary segregating populations (F_2 and BC_1), recombinant inbred lines (RILs), advanced backcross (AB) populations, backcross inbred lines (BILs), introgression lines (ILs) and near isogenic lines (NILs). Relevant molecular maps, markers, statistical methods and software are then described. With respect to application, basic principles of pyramiding multiple QTLs using MAS

[1] CNR–IGV Institute of Plant Genetics, Research Division Portici. Via Universita' 133, 80055 Portici (Naples) Italy.
[2] Department of Horticulture and Crop Science, The Ohio State University/Ohio Agricultural Research and Development Center, Wooster, OH 44691, USA.
*Corresponding author

and fine-mapping QTLs are presented, followed by a survey of each of the many tomato traits analyzed in these populations. Resistances to biotic stresses including viral, bacterial, fungal pathogens, and insect pests have all been important subjects of mapping studies. Research of abiotic stress tolerance has included cold, drought and salt. Numerous fruit traits influencing sensorial and nutritional quality such as fruit size, shape, color and firmness, as well as fruit primary and secondary metabolites are of critical importance to tomato improvement and have been extensively studied using QTL analysis. Finally, flowering and ripening time, flower morphology and reproductive barriers, leaf morphology, and seed weight QTL studies are reviewed. Given the advantages of using IL populations for the analysis of complex traits and their successful applications in tomato improvement, a particular attention is given to these genetic resources. Among others, we describe the use of the IL concept for studies of heterosis in tomato and related applications, for the resolution of integrated developmental networks, as well as for the application of integrated '-omics' tools aimed at the identification of candidate genes for key QTLs. The application of systems biology approaches combining transcriptomic, proteomic and metabolomic analyses along with the information derived from the tomato genome sequence will accelerate QTL cloning and continue to impact plant breeding methodologies.

Keywords: quantitative trait locus (QTL), mapping, introduction line (IL), quality, disease resistance, stress tolerance, morphology

6.1 Introduction

In crop plants many important traits such as yield, quality and stress tolerance show continuous variation, and result from the combined action of multiple, segregating genes interacting with each other and the environment. In early investigations, before molecular markers were available, the genetics of these complex traits was studied in general terms by 'quantitative genetics' using statistical techniques based on means, variances and covariances of relatives, rather than in terms of individual gene effects (Mather and Jinks 1982; Falconer 1989). These analyses provided general estimates including the approximate number of loci affecting a character, the average gene action (e.g., dominance, recessiveness) and the degree to which the various polygenes interacted with each other and the environment. However, they did not provide any information about the number and location of the genes underlying quantitative traits, termed polygenes by Mather (1949).

The development of molecular marker technology allowed the joint analysis of marker genotypes and phenotypic values in segregating populations. Thereafter it became possible to detect and locate the individual loci constituting polygenes, so-called quantitative trait loci (QTLs). The concepts for detecting QTLs or polygenes were developed more than 80 years ago when Sax (1923) reported the association of seed size (a quantitatively inherited trait) in beans with seed-coat pigmentation (a discrete monogenic trait). Subsequently, Thoday (1961) developed the idea of using single gene markers to systematically map and characterize individual polygenes controlling quantitative traits. However, the lack of appropriate genetic markers delayed the application of Thoday's idea until the early 1980s when isozyme markers started to be used as a general tool for mapping QTL in tomato (*Solanum lycopersicum*) (Tanksley et al. 1982; Vallejos and Tanksley 1983; Weller et al. 1988) and in maize (Edwards et al. 1987).

The first QTL mapping studies conducted in tomato mainly used morphological and isozyme markers along with primary segregating generations such as F_2 and backcross (e.g., BC_1) populations. Numerous quantitative traits were analyzed for fruit and plant characteristics. However, in all cases the number of informative isozyme markers was insufficient to cover the entire tomato genome. This made it not only difficult to estimate the precise position of the detected QTLs, but, in some cases, it was not even possible to determine their chromosomal positions (Tanksley et al. 1982; Vallejos and Tanksley 1983; Weller et al. 1988).

The introduction of DNA-based genetic markers, the first of which were restriction fragment length polymorphisms (RFLPs), overcame the problem of limited marker availability, and allowed the development of complete RFLP linkage maps for several organisms. In 1988 the first study was published in which an RFLP linkage map covering an entire genome (the tomato genome), was used to systematically map and characterize QTLs, thus demonstrating that quantitative traits could be resolved into Mendelian factors (Paterson et al. 1988). Since this pioneering study, QTL mapping in tomato has been applied to the analysis of numerous traits important to agriculture and biology, and for this purpose different populations and mapping strategies have been used. As will be illustrated in the following sections, during the past three decades tomato has proven to be an ideal system not only for the identification and map-based cloning of QTLs, but also for exploring the genetic bases of heterosis, and for the practical outcomes of QTL breeding based on the combined use of natural biodiversity and MAS approaches.

6.2 Populations used for QTL Mapping

QTL mapping assumes that alleles at marker loci will be in linkage disequilibrium (LD) (i.e., non-randomly associated) with alleles at physically linked QTLs (Tanksley 1993). Polymorphism at marker loci and in genes underlying the trait(s) of interest is an essential requirement for QTL mapping populations. For many self-pollinated crops such as tomato, the genetic variation within cultivated germplasm is very low (Miller and Tanksley 1990). Therefore, tomato QTL mapping studies have mostly involved interspecific crosses between cultivated germplasm and related wild species, although in a few cases *S. lycopersicum* intraspecific crosses have been successfully used (Tables 6-1–6-3; Causse et al. 2001; Saliba-Colombani et al. 2001; reviewed by Grandillo et al. 2011).

In regard to population structure, similar to other autogamous species, many QTL mapping studies conducted in tomato have used primary segregating populations such as F_2 or early backcross (BC) progenies. However, over time several other population structures have been developed and used for QTL analysis in tomato including recombinant inbred (RI) populations or lines (RILs), advanced backcross (AB) populations, backcross inbred lines (BILs) or inbred backcross (IBC) populations, and introgression lines (ILs).

To shorten the process of QTL identification, recent approaches making use of LD-based association analysis have been suggested. The method is based on the use of unrelated individuals with defined phenotypes, which are analyzed with many markers (Gupta et al. 2005). Chapter 4 provides an overview about the current status of molecular marker development and its potential use for the association mapping in tomato.

6.2.1 Primary Segregating Populations

F_2s and early backcross progenies (e.g., BC_1) offer the advantages of being easy and quick to develop, and also of having a high level of LD. Additional benefits of using F_2 over BC populations are the ability to detect recessive alleles and to obtain dominance estimates. These populations, however, suffer from several limitations: being transient populations, trait evaluation can be done only on a single plant basis, with no possibilty of replicating the trials over locations and time. In interspecific crosses, QTLs may not map accurately due to a high proportion of the wild species genome. With these populations estimation of epistasis is problematic, and mapped QTLs are not immediately available for application.

6.2.2 Recombinant Inbred Populations

RI populations are derived by inbreeding F_2 progeny by selfing or sibbing until they become homozygous lines. Due to the increased opportunity for meiotic recombination, RILs have relatively lower levels of LD. Because more recombination cycles have occurred, a lower number of QTLs are likely to be detected in a RIL relative to a primary segregating population

Table 6-1 Summary of disease (viral, bacterial, fungal) and insect resistance QTL studies conducted in tomato.

Type and source of resistance	Mapping population	No. plants, lines or families	Marker type[a]	No. markers[b]	No. QTLs	Chromosome	Ref
Viral							
Tomato yellow leaf curl virus (TYLCV)							
S. pimpinellifolium hirsute INRA	F_4 lines	11	RAPD	600	1	6	Chagué et al. 1997
Tomato mottle virus (ToMoV)							
S. chilense LA1932	F_2	na[c]	RAPD, MF	12, 2	2	6	Griffiths and Scott 2001
Bacterial							
Bacterial canker (*Clavibacter michiganensis* subsp. *michiganensis*)							
S. arcanum LA2157	three reciprocal BC_1	280, 199, 287	RFLP	73	5	1, 6–8, 10	Sandbrink et al. 1995
S. arcanum LA2157	F_2	324	RFLP	51	3	5, 7, 9	van Heusden et al. 1999
S. habrochaites LA0407	BILs (BC_2S_5), F_2 of selected BILs	64	RFLP, PCR-based	58, 5	2	2, 5	Kabelka et al. 2002; Coaker and Francis 2004
Bacterial spot (*Xanthomonas* sp.)							
S. lycopersicum cv. Hawaii 7998 (H7998)	BC_1	105	ISO, RFLP	11, 124	3	1 (two QTLs), 5	Yu et al. 1995
S. lycopersicum cv. Hawaii 7998 (H7998)	F_2, AB	na	SSR and SNP	37	2	1, 5	Yang et al. 2005

S. lycopersicum "cerasiforme" PI 114490	IBC	166	PCR-based	70	2	3, 11	Hutton et al. 2010
Bacterial wilt (*Ralstonia solanacearum*)							
S. lycopersicum "cerasiforme" (L285)	F_2	71	RFLP, RAPD	67, 12	3	6, 7, 10	Danesh et al. 1994
S. lycopersicum cv. Hawaii 7996 (H7996)	F_2	200	RFLP, RAPD, minisatellite	60, 13, 2	4	4 (two QTLs), 6, 11	Thoquet et al. 1996a
S. lycopersicum cv. Hawaii 7996 (H7996)	F_2, $F_{2:3}$	200	RFLP, RAPD, minisatellite	60, 13, 2	6	3, 4 (two QTLs), 6, 8, 10, 11	Thoquet et al. 1996b
S. lycopersicum cv. Hawaii 7996 (H7996)	F_3	3,500	RFLP	64	2	6 (two QTLs)	Mangin et al. 1999
S. lycopersicum cv. Hawaii 7996 (H7996)	$F_{2:3}$	200	RFLP, RAPD	76, 4	4 or 5	2, 6 (one or two QTLs), 8, 12	Wang et al. 2000
S. lycopersicum cv. Hawaii 7996 (H7996)	$F_{2:3}$, RILs (F_8)	189, 107[d]	RFLP	76	4	3, 4, 6, 8	Carmeille et al. 2006
Fungal							
Anthracnose (*Colletotrichum coccodes*)							
S. lycopersicum line 115-4	F_2	242	RAPD	42	several	several	Stommel and Zhang 2001
Black mold (*Alternaria alternata*)							
S. cheesmaniae LA0422	BC_1S_2 and BC_1S_3 lines from selected RILs	80, 151[e]	RFLP, PCR-based	na	5	2 (two QTLs), 3, 9, 12	Robert et al. 2001

Table 6-1 contd....

Table 6-1 contd....

Type and source of resistance	Mapping population	No. plants, lines or families	Marker type[a]	No. markers[b]	No. QTLs	Chromosome	Ref
Early blight (*Alternaria solani*)							
S. habrochaites PI 126445	BC_1, BC_1S_1	145	RFLP, RGA	141, 23	11, 13[d]	1–3, 5, 8–12	Foolad et al. 2002
S. habrochaites PI 126445	BC_1 (SGe)	820	RFLP, RGA	145, 34	7	3–6, 8, 10, 11	Zhang et al. 2003a
S. arcanum LA2157	F_2, $F_{2:3}$	176, 156[d]	SSR, SNP, AFLP	31, 14, 344	6	1, 2, 5–7, 9	Chaerani et al. 2007
Gray mold (*Botrytis cinerea*)							
S. habrochaites LYC4	F_2, BC_2S_1	174	AFLP, CAPS, SCAR	218, 51	3	1, 2, 4	Finkers et al. 2007a
S. habrochaites LYC4	ILs	30	AFLP, CAPS	457, 34	10	1–3, 4 (two QTLs), 6, 9 (two QTLs), 11, 12	Finkers et al. 2007b
S. lycopersicoides LA2951	ILs	58	CAPS, RFLP	100	7	1–5,9,11	Davis et al. 2009
Late blight (*Phytophthora infestans*)							
S. habrochaites LA2099	two reciprocal BC_1	213, 133[d]	RFLP	104, 98	15, 18[d]	all	Brouwer et al. 2004
S. habrochaites LA2099	NILs and sub-NILs for chr 4, 5, 11	400–500, 20–40[d]	RFLP, CAPS	8, 18, 14[f]	3	4, 5, 11	Brouwer and St Clair 2004
Powdery mildew (*Oidium lycopersici*)							
S. neorickii G1.1601	F_2, $F_{2:3}$	171	AFLP, PCR-based	318, 25	3	6, 12 (two QTLs)	Bai et al. 2003

Insect

2-tridecanone							
S. *habrochaites* LA0407	F_2	na	ISO, MF	na	5	na	Zamir et al. 1984
S. *habrochaites* PI 134417	F_2	74, 46[g]	RFLP	36–99	3	na	Nienhuis et al. 1987
Acylsugar level and composition							
S. *pennellii* LA0716	F_2	144	RFLP	150	5	2 (two QTLs), 3, 5, 11	Mutschler et al. 1996; Lawson et al. 1997
S. *pennellii* LA0716	F_2	231	RAPD, RFLP	na	6	2, 5–8, 12	Blauth et al. 1999

[a]ISO Isozyme, MF Morphological
[b]Per marker type or per population
[c]Not available
[d]Per population
[e]Successive field seasons
[f]Fine-mapped in NIL4, NIL5, and NIL11, respectively
[g]Two-phase screening

Table 6-2 Summary of abiotic stress tolerance QTL studies conducted in tomato.

Type and source of tolerance	Developmental stage/main specific traits[a]	Mapping population	No. plants, lines or families[b]	Marker type	No. markers[c]	No. traits or treatments	No. QTLs	Chromosome	Ref
Cold									
S. habrochaites	VG	BC$_1$	na[d]	ISO	11	1	3	6, 7, 12	Vallejos and Tanksley 1983
S. pimpinellifolium LA0722	SG	BC$_1$S$_1$	119	RFLP	151	1	3	1 (two QTLs), 4	Foolad et al. 1998b
S. habrochaites LA1778	VG/SW, RAU	BC$_1$	196	RFLP	89	7	10	1, 3, 5, 6 (three QTLs), 7, 9, 11, 12	Truco et al. 2000
Drought									
S. pennellii	VG/WUE	F$_3$, BC$_1$S$_1$	na	RFLP	17	1	3	Undetermined	Martin et al. 1989
S. pimpinellifolium LA0722	SG	BC$_1$S$_1$	30	RFLP	119	1	4	1, 8, 9, 12	Foolad et al. 2003a
S. pennellii LA0716	VG/WUE	ILs, sub-ILs of IL5-4	50, 43	STS, CAPS, AFLP, SSR	29	1	6	2, 3, 5, 7, 9, 12	Xu et al. 2008
Salt									
S. pennellii LA0716	VG/Na$^+$, Cl$^-$, K$^+$ accumulation	F$_2$	117	ISO	15	3	6	1, 2, 4–6, 12	Zamir and Tal 1987
S. pennellii LA0716	SG	F$_2$ (SGe)[e]	2,500, 1,700	ISO	16	2	5	1, 3, 7, 8, 12	Foolad and Jones 1993
S. pimpinellifolium L1	RS/TW, FN, FW	F$_2$	206	ISO, RAPD, RFLP	2, 2, 10	3	6	1–4, 10, 12	Bretó et al. 1994

S. pimpinellifolium L1	RS/TW, FN, FW	F$_2$	206	ISO, RFLP	2, 14	3	12	1–4, 10, 12	Monforte et al. 1996
S. pimpinellifolium (L1 and L5) and *S. cheesmaniae* L2	RS/TW, FN, FW, EA	three F$_2$	200, 150, 200	ISO, RFLP	3, 19	4	31	1–5,7, 9–12	Monforte et al. 1997a
S. pimpinellifolium (L1 and L5) and *S. cheesmaniae* L2	RS/TW, FN, FW, EA	three F$_2$	200, 150, 200	ISO, RFLP	3, 19	4/2	43	1–5, 7, 9–12	Monforte et al. 1997b
S. pennellii LA0716	SG	F$_2$ (SGe)	2,000	ISO, RFLP	16, 68	1	8	1–3, 7, 8, 9 (two QTLs), 12	Foolad et al. 1997
S. pennellii LA0716	SG	F$_2$ (SGe)	2,000	RAPD	53	1	8	1, 3, 5 (two QTLs), 6, 8, 9, unknown	Foolad and Chen 1998
S. pimpinellifolium LA0722	SG	BC$_1$S$_1$	119	RFLP	151	1	7	1 (two QTLs), 2, 5, 7, 9, 12	Foolad et al. 1998a
S. cheesmaniae L2	VG&RS/TW, FN, FW, EA, PHT, ID	F$_2$ and subpopulations	400	ISO, RFLP	3, 20	6	8	1, 2, 5, 7, 9, 12	Monforte et al. 1999
S. pimpinellifolium LA0722	VG	BC$_1$S$_1$	119	RFLP	151	1	5	1 (two QTLs), 3, 5, 9	Foolad and Chen 1999
S. pimpinellifolium LA0722	VG	BC$_1$ (SGe)	792	RFLP	115	1	5	1, 3, 5, 6, 11	Foolad et al. 2001
S. pimpinellifolium and *S. cheesmaniae* L2	VG&RS/TW, FN, FW, FRW, NFL, DTF, DRW, DFR, CI	two F$_7$-RILs	142, 116	RFLP, SSR, CG	153, 124	19	12, 23[f]	1–8, 10–12	Villalta et al. 2007
S. pimpinellifolium and *S. cheesmaniae* L2	VG/DSW, DLW, LA, K+ and Na+ concentration	two F$_8$-RILs	142, 116	RFLP, SSR, CG	153, 124	10	18, 25[b]	1, 3, 5–8, 11, 12	Villalta et al. 2008

Table 6-2 contd....

Table 6-2 contd....

Type and source of tolerance	Developmental stage/main specific traits[a]	Mapping population	No. plants, lines or families[b]	Marker type	No. markers[c]	No. traits or treatments	No. QTLs	Chromosome	Ref
S. cheesmaniae (L2) and *S. pimpinellifolium* (L5)	RS/FW, FN, TW, LNC, TN, LA, DLW	two F$_9$-RILs[g]	123, 100	RFLP, SSR, CG	153, 124	7	8	3, 9, 11	Estañ et al. 2009
S. pennellii LA0716	VG/Growth traits (PHT, STEM, LNO, DLW, DRW); Antioxidant content/activity AOX, PHE, FLA, SOD, CAT, APX, POX	ILs	52	na	na	12	125[h]	All	Frary et al. 2010

[a] AOX Antioxidant activity, APX Ascorbate peroxidase activity, CAT Catalase activity, DFR Days to first ripe fruit, DLW Dry leaf weight, DRW Dry root weight, DSW Dry stem weight, DTF Flowering time, EA Earliness, FLA Total flavonoid content, FN Fruit number, FRW Fresh root weight, FW Fruit weight, LA Leaf area, LNC Leaf sodium concentration, LNO Leaf number, NFL Number of flowers per inflorescence, PHE Total phenolic content, PHT Plant height, POX Gluthatione peroxidase activity, RAU Root ammonium uptake, RS Reproductive stage, SG Seed germination, SOD Superoxide dismutase, SW Shoot wilting, STEM Stem diameter, TN Transported sodium; TW Total fruit weight, VG Vegetative growth, WUE Water use efficiency
[b] Per population
[c] Per marker type
[d] Not available
[e] SGe = selective genotyping
[f] QTL detected for the six traits FW, FN, TW, Cl-, SF, NL, under both control and high salinity conditions
[g] both populations used as rootstocks
[h] Number of loci detected for antioxidant content under control and salt conditions

Table 6-3 Summary of plant, flower, fruit and yield QTL studies conducted in tomato.

Wild or donor parent	Main traits analyzed[a]	Mapping population	No. plants[b]	Marker type[c]	No. markers[d]	No. traits evaluated[e]	No. QTLs	Ref[f]
S. pennellii LA0716	FW, LFR, SE, SDW	BC$_1$	400	ISO	12	4	21	Tanksley et al. 1982
S. chmielewskii LA1028	B, FW, pH	BC$_1$	237	MF, ISO, RFLP	2, 5, 63	3	15	Paterson et al. 1988
S. pimpinellifolium CIAS27	AsA, B, DFL, DFR, FS, FW, HI, INL, LCN, LW, NDC, NFL, pH, PLH, PLZ, SDW, STW, SUG	F$_2$	1,700	MF, ISO	6, 4	18	85[g]	Weller et al. 1988
S. galapagense LA0483	B, FD, FW, pH	F$_2$	350	RFLP	71	4	29	Paterson et al. 1991
S. pennellii LA0716	BN, BRN, DFL, DTL, LFR, PLDW, PLFW, PLH, SNI, STD, TNI	F$_2$	432	RFLP	98	11	74	de Vicente and Tanksley 1993
S. chmielewskii	B, BY, CA, FC, FRU, FW, GLU, NFP, OA, pH, SUG, TS, Y	BILs (BC$_2$F$_5$)	64	RFLP	9	13	na[h]	Azanza et al. 1994
S. lycopersicum IVT KT$_1$ (breeding line containing S. pimpinellifolium and S. neorickii introgressions)	DFL, DFR, DFS, FW, LN, RT	F$_2$	690 (292)	RFLP	45	6	3	Lindhout et al. 1994
S. pennellii LA0716	B, BY, FW, PLW, PYG, Y	ILs/ILHs/ ILs x Tester	50	RFLP	375	6	104	Eshed and Zamir 1995; (Eshed and Zamir 1996; Eshed et al. 1996; Gur and Zamir 2004)

Table 6-3 contd....

Table 6-3 contd....

Wild or donor parent	Main traits analyzed[a]	Mapping population	No. plants[b]	Marker type[c]	No. markers[d]	No. traits evaluated[e]	No. QTLs	Ref[f]
S. galapagense LA0483	B, FW, SDW	F_8-RILs	97	MF, RFLP	2, 132	3	73[g]	Goldman et al. 1995
S. pimpinellifolium LA1589	ATI, ATL, ATW, B, DE, DFL, DFR, DTL, FC, FD, FS, FW, LCN, NFL, PLH, PT, RT, SDNF, SDW	BC_1	257	MF, RAPD, RFLP	2, 3, 115	19	54	Grandillo and Tanksley 1996a
S. pimpinellifolium LA1589	B, BYR, COV, DFR, FC, FER, FIR, FS, FW, GRO, MAT, pH, PUF, SCA, STR, VIS, Y, YR	$BC_2/$ BC_2F_1/BC_3	170	MF, RAPD, CAPS, RFLP	2, 3, 1, 115	21 (18)	87	Tanksley et al. 1996
S. habrochaites LA1777	BT, CI, FLZ, NFL, NFP, RL, SE, SI, UI	BC_1	149	RFLP	135	9	23	Bernacchi and Tanksley 1997
S. arcanum LA1708	B, BY, BYR, COV, DFR, ER, FC, FIR, FS, FW, FZ, GRO, HA, MAT, pH, PUF, SCA, SE, STR, VIS, Y, YG, YR	BC_3/BC_4	200	MF, CAPS, RFLP	1, 1, 205	35 (29)	166	Fulton et al. 1997
S. habrochaites LA1777	B, BYR, COV, FC, FIR, FS, FW, HA, MAT, NFL, pH, PUF, PYG, STR, VIS, Y, YR	BC_2/BC_3	200	RFLP	121	19	122	Bernacchi et al. 1998a
S. habrochaites LA1777 and *S. pimpinellifolium* LA1589	B, BY, COV, FC, FIR, FW, MAT, pH, STR, VIS, Y, YR	NILs	23	RFLP	na	12	22	Bernacchi et al. 1998b

S. pimpinellifolium LA0722	B, FD, FL, FS, FW, LYC, pH	BC_1/BC_1S_1	119	RFLP	151	7	59	Chen et al. 1999
S. pimpinellifolium LA1589 and S. pennelli LA0716 (IL2-5)	CF, FS	two F_2	82, 60	RFLP	82, 15	2	2	Ku et al. 1999
S. lycopersicum (line "Early cherry")	DFR, FW	F_2	76	RAPD	45	2	5	Doganlar et al. 2000a
S. neorickii LA2133	B, BYR, CAR, COV, ER, FC, FIR, FS, FW, HA, LYC, MAT, pH, PT, PUF, STR, STS, TA, TOA, VIS, Y, YR	BC_2/BC_3	170	MF, RFLP, PCR based	1, 131, 1	30	199	Fulton et al. 2000
S. lycopersicum "cerasiforme"	AV (18), B, CAR, ELA, FC, FD, FDW, FIR, FW, LYC, pH, SUG, TA	F_7-RILs	144	MF, RFLP, RAPD, AFLP	1, 84, 2, 16	32 (26)	81	Saliba-Colombani et al. 2001; (Causse et al 2002; Lecomte et al. 2004a; Chaib et al. 2006, 2007; Causse et al. 2007; Zanor et al. 2009)
S. lycopersicum "cerasiforme"	SA (12)	F_7-RILs	144	MF, RFLP, RAPD, AFLP	1, 84, 2, 16	12	49	Causse et al. 2001; (Causse et al. 2002; Bertin et al. 2003; Lecomte et al. 2004a; Chaib et al. 2006, 2007; Causse et al. 2007; Bertin et al. 2009; Zanor et al. 2009)
S. pimpinellifolium LA1589	FD, FL, FS, FW, LCN, SDNF, SDW	F_2	200 (114)	RFLP, CAPS	89, 1	7	30	Lippman and Tanksley 2001
S. pimpinellifolium LA1589	FS, OS	F_2	100	RFLP, SNP	108, 2	2	1	van der Knaap and Tanksley 2001

Table 6-3 contd....

Table 6-3 contd....

Wild or donor parent	Main traits analyzed[a]	Mapping population	No. plants[b]	Marker type[c]	No. markers[d]	No. traits evaluated[e]	No. QTLs	Ref[f]
S. habrochaites LA0407	STVM	BILs (BC$_2$S$_5$), F$_{2:3}$	64	RFLP and PCR-based	67	5	1	Coaker et al. 2002
S. pimpinellifolium LA1589	B, DFL, DFR, FC, FCR, FD, FER, FIR, FL, FS, FW, GRO, MAT, NFL, PUF, SDNF, SDW, STS	BILs (BC$_2$F$_6$)	196	MF, RFLP	1,126	22	71	Doganlar et al. 2002
S. habrochaites LA1777, *S. arcanum* LA1708, *S. neorickii* LA2133, *S. pimpinellifolium* LA0722	B, CA, FRT, FRU, GA, GLU, MA, pH, SA, SUC, TA	four AB	See reference	MF, RAPD, CAPS, RFLP	(See references)	16	222	Fulton et al. 2002a
S. pimpinellifolium LA1237 ("selfer") and LA1581 ("outcrosser")	FLM, NFL	F$_2$	147	RFLP	48	6 (4)	5	Georgiady et al. 2002
S. pimpinellifolium LA1589	EI, FS, PEI	F$_2$	85	RFLP	97	3	4	van der Knaap et al. 2002
S. pennellii LA0716	LM, LZ	ILs	58	RFLP	375	8	30	Holtan and Hake 2003
S. pennellii LA0716	CAR, FC, LYC	ILs	75	RFLP including CG	375	6	50	Liu et al. 2003b

S. pimpinellifolium LA1589	FS and FS-related traits, FW, LCN, NFL, SDNF	F_2	200	RFLP	93	10	50	van der Knaap and Tanksley 2003
S. pennellii LA0716	LCN	Several	See reference	See reference	See reference	2	4	Barrero and Tanksley 2004
S. pennellii LA0716	B, CA, FRU, FW, GLU, MA, pH, SUC, TA	ILs	75	RFLP, CG	375, 107	9	81	Causse et al. 2004
S. pennellii LA1657	B, BYR, ER, FC, FCR, FER, FIR, FS, FW, FZ, pH, PT, PUF, PYG, STR, VIS, Y, YR	BC_2/BC_2F_1	175	RFLP	150	25	84	Frary et al. 2004a
S. pennellii LA0716	LM, PTM, SPM	F_2	83	RFLP, SSR, COS	350, 10, 31	22 (18)	36	Frary et al. 2004b
S. habrochaites LA0407	FC	BILs (BC_2S_5), F_3, F_4	64	RFLP, PCR-based	58, 5	3	13	Kabelka et al. 2004
S. pennellii LA0716	B, FRU, FW, GLU, STA, SUC, Transcriptomic analysis	ILs (IL1-4, IL2-6, IL7-3, IL7-5, IL4-4, IL12-12)	6	RFLP	na	6	na	Baxter et al. 2005
S. habrochaites LA1777	HYI	ILs, BILs	71	RFLP	76	4	22	Moyle and Graham 2005
S. pennellii LA0716	AsA, CAR, LYC, PHE, TACW	Ils	76	RFLP	375	5	20	Rousseaux et al. 2005

Table 6-3 contd....

Table 6-3 contd....

Wild or donor parent	Main traits analyzed[a]	Mapping population	No. plants[b]	Marker type[c]	No. markers[d]	No. traits evaluated[e]	No. QTLs	Ref[f]
S. pennelli LA0716	PM (74) including: AsA, CA, FRU, GLU, SUC, Yield-related: B, FD, FL, FW, HI, PLW, SDNF, SDNP, Y	ILs	76	RFLP	375	83	889, 326[i]	Schauer et al. 2006
S. pennelli LA0716	ATI, ATL, ATW, B, BM, BY, FC, FD, FL, FN, FS, FW, NFL, NFLP, HI, OL, OS, OW, PLW, PT, SDL, SDNF, SDNP, SDS, SDW, SDWF, SDWI, SDWP, STIL, STIS, STIW, Y	ILs, ILHs	76	RFLP	375	35	841	Semel et al. 2006
S. pennellii	AV (23), CA, MA	ILs	74	RFLP	375	25 (24)	29	Tieman et al. 2006
S. pimpinellifolium LA1589	FS and FS-related attributes (14)	two F_2, BC_1	99, 130, 100	RFLP and PCR-based	111, 111, 108	15	36, 32, 27[l]	Brewer et al. 2007
S. pennellii LA0716 *S. habrochaites* PI24 *S. lycopersicum* "cerasiforme"	AsA, FDW, FW, SUC, TA	ILs, BC_2S_1, RILs	75, 130, 144	RFLP, AFLP, SSR	36, 138, 26	5	23	Stevens et al. 2007; (Stevens et al. 2008)
S. chmielewskii CH6047	DFL, LN	F_2	149	CAPS & SCAR, CG, SSR, AFLP	80, 11, 38, 136	2	8	Jiménez-Gómez et al. 2007

S. pimpinellifolium LA1589	FS and FS-related traits	three F$_2$	130, 106, 94	RFLP and PCR-based	111, 96, 97	14	20, 23, 20i	Gonzalo and van der Knaap 2008
S. habrochaites LYC4 (IL5-1 and IL5-2 lines) and S. habrochaites (IVT-line 1)	PAT, SE	two BC$_5$S$_1$, F$_2$	174, 183, 160	CAPS, COS, SSR	34	2	5	Gorguet et al. 2008
S. habrochaites LA1777	AV	ILs, BILs	89	RFLP	76	40 (27)	30	Mathieu et al. 2008
S. pennellii LA0716	HYI	ILs	75	RFLP	375	4	19	Moyle and Nakazato 2008
S. pennellii LA0716	PM (74) including: AsA, CA, FRU, GLU, SUC	ILs, ILHs	76	RFLP	375	74	332	Schauer et al. 2008
S. chmielewskii LA1840	CNP, CUTC, CZP, DMCP, DWP, FCR, FW, H4T, LN, LTA, LW, SDNF, SUC (two different fruit load conditions were tested)	ILs	20	COSII, SSR	130, 3	16 (14)	103	Prudent et al. 2009
S. chmielewskii LA1840	Metabolites, (two different fruit load conditions were tested)	ILs	23	COSII, SSR	130, 3	62	240 (HL), 128 (LL)m	Do et al. 2010

[a]**AsA** Ascorbic acid content, **ATI** Anther tube index (ATL/ATW), **ATL** Anther tube length, **ATW** Anther tube width, **AV** Aroma volatiles, **B** Soluble solids content (°Brix), **BM** Biomass, **BN** Total number of flower buds, **BT** Flower bud type, **BRN** Number of well developed branches, **BY** Brix x Total Yield, **BYR** Brix x Red Yield, **CA** Citric acid content, **CAR** β-carotene content, **CF** Constriction of the fruit, **CI** Corolla indentation, **CNP** Cell number of the pericarp, **COV** Cover, **CUTC** Cuticular conductance, **CZP** Cell size of the pericarp, **DE** Days to seedling emergence,

Table 6-3 contd....

Table 6-3 contd....

DFL Days to first flower (or differently measured), **DFR** Days to first ripe fruit (differently measured), **DFS** Days to first set fruit, **DMCP** Dry matter content of the pericarp, **DTL** Days to first true leaf, **DWP** Dry weight of the pericarp, **EI** Eccentricity index, **ELA** Elasticity, **ER** Epidermal reticulation, **FC** Fruit color (measured in diferent ways), **FCR** Fruit cracking, **FD** Fruit equatorial diameter, **FDW** Total dry matter weight of the fruit, **FER** Fertility, **FIR** Firmness, **FL** Fruit polar diameter, **FLM** Flower morphology, **FLZ** Flower size, **FN** Fruit number, **FRT** Fruitiness (measured by trained panel), **FRU** Fructose content, **FS** Fruit shape index (FL/FD), **FW** Fruit weight or fruit mass, **FZ** Fruit size, **GA** Glutamic acid, **GLU** Glucose content, **GRO** Growth, **H4T** Height of the 4th truss, **HA** Horticultural acceptability (differently measured), **HI** Harvest index, **HYI** Hybrid incompatibility, **INL** Internode length, **LCN** Number of locules, **LFR** Leaflet width/length ratio, **LM** Leaf morphology, **LN** Number of leaves to the first cluster, **LTA** Leaf total area, **LW** Leaf weight, **LYC** Lycopene content, **LZ** Leaf size, **MA** Malic acid content, **MAT** Maturity, **NDC** Number of nodes to the first cluster, **NFL** Number of flowers per cluster, **NFLP** Number of flowers per plant, **NFP** Number of fruits per plant, **OA** Organic acids, **OL** Ovary length, **OS** Ovary shape index (OL/OW), **OW** Ovary width, **PAT** Parthenocarpy, **PEI** Pericarp elongation index, **pH** Fruit pH, **PHE** Phenolics, **PLDW** Total dry weight of aerial portion of the plant, **PLFW** Total fresh weight of aerial portion of the plant, **PLH** Plant height (or differently measured), **PLW** Plant weight, **PLZ** Plant size, **PM** Primary metabolites, **PT** Pericarp thickness, **PTM** Petal morphology, **PUF** Puffiness, **PYG** Percent green yield, **RL** Inflorescence rachis length, **RT** Ripening time (or differently measured), **SA** Sensory attributes (measured by trained panel), **SCA** Sun scald, **SDL** Seed length, **SDNF** Numer of seeds per fruit, **SDNP** Number of seeds per plant, **SDS** Seed shape index (SDL/SDWI), **SDW** Seed weight, **SDWF** Seed weight per fruit, **SDWI** Seed width, **SDWP** Seed weight per plant, **SE** Stigma exsertion, **SI** Self incompatibility, **SNI** Number of internodes on the primary stem, **SPM** Sepal morphology, **STA** Starch content, **STD** Stem diameter at the first internode, **STIL** Stile length, **STIS** Stile shape index (STIL/STIW), **STIW** Stile width, **STR** Stem retention or stem release, **STS** Stem scar, **STVM** Stem vascular morphology, **STW** Stem width, **SUC** Sucrose content, **SUG** Sugar content, **TA** Total or titratable acidity, **TACW** Total antioxidant capacity, **TNI** Total number of internodes, **TOA** Total organic acids, **TS** Total solids, **UI** Unilateral incompatibility, **VIS** Viscosity, **Y** Total yield, **YG** Green yield, **YR** Red yield

[b]Number of plants selected for genotyping are indicated in parentheses

[c]CG Candidate genes, **ISO** Isozyme, **MF** Morphological, **COS** Conserved ortholog set (Fulton et al. 2002b), **COSII** Conserved ortholog set II (Wu et al. 2006)

[d]Per marker type or per population

[e]Number of traits for which QTL were identified are indicated in parentheses

[f]Related follow-up studies are indicated in parentheses

[g]Number of significant marker × trait associations

[h]Not available

[i]Fruit metabolism and yield-associated traits, respectively

[l]Per population

[m]HL = high load; LL = Low load

with equivalent marker density. A major advantage of RILs is their virtual homozygosity, which allows replication of experiments over time and locations. Moreover, with RI populations it is possible to obtain more precise estimates of the total phenotypic variance (Goldman et al. 1995). On the other hand, the development of RILs takes a relatively long time and can be a very difficult process when interspecific crosses are involved. In RI populations, both parental alleles are represented at 50% frequencies on average. The presence of a high percentage of the wild species genome in interspecific crosses can hinder the measurement of some quantitative traits. Compared to an F_2, RILs have increased statistical power in detecting additive QTLs; however, when all lines are homozygous it is no longer possible to detect non-additive QTLs.

In tomato, RI populations have been developed from interspecific crosses between the domesticated *S. lycopersicum* and the wild species *Solanum cheesmaniae* (Villalta et al. 2007, 2008), *Solanum galapagense* (Paran et al. 1995, 1997), *Solanum pennellii* (Foolad 2007), and *Solanum pimpinellifolium* (Graham et al. 2004; Foolad 2007; Villalta et al. 2007, 2008). These populations have been used to identify QTLs for fruit quality related traits as well as for abiotic stress tolerance (Goldman et al. 1995; Paran et al. 1997; Villalta et al. 2007, 2008). Moreover, an intraspecific RI population was developed from the cross between a small-fruited *S. lycopersicum* "cerasiforme" (cherry tomato) (Cervil) characterized by good taste and intense aroma, and a large-fruited *S. lycopersicum* line with a common taste (Levovil) (Saliba-Colombani et al. 2001). This population was used to identify QTLs for sensorial quality in fresh-market tomato, including physico-chemical characteristics, sensory traits and 12 aroma volatiles (Causse et al. 2001, 2002; Saliba-Colombani et al. 2001), and for vitamin C (ascorbic acid, AsA) content (Stevens et al. 2007).

6.2.3 Advanced Backcross Populations

AB populations are another type of segregating population that has been used for tomato QTL mapping. The use of AB populations was proposed by Tanksley and Nelson (1996) as a new breeding method designated 'Advanced backcross QTL analysis (AB-QTL)', with the purpose of integrating the process of QTL discovery with the development of improved varieties, by simultaneously identifying and transferring useful QTL alleles from unadapted (e.g., landraces, wild species) to elite germplasm. A cross is made between an elite cultivated recurrent parent and and an unadapted donor parent. Backcrosses to the recurrent parent are carried out before subsequent analyses of markers and traits (e.g., in BC_2 or BC_3). This results in a more uniform genome with increased similarity to the recurrent parent, facilitating trait evaluation. In addition, QTLs that rely on epistasis among donor parent alleles are less likely to be detected. The genetic background will facilitate discovery of additive QTLs, which will tend to behave as predicted when transferred into the cultivar. Development of near-isogenic lines (NILs) from AB populations can be accomplished using a small number of generations through employing markers. This permits favorable, wild QTL alleles to be quickly exploited via field testing, confirmation of QTL effects, and subsequent transfer into varieties.

In tomato, the AB-QTL analysis has been initially applied to five interspecific crosses generated using the same *S. lycopersicum* parent (E6203) and the following five wild species: *S. pimpinellifolium* (LA1589) (Tanksley et al. 1996), *S. arcanum* (LA1708) (Fulton et al. 1997), *S. habrochaites* (LA1777) (Bernacchi et al. 1998a, b), *S. neorickii* (LA2133) (Fulton et al. 2000), and *S. pennellii* (LA1657) (Frary et al. 2004a). These populations have been analyzed for numerous horticultural traits important for the tomato processing industry including total and red yield, fruit size and color, soluble solids content or brix, viscosity, firmness, and fruit pH (Table 6-3). The field trials were replicated in several locations worldwide, and wild QTL alleles with favorable effects were detected for more than 45% of traits evaluated across the five AB populations (Grandillo et al. 2008). Moreover, the first four AB-QTL populations have also been used to identify QTLs influencing sugars, organic acids and other biochemical properties possibly contributing to flavor (Fulton et al. 2002a). More recently, the AB-QTL method has been extended to the *S. chilense* accession LA1932, to explore the potentials of this wild relative as a source of useful QTL alleles for yield-related and fruit quality traits (Termolino et al. 2010). Results from this study have demonstrated that *S. chilense* also contains alleles capable of improving several traits of economic importance for processing tomatoes including brix, firmness and viscosity.

6.2.4 Backcross Inbred Lines

BILs represent another type of unbalanced population that has been used for QTL mapping in tomato. BILs are developed as follows—an F_1 population from two parental lines is backcrossed at least once to the recurrent parent; single-seed descent is then used to increase homozygosity within and heterogeneity among multiple lines. BIL populations have the major advantage of being permanent mapping resources that can be tested in replicated trials over time and locations. BILs can be a useful tool for breeding purposes as they may carry a relatively low number of introgressions from the donor parent. BILs have been developd from the AB-QTL mapping populations involving *S. pimpinellifolium* LA1589 (196 BC_2S_6 lines) (Doganlar et al. 2002; Brog et al. 2011), *S. habrochaites* LA1777 (Monforte and Tanksley 2000a), and *S. neorickii* LA2133 (Di Dato et al. 2010), from interspecific crosses involving *S. habrochaites* LA0407 (Kabelka et al. 2002), and from a high soluble solids introgression line LA1501 carrying segments introgressed from *Solanum chmielewskii* LA1028 (Azanza et al. 1994). BILs have been used for mapping QTLs for many agronomically important traits in tomato, including yield, fruit quality (Azanza et al. 1994; Doganlar et al. 2002; Di Dato et al. 2010; Tripodi et al. 2010; Brog et al. 2011) and disease resistance (Kabelka et al. 2002).

6.2.5 Introgression Lines

ILs are permanent lines that have been developed and widely used for QTL mapping studies in tomato. Unlike BILs that may retain multiple introgressions from the donor parent, each IL theoretically carries a single, possibly homozygous, marker-defined chromosomal segment originating from the donor parent, in an otherwise homogeneous recurrent parent genetic background. The entire donor genome can be represented in a set of ILs that are referred to as an IL library or an IL population. The map resolution of an IL population depends on the number, length, and overlap of adjacent introgressions, which define the so-called 'bins' (Pan et al. 2000; Zamir 2001). Comparisons among lines are used to assign genes or QTLs to specific bins. ILs allow one to focus on a specific region of interest for fine-mapping QTL (Paterson et al. 1990), or to screen for QTLs across entire genomes (Eshed and Zamir 1995; Zamir 2001; Lippman et al. 2007; Grandillo et al. 2008). Given the advantages of using IL populations for the analysis of complex traits and their numerous applications in tomato improvement, they are described in detail below (see Sect. 6.5).

6.3 Framework Maps and Markers

After the demonstration by Paterson et al. (1988) that RFLP linkage maps covering the entire tomato genome could be efficiently used to resolve quantitative traits into discrete Mendelian factors, RFLP mapping of complex traits in tomato was widely adopted. Most of the QTL mapping studies conducted in tomato have been based on RFLP markers (Tables 6-1–6-3; also see Chapter 4 and review by Grandillo et al. 2011). However, in some cases, RFLP maps have been integrated with PCR-based markers such as random amplified polymorphic DNAs (RAPDs), sequence characterized amplified regions (SCARs), simple sequence repeats (SSRs), expressed sequence tags (ESTs), amplified fragment length polymorphisms (AFLPs), cleaved amplified polymorphic sequences (CAPSs), and resistance gene analogs (RGAs) (Grandillo and Tanksley 1996a, b; Fulton et al. 2000; Lippman and Tanksley 2001; Saliba-Colombani et al. 2001; Zhang et al. 2003a; Brewer et al. 2007; Gonzalo and van der Knaap 2008). In other cases, the linkage maps were developed exclusively using PCR-based markers including single nucleotide polymorphisms (SNPs) (Stommel and Zhang 2001; Bai et al. 2003; Chaerani et al. 2007).

Efforts are being invested in developing maps with PCR-based markers, as they facilitate the application of molecular markers in tomato genetics and breeding. COSII markers, which are PCR-based markers developed from a set of single-copy conserved orthologous genes (COSII genes) in Asterid species (http://solgenomics.net/markers/cosii_markers.pl), have also been developed in tomato (Fulton et al. 2002b; Wu et al. 2006). Each COSII gene (representing a homologous set of Asterid unigenes) matches only one single-copy *Arabidopsis* gene, and has been annotated based on *Arabidopsis* gene ontology (GO). COSII markers and their corresponding universal primers are highly valuable for phylogenetic analyses as well as studies of synteny, genome evolution and genome organization.

Within the framework of the EU-SOL project (http://www.eu-sol. net/), a shared framework of COSII markers has been applied to anchor a diverse set of germplasm to the tomato high density molecular linkage map. Populations include ILs, BILs, and RILs derived from interspecific crosses of tomato and the six wild accessions *S. pennellii* LA0716, *S. habrochaites* LA1777, *S. neorickii* LA2133, *S. chmielweskii* LA1840, *S. galapagense* LA0483 and *S. pimpinellifolium* LA1589), and a diploid mapping population of potato (Tripodi et al. 2010; Brog et al. 2011). This will facilitate comparisons between functional maps of tomato and potato, QTL identification, and additional mapping and isolation of genes underlying target traits. It will also accelerate the rate of introgression breeding in tomato, and the exploitation of novel alleles using PCR-based markers.

6.4 QTL Mapping Approaches and Software

The basis for QTL identification is the statistical analysis of associations between the quantitative trait and the marker alleles segregating in a population. Various statistical methods can be used to determine these associations, and they are based on three broad classes: regression (Haley and Knott 1992; Whittaker et al. 1996), maximum-likelihood (Doerge et al. 1997), and Bayesian models (Sillanpää and Corander 2002).

An association between a single genetic marker and a phenotypic trait can be analyzed using 'single point analysis', also known as 'point analysis' (Tanksley 1993). This is the simplest method of QTL analysis. An estimate of the phenotypic effect is obtained by comparing mean phenotypes for different marker groups. Statistical significance is tested using standard linear regression (i.e., one-way ANOVA), assuming normal distribution of residual environmental variance. This type of analysis does not require a complete molecular linkage map; this is why single point analysis was used in the first QTL mapping studies conducted in tomato. However, the approach was also used in subsequent studies on AB populations or in conjunction with other statistical approaches (Fulton et al. 1997, 2000; Foolad et al. 1998a, 1999). This method has a major drawback in that accurate measures of QTL effects are possible only when they occur at marker loci. Phenotypic effects of loci not tightly linked to markers will be underestimated in proportion to the degree of recombination between marker and QTL. Futhermore, small phenotypic effects cannot be distinguished from distant linkage (Soller et al. 1976; Doerge et al. 1997). These problems can be minimized by increasing the number of segregating markers, preferably covering the genome at a distribution of one marker every 15 cM.

To overcome some of the limitations associated with single point analysis, interval mapping (IM) was developed (Lander and Botstein 1989). This method utilizes molecular linkage maps with broad genome coverage. QTL detection is based on maximum-likelihood parameter estimation and LOD scores (the log of the ratio of the likelihoods of there being one versus no QTL at a particular point). Recombination between markers and QTL are taken into account by simultaneously analyzing the effects of linked markers on quantitative traits. As a consequence, the chances of statistically detecting QTLs are greater, and the estimate of the QTL effect on the character is unbiased. QTL likelihood plots and estimates of the proportion of phenotypic variation explained (PVE) can be generated using the software MAPMAKER-QTL (Lincoln et al. 1992). The likely position of a QTL is indicated by the maximum LOD values in the profile, which exceed a pre-determined threshold. Because multiple tests are conducted, significance levels are adjusted accordingly. Map intervals corresponding to LOD = 1.0 decline along both sides of a QTL peak, and confidence intervals (CI) are set accordingly. Moreover, the multi-locus model of the program allows estimatation of the PVE by various combinations of significant QTLs. PVE values for IM will be approximately equal to R^2 values for single-point analysis when a QTL is tightly linked to a specific marker.

A limitation to the IM method is that multiple, linked QTL effects cannot be distinguished from each other (Zeng 1994). Linkage among QTLs can lead to biased estimates of both their locations and effects. A method that combined IM with multiple marker regression analysis, namely, composite interval mapping (CIM), was developed to address this issue (Jansen 1994; Zeng 1994). QTLs located on other intervals or chromosomes do not affect the statistical significance of the interval test. CIM using the software QTL Cartographer (Basten et al. 1997; Wang et al. 2005b) is among the most frequently applied QTL mapping method for populations derived from biparental crosses. However, even though multiple-QTL methods such as CIM can reduce the confidence interval of QTL locations and resolve two or more linked QTLs, the efficacy of these methods depends on whether the markers are evenly distributed on the map. Moreover, the localization of QTLs linked within less than 20 cM can still be inaccurate and fine-mapping approaches are envisaged as the only way to estimate the actual number of QTLs (Causse et al. 2001).

Another approach that has been proposed to identify marker-linked QTLs is 'distributional extreme analysis' or 'selective genotyping' (Lander and Botstein 1989; Tanksley 1993). With this method a larger population is evaluated for the phenotypic trait(s) of interest, but the genotyping is performed only on selected individuals belonging to the extreme tails of the distribution. The association of a QTL with a genetic marker is inferred when the allele frequency at any marker locus differs significantly between the extreme phenotypic classes. For analyzing a single quantitative trait this method remains valuable. Distributional extreme analyisis has been applied successfully in tomato to identify genetic markers associated with traits such as tolerance or resistance to abiotic and biotic stresses (Foolad and Jones 1993; Foolad et al. 1997, 2001; Foolad and Chen 1998; Zhang et al. 2003b).

The third class of statistical methods that have been widely explored for QTL mapping is Bayesian models (Bogdan et al. 2004; Wang et al. 2005a). However, partially due to the difficulty of choosing prior distributions, complexity of computation, and lack of user-friendly software, so far, their practical application has been limited (Li et al. 2007).

Other QTL mapping software programs that have been used in tomato include QGene (Nelson 1997) and MapQTL® (Van Ooijen et al. 2002; Chaerani et al. 2007). QGene 4.3.10 is an open-source Java program that implements either QTL, marker or trait analysis in a highly intuitive way (http://www.qgene.org/qgene/). This program offers most of the conventional QTL mapping methods including single-marker regression, IM, CIM, inclusive composite interval mapping (ICIM) (Li et al. 2007) and Bayesian interval mapping (BIM); it also allows their side-by-side comparisons. MapQTL® (Van Ooijen et al. 2002) allows application of the IM procedure as well as a multiple-QTL model (MQM) mapping procedure.

6.5 Introgression Line Mapping and Breeding Approach

QTL mapping studies have provided direct evidence for the genetic basis of transgressive variation, which is defined as the occurrence in segregating populations of individuals with phenotypes more extreme than those of both parents (Tanksley 1993). Tomato provides a good example in this respect, as many QTL mapping studies conducted for this crop reported the identification of QTLs with allelic effects opposite to those predicted based on the phenotypes of the parental lines of the cross (reviewed by Grandillo et al. 2008). The accumulation of such 'complementary QTLs' in certain progeny can explain the observed transgressive segregation. Interestingly, QTL mapping studies conducted using interspecific crosses have shown that, in spite of its general inferior phenotype, unadapted or exotic germplasm can represent a rich source of agronomically favorable complementary QTL alleles (Weller et al. 1988; de Vicente and Tanksley 1993; Eshed and Zamir 1995; Tanksley et al. 1996; Tanksley and McCouch 1997; Zamir 2001; Grandillo et al. 2008, 2011).

Considering the properties of IL populations and the potential of exotic germplasm as a source of untapped genetic variation, Zamir (2001) suggested to invest in the development of 'exotic libraries' in order to enhance the rate of introgression breeding. Each exotic library consists of a set of ILs originating from a cross between a recurrent cultivated parent and a donor exotic parent. For tomato, the first exotic library, covering the entire wild parent genome, was developd by Eshed and Zamir (1994). It consisted of 50 lines, each containing a single introgression from the self compatible (SC) green-fruited accession *S. pennellii* LA0716 in the genetic background of the processing tomato cultivar *S. lycopersicum* M-82. This population was used to identify QTLs for yield-associated traits including total yield, fruit weight and yield × brix, and to study their epistatic and environmental interactions (Eshed and Zamir 1995, 1996; Eshed et al. 1996). These studies highlighted the properties that make exotic libraries an efficient tool for the analysis of complex traits (Zamir 2001; Lippman et al. 2007; Grandillo et al. 2008). Since each IL in an exotic library contains a single introgression from the wild donor parent, their phenotype will generally resemble that of the cultivated recurrent parent; this reduces sterility problems and allows the lines to be measured for yield-related traits. Additional advantages of interspecific as well as intraspecific IL populations are that QTL detection is achieved using a very simple statistical procedure that compares phenotypes of each IL to the recurrent parent. One or more QTLs are detected on the introgressed segment when a significant difference for a particular trait is observed. Moreover, since all the phenotypic variation for the trait of interest is associated with the introgressed segment, the masking effects of major QTLs segregating in other regions of the genome are eliminated, thereby the ability of statistically identifying minor QTLs is increased. This explains why a relatively greater number of QTLs is oftentimes found when using IL populations versus other types of mapping populations (Eshed and Zamir 1995). The permanent nature of IL populations allows phenotypic measurements to be conducted on multiple replicates, thereby reducing environmental effects and increasing statistical power. Moreover, by replicating the trials in additional locations and over time, it becomes possible to estimate QTL × environment interactions (Eshed et al. 1996; Monforte et al. 2001; Liu et al. 2003b; Gur and Zamir 2004; Rousseaux et al. 2005).

In general, the initial map resolution of an IL population is relatively low. However, each IL can be used as the starting point for high-resolution mapping (see Sect. 6.8). Tight linkage of multiple QTLs affecting one or more trait(s) can be discerned from pleiotropy via high-resolution mapping (Eshed and Zamir 1996; Monforte and Tanksley 2000b; Monforte et al. 2001; Fridman et al. 2002; Frary et al. 2003; Chen and Tanksley 2004; Lecomte et al. 2004b; Stevens et al. 2008). Moreover, the identification of molecular markers more tightly linked to the QTL of interest is the basis for marker-assisted transferring of individual or a combination of QTLs into the desired genetic background.

ILs or QTL-NILs can also be used to more precisely estimate epistatic interactions (Eshed and Zamir 1996; Causse et al. 2007), QTL × genotype interactions (see Sect. 6.6) (Eshed and Zamir 1995; Eshed et al. 1996; Monforte et al. 2001; Gur and Zamir 2004; Lecomte et al. 2004a; Chaïb et al. 2006), and for the positional cloning of key genes underlying quantitative traits (see Sect. 6.8) (Alpert and Tanksley 1996; Frary et al. 2000; Fridman et al. 2000, 2004; Liu et al. 2002; Chen et al. 2007; Cong et al. 2008).

The *S. pennellii* (LA0716) IL population presently consists of 76 lines, which cover the entire tomato genome delimiting 107 bins, each with an average length of 12 cM (Liu et al. 1999; Pan et al. 2000; Liu et al. 2003b; http://solgenomics.net/). As part of the EU-SOL project, this IL library has been expanded through the addition of >400 sub-ILs; this will significantly improve mapping resolution (Lippman et al. 2007). The *S. pennellii* ILs have been phenotyped for hundreds of traits allowing the identification of >2,700 QTLs (Lippman et al. 2007) (Tables 6-1–6-3). Repeated measurements have been conducted for 20 yield-associated, fruit morphology and biochemical traits, and the resulting data are available in silico through the search engine 'Real Time QTL' (http://zamir.sgn.cornell.edu; Gur et al. 2004). Some of the analyzed quantitative traits include: fruit weight, fruit shape, soluble solids content, pH and yield (Eshed and Zamir 1994, 1995, 1996; Eshed et al. 1996; Monforte et al. 2001; Causse et al. 2004; Gur and Zamir 2004; Semel et al. 2006; Gur et al. 2011), leaf morphology (Holtan and Hake 2003), harvest index and earliness (Gur et al. 2010), carotenoid content in relation to fruit color (Liu et al. 2003b), fruit nutritional and antioxidant content (Rousseaux et al. 2005; Schauer et al. 2006; Stevens et al. 2007; Almeida et al. 2011), fruit primary metabolites (Causse et al. 2004; Baxter et al. 2005; Schauer et al. 2006, 2008; Gur et al. 2010), aroma compounds (Tadmor et al. 2002; Tieman et al. 2006), water use efficiency (WUE) (Xu et al. 2008), salt tolerance and antioxidant response (Frary et al. 2010), trichome specialized metabolites (Schilmiller et al. 2010), as well as enzyme activity (Steinhauser et al. 2011) and postzygotic interspecific barriers (Moyle and Nakazato 2008).

The *S. pennellii* IL population has been used to explore the potential of the 'candidate gene approach' to identify candidate genes for QTLs influencing the intensity of tomato fruit color (Liu et al. 2003b), tomato fruit size and composition (Causse et al. 2004), as well as fruit AsA content (Stevens et al. 2008). While no co-locations were initially found between candidate genes and fruit color QTLs (Liu et al. 2003b), Causse et al. (2004) and Stevens et al. (2008) observed several apparent links. The *S. pennelii* ILs have also been used to find associations between trascriptomic changes and fruit composition (Baxter et al. 2005; Di Matteo et al. 2010; Lee et al. 2011).

In addition to the *S. pennellii* LA0716 IL population, ILs have been developed from other wild relatives of tomato including *S. habrochaites* LA1777 (Monforte and Tanksley 2000a), *S. habrochaites* LA0407 (Francis et al. 2001), *S. lycopersicoides* LA2951 (Canady et al. 2005), and *S. chmielewskii* LA1840 (Prudent et al. 2009).

The first set of *S. habrochaites* LA1777 ILs was developed by Monforte and Tanksley (2000a) from an AB-QTL population (Bernacchi et al. 1998a, b), and consisted of 99 NILs and BILs, which provided a coverage of approximately 85% of the wild donor genome. The lines are highly variable for numerous traits including yield, leaf morphology and trichome density, as well as fruit traits such as shape, size, color, biochemical composition and flavor volatiles; favorable wild QTL alleles have been identified for several of these traits (Monforte and Tanksley 2000b; Van der Hoeven et al. 2000; Monforte et al. 2001; Yates et al. 2004; Mathieu et al. 2008; Grandillo et al. unpublished data) (see Sects. 6.9.4 and 6.9.5). The lines have also been used to identify QTLs associated with postzygotic interspecific barriers—primarily pollen and seed sterility in hybrids (Moyle and Graham 2005) (see Sect. 6.9.7). Within the framework of the EU-SOL project, a new set of *S. habrochaites* LA1777 ILs have been developed and anchored to a shared framework of ca. 130 COSII markers. This has allowed a better genome coverage and an increased mapping resolution of the population, with each line carrying a single wild introgression at the given marker density (Tripodi et al. 2010; Grandillo et al. unpublished data; http://www.eu-sol.net/).

6.5.1 Introgression Line Heterosis

Heterosis has played an important role in crop improvement, particularly for quantitative traits such as growth and yield. Two types of heterosis have been recognized—mid-parent and best-parent. In the former, an F_1 hybrid potentially exceeds the mean of both parents, while in the latter it can exceed the superior parent (East and Jones 1919). This so-called hybrid vigor is associated with heterozygosity, although the genetic mechanism of this complex phenomenon is not understood. Possible genetic explanations include: dominance, true overdominance (ODO), pseudoODO (i.e., nearby loci at which alleles having dominant or partially dominant advantageous effects are in repulsion linkage phase), and certain types of epistasis. However, the importance of loci with ODO effects remains to be determined.

Exotic libraries provide very efficient means with which to explore the underlying genetic mechanisms of heterosis. The influence of heterozygosity on phenotypic traits can be analyzed by crossing homozygous ILs to different tester lines, facilitating the analysis of the effects of heterozygosity on phenotype (Semel et al. 2006). Maximal genotypic and phenotypic diversity of interspecific ILs facilitates characterization of the lines for a broad range of traits. Another advantage of IL populations is that their genetic structure allows the assessment of the contribution of ODO effects to heterosis while excluding epistasis. For these reasons, Semel et al. (2006) evaluated the 76 *S. pennellii* LA0716 ILs and their hybrids (introgression line hybrids, ILHs), for 35 different traits related to yield and fitness. They found that while dominant and recessive QTLs were detected for all analyzed traits, ODO QTLs were identified only for the reproductive traits. These results suggested that the true ODO model, involving a single functional Mendelian locus, was a more likely explanation for the heterosis observed in the ILs than the pseudoODO model involving linked loci with dominant alleles in repulsion. The *S. pennelli* ILs and ILHs have also been characterized for primary metabolic traits (Schauer et al. 2008). The analysis, conducted over multiple years, indicated a lack of heterosis for primary metabolite content in the ILs. Very few ODO QTLs were reported, and a similar mode of inheritance was found between the metabolite QTLs and the other nonreproductive morphological traits, which was consistent with the hypothesis proposed by Semel et al. (2006).

6.6 Interactions between QTL and Genetic Background

The influence of a gene on a polygenic trait may vary depending on epistatic interactions with specific genes or the total genetic background. With respect to quantitative traits, epistasis is defined as the deviation from the sum of the independent effects of the individual genes (Falconer 1989). In classical quantitative genetics, the genetic variance attributable to epistasis is designated as genotype-by-genotype interaction (Falconer 1989). The potential number of epistatic interactions (two-locus, three-locus, etc.) is quite vast. Molecular markers underlying individual QTLs can in principle be used to precisely measure epistatic interactions among QTLs. However, in primary segregating populations (e.g., F_2, early BC), as well as in RI populations, the statistical power to detect significant interactions between QTLs is poor, and extraordinarily large population sizes would be required in order to permit the detection of epistasis (Tanksley 1993).

Despite the above-mentioned limitations, significant epistasis was observed for several quantitative traits in tomato including salt tolerance

(Bretó et al. 1994; Foolad et al. 1998a; Foolad and Chen 1999), fruit shape (van der Knaap et al. 2002), locule number (Lippman and Tanksley 2001), fruit color (Kabelka et al. 2004), soluble solids (Monforte et al. 2001), fruit firmness, and aroma (Saliba-Colombani et al. 2001; Causse et al. 2002).

NILs containing single QTL and combinations of QTLs for a given trait provide a much more efficient tool to reveal and characterize epistatic interactions (Eshed and Zamir 1996; Coaker and Francis 2004; Causse et al. 2007). Eshed and Zamir (1996) detected epistasis at a high frequency (28%), and the majority of interactions they found were less-than-additive. Relative to pairs of QTLs, sets of three QTLs showed increased epistatic effects. Coaker and Francis (2004) fine-mapped two major QTLs for resistance to bacterial canker, and found additive-by-additive epistasis in populations in which a large proportion of variance in resistance was associated with the two QTLs. Causse et al. (2007) evaluated six traits related to tomato development and fruit quality in 13 QTL-NILs. These lines shared an *S. lycopersicum* large-fruited genetic background introgressed with one to five fragments from a cherry tomato parent. Although a large fraction of the variation of the six fruit quality traits was under additive control, interactions leading to less-than-additive effects were common. This study showed that epistasis was detected more frequently in NILs than in the RIL population derived from the same cross, confirming the low statistical power that RIL populations provide in detecting epistasis.

ILs or NILs can also be used to study QTL interactions with the overall genetic background. This can be examined by transferring the same QTL into several genetic backgrounds (Lecomte et al. 2004a; Chaïb et al. 2006, 2007; Zanor et al. 2009) or by crossing the same IL to different testers (Eshed and Zamir 1995; Eshed et al. 1996). For example, Lecomte et al. (2004a), Chaïb et al. (2006, 2007) and Zanor et al. (2009) found significant interactions between fruit quality QTLs and genetic background. From a breeder's perspective, these results illustrate the need to introgress and test the effects of QTLs in several recipients before they can be used for crop improvement.

6.7 MAS and QTL Pyramiding

Results from QTL mapping studies in tomato and other crops indicated that the MAS transfer of multiple favorable QTLs, or QTL pyramiding, is the most promising approach to produce new varieties improved for yield-associated phenotypes as well as for other agriculturally important quantitative traits (Lawson et al. 1997; Ashikari and Matsuoka 2006).

An example of the successful application of this strategy is the study conducted by Gur and Zamir (2004) in which a multiple-introgression line was developed by pyramiding three independent yield-promoting genomic regions derived from the drought tolerant green-fruited wild accession

S. pennellii LA0716 into the genetic background of the cultivated recipient genotype M82. The resulting hybrids had yields 50% higher than leading commercial varieties when tested in multiple environments and irrigation regimes. The introduction of the *S. pennellii* introgressions into processing tomato lines resulted in the development of a leading hybrid variety, AB2 (Lippman et al. 2007).

These results clearly demonstrate that IL heterosis and breeding have moved from a mere scientific level to practical application, and that the use of MAS QTL pyramiding in conjuction with the exploitation of exotic germplasm can enrich the genetic basis of cultivated varieties and allow plant breeders to continue raising the ceiling in plant improvement.

6.8 Mendelizing QTLs

Mapping of traits and the identification of closely linked markers allows the utilization of MAS to improve cultivars. When cloning of the underlying gene is desired, a QTL must be introgressed into a homozygous background such that only the locus of interest segregates and the QTL is Mendelized. The *S. pennellii* ILs that each harbor a specific segment of *S. pennellii* in a homozygous cultivated tomato background (Eshed and Zamir 1995) have been extremely useful for the cloning of genes underlying QTLs as well as single locus traits. Once a QTL has been found in an F_2 or BC population, the cultivated parent is crossed to the IL that harbors the *S. pennellii* genome spanning the QTL. Fine-mapping can then be conducted in the resulting F_2 population. Several QTLs have been cloned using the ILs such as *fw2.2*, *brix9-2-5*, *ovate*, *style2.1* and *fas* (Frary et al. 2000; Fridman et al. 2000; Liu et al. 2002; Chen et al. 2007; Cong et al. 2008). An attempt was also made to clone *sun* using the ILs. Unfortunately, *sun* mapped inside a paracentric inversion within the *S. pennellii* genome; this precluded map-based cloning using that resource (van der Knaap et al. 2004).

However, the ILs are not appropriate for attempting to clone every QTL. For many QTLs, the effect of the genetic background can be significant. Crosses to a particular IL accession may result in phenotypic segregation that is only partially explained by the locus of interest. This can make scoring of the trait excessively cumbersome. Moreover, cloning of major QTLs often requires the presence of other loci to enhance effect and improve the ease of scoring the alleles at the major locus. In those situations, it is important to develop ILs from the starting F_2, RIL or BC populations that demonstrated the QTL effect.

After a QTL is Mendelized, fine-mapping of the locus can begin. Plants that are recombinant between flanking markers are identified. These primary recombinants serve as the parents of families that are progeny-tested to determine whether families segregate or are fixed for the QTL.

Alternatively, the primary recombinants are developed into a collection of substitution lines that can be grown in different locations for phenotypic evaluations. At the start of fine-mapping, QTL regions are usually relatively large (20 cM is not uncommon). Therefore, fine-mapping often occurs in stages, during which the QTL interval becomes progressively shorter. This results in confirmation of the QTL simultaneously to locus placement within a narrower genetic interval. With genomic sequences in hand, the researcher can quickly determine the physical distance of the interval and the number of candidate genes. For some traits, it may be possible to predict the underlying gene from the available candidate genes. When the region recombines well and the number of candidate genes is large, it is best to identify additional recombinant plants to narrow down the region, reducing the number of candidate genes that need to be studied. Without genomic sequences in hand, the researcher will typically aim to locate the QTL to a region of less than 1 cM. If the region recombines well, a bacterial artificial chromosome (BAC) clone of 100 kb may span a genetic distance of 1 cM, which implies that the interval is spanned by one clone. Markers flanking the locus are used to screen BAC libraries for positive clones. BAC end sequences are converted into markers to determine the genetic-to-physical distance at the locus and whether the gene is on or adjacent to the BAC clone. BAC ends can also be converted into probes for continued chromosome walking when needed. Once a BAC that harbors the gene is identified, candidate gene(s) will be predicted from the sequence analysis of the clone. Final confirmation that the gene has been identified is usually accomplished by stable genetic transformation.

6.9 QTL Mapping on a Trait Basis

In tomato, mapping studies have been conducted to identify QTLs for numerous agriculturally and biologically important traits including resistance/tolerance to biotic and abiotic stresses, yield-related traits, sensory and nutritional quality of the fruit (see reviews by Foolad 2007; Labate et al. 2007; Grandillo et al. 2011). Here we review the major results for the majority of analyzed traits (Tables 6-1–6-5).

6.9.1 Disease Resistance

Hundreds of pathogens of diverse etiology are known to infect tomato (Jones et al. 1991). QTL mapping and MAS can aid in the transfer of polygenically based, horizontal (field or race-nonspecific) resistance for which phenotypic selection can be intractable. Since the pioneering study by Rick and Fobes (1974) that reported an allozyme marker to be associated with root-knot nematode (*Meloidogyne incognita*) resistance, numerous studies have

Table 6-4 Fruit weight QTLs.

Fruit weight locus	Ref
fw1.1	Grandillo and Tanksley 1996a; Chen et al. 1999; Lippman and Tanksley 2001; van der Knaap and Tanksley 2003
fw1.2	Grandillo and Tanksley 1996a; Lippman and Tanksley 2001
fw2.1	Grandillo and Tanksley 1996a; Lippman and Tanksley 2001; Saliba-Colombani et al. 2001
fw2.2	Alpert et al. 1995; Grandillo and Tanksley 1996a; Chen et al. 1999; Lippman and Tanksley 2001; Saliba-Colombani et al. 2001; van der Knaap and Tanksley 2003
fw3.1	Chen et al. 1999
fw3.2	Chen et al. 1999; Lippman and Tanksley 2001; Saliba-Colombani et al. 2001; van der Knaap and Tanksley 2003
fw4.1	Chen et al. 1999
fw5.2	van der Knaap and Tanksley 2003
fw6.2	van der Knaap and Tanksley 2003
fw6.3	Chen et al. 1999
fw7.1	Chen et al. 1999
fw7.2	Chen et al. 1999; van der Knaap and Tanksley 2003
fw8.1	Grandillo and Tanksley 1996a; Chen et al. 1999
fw8.2	Grandillo and Tanksley 1996a
fw9.2	Chen et al. 1999
fw11.2	Chen et al. 1999
fw11.3	Grandillo and Tanksley 1996a; Lippman and Tanksley 2001; Saliba-Colombani et al. 2001; van der Knaap and Tanksley 2003
fw12.1	Chen et al. 1999; Saliba-Colombani et al. 2001

Table 6-5 Fruit shape QTLs.

Fruit shape locus	Ref
ovate	Ku et al. 1999; Gonzalo and van der Knaap 2008
lc	Lippman and Tanksley 2001; van der Knaap and Tanksley 2003; Barrero and Tanksley 2004; Lecomte et al. 2004b
sun	van der Knaap and Tanksley 2001; Brewer et al. 2007; Gonzalo and van der Knaap 2008
fs8.1	Grandillo et al. 1996; Ku et al. 2000; Brewer et al. 2007; Gonzalo and van der Knaap 2008
fas	Lippman and Tanksley 2001; Barrero and Tanksley 2004

been conducted to identify molecular markers linked to major genes for vertical resistance and to QTLs for horizontal resistance. This includes a wide variety of viral, bacterial, fungal and nematode diseases in tomato (Table 6-1; see also review by Foolad 2007). In most cases, QTLs for disease resistance have been mapped using interspecific populations, with the source of the resistant allele being the resistant wild parent. However, in a few cases either the source of the resistance was a cultivated tomato or transgressive resistant QTLs were identified in the susceptible cultivated parent. The most frequently used population structures have been F_2s and early BCs; however, several studies have used BILs, RILs and ILs. In many cases multiple QTLs have been identified, with some of the QTLs having a major effect on resistance.

Despite the numerous QTL mapping studies conducted so far, the use of MAS to improve quantitative resistance to tomato diseases remains limited (Foolad 2007). A few attempts made to date have clearly indicated that further refinement and characterization of the target QTLs is necessary before they can be transferred into the desired cultivated background. This is particularly true when the origin of the QTL allele is a wild species (Robert et al. 2001; Brouwer and St Clair 2004).

6.9.1.1 Viral Resistance

A major QTL for resistance to the tomato yellow leaf curl virus (TYLCV) was identified using bulked segregant analysis (BSA) and RAPD markers (Chagué et al. 1997). The resistance originated from the *S. pimpinellifolium* parent and the identified QTL, which explained up to 27.7% of the resistance, mapped to a region of chromosome 6 where a major TYLCV tolerance locus (*Ty-1*, originally introgressed from *Solanum chilense*) (Zamir et al. 1994) plus several major resistance loci (including *Mi*, *Cf-2*, *Cf-5*, *Ol-1* and *Meu-1*) had previously been mapped.

RAPD markers have been used to identify QTLs for oligogenic resistance to tomato mottle virus (ToMoV) (Griffiths and Scott 2001). Twelve RAPD markers that were associated with the ToMoV resistance segregated into two linked regions flanking either side of the two morphological markers *self-pruning (sp)* and *potato leaf (c)* located on the long arm of chromosome 6.

6.9.1.2 Bacterial Resistance

QTL mapping studies have been conducted to identify molecular markers associated with resistances to three major bacterial diseases of tomato: bacterial canker, caused by the gram-positive bacterium *Clavibacter michiganensis* ssp. *michiganensis* (Cmm), bacterial spot caused by *Xanthomonas* sp., and bacterial wilt caused by the soilborne bacterium *Ralstonia solanacearum*.

For bacterial canker, van Heusden et al. (1999) identified three QTLs on chromosomes 5, 7 and 9, which showed a substantial influence on resistance to *Cmm*. Their effects were additive and the resistant alleles originated from the resistant parent *S. arcanum* LA2157. The strongest effect QTL, mapping on chromosome 7, had also been identified by Sandbrink et al. (1995) using an *S. arcanum* intraspecific population, and apparently a bacterial wilt resistance gene had also been mapped to the same position by Danesh et al. (1994). Two major QTLs for resistance to *Cmm* have been identified using a BIL population derived from a *S. lycopersicum* x *S. habrochaites* LA0407 cross (Kabelka et al. 2002). These QTLs were subsequently fine-mapped, and an additive-by-additive epistasis between them was confirmed (Coaker and Francis 2004).

For bacterial spot, Yu et al. (1995) used an interspecific BC_1 population derived from a cross between the resistant cv. Hawaii 7998 and a susceptible wild accession of *S. pennellii*. Hypersensitive resistance to race T1 found in the cv. Hawaii 7998 was associated with at least three QTLs, two mapping on chromosome 1 (*Rx1* and *Rx2*), and one on chromosome 5 (*Rx3*). More recently, Yang et al. (2005) used an intraspecific F_2 population derived from a cross between the cv. Hawaii 7998 (H 7998) and an elite breeding line, Ohio 88119. Due to the low level of polymorphism between the two parents, a limited number of markers were available for mapping, and only two of the previously identified QTLs could be detected (*Rx1* and *Rx3*). Resistance to multiple races, including race T4, has been described in the *S. lycopersicum* "cerasiforme" accession PI 114490, and major and minor resistance QTLs to bacterial spot have been identified in an elite IBC population (Hutton et al. 2010).

Several QTLs conferring resistance to bacterial wilt race 1 strains (phylotypes I and II) have been identified in tomato populations derived from various resistance sources (Danesh et al. 1994; Thoquet et al. 1996a, b; Mangin et al. 1999; Wang et al. 2000). A major QTL, accounting for 20% to 50% of PVE for resistance in the field, was mapped to chromosome 6. The experiments included various strains of the pathogen (Danesh et al. 1994; Thoquet et al. 1996a, b; Mangin et al. 1999). Mangin et al. (1999), by using a large F_2 population, temporal analysis and two different statistical approaches, showed that the broad LOD peak previously observed on chromosome 6 was more likely due to two QTLs 30 cM apart. In another study, Wang et al. (2000) identified a new major locus on chromosome 12, for which the allele carried by tomato cultivar Hawaii 7996 gave resistance to the bacterial strain Pss4 endemic to Taiwan. In this population, the QTL mapping on chromosome 6 had a minor effect. Four QTLs for partial resistance to a race 3-phylotype II strain of *R. solanacearum* were identified in F_2 progeny of a cross between the partially resistant *S. lycopersicum* cv. Hawaii 7996 and a highly susceptible cultivar (Carmeille et al. 2006). The comparison of these QTLs with those detected in the same study and in previous studies against other races of the pathogen (race 3-phylotype II strain JT516 and race 1-phylotype I and II strains), revealed the possible occurrence of some phylotype-specific resistance QTLs in Hawaii 7996.

6.9.1.3 Fungal Resistance

Major and minor QTLs have been identified for resistances to important fungal diseases of cultivated tomato, including: anthracnose (*Colletotrichum coccodes*), blackmold (*Alternaria alternata*), early blight (*Alternaria solani*), grey mold (*Botrytis cinerea*), late blight (*Phytophthora infestans*), and powdery mildew (*Oidium lycopersici*).

A study conducted by Stommel and Zhang (2001) aimed to identify QTLs for resistance to anthracnose. Putative RAPD markers linked to loci that influenced resistance were identified, however, the phenotypic variation for anthracnose resistance explained by any individual QTL was very low, ranging from 1.7% to 5.1%

QTLs for resistance to blackmold have been mapped in progeny derived from a cross between a susceptible cv. VF145B-7879 and the resistant *S. cheesmaniae* LA0422 (Robert et al. 2001). Five of the mapped QTLs were selected for further characterization and the analysis allowed the identifiction of one QTL on chromosome 2 that had the largest positive effect on blackmold resistance, alone and in combination with other QTLs, and was also associated with earliness, a positive horticultural trait.

For early blight, 14 resistance QTLs were detected in BC populations (BC_1 and BC_1S_1) deriving from a susceptible tomato breeding line and a resistant *S. habrochaites* accession (PI 126445) (Foolad et al. 2002; Zhang et al. 2003a). The individual QTL effects ranged from about 10% to 25% of the total phenotypic variation, and the combination of five to six complementary QTLs explained the majority (>50%) of the variation (Foolad et al. 2002). Four of these QTLs (detected as major QTLs in both studies) were considered as the most reliable, and therefore useful, in marker-assisted breeding programs. Some of the QTLs for early blight resistance colocalized with several RGAs, raising the possibility of isolating early blight resistance genes by cloning. In a subsequent study, six QTLs for resistance to early blight were found in the resistant wild accession *S. arcanum* LA2157, by using two different evaluation criteria (Chaerani et al. 2007). The PVE for each QTL ranged from 7% to 16%.

For grey mold, three resistance QTLs, explaining 7% to 15% of the total phenotypic variation, were identified in an F_2 population and confirmed in BC_2S_1 families deriving from a *S. lycopersicum* cv. Moneymaker x *S. habrochaites* LYC4 cross (Finkers et al. 2007a). The evaluation of an IL population of 30 BC_5S_2 lines developed from the same cross allowed the verification of the three previously mapped QTLs for resistance to *B. cinerea* and the identification of seven additional QTLs (Finkers et al. 2007b). The

S. lycopersicoides accession LA2951 exhibits high foliar resistance to gray mold, and five putative resistance QTLs were identified in a population of 58 *S. lycopersicoides* LA2951 ILs, with the major resistance QTL mapping on the long arm of chromosome 1 (Davis et al. 2009).

Reciprocal *S. lycopersicum* (E) x *S. habrochaites* LA2099-MD1 (H) backcross populations have been used to identify QTLs for quantitative resistance to late blight by using three types of replicated disease assays (detached-leaflet, whole-plant, and field) (Brouwer et al. 2004). In BC and BC-H populations, totals of 15 and 18 QTLs were mapped, respectively. Eight QTLs were highly consistent across replicated trials using various assay techniques. The resistance QTLs were mapped on all 12 chromosomes, and the resistance allele originated from the resistant wild parent in all but one case. Some resistance QTLs detected in tomato coincided with the chromosomal locations of previously mapped *P. infestans* resistance QTLs and R genes in potato, suggesting functional conservation within the Solanaceae. The effects of three of these resistance QTLs, mapping on chromosomes 4, 5 and 11, were confirmed with the development and characterization of NILs and sub-NILs (Brouwer and St Clair 2004).

For quantitative resistance to powdery mildew, Bai et al. (2003) identified three QTLs in an F_2 population derived from a cross between the susceptible cv. Moneymaker and the resistant *S. neorickii* accession G11601. The three QTLs mapped on chromosomes 6 and 12 (two QTLs), and explained 16.1%, 29.5%, and 22.3% of the total phenotypic variation, respectively. In combination, the three QTLs explained 68% of the phenotypic variation. All QTLs were confirmed by testing F_3 progenies, and PCR-based markers were developd for MAS. Interestingly, the authors provided evidence for colocalization of two QTLs with R genes, *Ol-1/Ol-3* on chromosome 6 and *Lv* on chromosome 12, involved in tomato powdery mildew resistance.

6.9.2 Insect Resistance

Cultivated tomato is susceptible to numerous arthropod pests (Kennedy 2003), and resistances have been reported in the wild species *S. habrochaites*, *S. peruvianum* and *S. pennellii* (see reviews by Labate et al. 2007). Methyl-chetones, such as 2-tridecanone (2TD), and sesquiterpenes have been found to be associated with pest resistance in *S. habrochaites,* whereas acylsugars exuded by type IV glandular trichomes are the allochemicals related to insect resistances found in many *S. pennellii* accessions (references in Labate et al. 2007).

By using an F_2 population derived from the cross *S. lycopersicum* x *S. habrochaites* (PI 134417), Zamir et al. (1984) identified five independent loci (three isozyme and two morphological markers) associated with expression of 2TD (Table 6-1). In a subsequent study, Nienhuis et al. (1987)

used an F_2 population derived from an *S. lycopersicum* x *S. habrochaites* LA0407 cross and identified RFLP loci on three linkage groups associated with expression of 2TD.

Using an interspecific F_2 population created by the cross *S. lycopersicum* x *S. pennellii* LA0716, Mutschler et al. (1996) identified five putative QTLs associated with acylsugar accumulation, two on chromosome 2 and one each on chromosomes 3, 4, and 11. The major effects on both acylsucroses and total acylsugars were associated with the regions on chromosomes 2 and 3. The observed range in PVE was ~7% to 16% depending on the trait and analytical method. Lawson et al. (1997) targeted these five QTLs for a MAS program aimed at transferring the acyl-sugar-mediated insect resistance from *S. pennellii* into cultivated tomato. The obtained multi-QTL line accumulated acylsugars but at levels lower than the interspecific F_1 control, suggesting that there are additional QTLs, still unidentified, necessary for higher levels of acylsugar accumulation.

Blauth et al. (1999) used an intraspecific F_2 population derived from a cross between *S. pennellii* LA0716 and *S. pennelli* LA1912 to analyze the genetic basis of acylsugar fatty acid composition. Six QTLs were detected which together explained 23% to 60% of the variance observed for each of the nine segregating fatty acid constituents.

More recently, to determine the biological roles and chemical contents of individual glandular trichome types in *Solanum*, McDowell et al. (2011) have investigated the metabolic and transcriptomic profiles of glandular trichome types in *S. lycopersicum* (glandular tricome types I, VI and VII), *S. habrochaites* (types I, IV, and VI), *S. pennellii* (types IV and VI), *S. arcanum* (type VI) and *S. pimpinellifolium* (type VI). This study showed that glandular trichomes of cultivated tomato and wild tomato relatives produce a variety of structurally diverse volatile and non-volatile specialized ('secondary') metabilites, including terpenes, flavonoids and acylsugars. A genetic screen of leaf trichome and surface metabolite extracts of the *S. pennellii* LA0716 IL population allowed the identification of *S. pennellii* regions influencing mono- and sesquiterpenes or only sesquiterpenes, as well as the quality or quantity of acylsugars (Schilmiller et al. 2010).

6.9.3 Abiotic Stress Tolerance

Wild tomato species harbor a valuable pool of alleles conferring tolerance to abiotic stresses such as drought, salt, and temperature extremes. This is critical for continued crop improvement (Foolad 2007). With the exception of heat tolerance, traditional breeding for abiotic tolerance has been intractable to date, due to the complex nature of these traits combined with the difficulty in developing reliable field screening techniques (Foolad 2007). Moreover, the stage-specific nature of abiotic stress tolerance renders it impracticable to carry out simultaneous or sequential screening. Rather, specific developmental stages throughout the ontogeny of the plant might be measured as different traits and evaluated with different screening procedures (Foolad 2007). Consequently, the identification of molecular markers tightly linked to QTLs for abiotic stress tolerance is considered a more promising approach to facilitate selection and breeding of these complex traits.

In tomato, QTL mapping studies have focused on salt, drought and cold tolerance, while much less effort has been invested in identifying QTLs conferring tolerance to high temperatures or other stresses (Table 6-2) (Foolad 2004, 2007).

6.9.3.1 Cold Tolerance

Cultivated tomato is sensitive to low temperatures (~0–15°C) during all developmental stages including germination and emergence (Foolad and Lin 1998). Foolad et al. (1998b) identified two putative QTLs for cold tolerance during seed germination, in an *S. lycopersicum* and *S. pimpinellifolium* (LA0722) BC_1 population. The QTLs mapped on chromosomes 1 (two QTL) and 4, with favorable QTL alleles contributed by *S. pimpinellifolium* on chromosome 1. Individual QTLs explained 11.9% to 33.4% of total phenotypic variation. Two QTLs were confirmed in a follow-up study of a larger BC_1 population and an F_9 RIL population derived from the same cross (Foolad 2007).

Cold tolerance during vegetative growth has been the subject of two QTL mapping studies. In a BC_1 population between a cold sensitive *S. lycopersicum* line and a cold tolerant *S. habrochaites* accession, Vallejos and Tanksley (1983) identified three QTLs (on chromosomes 6, 7 and 12) associated with chilling tolerance, using isozyme markers and differential growth at 5°C for three weeks as criteria. Truco et al. (2000) identified QTLs for shoot wilting and root ammonium uptake under chilling temperatures in a BC_1 population from an *S. lycopersicum* x *S. habrochaites* (LA1778) cross. At least three QTLs for chilling-induced shoot wilting, and four QTLs for recovery from wilting, were identified. A minimum of two QTLs seemed to be involved in ammonium uptake, and, unexpectedly, for both QTLs the *S. habrochaites* allele contributed to decreased ammonium uptake after chilling.

6.9.3.2 Drought Tolerance

Less attention has been given to studying drought tolerance in tomato compared to salt and cold tolerance. Plant WUE will affect yield when water is a limiting factor. WUE is defined as the rate of carbon gained to water transpired. Relative proportion of ^{13}C to ^{12}C in plant organic matter (known as stable isotope discrimination or δ^{13} C) can potentially serve as an indirect indicator though which WUE can be improved in C3 plants. This would be accomplished by the elimination of undesirable linkages between WUE and yield.

The drought tolerant *S. pennellii* wild species was found to have greater WUE and less negative δ^{13} C than cultivated tomato (Martin et al. 1999). Martin et al. (1989) identified three RFLP markers that explained a large proportion of the genetic variance for δ^{13} C in F_3 and BC_1S_1 tomato populations derived from an *S. lycopersicum* x *S. pennellii* cross. Recently, the use of the *S. pennelli* LA0716 IL population has allowed the identification of a dominant QTL for δ^{13} C, QWUE5.1, in the *S. pennellii* IL5-4, for which the wild allele determined high δ^{13} C (small negative value) (Xu et al. 2008). Sub-ILs developed from IL5-4 allowed fine mapping of QWUE5.1 to an interval of approximately 2.2 cM in length.

Drought tolerance during seed germination was analyzed by Foolad et al. (2003a) in an *S. lycopersicum* x *S. pimpinellifolium* (LA0722) BC_1 population, using a selective genotyping approach. Four QTLs for rapid seed germination under drought stress were identified on chromosomes 1, 8, 9 and 12. The two QTLs located on chromosomes 1 and 9 had larger effects and the drought tolerant alleles originated from the *S. pimpinellifolium* donor parent. The QTL mapping results were subsequently confirmed in an F_9 RIL population derived from the same interspecific cross (Foolad 2007).

Responses to various abiotic stresses during seed germination have been analyzed by QTL mapping (Foolad 1999b; Foolad et al. 1999, 2003b, 2007). These studies indicated that genetic relationships among cold, salt, and drought tolerance during seed germination exist, and that similar physiological mechanism(s) may contribute to rapid seed germination under different conditions. Seven QTLs, mapped on chromosomes 1, 4, 9 and 11, were identified by measuring effects on germination rate under three conditions (Foolad et al. 2007). The authors suggested that these QTLs can be considered as the most useful for MAS improvement of tomato seed germinability under different abiotic stresses.

6.9.3.3 Salt Tolerance

S. lycopersicum is a mesophytic species, and most commercial cultivars are moderately sensitive to salinity at all stages of plant development, with seed germination and early seedling growth being the most vulnerable stages (Foolad 2004, 2007). Genetic resources for salt tolerance have been identified within related wild tomato species and primitive cultivars including *S. pimpinellifolium*, *S. chilense*, *S. cheesmaniae*, *S. galapagenese*, *S. pennellii* and *S. peruvianum* (reviewed by Foolad 2004). Most studies have focused on specific developmental stages; in only a few instances has salt tolerance been compared at different developmental stages (Foolad 1999a; Zhang et al. 2003b).

QTLs conferring salt tolerance during seed germination have been identified in mapping populations involving two salt tolerant wild accessions: *S. pennellii* LA0716 (Foolad and Jones 1993; Foolad et al. 1997; Foolad and Chen 1998), and *S. pimpinellifolium* LA0722 (Foolad et al. 1998a). The QTL mapping results obtained in the two interspecific mapping populations indicated that salt tolerance during seed germination in tomato was controlled by a few major QTLs acting together with a number of smaller effect QTLs. For example, in the *S. pimpinellifolium* LA0722 study, seven putative QTLs were identified, and the *S. pimpinellifolium* accession had favorable QTL alleles at six of them (Foolad et al. 1998a). The PVE of individual QTL ranged from 6.5% to 15.6%, and the identified QTLs together could explain 44.5% of the total phenotypic variation (Foolad et al. 1998a). A comparison of the QTLs identified in the *S. pennellii* F_2 and in the *S. pimpinellifolium* BC_1S_1 mapping populations indicated that some of the QTLs for salt tolerance during germination in tomato were conserved across species, while other QTLs were species-specific (Foolad et al. 1997; Foolad and Chen 1998; Foolad et al. 1998a).

The *S. pimpinellifolium* LA0722 BC_1 population has also been used to identify QTLs conferring salt tolerance during the vegetative stage (Foolad and Chen 1999; Foolad et al. 2001). Foolad and Chen (1999) detected five QTLs on chromosomes 1 (two QTLs), 3, 5 and 9. Each QTL accounted for between 5.7% and 17.7% of the total phenotypic variance. In all cases the positive QTL alleles were derived from the salt tolerant parent, and digenic epistatic interactions were also identified. Some of these QTLs were subsequently confirmed using the selective genotyping approach (Foolad et al. 2001). QTL mapping studies have been conducted to identify QTLs for salt tolerance during the reproductive stage in tomato (Bretó et al. 1994; Monforte et al. 1996, 1997a, b, 1999; Villalta et al. 2007). Bretó et al. (1994) and Monforte et al. (1996) used an F_2 population derived from a *S. lycopersicum* cv. Madrigal x *S. pimpinellifolium* (line L1) cross to identify QTLs affecting average fruit weight, fruit number and total fruit weight under salinity. These studies showed that fruit weight was a key trait for salt tolerance. Moreover, MAS provided a very powerful tool, relative to phenotypic selection, with which to develop salt tolerant breeding lines (Monforte et al. 1996). *S. lycopersicum* x *S. cheesmaniae* and *S. lycopersicum* x *S. pimpinellifolium* F_2 populations were analyzed for salt tolerance QTLs (Monforte et al. 1997a, b). In the *S. lycopersicum* x *S. pimpinellifolium* crosses several salt tolerence QTLs, measured in terms of yield, mapped to loci known to influence fruit weight (Monforte et al. 1997a). Two populations of RILs were developed from these same interspecific F_2 populations, and were used to identify QTLs for salt tolerance under different salinity levels in terms of G x E interactions (Villalta et al. 2007, 2008). Villalta et al. (2007) studied the effects of salinity on 19 traits including fruit weight, fruit number, total fruit weight, flowering time and chloride (Cl⁻) concentration. For the six traits associated with yield, flowering time, and Cl⁻ concentration, 15 QTLs were identified in both populations under high salinity conditions, while 15 and 16 QTLs were identified under control conditions, in the *S. pimpinellifolium* and in the *S. cheesmaniae* populations, respectively. For both populations, eight QTLs were detected under both control and high salinity conditions. The contributions of individual QTLs were, in general, low. Leaf Cl⁻ concentration was the trait most affected by salinity, followed by flowering time.

In a subsequent study, the two RIL populations (F_8 lines) were used to identify QTLs for Na^+ and K^+ in stems and leaves (Villalta et al. 2008). A major QTL controlling Na^+ (and K^+ under salinity) concentration in aerial parts of the plant was identified on chromosome 7. The results indicated that none of the Na^+ transporters or regulatory proteins tested in that study colocalized with the QTL. In an earlier study, Zamir and Tal (1987) used isozyme markers in an *S. lycopersicum* × *S. pennellii* LA0716 F_2 population, and identified four QTLs having a similar effect on Na^+ and Cl⁻ uptake, and two other QTLs that affected K^+ uptake.

An alternative approach that could be pursued to improve tomato crop productivity under salinity is by grafting cultivars onto salt tolerant wild relatives. Towards this aim, the rootstock effects on fruit yield (Estañ et al. 2009) and on different physiological traits (Asins et al. 2010) were analyzed in the shoot of a grafted tomato variety under moderate salinity (75 mM NaCl), using as rootstocks the two interspecific RIL populations previously analyzed by Villalta et al. (2007, 2008). A minimum of eight QTLs were identified that contributed to this ST rootstock effect on yield, with the most relevant component being the number of fruits (Estañ et al. 2009). The co-localization of scion water content and yield QTLs in one of the two populations, suggested that the improvement of fruit yield under salinity observed by grafting was in part explained by the rootstock's ability to reduce the perturbations in scion water status and leaf fresh weight (Asins et al. 2010).

6.9.4 Fruit Yield

Improving yield performance is a major objective for most breeding programs. For tomato, total fruit yield is estimated by the total weight of the fruits harvested per unit area (measured in kg/m²). In some cases, total yield has been analyzed in conjunction with red and green yields. For processing tomatoes, a biological and agricultural estimate of the productivity is given by the total sugar production per unit area (brix × yield), measured in g/m^2 (Eshed and Zamir 1995; Gur and Zamir 2004). Yield and agricultural yield were evaluated in five AB QTL mapping studies involving five different wild tomato accessions (Tanksley et al. 1996; Fulton et al. 1997, 2000; Bernacchi et al. 1998a; Frary et al. 2004a), in *S. chmielewskii* BILs (Azanza et al. 1994), in the *S. pennellii* LA0716 ILs (Eshed and Zamir 1995, 1996; Eshed et al. 1996; Gur and Zamir 2004; Schauer et al. 2006, 2008; http://zamir.sgn. cornell.edu) and in the *S. habrochaites* LA1777 ILs (Monforte and Tanksley 2000b; Monforte et al. 2001; Grandillo et al. unpublished data) (Table 6-3). Moreover, yield has been analyzed in interspecific F_2 and RI populations under salt-stress conditions (Bretó et al. 1994; Monforte et al. 1996, 1997a, b; Villalta et al. 2007). Taken together, the results obtained in the five AB-QTL mapping studies (Tanksley et al. 1996; Fulton et al. 1997, 2000; Bernacchi ct al. 1998a; Frary et al. 2004a) and those obtained by Eshed and Zamir (1995) and Eshed et al. (1996) on the *S. pennellii* LA0716 IL population, indicated that at least 30 QTLs exist for fruit yield in tomato, located on all 12 tomato chromosomes. Most of the agricultural yield QTLs co-localized with the total or red yield QTLs. The highest number of 12 QTLs identified in a single mapping population was reported for red yield by Fulton et al. (1997) in the *S. arcanum* LA1708 AB population, and for total yield by Bernacchi et al. (1998a) in the *S. habrochaites* LA1777 AB population. For agricultural yield, the highest number of 14 QTLs was reported by Eshed and Zamir (1995) for the *S. pennellii* LA0716 IL population. Also for yield, major QTLs were identified together with minor QTLs. Many of the identified yield QTLs were detected in more than one mapping population and species, and can be considered potentially conserved QTLs. As expected based on the parental phenotypes, most of the wild alleles had a negative effect on yield. However, in some cases favorable yield QTL alleles of wild origin were identified (Eshed and Zamir 1995; Fulton et al. 2000; as reviewed by Grandillo et al. 2008, 2011). The practical value of these transgressive wild QTL alleles for the improvement of cultivated tomato germplasm has been demonstrated by Gur and Zamir (2004) (see Sect. 6.7).

Recently, a reciprocal grafting approach was used by Gur et al. (2011) on a subset of yield-improving *S. pennellii* LA0716 ILs to distinguish between root- and shoot-related yield QTLs. The results showed that yield QTLs from wild tomato are predominantly expressed by the shoot. Moreover, the study provided a preliminary characterization of the metabolic factors that may be responsible for the unique observation of a root-expressed QTL.

6.9.5 Fruit Quality

Numerous traits influence tomato fruit quality, and most of them are polygenically inherited. Fruit soluble solids, pH, and paste viscosity are the most important quality traits in processing tomatoes, while wholesale and retail distributors of fresh market tomatoes require extended shelf-life and firmness. In order to satisfy consumers' expectations it is important to also improve sensory quality, which, for fresh market tomatoes, is correlated with visual aspect (size, shape and color), texture (firmness, mealiness, juiciness), taste and aroma (both named flavor, since they are sometimes difficult to distinguish). The characteristic flavor of tomato fruits depends on a complex mixture of sugars, acids, amino acids, minerals and volatile compounds (Baldwin et al. 1991a, b). The nutritional quality of tomato fruit is also of increasing interest, since this vegetable represents a relevant source of antioxidants (mainly AsA and carotenoids) that have been shown to reduce the risks of certain types of cancers and cardiovascular diseases (Giovannucci 1999; Willcox et al. 2003).

For most of these quality traits, QTLs have been identified in different experiments and/or using different mapping populations (Tables 6-3–6-5). Due to space limitations, here we will review the QTL mapping results obtained for some of the key quality traits that have been analyzed in tomato. We must mention that, for this crop, QTLs have been detected for numerous other fruit quality traits including fruit viscosity, puffiness, cracking, epidermal reticulation, sunscald, pericarp thickness, stem retention, and stem scar (Table 6-3; see review by Foolad 2007; Tanksley et al. 1996; Fulton et al. 1997, 2000; Bernacchi et al. 1998a; Monforte et al. 2001; Frary et al. 2004a).

6.9.5.1 Fruit Weight

Fruit weight has been analyzed in most of the QTL mapping studies conducted in tomato (Tables 6-3 and 6-4). This trait is controlled by many major and minor QTLs, and a set of at least 28 QTLs controlling the variation in fruit weight have been identified in a synthesis of 17 studies involving seven different wild species (Grandillo et al. 1999; Tanksley 2004). Studies conducted with populations derived from crosses between *S. lycopersicum* and close relatives of cultivated tomato, the weedy cherry tomato and the red-fruited wild species *S. pimpinellifolium*, revealed 18 QTLs controlling fruit weight (Grandillo and Tanksley 1996a; Chen et al. 1999; Lippman and Tanksley 2001; Saliba-Colombani et al. 2001; van der Knaap and Tanksley 2003). QTLs that were found in three or more studies were *fw1.1, fw2.1, fw2.2, fw3.2* and *fw11.3*. The effect of these QTLs differed among the populations; however, *fw2.2* exhibited the largest effect in the majority of studies. The other fruit weight loci were restricted to one or two studies (Table 6-4). The gene underlying *fw2.2* was isolated and found to encode a plant-specific protein that is associated with cell cycle regulators (Frary et al. 2000; Cong and Tanksley 2006). Expression levels of *FW2.2* were negatively correlated with fruit growth. The most dramatic morphological differences within the fruit appeared in the placenta tissues of the organ (Cong et al. 2002; Liu et al. 2003a).

6.9.5.2 Fruit Shape

Cultivated tomato displays a range of fruit shapes (see Chapter 3). Shape QTLs identified in populations derived from crosses with cultivated accessions and *S. pimpinellifolium* or cherry tomato were found on chromosome 2 (*ovate* and *lc*) (Ku et al. 1999; Lippman and Tanksley 2001; van der Knaap and Tanksley 2003; Barrero and Tanksley 2004; Lecomte et al. 2004b; Gonzalo and van der Knaap 2008), chromosome 7 (*sun*) (van der Knaap and Tanksley 2001; Brewer et al. 2007; Gonzalo and van der Knaap 2008), chromosome 8 (*fs8.1*) (Grandillo and Tanksley 1996a; Brewer et al. 2007; Gonzalo and van der Knaap 2008), and chromosome 11 (*f* or *fas*) (Lippman and Tanksley 2001; Barrero and Tanksley 2004) (Table 6-5, Fig. 6-1). Fruit elongation is controlled by *sun, ovate,* and *fs8.1* whereas locule number is controlled mainly by *lc* and *fas* that interact epistatically. The genes underlying *sun, ovate, fas* and *lc* have been cloned (Liu et al. 2002; Cong et al. 2008; Xiao et al. 2008; Muños et al. 2011). *SUN* encodes a member of the IQD family of plant proteins and imparts elongated fruit shape after anthesis (Xiao et al. 2008). It is hypothesized that *SUN* may affect hormone or secondary metabolite levels resulting in altered organ shape (Xiao et al. 2008). The *OVATE* gene encodes a protein that negatively regulates plant growth (Liu et al. 2002). Ovate family proteins (OFPs) act as transcriptional repressors via interactions with BELL and KNOX homeodomain transcription factors. In addition, *AtOFP1* expression led to down-regulation of GA20 oxidase-1 expression thereby controlling plant growth and development via modulation of gibberellin levels (Hackbusch et al. 2005; Wang et al. 2007). *FAS* encodes a YABBY protein involved in organ polarity and carpel number (Cong et al. 2008). Map-based cloning of the *lc* QTL showed that two SNPs immediately downstream of *WUSCHEL* control the trait; however, the gene responsible for *lc* and its role during domestication remains unknown (Muños et al. 2011).

Additional fruit shape QTLs were found on the middle of chromosomes 1 and 9, top of 2 and 12, and bottom of 3, 5, 6, 7 and 11, of which each affected two or more attributes of shape (Brewer et al. 2007; Gonzalo and van der Knaap 2008). Some of the shape QTLs overlapped with fruit size QTLs, notably *fw1.1, fw3.2* and the overlapping loci *fw11.3* and *fas*, and thus may represent genes with pleiotropic effects on morphology.

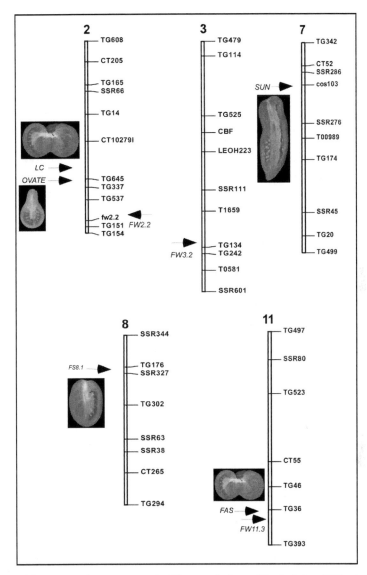

Figure 6-1 Five tomato shape genes control the morphology of the fruit. In addition to *FW2.2* controlling fruit weight, *OVATE* and *LC* map to chromosome 2. *OVATE* controls obovoid and ellipsoid shape whereas *LC* controls locule number and flat shape. *SUN* maps to chromosome 7 and controls the shape of very elongated fruit. *FS8.1* maps to chromosome 8. While the underlying molecular basis is unknown, the gene controls fruit elongation and is found in the cultivar Rio Grande. *FAS* maps to chromosome 11, very closely to a fruit weight locus, *fw11.3*. Like *LC*, *FAS* controls locule number leading to flat shape (Huang and Van der Knaap 2011; Rodriguez et al. 2011).

Color image of this figure appears in the color plate section at the end of the book.

6.9.5.3 Fruit Color

The color of tomato fruit is one of the most important and complex traits contributing to fruit organoleptic and nutritional quality (Francis 1995). The final color is conditioned by pigment types and concentrations, and is genetically and environmentally regulated. Red color of tomato fruit is mainly determined by the carotenoid lycopene. More than 20 genes that influence the type, amount or distribution of tomato fruit carotenoids, and which are either directly or indirectly involved in the carotenoid pathway, have been characterized (see review by Labate et al. 2007). Although these monogenic mutants can substantially influence fruit pigmentation, in practice they have not made major contributions to the development of commercial cultivars with improved fruit color.

In order to analyse the genetic basis of quantitative variation of tomato fruit color, QTL mapping studies have been conducted using an intraspecific population (Saliba-Colombani et al. 2001; Causse et al. 2002) as well as interspecific populations involving six wild species: *S. pimpinellifolium* (Grandillo and Tanksley 1996a; Tanksley et al. 1996; Chen et al. 1999; Doganlar et al. 2002), *S. habrochaites* (Bernacchi et al. 1998a, b; Monforte et al. 2001; Kabelka et al. 2004; Yates et al. 2004), *S. pennellii* (Monforte et al. 2001; Liu et al. 2003b; Frary et al. 2004a), *S. peruvianum* (Fulton et al. 1997; Monforte et al. 2001; Yates et al. 2004), and *S. neorickii* (Fulton et al. 2000) (Table 6-3). Trait evaluations have been performed using various methods, including visual evaluation of internal and/or external color of fresh fruit, measurements of fresh fruit using a chromometer, analytical measurements of processed products, and estimates of lycopene and carotenoid content.

These studies have revealed numerous fruit color QTLs, which map to all 12 tomato chromosomes. Interestingly, QTL alleles enhancing the red color of the fruit have been found not only in the red-fruited *S. pimpinellifolium* and *S. lycopersicum* "cerasiforme" (cherry tomato) (Saliba-Colombani et al. 2001), but also in the green-fruited wild species *S. habrochaites*, *S. arcanum*, *S. neorickii* and *S. pennellii*. For example, in the *S. pennellii* LA0716 IL population the wild allele increased red color of the fruit for 10 of 16 QTLs (63%) that affected red color intensity (Liu et al. 2003b). It is notable that some of these wild fruit color QTL alleles substantially enhanced the intensity of the red color once introgressed into a *S. lycopersicum* genetic background. This was the case, for example, for the fruit color QTLs mapped at the bottom of chromosome 4: the NILs IL4-4, TA1160 and TA517, carrying introgressions for this region from the *S. pennellii*, *S. arcanum* and *S. habrochaites* wild relatives, respectively, showed increases in the intensity of red color that ranged from 52% for IL4-4 to 84% for TA1160 (Monforte et al. 2001). These results are of particular interest if one considers that all the three wild species are green-fruited, and lack an active enzyme for the last step of the pathway and cannot synthesize lycopene, suggesting that these fruit color wild QTL alleles might alter earlier steps of the pathway.

The complexity of the variation present in the clade *Solanum* sect. *Lycopersicon* for fruit color explains why the candidate gene approach does not seem to be an efficient tool for the identification of sequences that regulate quantitative variation of red color and pigment accumulation in tomato fruit (Liu et al. 2003b).

6.9.5.4 Fruit Firmness

Firmness is an important trait for fruit quality as it relates to fruit texture and shelf life, which are both considered critical characteristics for the consumer's perception of fruit quality (Causse et al. 2003, 2010; Sinesio et al. 2010). Tomato fruit firmness and viscosity depend on insoluble solids content. Although the genetic basis of fruit ripening and maturation has been dissected using long shelf-life mutants such as *rin*, *nor* and *Cnr* (Giovannoni 2004), increased knowledge of the genetic control of fruit firmness will be necessary in order to ensure breeding progress.

Towards this aim, several QTL mapping studies have been conducted in tomato that have analyzed fruit firmness, allowing the identification of numerous QTLs (Table 6-3; also see review by Labate et al. 2007). Two studies used the intraspecific RIL population derived from the cherry tomato × large tomato cross, and identified a total of two and five QTLs for fruit firmness measured by using a penetrometer (Saliba-Colombani et al. 2001) and by a taste panel (Causse et al. 2002), respectively. Two QTLs explained a large portion of the total phenotypic variation, with PVE values of 33.3% and 41% for firmness measured mechanically and by taste, respectively. In both cases the QTL alleles increasing firmness were derived from the cherry tomato parent. The other five QTLs explained less than 13% of the total phenotypic variation. Fruit firmness, along with other fruit quality traits, has been further analyzed in QTL-NILs derived from the same intraspecific cross (Lecomte et al. 2004a; Chaïb et al. 2006; Bertin et al. 2009)

Firmness was evaluated also in the five interspecific AB populations (Tanksley et al. 1996; Fulton et al. 1997, 2000; Bernacchi et al. 1998a; Frary et al. 2004a), in the *S. pimpinellifolium* BIL population (Doganlar et al. 2002), and in experiments conducted using NILs (Frary et al. 2003; Yates et al. 2004; Chaïb et al. 2006). In all AB populations and in the *S. pimpinellifolium* BIL population, firmness was measured by touching. None of the fruit firmness QTLs detected in these studies explained more than 16% of the total phenotypic variation. The largest number of fruit firmness QTLs (12) was detected in the *S. lycopersicum* x *S. neorickii* AB population, and for six of these QTLs the wild allele increased the trait (Fulton et al. 2000).

On average across the five studies, for 25% of QTLs the favorable allele originated from the wild species.

A comparison of the fruit firmness QTLs identified in the different studies indicated that in some instances QTLs detected by different methods co-localized (see review by Labate et al. 2007). Also, it is worth noting that on chromosomes 2, 5, and 10, QTLs for fimrness co-localized with the loci *rin*, *nor*, and *Cnr* (references in Labate et al. 2007).

6.9.5.5 *Primary Metabolic Traits*

To explore the genetic basis of tomato fruit primary metabolic traits, Schauer et al. (2006) realized a comprehensive metabolic profiling and phenotyping of the 76 *S. pennellii* LA0716 ILs (Table 6-3). They quantified 74 metabolites of known chemical structure including most plant amino and organic acids, sugars, sugar alcohols, fatty acids and vitamins C (AsA) and E (α-tocopherol). This analysis allowed the identification of ~890 fruit pericarp metabolite QTLs and 326 QTLs that modified yield-associated traits that were stable over two independent harvests. The ~890 metabolite QTLs mapped to only 76 loci, which clearly indicated that several of the analyzed traits co-localized. Interestingly, at least 50% of the metabolic loci were associated with QTLs that modified whole-plant yield-associated traits, and although many QTLs were found in the ILs for increased metabolite content, in most cases these metabolite enhancing QTLs were associated with a reduction in yield (Schauer et al. 2006). The results showed that the central hub in fruit metabolism is harvest index, which is a measure of the sink-source relationship between the reproductive and vegetative tissues. The analysis extended to an additional year of harvest and to the ILH, allowed the evaluation of the stability, heritability and mode of inheritance of the previously identified QTLs (Schauer et al. 2008). Although the majority of metabolites were influenced by a combination of genetic and environmental factors, relatively few of them could be defined as environmentally determined. Most of the primary metabolism QTLs were either dominantly or additively inherited, and a very limited number showed an overdominant pattern of inheritance. Although there was little evidence of heterosis for the metabolite QTLs, the analysis conducted on the heterozygous lines did not reveal any strong negative association between metabolite content and yield. The uncoupling of the metabolic and morphological traits observed in the ILHs was suggested to be due to the reduced fertility problems and range of fruit sizes displayed by the heterozygous lines compared to the homozygous counterparts.

Recently, the influence of fruit load on primary metabolite composition of tomato pericarp has been evaluated in 23 *S. chmilelewskii* LA1840 ILs (Do et al. 2010). The ILs were grown under high (HL) and low (LL) fruit loads, and the analysis of 62 metabolite traits allowed the identification of 240 and 128 metabolite QTLs under HL and LL coltivation, respectively. The data have been discussed in the context of trascriptomic data collected on a subset of the ILs, and, among others, the study demonstrated that the aminoacid content was strongly influenced by cultivation conditions (Prudent et al. 2009, 2010).

Several of the traits analyzed by Schauer et al. (2006, 2008) and by Do et al. (2010), such as sugars, acids and AsA, were addressed in other QTL mapping studies, and will be described in greater detail below.

6.9.5.6 Sugar and Acid Content

Sugar and organic acids represent about 60% of tomato dry matter and both are key components contributing to tomato flavor (Davies and Hobson 1981). Sugars constitute approximately 55% to 65% of the total soluble solids fraction and 50% of the total solids in tomato fruit. Commercial hybrid cultivars generally range from 4.5% to 6.0% soluble solids, whereas wild tomato species can reach concentrations near 15% (Hewitt and Garvey 1987). The storage sugars glucose and fructose accumulate in *S. lycopersicum* and *S. pimpinellifolium* fruit, while green-fruited wild tomato species accumulate sucrose in addition to glucose and fructose (Davies 1966).

Besides the comprehensive QTL mapping study conducted for fruit primary metabolites by Schauer et al. (2006, 2008), sugar content and related traits (brix, fructose, glucose, or sucrose content) have been analyzed in numerous other QTL mapping studies involving eight different species (Table 6-3): *S. cheesmaniae* (Paterson et al. 1991; Goldman et al. 1995), *S. chmielewskii* (Azanza et al. 1994; Osborn et al. 1987; Paterson et al. 1988, 1990; Prudent et al. 2009; Do et al. 2010), *S. habrochaites* (Bernacchi et al. 1998a), *S. neorickii* (Fulton et al. 2000), *S. pimpinellifolium* (Grandillo and Tanksley 1996a; Tanksley et al. 1996; Chen et al. 1999; Doganlar et al. 2002), *S. pennellii* (Eshed and Zamir 1995; Causse et al. 2004; Frary et al. 2004a), *S. arcanum* (Fulton et al. 1997), and *S. lycopersicum* "cerasiforme" (cherry tomato) (Saliba-Colombani et al. 2001; Georgelis et al. 2004; Lecomte et al. 2004a,b; Chaïb et al. 2006; Causse et al. 2007; Bertin et al. 2003, 2009).

The mapping results based on 14 populations were previously summarized (see review by Labate et al. 2007). QTLs were positioned on the tomato reference map (Tanksley et al. 1992) based on the nearest marker, resulting in a total of 95 QTLs concentrated in 56 regions. The lowest number of three QTLs was detected in the *S. lycopersicum* x *S. habroachaites* AB population (Bernacchi et al. 1998a), while up to 18 and 21 QTLs were identified for the *S. lycopersicum* x *S. arcanum* AB population (Fulton et al. 1997) and the *S. pennellii* LA0716 IL population (Eshed and Zamir 1995), respectively. Wild species alleles enhanced the trait value for most of the

QTLs. QTLs were detected in multiple populations for 28 regions, possibly indicating equivalency among QTLs (references in Labate et al. 2007).

The stability of QTLs involved in sugar concentration partly depends on environmental variations such as the year of growing periods (Chaïb et al. 2006), as well as the changes in carbon availability within the plant that can derive from different fruit loads (Prudent et al. 2009). Dissecting complex traits into elementary physiological processes may clarify the genetic control of quality traits and facilitate the identification of candidate genes. To identify the growth processes involved in QTLs for tomato fruit size and composition, a combination of ecophysiological modelling and QTL analysis was used on four QTL-NILs derived from the tasty cherry tomato × normal tasty large-fruited tomato cross (Bertin net al 2009). QTLs for fruit composition reflected differences in water accumulation rather than in sugar accumulation, with the exception of one line for which the upregulation of sucrose unloading and hexose transport and/or starch synthesis was suggested. Along this line, a model-based approach followed by genetic analysis was applied to a set of 20 *S. chmielewskii* ILs to identify key elementary processes underlying sugar concentration in tomato fruit including the assimilate supply, the transformation of sugars into other compounds, and their dilution by water uptake (Prudent et al. 2011). This work allowed uncoupling genetic from physiological relationships among processes, and thus provided new insights towards understanding tomato fruit sugar assimilation.

A major wild-species QTL, *Brix9-2-5*, that increased sugar yield of tomatoes by more than 20%, was isolated using map-based cloning (Fridman et al. 2000). The gene responsible for the QTL was the *lin5* gene, which encodes for an apoplastic invertase that modulates sugar partitioning to the fruit. The functional polymorphism of *Brix9-2-5* was identified as an amino acid change close to the catalytic site of the invertase protein, thus introducing the concept of a quantitative trait nucleotide (QTN, Fridman et al. 2004). High-resolution QTL mapping of the 9-cM *S. pennellii* introgression carrying *Brix9-2-5*, allowed the identification of another pleiotropic QTL for brix (*PW9-2-5*, 0.3 cM distant) that interacted epistatically with the genetic background (Fridman et al. 2002). *PW9-2-5* conferred an altered growth habit resulting in increased plant weight, yield, and brix.

Total acid content is very important in tomato not only for its contribution to flavor, but also because it can change the volatility of aromatic substances, and a low pH increases the safety of the processed product (Bucheli et al. 1999). Total acid content, pH, and/or the concentration of the various acids have been analyzed in several QTL mapping studies (Eshed and Zamir 1995; Causse et al. 2002, 2004; Fulton et al. 2002a; Chaïb et al. 2006; Lecomte et al. 2004a; Schauer et al. 2006, 2008; Bertin et al. 2009).

Fulton et al. (2002a) examined four AB populations each carrying introgressions from a different wild species (*S. pimpinellifolium, S. arcanum, S. habrochaites and S. neorickii*) for total acids, pH, citric acid, malic acid and glutamic acid. For these five traits a total of 102 QTLs were identified, and the PVE for individual QTLs ranged from 3% to 18%. For 51% of the QTLs the wild parent alleles were associated with an increase in the acidity-related trait value.

In addition, total acidity and/or pH were mapped in an intraspecific population derived from crosses between large tomato and cherry tomato (Causse et al. 2002; Georgelis et al. 2004), in derived QTL-NILs (Chaïb et al. 2006; Lecomte et al. 2004a; Causse et al. 2007) and in the *S. pennellii* IL population (Eshed and Zamir 1995). Citric acid, malic acid and glutamic acid were analyzed in the *S. pennellii* ILs by Causse et al. (2004) and/or by Schauer et al. (2006, 2008). In total, only a few regions were identified that influenced both acid and sugar content.

6.9.5.7 Nutritional and Antioxidant Compounds

AsA, tocopherol, carotenoids, and phenolic compounds are the main components of antioxidant activity in fruits and vegetables. Tomato is a valuable dietary source of water-soluble antioxidants such as AsA and lipid-soluble antioxidants such as carotenoids. Tomatoes and tomato-based products such as pastes and sauces play an important role in nutrition throughout the world (Willcox et al. 2003); there is keen interest in the genetic improvement of tomato antioxidant levels.

QTLs for the content of the carotenoids lycopene and ß-carotene were identified in several mapping studies (Table 6-3) (Chen et al. 1999; Fulton et al. 2000; Saliba-Colombani et al. 2001; Causse et al. 2002; Liu et al. 2003b; Yates et al. 2004; Rosseaux et al. 2005). Chen et al. (1999) used a backcross population involving *S. pimpinellifolium* LA0722 and identified eight QTLs for lycopene content. The PVE for single QTLs ranged from 5.2% to 12.4%. Consistent with the parental phenotypes, the wild QTL allele conferred an increase of lycopene content at all but one minor QTLs.

Saliba-Colombani et al. (2001) used a large tomato x cherry tomato RI population and identified two QTLs for lycopene, mapping on chromosomes 4 and 11, and three QTLs for ß-carotene, mapping on chromosomes 2, 3 and 8. The PVE for each QTL ranged from 5.6% to 17.8%, and in all cases the favorable allele derived from the cherry tomato parent. Only the lycopene QTL detected on chromosome 4 mapped close to a lycopene QTL previously identified by Chen et al. (1999).

Liu et al. (2003b) and Rousseaux et al. (2005) analyzed lycopene and ß-carotene content in 75 and 50 *S. pennellii* LA0716 ILs, respectively. Liu et al. (2003b) identified five QTLs for lycopene content on bins 2K, 3C, 5A, 6A and 12C, while two QTLs were mapped for ß-carotene content on bins 10E and 12C. For three lycopene QTLs and for both ß-carotene QTLs, the wild allele increased the trait value. These authors found lycopene content and the mean intensity of the internal fruit color of the fruits to be positively correlated ($r = 0.69$).

Rousseaux et al. (2005) identified four QTLs for lycopene content on chromosomes 3, 6 and 12, and two for ß-carotene content on chromosomes 6 and 12. For none of the four QTLs for lycopene content did the wild allele enhance the trait; however, the wild allele increased the levels of the compound for two ß-carotene QTLs. In this study, the QTLs found for lycopene and ß-carotene concentrations were concordant with known carotenoid mutations for yellow (*r*) and orange (*B* and *Del*) fruits that have been positioned on bins 3C, 6E and 12C, respectively. Liu et al. (2003b) identified a QTL for reduced lycopene concentration on bin 3C, and for reduced lycopene and increased ß-carotene concentrations on bin 12C. However, no QTL for lycopene and ß-carotene concentrations co-localized with the *B* mutation on bin 6E. In this same population, Rousseaux et al. (2005) identified nine QTLs for total phenolic concentration.

Several QTL mapping studies have been conducted to analyze the genetic basis of quantitative variation of AsA content in tomato fruit (Rosseaux et al. 2005; Schauer et al. 2006, 2008; Stevens et al. 2007, 2008). Rosseaux et al. (2005) used the population of 50 *S. pennellii* ILs and identified six QTLs for AsA content (total ascorbate) in three years of field experiments. An *S. pennellii* QTL allele on chromosome 12 increased ascorbate content 44% relative to *S. lycopersicum* M82 levels, even though *S. pennellii* had intrinsically low levels of ascorbate. A set of 76 *S. pennellii* ILs were used for detection of 17 QTLs for oxidized AsA (dehydroascorbate) content in peeled tomato pericarps (Schauer et al. 2006). A more comprehensive study of the genetic basis of acorbic acid content in tomato fruits was conducted by Stevens et al. (2007). The authors evaluated the variation in total AsA (i.e., reduced + oxidized forms) content of the whole fruit in three segregating populations: 75 *S. pennellii* LA0716 ILs, 130 BC_2S_1 families derived from an *S. lycopersicum* accession Ferum x *S. habrochaites* PI 247087 cross, and 144 RILs derived from the cross between a small-fruited cherry tomato and a large-fruited line. A minimum of 23 AsA QTLs (19 when expressed relative to fresh weight and 13 to dry matter weight) covering 15 bins were identified, with major effects found on bins 2K, 8B, 9D, 9J, 10D and 12B. In most cases the wild allele increased AsA content, although some favorable QTL alleles were found in the *S. lycopersicum* parent. Taken together, the results by Schauer et al. (2006) and by Stevens et al. (2007) indicated that at least 30 QTLs exist for AsA content. A comparison of the mapping results obtained in the different studies showed only a few QTLs to be held in common (Stevens et al. 2007). This could be explained by the fact that different AsA forms were measured, and that different sampling and extraction procedures were used.

A complete set of candidate genes involved in tomato AsA biosynthesis and metabolism have been mapped on the *S. pennellii* ILs (Stevens et al. 2007). This allowed the identification of several co-locations between genes and QTLs for AsA content (Stevens et al. 2007). Interestingly, the QTL mapped on bin 9-D co-localized with a monodehydroascorbate reductase (*MDHAR*) gene. This QTL was targeted for fine mapping and was found to be composed of several QTLs with sometimes opposing effects (Stevens et al. 2008). This study confirmed the *MDHAR* gene as a candidate component of the QTL, and also showed that part of the AsA QTL on chromosome 9 may be under environmental control.

A comparative microarray analysis conducted on fruits from *S. pennellii* LA0716 IL 12-4 (line with high levels of AsA) and cv M82 (control parent with low levels of AsA) suggested that the higher content of AsA in IL12-4 was mainly achieved by increasing flux through the L-galactonic acid pathway, which is driven by pectin degradation and may be triggered by ethylene (Di Matteo et al. 2010).

QTLs for α-tocopherol content in ripe fruit have been identified in the *S. pennellii* LA0716 IL population (Schauer et al. 2006). Two QTLs explaining variation in the α-tocopherol fruit content were mapped on chromosomes 6 and 9. Additional independent experiments available at the Tomato Functional Genomics database (Fei et al. 2006; http://ted.bti.cornell.edu/) have also reported differences in tocopherol content associated with the same genomic regions. Almeida et al. (2011) used a subset of the *S. pennellii* LA0716 IL population to identify QTLs associated to variations of tocopherol isoforms (α β, γ, and δ). This study confirmed the previously identified vitamine E QTLs on chromosomes 6 and 9, and novel QTLs were mapped on chromosomes 7 and 8. Moreover, integrated metabolic, genetic and genomic analyses allowed the idetification of 16 candidate loci putatively affecting tocopherol content in tomato fruit.

6.9.5.8 Volatile Compounds

Aroma volatiles along with sugars and acids contribute to the flavor of fresh tomato fruits (Buttery 1993). Although more than 400 aroma volatiles have been found in tomato fruit, only a subset of approximately 15–20 volatiles are present in sufficient quantities to impact tomato flavor, i.e., those that exceed odor perception thresholds (Petro-Turza 1987; Tandon et al. 2000; Klee 2010). These volatiles were synthesized from different precursors, including carotenoids, amino acids and lipids. Some of these volatiles were perceived positively while others contributed negatively to flavor (Baldwin et al. 1998). Despite the recognized importance of volatiles for tomato fruit flavor, our knowledge of the genetic and molecular bases of these compounds remains limited.

Three QTL mapping studies were conducted in tomato to identify the loci influencing volatile compounds of the fruit: one study used the intraspecific RI population derived from a cross between a cherry tomato line with a good overall aroma intensity and an inbred large-fruited line with a common taste (RI-cherry) (Saliba-Colombani et al. 2001), a second study used the *S. pennellii* LA0716 IL population (74 lines) to exploit the broad biochemical diversity of this population (Tieman et al. 2006), and a third study (Mathieu et al. 2008) used a set of 89 *S. habrochaites* LA1777 ILs and BILs available from the C. M. Rick Tomato Genetics Resource Center (TGRC, http://tgrc.ucdavis.edu) that were originally developed by Monforte and Tanksley (2000a) (Table 6-3). Each of the three studies analyzed slightly different subsets of volatiles.

Saliba-Colombani et al. (2001) measured 18 volatiles in the RI-cherry population and identified 29 QTLs for 12 of the quantified compounds. The QTLs were mainly localized on chromosomes 4, 8 and 9, sugesting the involvement of these regions in aroma-volatile metabolism. Major QTLs (PVE>30%) were identified for six aroma volatiles. In some cases, most of the variation was explained by one major QTL, e.g., orthomethoxyphenol and 3-methylpentan-1-ol contents for which the two QTLs *myp9.1* and *mno4.1* explained 87% and 63% of the phenotypic variation, respectively. In contrast, for other volatile compounds such as eugenol and 3-(methylthio) propanal, the variation in content was explained by the action of three and five QTLs, respectively.

Tieman et al. (2006) measured 23 volatiles in the *S. pennellii* LA0716 IL population and identified a minimum of 25 loci that altered content of one or more volatiles. Although ten volatiles were analyzed in both studies, only three QTLs were detected within the same regions. These differences between the two studies could be partly explained by the strong variation observed across years or environments (Tieman et al. 2006). In both studies, QTLs for several volatiles frequently co-localized. In most of the cases the clustered QTLs corresponded to volatiles derived from different metabolic pathways, suggesting a regulatory gene acting on several pathways. However, in a few cases the co-localizations involved QTLs for volatiles derived from the same pathway, suggesting the action of a gene within a single pathway.

Mathieu et al. (2008) characterized the population of *S. habrochaites* LA1777 ILs over multiple seasons and locations, and at least 30 QTLs affecting the emission of one or more of 24 different volatiles were identified. Some of the *S. habrochaites*-derived QTLs detected in this study localized to similar positions as previously mapped QTLs in the RI-cherry population (Saliba-Colombani et al. 2001) or in the *S. pennellii* LA0716 IL population (Tieman et al. 2006). It is worth noting that the rather large segments defined by some of the ILs do not exclude multiple loci affecting the same volatiles. The biochemical data obtained for the *S. habrochaites* IL population were combined with those previously obtained for the *S. pennelli* IL population, and the combined analysis allowed construction of a correlation database, and derivation of a metabolite tree for the major flavor associated tomato volatiles (Mathieu et al. 2008). This metabolic tree provides useful information about metabolic relationships and the potential limiting steps in the biosynthesis of a number of volatiles.

6.9.5.9 Sensory Traits

Sensory analysis, performed by trained panelists, is a highly valid method with which to study sensorial characteristics, particularly aroma and texture. The cherry tomato × large-fruited *S. lycopersicum* RI population has been used to study the genetic control of sensory quality through all of its components (Saliba-Colombani et al 2001; Causse et al. 2001, 2002) (Table 6-3). The population was characterized for physical and chemical traits such as color, weight, firmness, pH and titratable acidity, sugars and soluble solids content, concentration of 12 aroma volatiles (see above), as well as for sensory traits based on a panel of 56 judges. Judges were trained to quantify their sensory perceptions of taste (sourness, sweetness), aroma (aroma intensity, lemon, candy, citrus, pharmaceutical aroma) and texture (meltiness, firmness, mealiness, juiciness) and skin. A total of 38 significant QTLs were identified for 12 sensory attributes by CIM analysis, with a range of one to five QTL(s) detected per attribute. PVE ranged from 9% to 45%, with some QTLs (e.g., those for six aroma volatiles) associated with >20% of total phenotypic variation (Saliba-Colombani et al 2001).

Surprisingly, most QTLs were grouped within a few clusters (Causse et al. 2002). The favorable alleles for all fruit quality traits originated from the cherry tomato line (Causse et al. 2001, 2002). Despite their low heritability, QTLs for sensory traits were detected, and were often consistent with the location of QTLs using mechanical measurements (see Labate et al. 2007). For some of the most interesting regions QTL-NILs have been developed and further characterized (Lecomte et al. 2004a; Chaib et al. 2006, 2007; Zanor et al. 2009). Zanor et al. (2009) conducted an extensive profiling of both central primary metabolism and volatile organic components of the fruit in QTL-NILs known to possess characteristic sensory properties, as assessed by a trained tasting panel (Chaïb et al, 2006). This study allowed the identification of a number of QTLs which influenced the levels of primary metabolites and/or volatile organic components and to determine their co-location with previously defined sensory QTLs. In addition, correlation analyses revealed that there was a relatively weak association between the levels of primary metabolites and volatile organic compounds, with the exception of strong associations observed between the levels of sucrose and those of a number of volatile organic compounds. On the contrary, a number of interesting associations were found between sensorial properties and either the primary metabolites (e.g., sourness-alanine) or the volatile organic compounds (e.g., pharmaceutical aroma-guaiacol).

6.9.6 Flowering and Ripening Time

Artificial selection of cultivated tomato has resulted in the generation of early-flowering, day-length-insensitive cultivars suitable for production within different environments; in comparison, the wild progenitors of the crop require additional time and specific photoperiods in order to flower (Atherton and Harris 1986). Several studies have been conducted to identify QTLs associated with flowering time variation in tomato using experimental populations derived from interspecific crosses between cultivated tomato and *S. pimpinellifolium* (Weller et al. 1988; Grandillo and Tanklsey 1996a; Doganlar et al. 2002), *S. pennellii* (de Vicente and Tanksley 1993), IVT-KT1 (an early flowering hybrid line derived from *S. lycopersicum, S. pimpinellifolium* and *S. neorickii*, Lindhout et al. 1994), and *S. chmielewskii* (Jiménez-Gómez et al. 2007) (Table 6-3). Flowering time has been mainly scored as number of days to flowering (DTF) or the number of leaves produced by the plant before differentiation of the first inflorescence (LN). Taken together, these QTL studies identified at least nine genomic regions affecting DTF in tomato, and for all QTLs, except for the one mapped on chromosome 3, the *S. lycopersicum* allele reduced DTF. The trait LN has been analyzed in three interspecific mapping populations, and the results suggested the presence of at least eight regions affecting LN in tomato (Weller et al. 1988; Lindhout et al. 1994; Jiménez-Gómez et al. 2007). For the QTLs identified on chromosomes 1 and 3, the wild-species alleles reduced LN in comparison to *S. lycopersicum*.

QTL affecting DTF and LN co-localized at five of the 13 chromosomal regions detected, suggesting that genetic control of the two traits overlapped to a certain extent (Jiménez-Gómez et al. 2007). QTLs for other partially related traits, such as days to fruit set and days to fruit ripening, were identified in previous studies (Weller et al. 1988; Lindhout et al. 1994; Grandillo and Tanksley 1996a; Tanksley et al. 1996; Fulton et al. 1997; Monforte et al. 1999; Doganlar et al. 2000a, 2002). Jiménez-Gómez et al. (2007) showed that these QTLs co-localized with seven of the 13 regions containing DTF or LN QTLs, indicating that flowering time control might share common genetic components with ripening time traits. Moreover, three flowering time genes on the genetic map, *PHYB2, FALSIFLORA*, and a tomato *FLC-like* sequence, co-localized with some of the identified QTLs, suggesting their role as candidate genes that might have been targets of selection during domestication of tomato.

6.9.7 Flower Morphology and Reproductive Barriers

The ability of plants to develop new mating systems is an important aspect of evolution. The tomato clade (*Solanum* sect. *Lycopersicon*) and its four allied species (*S. ochrantum*, *S. juglandifolium*, *S. lycopersicoides* and *S. sitiens*) represent an ideal system for genetic studies of self-incompatibility (SI), unilateral incongruity or incompatibility (UI) and floral variation associated with mating system. In fact, cultivated tomato (*S. lycopersicum*) and its related wild tomato relatives cover the full range of mating systems, from autogamous to strict allogamy, with variations between and within species , and associated floral traits (Chapter 1; reviews by Peralta et al. 2008; Bedinger et al. 2011; Grandillo et al. 2011). *S. lycopersicum* and the other red- or orange-fruited species are all self-compatible (SC). Genetically controlled SI—intraspecific pollen recognition systems—is a common mechanism in many plants to prevent self-fertilization and hence promote outcrossing. Mutations causing loss of SI seem to represent the first steps in the evolution of autogamy from obligate allogamy (Takayama and Isogai 2005). In the Solanaceae SI specificity is controlled by a polymorphic *S* locus, encoding S-ribonucleases (S-RNases) expressed in the pistil (McClure et al. 1989) and S-locus F-box (SLF) proteins expressed in pollen (Sijacic et al. 2004), so that "self" pollen is rejected on pistils with matching *S* alleles. UI is a barrier that occurs when pollen of one species is rejected on pistils of a related species, but not vice versa (De Nettancourt 1977). In the Solanaceae pollen of *S. lycopersicum* (SC) is rejected on styles of almost all the green-fruited species, but not on styles of other red- or orange-fruited species (reviewed by Mutschler and Liedl 1994). Thus UI in the Solanaceae is mostly consistent with the "SI × SC rule" with some exceptions (Bedinger et al. 2011; Li et al. 2010). This would suggest that there is a relationship between the ability to reject self-pollen and the ability to reject foreign (interspecific) pollen, although several other observations demonstrate that there are also differences between UI and SI (Liedl et al. 1996; Bedinger et al. 2011; Covey et al. 2010; Li et al. 2010).

While the molecular mechanisms underlying SI have been well studied, much less known are the mechanisms underlying UI. Several QTLs underlying pistil-side unilateral and self-incompatibilities were identified in a BC_1 population deriving from an *S. lycopersicum* × *S. habrochaites* LA1777 (an SI accession) cross (Bernacchi and Tanksley 1997); the major QTL for both types of pollen rejection was mapped at or near the *S* locus on chromosome 1, which controls SI specificity (Tanksley and Loaiza-Figueroa

1985). The minor UI QTLs mapped on chromosomes 3 and 12 modulated this reproductive trait.

The genetics of pollen-side UI in the tomato clade, or any other system, has received less attention, and little data is available. Two to three genetic loci from *S. pennellii* have been identified that are necessary for the pollen to overcome incompatibility on pistils of *S. lycopersicum* × *S. lycopersicoides* and *S. lycopersicum* × *S. sitiens* hybrids (Chetelat and Deverna 1991; Pertuze et al. 2003). One factor, *ui1.1*, maps at or near the S locus on chromosome 1, and the other, *ui6.1*, is located on the short arm of chromosome 6. Map-based cloning of *ui6.1* showed that it encodes for a CUL1 protein and that the pollen-expressed *Cullin1* gene interacts with a gene at or near the S locus to control UI (Li and Chetelat 2010).

The absence of SI does not guarantee self-fertilization, and changes in floral morphology are also required to promote autogamy over allogamy (Kalisz et al. 1999). Most major QTLs for several floral traits important to pollination biology (e.g., number and size of flowers) have also been mapped to the S locus region, suggesting a gene complex controlling both genetic and morphological mechanisms of reproduction (de Vicente and Tanksley 1993; Bernacchi and Tanksley 1997). One key morphological trait is the degree to which the female stigma surface is either exserted above the anthers (promoting outcrossing) or recessed below the anthers (promoting self-fertilization) (Chen and Tanksley 2004). The cultivated tomato is autogamous and normally has flowers with recessed stigmas, whereas most wild tomato species are allogamous and have flowers with highly exserted stigmas (Rick 1979).

A few QTL mapping studies for stigma exsertion in tomato have been reported (Bernacchi and Tanksley 1997; Fulton et al. 1997; Georgiady et al. 2002). The studies conducted using crosses between wild SI species and SC cultivated tomato demonstrated that a single major QTL on chromosome 2, designated *stigma exsertion 2.1* (*se2.1*), explained most of the structural changes that accompanied the evolutionary transition from allogamous to autogamous flowers (Bernacchi and Tanksley 1997; Fulton et al. 1997). The study by Georgiady et al. (2002) involving a cross between allogamous and autogamous *S. pimpinellifolium* accessions identified a QTL of large effect on chromosome 8, and three QTLs affecting aspects of anther length on different chromosomes. Fine mapping studies have shown that *se2.1* is a complex locus composed of at least five closely linked genes: three controlling stamen length, one controlling style length, and one conditioning anther dehiscence (Chen and Tanksley 2004). Of these five loci, the locus controlling style length (*Style2.1*) accounted for the greatest change in stigma exsertion. Map-based cloning of *Style2.1* demonstrated that this gene encodes a putative transcription factor that regulates cell elongation in developing styles (Chen et al. 2007). The transition from allogamy to autogamy was accompanied by a mutation in the *Style2.1* promoter that results in a down-regulation of *Style2.1* expression during flower development.

In some cases stigma exsertion and lack of stylar UI barriers are not suffient to avoid reproductive isolation, suggesting that postzygotic barriers—primarily pollen and seed sterility in hybrids—also contribute to reproduction isolation in the tomato clade. Using the two populations of *S. pennellii* LA0716 ILs and *S. habrochaites* LA1777 ILs and BILs, it was found that hybrid pollen and seed infertility were each based on a modest number of QTLs, male (pollen) and other (seed) incompatibility factors were roughly comparable in number, and seed-infertility QTLs acted additively or recessively (Moyle and Graham 2005; Moyle and Nakazato 2008). In all cases sterility effects were quantitiative, and individual QTLs reduced fertility by 25–55% (pollen) to 40–80% (seed) in comparison with intraspecific leves. The co-localization of QTLs for pollen and seed sterility from the two *Solanum* studies suggested a shared evolutionary history for these QTLs, a nonrandom genomic distribution of loci causing sterility, and/or a tendency of certain genes to be involved in hybrid sterility. Results deriving from additional QTL mapping of pollen and/or seed sterility between cultivated tomato and other wild tomato relatives are consistent with the progressive accumulation of postzygotic QTLs with increasing divergence between species (Bedinger et al. 2011; Canady et al. 2005; Moyle and Nakazato 2008, 2010). In this respect, the recent study by Moyle and Nakazato (2010) provided direct empirical support for the Dobzhansky-Muller model of hybrid incompatibility, and the snowball prediction.

6.9.8 Leaf, Sepal, and Petal Morphology

There exists a great diversity in leaf and flower morphology among cultivated and wild tomato species. A few QTL mapping studies have analyzed the differences in leaf and/or floral morphology that characterize two highly divergent tomato species: cultivated tomato, *S. lycopersicum*, and its wild relative *S. pennellii* (de Vicente and Tanksley 1993; Holtan and Hake 2003; Frary et al 2004b). *S. pennelli* is a xerophytic species adapted to arid conditions, and its leaves are relatively small, consisting of small, rounded, thick leaflets attached to the rachis on short petiolules. In contrast, the leaves of *S. lycopersicum* are larger and characterized by numerous elongated leaflets with narrowed tips on elongated petiolules.

Holtan and Hake (2003) used the population of 50 *S. pennellii* LA0716 ILs to identify QTLs for leaf dissection traits. Eight traits were analyzed and a total of 30 QTLs were identified, of which 22 primarily affected leaf dissection and eight primarily affected leaf size. Five of the detected QTLs resulted in transgressive phenotypes with drastically increased dissection relative to both parent lines.

Frary et al. (2004b) analyzed an *S. lycopersicum* × *S. pennellii* F_2 population for 18 traits related to the lateral leaflets of tomato's compound leaves and the most leaflike of the floral organs, the sepals and petals, which together comprise the perianth. This comparative study aimed at determining whether the same QTLs underly aspects of both vegetative and floral morphology. A total of 36 highly significant QTLs were identified for the 18 traits, and some of them explained a large portion of the total phenotypic variation (up to 54.2%). A pair of leaflet trait QTLs identified in this study may correspond to leaflet shape QTLs previously identifed by de Vicente and Tanksley (1993). While little or no overlap in QTL positioning between different organs was observed in tomato, in contrast, common QTLs for control of leaflet shape between tomato and eggplant were detected (Frary et al. 2004b). As the authors stated, these findings suggested that the gencs determining the size and shape of leaflets, sepals, and petals in tomato are organ specific, but that there has been some conservation in the genes controlling leaf morphology within the Solanaceae.

6.9.9 Seed Weight

The seeds of cultivated tomato are much larger than those of its wild relatives (Doganlar et al. 2000b). This increase in seed size is a result of the selection for yield, uniform germination, and seedling vigor that has been exerted during domestication and breeding. Several QTL mapping studies have been conducted to identify seed-weight QTLs in tomato (Table 6-3). The mapping results obtained with seven different interspecific populations involving five wild relatives have been summarized by Doganlar et al. (2000b). The data showed that overall, a total of 24 seed-weight QTLs were identified covering all 12 tomato chromosomes. The largest number of 14 QTLs was identified in an *S. lycopersicum* x *S. galapagense* LA0483 F_8RIL population (Goldman et al. 1995). One seed-weight QTL (*sw4.1*) was detected in all five wild species, whereas 11 QTLs were detected in two or more different species, and 12 were identified in only one species. The PVE for each QTL was determined in three studies, and varied from 3% to 24.5% (Goldman et al. 1995; Grandillo and Tanksley 1996a; Doganlar et al. 2000b). Among all the identified seed-weight QTLs, *sw4.1* had the largest effect on the trait accounting for 8.4% to 24.5% of the PVE in segregating populations, and was independent from fruit-weight QTLs. Moreover, this QTL is conserved across tomato species and potentially plays a role in the evolution and domestication of cultivated tomato (Doganlar et al. 2000b). For these reasons, *Sw4.1* has been targeted for map-based cloning (Doganlar et al. 2000b). Through applying a combination of genetic, developmental, molecular and transgeneic techniques, Orsi and Tanksley (2009) have identified the *Sw4.1* QTL as an ABC transporter gene.

6.10 Conclusions and Future Prospects

We have reviewed more than 25 years of research conducted to identify the genetic and molecular basis of quantitative variation in tomato. Following the demonstration by Paterson et al. (1988) that whole genome molecular maps could be used to resolve complex traits into Mendelian factors, QTL mapping studies in tomato flourished and were extended to hundreds of traits involved in plant morphology, adaptation, yield, metabolism and gene expression. The outcome of these studies has been the identification of thousands of QTLs, many of which are potentially of value for crop improvement, and whose molecular basis is waiting to be deciphered.

The reduced polymorphism that characterizes cultivated tomato has fostered the extensive utilization of interspecific mapping populations. Since the earliest studies, it has been clear that wild relatives can provide a rich source of valuable QTL alleles for the improvement of modern cultivars, including traits for which the wild species do not show a favorable phenotype. These findings have changed the paradigm of plant breeding from selection of phenotypes towards selection of alleles (Tanksley and McCouch 1997). The development of permanent mapping populations deriving from interspecific crosses, such as 'exotic libraries' or IL libraries, have provided new tools for efficient analysis of complex traits in tomato, and for the exploitation of tomato natural biodiversity in MAS breeding programs (Zamir 2001; Lippman et al. 2007; Grandillo et al. 2008, 2011). The IL approach allowed the map based cloning of the first QTL to be isolated (Frary et al. 2000; Fridman et al. 2000), and, in addition, it has proven to be an ideal system within which to explore the genetic basis of heterosis for 'real-world' applications, as shown by the development of a new leading hybrid of processing tomato (Lippman et al. 2007).

The genomics tools and resources that are increasingly available will greatly improve the efficiency of QTL mapping and cloning of the corresponding genes. Comparative mapping and the use of COS markers will facilitate synteny studies among the Solanaceae crops and the microsynteny with *Arabidopsis*; this should accelerate gene and QTL characterization, and gradually increase our knowledge of tomato gene function (Fulton et al. 2002b). The more than 330,000 ESTs available for tomato in public databases (see Chapter 9), e.g., dbEST: (http://www.ncbi.nlm.nih.gov/dbEST/), SGN (Mueller et al. 2005; http://solgenomics.net/) and SolEST db (D'Agostino et al. 2009; http://biosrv.cab.unina.it/solestdb/) represent a valuable source of unbiased SNPs, which will improve genome mapping, MAS, and positional cloning within *S. lycopersicum* intraspecific germplasm (Labate et al. 2009). From the EST data more than 30,000 unigenes have been defined (Van der Hooven et al. 2002; http://solgenomics.net/; http://www.ncbi.nlm.nih.gov) that represent a large set of candidate sequences for numerous important quantitative traits. Additional candidate genes for QTLs of interest can be identified by applying high-throughput expression technologies such as transcriptional, proteomic and metabolic profiling to selected ILs or QTL-NILs (see Chapters 12 and 13) (Baxter et al. 2005; Faurobert et al. 2007; Li et al. 2005; Overy et al. 2005; Lee et al. 2011).

A high quality draft sequence of the 950 Mb tomato genome is available, as are the sequences of related species (http://solgenomics.net/; see Chapter 10), including that of the potato genome (The Potato Genome Sequencing Consortium et al. 2011). The information derived from the tomato genome sequence will further facilitate linking the Solanaceae to each other and to other families, and accelerate candidate gene identification. Therefore, although only a handful of QTLs have been cloned in tomato to date, the application of systems biology approaches combining transcriptomic, proteomic and metabolomic analyses with the tomato genome sequence will increase the number of QTLs for which the molecular bases will be known.

In order to take full advantage of the potential of genomic-assisted comparative QTL studies, it is necessary to reach an integrative view of the large amount of data that is being generated. This will allow linkages among genome information (including gene content, expression and function), QTLs and plant breeding (see Chapter 14). Given the major role that extensive and precise phenotyping will play in the future, quantitative genetics is now facing the bioinformatic challenge of presenting, in silico, detailed information concerning the components of genotypic and phenotypic variation in a unified ontology-based genomic framework (Gur et al. 2004; Lippman et al. 2007).

The increasing knowledge deriving from the many 'omics' tools will continue to impact plant breeding methodologies (Bai and Lindhout 2007). Knowing the map position of QTLs of agronomic interest and the markers tightly linked to them, the allelic variation at the loci can be assessed using marker haplotypes on available germplasm collections; extensive phenotyping can then be used to determine the phenotypic contribution of each allele. This knowledge will enable the breeder to select the best combinations of alleles and to design programs to combine traits in new, superior genotypes following the 'breeding by design' concept (Peleman and Van der Voort 2003).

Acknowledgments

The authors are very thankful to Joanne Labate for critical reading and editing of the manuscript. Work in the laboratory of S Grandillo was supported by the European Union (EU) program EU-SOL (contract PL 016214-2 EU-SOL) and in part by the Italian MIUR project GenoPOM. This work was also in part supported by the 'International Exchange Program of the University of Naples Federico II' to S Grandillo. Contribution n. 370 from CNR-IGV, Institute of Plant Genetics, Research Division Portici.

References

Almeida J, Quadrana L, Asís R, Setta N, de Godoy F, Bermúdez L, Otaiza SN, Corrêa da Silva JV, Fernie AR, Carrari F, Rossi M (2011) Genetic dissection of vitamin E biosynthesis in tomato J Exp Bot 62(11): 3781–3798.

Alpert K, Tanksley S (1996) High-resolution mapping and isolation of a yeast artificial chromosome contig containing *fw2.2*: A major fruit weight quantitative trait locus in tomato. Proc Natl Acad Sci USA 93: 15503–15507.

Alpert K, Grandillo S, Tanksley SD (1995) *fw2.2*: a major QTL controlling fruit weight is common to both red- and green-fruited tomato species. Theor Appl Genet 91: 994–1000.

Asins MJ, Bolarín MC, Pérez-Alfocea F, Estañ MT, Martínez-Andújar C, Albacete A, Villalta I, Bernet GP, Dodd IC, Carbonell EA (2010) Genetic analysis of physiological components of salt tolerance conferred by *Solanum* rootstocks. What is the rootstock doing for the scion? Theor Appl Genet 121(1): 105–115.

Ashikari M, Matsuoka M (2006) Identification, isolation and pyramiding of quantitative trait loci for rice breeding. Trends Plant Sci 11(7): 344–350.

Atherton JG, Harris GP (1986) Flowering. In: Atherton JG, Harris GP (eds) The Tomato Crop: A Scientific Basis for Improvement. Chapman and Hall, London, UK, pp 167–200.

Azanza F, Young TE, Kim D, Tanksley SD, Juvik JA (1994) Characterization of the effect of introgressed segments of chromosome 7 and 10 from *Lycopersicon chmielewskii* on tomato soluble solids, pH, and yield. Theor Appl Genet 87: 965–972.

Bai YL, Lindhout P (2007) Domestication and breeding of tomatoes: what have we gained and what can we gain in the future? Ann Bot 100: 1085–1094.

Bai Y, Huang CC, van der Hulst R, Meijer-Dekens F, Bonnema G, Lindhout P (2003) QTLs for tomato powdery mildew resistance (*Oidium lycopersici*) in *Lycopersicon parviflorum* G1.1601 co-localize with two qualitative powdery mildew resistance genes. Mol Plant Microbe Interact 16: 169–176.

Baldwin EA, Nisperos-Carriedo MO, Baker R, Scott JW (1991a) Quantitative analysis of flavour parameters in 6 Florida tomato cultivars (*Lycopersicon esculentum* Mill). J Agri Food Chem 39: 1135–1140.

Baldwin EA, Nisperos-Carriedo MO, Moshonas MG (1991b) Quantitative analysis of flavor and other volatiles and for certain constituents of two tomato cultivars during ripening. J Am Soc Hort Sci 116: 265–269.

Baldwin EA, Scott JW, Einstein MA, Malundo TMM, Carr BT, Shewfelt RL, Tandon KS (1998) Relationship between sensory and instrumental analysis for tomato flavour. J Am Soc Hort Sci 123: 906–915.

Barrero LS, Tanksley SD (2004) Evaluating the genetic basis of multiple-locule fruit in a broad cross section of tomato cultivars. Theor Appl Genet 109: 669–679.

Basten CJ, Weir BS, Zeng ZB (1997) QTL cartographer. A reference manual and tutorial for QTL mapping. Department of Statistics, North Carolina State University, Raleigh, NC, USA.

Baxter CJ, Sabar M, Quick WP, Sweetlove LJ (2005) Comparison of changes in fruit gene expression in tomato introgression lines provides evidence of genome-wide transcriptional changes and reveals links to mapped QTLs and described traits. J Exp Bot 56: 1591–1604.

Bedinger PA, Chetelat RT, McClure B, Moyle LC, Rose JK, Stack SM, van der Knaap E, Baek YS, Lopez-Casado G, Covey PA, Kumar A, Li W, Nunez R, Cruz-Garcia F, Royer S

(2011) Interspecific reproductive barriers in the tomato clade: opportunities to decipher mechanisms of reproductive isolation. Sex Plant Reprod 24(3): 171–187.

Bernacchi D, Tanksley SD (1997) An interspecific backcross of *Lycopersicon esculentum* × *L. hirsutum*: linkage analysis and a QTL study of sexual compatibility factors and floral traits. Genetics 147: 861–877.

Bernacchi D, Beck-Bunn T, Eshed Y, Lopez J, Petiard V, Uhlig J, Zamir D, Tanksley S (1998a) Advanced backcross QTL analysis in tomato. I. Identification of QTLs for traits of agronomic importance from *Lycopersicon hirsutum*. Theor Appl Genet 97: 381–397.

Bernacchi D, Beck-Bunn T, Emmatty D, Eshed Y, Inai S, Lopez J, Petiard V, Sayama H, Uhlig J, Zamir D, Tanksley S (1998b) Advanced backcross QTL analysis of tomato. II. Evaluation of near-isogenic lines carrying single-donor introgressions for desirable wild QTL-alleles derived from *Lycopersicon hirsutum* and *L. pimpinellifolium*. Theor Appl Genet 97: 170–180 and 1191–1196.

Bertin N, Borel C, Brunel B, Cheniclet C, Causse M (2003) Do genetic make-up and growth manipulation affect tomato fruit size by cell number, or cell size and DNA endoreduplication? Ann Bot 92(3): 415–424.

Bertin N, Causse M, Brunel B, Tricon D, Génard M (2009) Identification of growth processes involved in QTLs for tomato fruit size and composition. J Exp Bot 60(1): 237–248.

Blauth SL, JC Steffens, GA Churchill, MA Mutschler (1999) Identification of QTLs controlling acylsugar fatty acid composition in an intraspecific population of *Lycopersicon pennellii* (Corr.) D'Arcy. Theor Appl Genet 99: 373–381.

Bogdan M, Ghosh JK, Doerge RW (2004) Modifying the Schwarz Bayesian information criterion to locate multiple interacting quantitative trait loci. Genetics 167: 989–999.

Bretó MP, Asins MJ, Carbonell EA (1994) Salt tolerance in *Lycopersicon* species. III. Detection of quantitative trait loci by means of molecular markers. Theor Appl Genet 88: 395–401.

Brewer MT, Moyseenko JB, Monforte AJ, van der Knaap E (2007) Morphological variation in tomato fruit: a comprehensive analysis and identification of loci controlling fruit shape and development. J Exp Bot 58: 1339–1349.

Brog M, Tripodi P, Cammareri M, Osorio-Algar S, Fraser P, Fernie A, Grandillo S, Zamir D (2011) Towards phenomics of the sequenced genomes of the cultivated tomato and its wild ancestor *Solanum pimpinellifolium*. Proceedings of the Joint Meeting AGI-SIBV-SIGA Assisi, Italy, 19–22 September, 2011 ISBN 978-88-904570-2-9.

Brouwer DJ, St Clair DA (2004) Fine mapping of three quantitative trait loci for late blight resistance in tomato using near isogenic lines (NILs) and sub-NILs. Theor Appl Genet 108: 628–638.

Brouwer DJ, Jones, ES, St Clair DA (2004) QTL analysis of quantitative resistance to *Phytophthora infestans* (late blight) in tomato and comparison with potato. Genome 47: 475–492.

Bucheli P, Voirol E, de la Torre R, Lopez J, Rytz A, Tanksley SD, Petiard V (1999) Definition of nonvolatile markers for flavor of tomato (*Lycopersicon esculentum* Mill.) as tools in selection and breeding. J Agri Food Chem 47: 659–664.

Buttery RG (1993) Quantitative and sensory aspects of flavour of tomato and other vegetables and fruits. In: Acree TE, Teranishsi R (eds) Flavor Science: Sensible Principles and Techniques. American Chemical Society, Washington DC, USA, pp 259–286.

Canady MA, Meglic V, Chetelat RT (2005) A library of *Solanum lycopersicoides* introgression lines in cultivated tomato. Genome 48: 685–697.

Carmeille A, Caranta EC, Dintinger EJ, Prior P, Luisetti EJ, Besse EP (2006) Identification of QTLs for *Ralstonia solanacearum* race 3-phylotype II resistance in tomato. Theor Appl Genet 114: 110–121.

Causse M, Saliba-Colombani V, Lesschaeve I, Buret M (2001) Genetic analysis of organoleptic quality in fresh market tomato. 2. Mapping QTLs for sensory attributes. Theor Appl Genet 102: 273–283.

Causse M, Saliba-Colombani V, Lecomte L, Duffé P, Rousselle P, Buret M (2002) QTL analysis of fruit quality in fresh market tomato: a few chromosome regions control the variation of sensory and instrumental traits. J Exp Bot 53: 2089–2098.

Causse M, Buret M, Robini K, Verschave P (2003) Inheritance of nutritional and sensory quality traits in fresh market tomato and relation to consumer preferences. J Food Sci 68: 2342–2350.

Causse M, Duffe P, Gomez MC, Buret M, Damidaux R, Zamir D, Gur A, Chevalier C, Lemaire-Chamley M, Rothan C (2004) A genetic map of candidate genes and QTLs involved in tomato fruit size and composition. J Exp Bot 55: 1671–1685.

Causse M, Chaïb J, Lecomte L, Buret M, Hospital F (2007) Both additivity and epistasis control the genetic variation for fruit quality traits in tomato. Theor Appl Genet 115(3): 429–442.

Causse M, Friguet C, Coiret C, Lépicier M, Navez B, Lee M, Holthuysen N, Sinesio F, Moneta E, Grandillo S (2010) Consumer preferences for fresh tomato at the European scale: a common segmentation on taste and firmness. J Food Sci: 75(9): S531–541.

Chaerani R, Smulders MJ, van der Linden CG, Vosman B, Stam P, Voorrips RE (2007) QTL identification for early blight resistance (*Alternaria solani*) in a *Solanum lycopersicum* × *S. arcanum* cross. Theor Appl Genet 114(3): 439–450.

Chagué V, Mercier JC, Guénard M, Courcel Ad, Vedel F (1997) Identification of RAPD markers linked to a locus involved in quantitative resistance to TYLCV in tomato by bulked segregant analysis. Theor Appl Genet 95: 671–677.

Chaïb J, Lecomte L, Buret M, Causse M (2006) Stability over genetic backgrounds, generations and years of quantitative trait locus (QTLs) for organoleptic quality in tomato. Theor Appl Genet 112: 934–944.

Chaïb J, Devaux MF, Grotte MG, Robini K, Causse M, Lahaye M, Marty I (2007) Physiological relationships among physical, sensory, and morphological attributes of texture in tomato fruits. J Exp Bot 58(8): 1915–1925.

Chen KY, Tanksley SD (2004) High-resolution mapping and functional analysis of *se2.1*: a major stigma exsertion quantitative trait locus associated with the evolution from allogamy to autogamy in the genus *Lycopersicon*. Genetics 168: 1563–1573.

Chen FQ, Foolad MR, Hyman J, St Clair DA, Beelman RB (1999) Mapping of QTLs for lycopene and other fruit traits in a *Lycopersicon esculentum* x *L. pimpinellifolium* cross and comparison of QTLs across tomato species. Mol Breed 5: 283–299.

Chen KY, Cong B, Wing R, Vrebalov J, Tanksley SD (2007) Changes in regulation of a transcription factor lead to autogamy in cultivated tomatoes. Science 318: 643–645.

Chetelat RT, DeVerna JW (1991) Expression of unilateral incompatibility in pollen of *Lycopersicon pennellii* is determined by major loci on chromosomes 1, 6 and 10. Theor Appl Genet 82: 704–712.

Coaker GL, Francis DM (2004) Mapping, genetic effects, and epistatic interaction of two bacterial canker resistance QTLs from *Lycopersicon hirsutum*. Theor Appl Genet 108: 1047–1055.

Coaker GL, Meulia T, Kabelka EA, Jones AK, Francis DM (2002) A QTL controlling stem morphology and vascular development in *Lycopersicon esculentum* x *Lycopersicon hirsutum* (Solanaceae) crosses is located on chromosome 2. Am J Bot 89: 1859–1866.

Cong B, Tanksley SD (2006) Fw2.2 and cell cycle control in developing tomato fruit: A possible example of gene co-option in the evolution of a novel organ. Plant Mol Biol 62: 867–880.

Cong B, Liu J, Tanksley SD (2002) Natural alleles at a tomato fruit size quantitative trait locus differ by heterochronic regulatory mutations. Proc Natl Acad Sci USA 99: 13606–13611.

Cong B, Barrero LS, Tanksley SD (2008) Regulatory change in YABBY-like transcription factor led to evolution of extreme fruit size during tomato domestication. Nat Genet 40: 800–804.

Covey PA, Kondo K, Welch L, Frank E, Sianta S, Kumar A, Nuñez R, Lopez-Casado G, van der Knaap E, Rose JK, McClure BA, Bedinger PA (2010) Multiple features that distinguish unilateral incongruity and self-incompatibility in the tomato clade. Plant J 64(3): 367–378.

D'Agostino N, Traini A, Frusciante L, Chiusano ML (2009) SolEST database: a "one-stop shop" approach to the study of Solanaceae transcriptomes. BMC Plant Biology 9: 142.

Danesh D, Aarons S, McGill GE, Young ND (1994) Genetic dissection of oligogenic resistance to bacterial wilt in tomato. Mol Plant-Microbe Interact 7: 464–471.

Davies JN (1966) Occurrence of sucrose in the fruit of some species of *Lycopersicon*. Nature 209: 640–641.

Davies JN, Hobson GE (1981) The constituents of tomato fruit—The influence of environment, nutrition and genotype. Crit Rev Food Sci Nutr 15: 205–280.

Davis J, Yu D, Evans W, Gokirmak T, Chetelat RT, Stotz HU (2009) Mapping of loci from *Solanum lycopersicoides* conferring resistance or susceptibility to *Botrytis cinerea* in tomato. Theor Appl Genet 119: 305–314.

De Nettancourt D (1977) Incompatibility in Angiosperms. Springer-Verlag, Berlin.

de Vicente MC, Tanksley SD (1993) QTL analysis of transgressive segregation in an interspecific tomato cross. Genetics 134: 585–596.

Di Dato F, Brog M, Tripodi P, Zamir D, Grandillo S (2010) QTL analysis of *Solanum neorickii* (LA2133) BILs anchored to a common set of COSII markers. Proceedings of the 54th Italian Society of Agricultural Genetics Annual Congress Matera, Italy, September 27–30, 2010 ISBN 978-88-904570-0-5.

Di Matteo A, Sacco A, Anacleria M, Pezzotti M, Delledonne M, Ferrarini A, Frusciante L, Barone A (2010) The ascorbic acid content of tomato fruits is associated with the expression of genes involved in pectin degradation. BMC Plant Biol 10: 163.

Do PT, Prudent M, Sulpice R, Causse M, Fernie AR (2010) The influence of fruit load on the tomato pericarp metabolome in a *Solanum chmielewskii* introgression line population. Plant Physiol: 154(3): 1128–1142.

Doerge RW, Zeng ZB, Weir BS (1997) Statistical issues in the search for genes affecting quantitative traits in experimental populations. Stat Sci 12: 195–219.

Doganlar S, Tanksley SD, Mutschler MA (2000a) Identification and molecular mapping of loci controlling fruit ripening time in tomato. Theor Appl Genet 100(2): 249–255.

Doganlar S, Frary A, Tanksley SD (2000b) The genetic basis of seed-weight variation: tomato as a model system. Theor Appl Genet 100: 1267–1273.

Doganlar S, Frary A, Ku H-M, Tanksley SD (2002) Mapping quantitative trait loci in inbred backcross lines of *Lycopersicon pimpinellifolium* (LA1589). Genome 45: 1189–1202.

East EM, Jones DF (1919) Inbreeding and outbreeding: Their genetic and sociological significance. Monographs on Experimental Biology. Loeb J, Morgan TH, Osterhout WJV (eds). JB Lippincott Co, Philadelphia and London, 285 pp.

Edwards MD, Stuber CW, Wendel JF (1987) Molecular-marker facilitated investigations of quantitative-trait loci in maize. I. Numbers, genomic distribution and types of gene action. Genetics 116: 113–125.

Eshed Y, Zamir D (1994) Introgressions from *Lycopersicon pennellii* can improve the soluble solids yield of tomato hybrids. Theor Appl Genet 88: 891–897.

Eshed Y, Zamir D (1995) An introgression line population of *Lycopersicon pennellii* in the cultivated tomato enables the identification and fine mapping of yield-associated QTL. Genetics 141: 1147–1162.

Eshed Y, Zamir D (1996) Less-than-additive epistatic interactions of quantitative trait loci in tomato. Genetics 143: 1807–1817.

Eshed Y, Gera G, Zamir D (1996) A genome-wide search for wild-species alleles that increase horticultural yield of processing tomatoes. Theor Appl Genet 93: 877–886.

Estañ MT, Villalta I, Bolarín MC, Carbonell EA, Asins MJ (2009) Identification of fruit yield loci controlling the salt tolerance conferred by *Solanum* rootstocks. Theor Appl Genet 118: 305–312.

Falconer DS (1989) Introduction to Quantitative Genetics, 3rd edn. Longman Scientific & Technical, Essex, UK.

Faurobert M, Mihr C, Bertin N, Pawłowski T, Negroni L, Sommerer N, Causse M (2007) Major proteome variations associated with cherry tomato pericarp development and ripening. Plant Physiol 43: 1327–1346.

Fei Z, Tang X, Alba R, Giovannoni J (2006) Tomato Expression Database (TED): a suite of data presentation and analysis tools. Nucleic Acids Res 1: 34.

Finkers R, van den Berg P, van Berloo R, ten Have A, van Heusden AW, van Kan JA, Lindhout P (2007a) Three QTLs for *Botrytis cinerea* resistance in tomato. Theor Appl Genet 114(4): 585–593.

Finkers R, van Heusden AW, Meijer-Dekens F, van Kan JAL, Maris P, Lindhout P (2007b) The construction of a *Solanum habrochaites* LYC4 introgression line population and the identification of QTLs for resistance to *Botrytis cinerea*. Theor Appl Genet 114: 1071–1080.

Foolad MR (1999a) Comparison of salt tolerance during seed germination and vegetative growth in tomato by QTL mapping. Genome 42: 727–734.

Foolad MR (1999b) Genetics of salt tolerance and cold tolerance in tomato: quantitative analysis and QTL mapping. Plant Biotechnol 16: 55–64.

Foolad MR (2004) Recent genetic advances in genetics of salt tolerance in tomato. Plant Cell Tiss Org Cult 76: 101–119.

Foolad MR (2007) Genome mapping and molecular breeding of tomato. Int J Plant Genomics doi: 10.1155/2007/64358.

Foolad MR, Chen FQ (1998) RAPD markers associated with salt tolerance in an interspecific cross of tomato (*Lycopersicon esculentum x L. pennellii*). Plant Cell Rep 17: 306–312.

Foolad MR, Chen FQ (1999) RFLP mapping of QTLs conferring salt tolerance during vegetative stage in tomato. Theor Appl Genet 99: 235–243.

Foolad MR, Jones RA (1993) Mapping salt-tolerance genes in tomato (*Lycopersicon esculentum*) using trait-based marker analysis. Theor Appl Genet 87: 184–192.

Foolad MR, Lin GY (1998) Genetic analysis of low temperature tolerance during germination in tomato, *Lycopersicon esculentum* Mill. Plant Breed 117: 171–176.

Foolad MR, Stoltz T, Dervinis C, Rodriguez RL, Jones RA (1997) Mapping QTLs conferring salt tolerance during germination in tomato by selective genotyping. Mol Breed 3: 269–277.

Foolad MR, Chen FQ, Lin GY (1998a) RFLP mapping of QTLs conferring salt tolerance during germination in an interspecific cross of tomato. Theor Appl Genet 97: 1133–1144.

Foolad MR, Chen FQ, Lin GY (1998b) RFLP mapping of QTLs conferring cold tolerance during seed germination in an interspecific cross of tomato. Mol Breed 4: 519–529.

Foolad MR, Lin GY, Chen FQ (1999) Comparison of QTLs for seed germination under non-stress, cold stress and salt stress in tomato. Plant Breed 118: 167–173.

Foolad MR, Zhang LP, Lin GY (2001) Identification and validation of QTLs for salt tolerance during vegetative growth in tomato by selective genotyping. Genome 44: 444–454.

Foolad MR, Zhang LP, Khan AA, Nino-Liu D, Lin GY (2002) Identification of QTLs for early blight (*Alternaria solani*) resistance in tomato using backcross populations of a *Lycopersicon esculentum x L. hirsutum* cross. Theor Appl Genet 104: 945–958.

Foolad MR, Zhang LP, Subbiah P (2003a) Genetics of drought tolerance during seed germination in tomato: inheritance and QTL mapping. Genome 46: 536–545.

Foolad MR, Zhang LP, Subbiah P (2003b) Relationships among cold, salt and drought tolerance during seed germination in tomato: inheritance and QTL mapping. Acta Hort 618: 47–57.

Foolad MR, Subbiah P, Zhang LP (2007) Common QTL affect the rate of tomato seed germination under different stress and nonstress conditions. Int J Plant Genom doi: 10.1155/2007/97386.

Francis FJ (1995) Quality as influenced by color. Food Qual Prefer 6: 149–155.

Francis DM, Kabelka E, Bell J, Franchino B, St. Clair D (2001) Resistance to bacterial canker in tomato (*Lycopersicon hirsutum* LA407) and its progeny derived from crosses to *L. esculentum*. Plant Dis 85: 1171–1176.

Frary A, Nesbitt TC, Frary A, Grandillo S, van der Knaap E, Cong B, Liu J, Meller J, Elber R, Alpert KB, Tanksley SD (2000) *fw-2.2*: a quantitative trait locus key to the evolution of tomato fruit size. Science 289: 85–88.

Frary A, Doganlar S, Frampton A, Fulton T, Uhlig J, Yates H, Tanksley S (2003) Fine mapping of quantitative trait loci for improved fruit characteristics from *Lycopersicon chmielewskii* chromosome 1. Genome 46: 235–243.

Frary A, Fulton TM, Zamir D, Tanksley SD (2004a) Advanced backcross QTL analysis of a *Lycopersicon esculentum* × *L. pennellii* cross and identification of possible orthologs in the Solanaceae. Theor Appl Genet 108: 485–496.

Frary A, Fritz LA, Tanksley SD (2004b) A comparative study of the genetic bases of natural variation in tomato leaf, sepal, and petal morphology. Theor Appl Genet 109: 523–533.

Frary A, Göl D, Keleş D, Okmen B, Pinar H, Siğva HO, Yemenicioğlu A, Doğanlar S (2010) Salt tolerance in *Solanum pennellii*: antioxidant response and related QTL. BMC Plant Biol 10: 58.

Fridman E, Pleban T, Zamir D (2000) A recombination hotspot delimits a wild-species quantitative trait locus for tomato sugar content to 484 bp within an invertase gene. Proc Natl Acad Sci USA 97: 4718–4723.

Fridman E, Liu YS, Carmel-Goren L, Gur A, Shoresh M, Pleban T, Eshed Y, Zamir D (2002) Two tightly linked QTLs modify tomato sugar content via different physiological pathways. Mol Genet Genom 266: 821–826.

Fridman E, Carrari F, Liu YS, Fernie AR, Zamir D (2004) Zooming in on a quantitative trait for tomato yield using interspecific introgressions. Science 305: 1786–1789.

Fulton TM, Beck-Bunn T, Emmatty D, Eshed Y, Lopez J, Petiard V, Uhlig J, Zamir D, Tanksley SD (1997) QTL analysis of an advanced backcross of *Lycopersicon peruvianum* to the cultivated tomato and comparisons with QTLs found in other wild species. Theor Appl Genet 95: 881–894.

Fulton TM, Grandillo S, Beck-Bunn T, Fridman E, Frampton A, Lopez J, Petiard V, Uhlig J, Zamir D, Tanksley SD (2000) Advanced backcross QTL analysis of a *Lycopersicon esculentum* × *L. parviflorum* cross. Theor Appl Genet 100: 1025–1042.

Fulton TM, Bucheli P, Voirol E, López J, Pétiard V, Tanksley SD (2002a) Quantitative trait loci (QTL) affecting sugars, organic acids and other biochemical properties possibly contributing to flavor, identified in four advanced backcross populations of tomato. Euphytica 127: 163–177.

Fulton TM, van der Hoeven R, Eanetta NT, Tanksley SD (2002b) Identification, analysis, and utilization of conserved ortholog set markers for comparative genomics in higher plants. Plant Cell 14: 1457–1467.

Georgelis N, Scott JW, Baldwin EA (2004) Relationship of tomato fruit sugar concentration with physical and chemical traits and linkage of RAPD markers. J Am Soc Hort Sci 129: 839–845.

Georgiady MS, Whitkus RW, Lord EM (2002) Genetic analysis of traits distinguishing outcrossing and self-pollinating forms of currant tomato, *Lycopersicon pimpinellifolium* (Jusl.) Mill. Genetics 161: 333–344.

Giovannoni JJ (2004) Genetic regulation of fruit development and ripening. Plant Cell 16: S170–S180.

Giovannucci E (1999) Tomatoes, tomato-based products, lycopene, and cancer: Review of the epidemiologic literature. J Natl Canc Inst 91: 317–331.

Goldman IL, Paran I, Zamir D (1995) Quantitative trait locus analysis of a recombinant inbred line population derived from a *Lycopersicon esculentum* × *L. cheesmanii* cross. Theor Appl Genet 90: 925–932.

Gonzalo MJ, van der Knaap E (2008) A comparative analysis into the genetic bases of morphology in tomato varieties exhibiting elongated fruit shape. Theor Appl Genet 116: 647–656.

Gorguet B, Eggink PM, Ocaña J, Tiwari A, Schipper D, Finkers R, Visser RGF, van Heusden AW (2008) Mapping and characterization of novel parthenocarpy QTLs in tomato. Theor Appl Genet 116: 755–767.

Graham EB, Frary A, Kang JJ, Jones CM, Gardener RG (2004) A recombinant inbred line mapping population derived from a *Lycopersicon* esculentum × *L. pimpinellifolium* cross. Tomato Genet Coop Rep 54: 22–25.

Grandillo S, Tanksley SD (1996a) QTL analysis of horticultural traits differentiating the cultivated tomato from the closely related species *Lycopersicon pimpinellifolium*. Theor Appl Genet 92: 935–951.

Grandillo S, Tanksley SD (1996b) Genetic analysis of RFLPs, GATA microsatellites and RAPDs in a cross between *L. esculentum* and *L. pimpinellifolium*. Theor Appl Genet 92: 957–965.

Grandillo S, Ku HM, Tanksley SD (1996) Characterization of *fs8.1*, a major QTL influencing fruit shape in tomato. Mol Breed 2: 251–260.

Grandillo S, Ku HM, Tanksley SD (1999) Identifying the loci responsible for natural variation in fruit size and shape in tomato. Theor Appl Genet 99: 978–987.

Grandillo S, Tanksley SD, Zamir D (2008) Exploitation of natural biodiversity through genomics. In: Varshney RK, Tuberosa R (eds) Genomics assisted crop improvement: vol I Genomics approaches and platforms.) Springer, Netherlands, pp 121–150.

Grandillo S, Chetelat R, Knapp S, Spooner D, Peralta I, Cammareri M, Perez O, Termolino P, Tripodi P, Chiusano Ml, Ercolano MR, Frusciante L, Monti L, Pignone D (2011) *Solanum* sect. *Lycopersicon*. In: Kole C (ed) Wild Crop Relatives: Genomic and Breeding Resources, Vol 5 Vegetables. Springer, Netherlands, pp 129–215.

Griffiths PD, Scott JW (2001) Inheritance and linkage of *Tomato mottle virus* resistance genes derived from *Lycopersicon chilense* accession LA 1932. J Am Soc Hort Sci 126: 462–467.

Gupta PK, Rustgi S, Kulwal PL (2005) Linkage disequilibrium and association studies in higher plants: present status and future prospects. Plant Mol Biol 57(4): 461–485.

Gur A, Zamir D (2004) Unused natural variation can lift yield barriers in plant breeding. PLoS Biol 2(10): e245.

Gur A, Semel Y, Cahaner A, Zamir D (2004) Real Time QTL of complex phenotypes in tomato interspecific introgression lines. Trends Plant Sci 9: 107–109.

Gur A, Osorio S, Fridman E, Fernie AR, Zamir D (2010) *hi2-1*, a QTL which improves harvest index, earliness and alters metabolite accumulation of processing tomatoes. Theor Appl Genet 121(8): 1587–1599.

Gur A, Semel Y, Osorio S, Friedmann M, Seekh S, Ghareeb B, Mohammad A, Pleban T, Gera G, Fernie AR, Zamir D (2011) Yield quantitative trait loci from wild tomato are predominately expressed by the shoot. Theor Appl Genet 122(2): 405–420.

Hackbusch J, Richter K, Muller J, Salamini F, Uhrig JF (2005) A central role of *Arabidopsis thaliana* ovate family proteins in networking and subcellular localization of 3-aa loop extension homeodomain proteins. Proc Natl Acad Sci USA 102: 4908–4912.

Haley CS, Knott SA (1992) A simple regression method for mapping quantitative loci in line crosses using flanking markers. Heredity 69: 315–324.

Hewitt JD, Garvey TC (1987) Wild sources of high soluble solids in tomato. In: Nevins DJ, Jones RA (eds) Plant Biology. Vol 4: Tomato Biotechnology, AR Liss, New York, USA, pp 45–54.

Holtan HE, Hake S (2003) Quantitative trait locus analysis of leaf dissection in tomato using *Lycopersicon pennellii* segmental introgression lines. Genetics 165: 1541–1550.

Huang Z, van der Knaap E (2011) Tomato *fruit weight 11.3* maps close to *fasciated* on the bottom of chromosome 11. Theor Appl Genet 123: 465–474.

Hutton SF, Scott JW, Yang W, Sim SC, Francis DM, Jones JB (2010) Identification of QTL associated with resistance to bacterial spot race T4 in tomato. Theor Appl Genet 121(7): 1275–1287.

Jansen RC (1994) High resolution of quantitative traits into multiple loci via interval mapping. Genetics 136: 1447–1455.

Jiménez-Gómez JM, Alonso-Blanco C, Borja A, Anastasio G, Angosto T, Lozano R, Martínez-Zapater JM (2007) Quantitative genetic analysis of flowering time in tomato. Genome 50: 303–315.

Jones JB, Jones JP, Stall RE, Zitter TA (1991) Compendium of tomato diseases. The American Phytopathological Society, St Paul, MN, USA.

Kabelka E, Franchino B, Francis DM (2002) Two loci from *Lycopersicon hirsutum* LA407 confer resistance to strains of *Clavibacter michiganensis* subsp. *michiganensis*. Phytopathology 92: 504–510.

Kabelka E, Yang WC, Francis DM (2004) Improved tomato fruit color within an inbred backcross line derived from *Lycopersicon esculentum* and *L. hirsutum* involves the interaction of loci. J Am Soc Hort Sci 129: 250–257.

Kalisz S, Vogler D, Fails B, Finer M, Shepard E, Herman T, Gonzales R (1999) The mechanism of delayed selfing in *Collinsia verna* (Scrophulariaceae). Am J Bot 86: 1239–1247.

Kennedy GG (2003) Tomato, pests, parasitoids, and predators: Tritrophic interactions involving the genus *Lycopersicon*. Ann Rev Entomol 48: 51–72.

Klee HJ (2010) Improving the flavor of fresh fruits: genomics, biochemistry, and biotechnology. New Phytol 187(1): 44–56.

Ku HM, Doganlar S, Chen KY, Tanksley SD (1999) The genetic basis of pear-shaped tomato fruit. Theor Appl Genet 99: 844–850.

Ku HM, Grandillo S, Tanksley SD (2000) *fs8.1*, a major QTL, sets the pattern of tomato carpel shape well before anthesis. Theor Appl Genet 101: 873–878.

Labate JA, Grandillo S, Fulton T, Muños S, Caicedo AL, Peralta I, Ji Y, Chetelat RT, Scott JW, Gonzalo MJ, Francis D, Yang W, van der Knaap E, Baldo AM, Smith-White B, Mueller LA, Prince JP, Blanchard NE, Storey DB, Stevens MR, Robbins MD, Wang JF, Liedl BE, O'Connell MA, Stommel JR, Aoki K, Iijima Y, Slade AJ, Hurst SR, Loeffler D, Steine MN, Vafeados D, McGuire C, Freeman C, Amen A, Goodstal J, Facciotti D, Van Eck J, Causse M (2007) Tomato. In: Kole C (ed) Genome mapping and molecular breeding in plants. Vol 5: Vegetables. Springer, Berlin, Heidelberg, Germany, pp 1–96.

Labate JA, Robertson LD, Wu F, Tanksley SD, Baldo AM (2009) EST, COSII, and arbitrary gene markers give similar estimates of nucleotide diversity in cultivated tomato (*Solanum lycopersicum* L.). Theor Appl Genet 118: 1005–1014.

Lander ES, Botstein D (1989) Mapping Mendelian factors underlying quantitative traits using RFLP linkage maps. Genetics 121: 185–199.

Lawson DM, Lunde CF, Mutschler MA (1997) Marker-assisted transfer of acylsugar-mediated pest resistance from the wild tomato, *Lycopersicon pennellii*, to the cultivated tomato, *Lycopersicon esculentum*. Mol Breed 3: 307–317.

Lecomte L, Duffé P, Buret M, Servin B, Hospital F, Causse M (2004a) Marker-assisted introgression of five QTLs controlling fruit quality traits into three tomato lines revealed interactions between QTLs and genetic backgrounds. Theor Appl Genet 109: 658–668.

Lecomte L, Saliba-Colombani V, Gautier A, Gomez-Jimenez MC, Duffé P, Buret M, Causse M (2004b) Fine mapping of QTLs of chromosome 2 affecting the fruit architecture and composition of tomato. Mol Breed 13: 1–14.

Li W, Chetelat RT (2010) A pollen factor linking inter- and intraspecific pollen rejection in tomato. Science 330(6012): 1827–1830.

Li W, Royer S, Chetelat RT (2010) Fine mapping of *ui6.1*, a gametophytic factor controlling pollen-side unilateral incompatibility in interspecific solanum hybrids. Genetics 185(3): 1069–1080.

Li Z-K, Fu B-Y, Gao Y-M, Xu J-L, Ali J, Lafitte HR, Jiang Y-Z, Rey JD, Vijayakumar CHM, Maghirang R, Zheng T-Q, Zhu L-H. (2005) Genome-wide introgression lines and their use in genetic and molecular dissection of complex phenotypes in rice (*Oryza sativa* L.). Plant Mol Biol 59: 33–52.

Li H, Ye G, Wang J (2007) A modified algorithm for the improvement of composite interval mapping. Genetics 175: 361–374.

Liedl BE, Mccormick S, Mutschler MA (1996) Unilateral incongruity in crosses involving *Lycopersicon pennellii* and *L. esculentum* is distinct from self-incompatibility in expression, timing and location. Sex Plant Reprod 9: 299–308.

Lincoln S, Daly M, Lander E (1992) Mapping genes controlling quantitative traits with MAPMAKER/QTL 1.1. Whitehead Institute Technical Report, 2nd edn.

Lindhout P, Heusden S, Pet G, Ooijen JW, Sandbrink H, Verkerk R, Vrielink R, Zabel P (1994) Perspectives of molecular marker assisted breeding for earliness in tomato. Euphytica 79: 279–286.

Lippman Z, Tanksley SD (2001) Dissecting the genetic pathway to extreme fruit size in tomato using a cross between the small-fruited wild species *Lycopersicon pimpinellifolium* and *L. esculentum* var. Giant Heirloom. Genetics 158: 413–422.

Lippman ZB, Semel Y, Zamir D (2007) An integrated view of quantitative trait variation using tomato interspecific introgression lines. Curr Opin Genet Dev 17: 545–552.

Liu YS, Zamir D (1999) Second generation *L. pennellii* introgression lines and the concept of bin mapping. Tomato Genet Coop Rep 49: 26–30.

Liu J, Van Eck J, Cong B, Tanksley SD (2002) A new class of regulatory genes underlying the cause of pear-shaped tomato fruit. Proc Natl Acad Sci USA 99: 13302–13306.

Liu J, Cong B, Tanksley SD (2003a) Generation and analysis of an artificial gene dosage series in tomato to study the mechanisms by which the cloned quantitative trait locus *fw2.2* controls fruit size. Plant Physiol 132: 292–299.

Liu J, Gur A, Ronen G, Causse M, Damidaux R, Buret M, Hirschberg J, Zamir D (2003b) There is more to tomato fruit colour than candidate carotenoid genes. Plant Biot J 1: 195–207.

Mangin B, Thoquet P, Olivier J, Grimsley NH (1999) Temporal and multiple quantitative trait loci analyses of resistance to bacterial wilt in tomato permit the resolution of linked loci. Genetics 151: 1165–1172.

Martin B, Nienhuis J, King G, Schaefer A (1989) Restriction fragment length polymorphisms associated with water use efficiency in tomato. Science 243: 1725–1728.

Martin B, Tauer CG, Lin RK (1999) Carbon isotope discrimination as a tool to improve water-use efficiency in tomato. Crop Sci 39: 1775–1783.

Mather K (1949) Biometrical genetics, the study of continuous variation. Methuen & Co, London, UK.

Mather K, Jinks JL (1982) Biometrical Genetics, 3rd edn. Chapman and Hall, London, UK.

Mathieu S, Dal Cin V, Fei Z, Li H, Bliss P, Taylor MG, Klee HJ, Tieman DM (2008) Flavour compounds in tomato fruits: identification of loci and potential pathways affecting volatile composition. J Exp Bot 60: 325–337.

McClure BA, Haring V, Ebert PR, Anderson MA, Simpson RJ, Sakiyama F, Clarke AE (1989) Style self-incompatibility gene products of *Nicotiana alata* are ribonucleases. Nature 342: 955–957.

McDowell ET, Kapteyn J, Schmidt A, Li C, Kang JH, Descour A, Shi F, Larson M, Schilmiller A, An L, Jones AD, Pichersky E, Soderlund CA, Gang DR (2011) Comparative functional genomic analysis of *Solanum* glandular trichome types. Plant Physiol 155(1): 524–539.

Miller JC, Tanksley SD (1990) RFLP analysis of phylogenetic relationships and genetic variation in the genus *Lycopersicon*. Theor Appl Genet 80: 437–448.

Monforte AJ, Tanksley SD (2000a) Development of a set of near isogenic and backcross recombinant inbred lines containing most of the *Lycopersicon hirsutum* genome in a *L. esculentum* genetic background: A tool for gene mapping and gene discovery. Genome 43: 803–813.

Monforte AJ, Tanksley SD (2000b) Fine mapping of a quantitative trait locus (QTL) from *Lycopersicon hirsutum* chromosome 1 affecting fruit characteristics and agronomic traits: breaking linkage among QTLs affecting different traits and dissection of heterosis for yield. Theor Appl Genet 100: 471–479.

Monforte AJ, Asins MJ, Carbonell EA (1996) Salt tolerance in *Lycopersicon* species. IV. High efficiency of marker-assisted selection to obtain salt-tolerant breeding lines. Theor Appl Genet 93: 765–772.

Monforte AJ, Asìns MJ, Carbonell EA (1997a) Salt tolerance in *Lycopersicon* species. V. Does genetic variability at quantitative trait loci affect their analysis? Theor Appl Genet 95: 284–293.

Monforte AJ, Asìns MJ, Carbonell EA (1997b) Salt tolerance in *Lycopersicon* species. VI. Genotype by salinity interaction in quantitative trait loci detection: constitutive and response QTLs. Theor Appl Genet 95: 706–713.

Monforte AJ, Asìns MJ, Carbonell EA (1999) Salt tolerance in *Lycopersicon* spp. VII. Pleiotropic action of genes controlling earliness on fruit yield. Theor Appl Genet 98: 593–601.

Monforte AJ, Friedman E, Zamir D, Tanksley SD (2001) Comparison of a set of allelic QTL-NILs for chromosome 4 of tomato: Deductions about natural variation and implications for germplasm utilization. Theor Appl Genet 102: 572–590.

Moyle LC, Graham EB (2005) Genetics of hybrid incompatibility between *Lycopersicon esculentum* and *L. hirsutum*. Genetics 169: 355–373.

Moyle LC, Nakazato T (2008) Comparative genetics of hybrid incompatibility: sterility in two Solanum species crosses. Genetics 179: 1437–1453.

Mueller AL, Solow TH, Taylor N, Skwarecki B, Buels R, Bins J, Lin C, Wright MH, Ahrens R, Wang Y, et al. (2005) The SOL Genomics Network (SGN): a comparative resource for Solanaceous biology and beyond. Plant Physiol 138: 1310–1317.

Muños S, Ranc N, Botton E, Bérard A, Rolland S, Duffé P, Carretero Y, Le Paslier MC, Delalande C, Bouzayen M, Brunel D, Causse M (2011) Increase in tomato locule number is controlled by two single-nucleotide polymorphisms located near *WUSCHEL*. Plant Physiol 156(4): 2244–2254.

Mutschler MA, Liedl BE (1994) Interspecific crossing barriers in *Lycopersicon* and their relationship to self-incompatibility. In Williams G, Clarke AE, Knox BR (eds) Genetic control of self-incompatibility and reproductive development in flowering plants, Kluwer, The Netherlands, pp 164–188.

Mutschler MA, Doerge RW, Liu SC, Kuai JP, Liedl BE, Shapiro JA (1996) QTL analysis of pest resistance in the wild tomato *Lycopersicon pennellii*: QTLs controlling acylsugar level and composition. Theor Appl Genet 92: 709–718.

Nelson JC (1997) QGENE: software for marker-based genomic analysis and breeding. Mol Breed 3: 229–235.

Nienhuis J, Helentjaris T, Slocum M, Ruggero B, Schaefer A (1987) Restriction fragment length polymorphism analysis of loci associated with insect resistance in tomato. Crop Sci 27: 797–803.

Orsi CH, Tanksley SD (2009) Natural variation in an ABC transporter gene associated with seed size evolution in tomato species. PLoS Genet 5: e1000347.

Osborn TC, Alexander DC, Fobes JF (1987) Identification of restriction fragment length polymorhisms linked to genes controlling soluble solids content in tomato. Theor Appl Genet 73: 350–356.

Overy SA, Walker HJ, Malone S, Howard TP, Baxter CJ, Sweetlove LJ, Hill SA, Quick WP (2005) Application of metabolite profiling to the identification of traits in a population of tomato introgression lines. J Expt Bot 56: 287–296.

Pan Q, Liu YS, Budai-Hadrian O, Sela M, Carmel-Goren L, Zamir D, Fluhr R (2000) Comparative genetics of nucleotide binding site-leucine rich repeat resistance gene homologues in the genomes of two dicotyledons: tomato and Arabidopsis. Genetics 155: 309–322.

Paran I, Goldman I, Tanksley SD, Zamir D (1995) Recombinant inbred lines for genetic mapping in tomato. Theor Appl Genet 90: 542–548.

Paran I, Goldman I, Zamir D (1997) QTL analysis of morphological traits in a tomato recombinant inbred line population. Genome 40: 242–248.

Paterson AH, Lander ES, Hewitt JD, Peterson S, Lincoln SE, Tanksley SD (1988) Resolution of quantitative traits into Mendelian factors by using a complete linkage map of restriction fragment length polymorphisms. Nature 335: 721–726.

Paterson AH, DeVerna JW, Lanini B, Tanksley SD (1990) Fine mapping of quantitative trait loci using selected overlapping recombinant chromosomes, in an interspecies cross of tomato. Genetics 124: 735–742.

Paterson AH, Damon S, Hewitt JD, Zamir D, Rabinovitch HD, Lincoln SE, Lander ES, Tanksley SD (1991) Mendelian factors underlying quantitative traits in tomato: comparison across species, generations, and environments. Genetics 127: 181–197.

Peleman JD, van der Voort JR (2003) Breeding by deisgn. Trends Plant Sci 8: 330–334.

Peralta IE, Spooner DM, Knapp S (2008) Taxonomy of wild tomatoes and their relatives (*Solanum* sections *Lycopersicoides, Juglandifolia, Lycopersicon*; Solanaceae). Syst Bot Monogr 84: 1–186.

Pertuzé RA, Ji Y, Chetelat RT (2003) Transmission and recombination of homeologous *Solanum sitiens* chromosomes in tomato. Theor Appl Genet 107: 1391–1401.

Petro-Turza M (1987) Flavor of tomato and tomato products. Food Rev Int 2: 309–351.

Prudent M, Causse M, Génard M, Tripodi P, Grandillo S, Bertin N (2009) Genetic and physiological analysis of tomato fruit weight and composition: influence of carbon availability on QTL detection. J Exp Bot 60: 923–937.

Prudent M, Bertin N, Génard M, Muños S, Rolland S, Garcia V, Petit J, Baldet P, Rothan C, Causse M (2010) Genotype-dependent response to carbon availability in growing tomato fruit. Plant Cell Environ. 33(7): 1186–1204.

Prudent M, Lecomte A, Bouchet JP, Bertin N, Causse M, Génard M (2011) Combining ecophysiological modelling and quantitative trait locus analysis to identify key elementary processes underlying tomato fruit sugar concentration. J Exp Bot (3): 907–919.

Rick C (1979) Biosystematic studies in *Lycopersicon* and closely related species of *Solanum*. In: Hawkes JG, Lester RN, Skelding AD (eds) The Biology and Taxonomy of the Solanaceae, Linnean Soc Symposium Series No 7, pp 667–678.

Rick CM, Fobes JF (1974) Association of an allozyme with nematode resistance. Tomato Genet Coop Rep 24: 25.

Robert VJM, West MAL, Inai S, Caines A, Arntzen, Smith KJ, St Clair DA (2001) Marker-assisted introgression of blackmold resistance QTL alleles from wild *Lycopersicon cheesmanii* to cultivated tomato (*L. esculentum*) and evaluation of QTL phenotypic effects. Mol Breed 8: 217–233.

Rodriguez GR, Munos S, Anderson C, Sim S-C, Michel A, Causse M, McSpadden Gardener BB, Francis D, van der Knaap E (2011) Distribution of *SUN, OVATE, LC*, and *FAS* in the tomato germplasm and the relationship to fruit shape diversity. Plant Physiol 156: 275–285.

Rousseaux MC, Jones CM, Adams D, Chetelat R, Bennett A, Powel A (2005) QTL analysis of fruit antioxidants in tomato using *Lycopersicon pennellii* introgression lines. Theor Appl Genet 111: 1396–1408.

Saliba-Colombani V, Causse M, Langlois D, Philouze J, Buret M (2001) Genetic analysis of organoleptic quality in fresh market tomato. 1. Mapping QTLs for physical and chemical traits. Theor Appl Genet 102: 259–272.

Sandbrink JM, van Ooijen J, Purimahua CC, Vrielink M, Verkerk R, Zabel P, Lindhout P (1995) Localization of genes for bacterial canker resistance in *Lycopersicon peruvianum* using RFLPs. Theor Appl Genet 90: 444–450.

Sax K (1923) Association of size differences with seed-coat pattern and pigmentation in *Phaseolus vulgaris*. Genetics 8: 552–560.

Schauer N, Semel Y, Roessner U, Gur A, Balbo I, Carrari F, Pleban T, Perez-Melis A, Bruedigam C, Kopka J, Willmitzer L, Zamir D, Fernie AR (2006) Comprehensive metabolic profiling and phenotyping of interspecific introgression lines for tomato improvement. Nat Biotechnol 24: 447–454.

Schauer N, Semel Y, Balbo I, Steinfath M, Repsilber D, Selbig J, Pleban T, Zamir D, Fernie AR (2008) Mode of inheritance of primary metabolic traits in tomato. Plant Cell 20: 509–523.

Schilmiller A, Shi F, Kim J, Charbonneau AL, Holmes D, Daniel Jones A, Last RL (2010) Mass spectrometry screening reveals widespread diversity in trichome specialized metabolites of tomato chromosomal substitution lines. Plant J 62(3): 391–403.

Semel Y, Nissenbaum J, Menda N, Zinder M, Krieger U, Issman N, Pleban T, Lippman Z, Gur A, Zamir D (2006) Overdominant quantitative trait loci for yield and fitness in tomato. Proc Natl Acad Sci USA 103: 12981–12986.

Sijacic P, Wang X, Skirpan AL, Wang Y, Dowd PE, McCubbin AG, Huang S, Kao TH (2004) Identification of the pollen determinant of S-RNase-mediated self-incompatibility. Nature 429(6989): 302–305.

Sillanpää MJ, Corander J (2002) Model choice in gene mapping: what and why. Trends Genet 18: 302–307.

Sinesio F, Cammareri M, Moneta E, Navez B, Peparaio M, Causse M, Grandillo S (2010) Sensory quality of fresh French and Dutch market tomatoes: a preference mapping study with Italian consumers. J Food Sci: 75(1): S55–67.

Soller M, Brody T, Genizi A (1976) On the power of experimental design for detection of linkage between marker loci and quantitative loci in crosses between inbred lines. Theor Appl Genet 47: 35–39.

Steinhauser MC, Steinhauser D, Gibon Y, Bolger M, Arrivault S, Usadel B, Zamir D, Fernie AR, Stitt M (2011) Identification of enzyme activity quantitative trait loci in a *Solanum lycopersicum* x *Solanum pennellii* introgression line population. Plant Physiol 157(3): 998–1014.

Stevens R, Buret M, Duffé P, Garchery C, Baldet P, Rothan C, Causse M (2007) Candidate genes and quantitative trait loci affecting fruit ascorbic acid content in three tomato populations. Plant Physiol 143: 1943–1953.

Stevens R, Page D, Gouble B, Garchery C, Zamir D, Causse M (2008) Tomato fruit ascorbic acid content is linked with monodehydroascorbate reductase activity and tolerance to chilling stress. Plant Cell Environ 31: 1086–1096.

Stommel JR, Zhang Y (2001) Inheritance and QTL analysis of anthracnose resistance in the cultivated tomato (*Lycopersicon esculentum*). Acta Hort 542: 303–310.

Tadmor Y, Fridman E, Gur A, Larkov O, Lastochkin E, Ravid U, Zamir D, Lewinsohn E (2002) Identification of *malodorous*, a wild species allele affecting tomato aroma that was selected against during domestication. J Agri Food Chem 50: 2005–2009.

Takayama S, Isogai A (2005) Self-incompatibility in plants. Annu Rev Plant Biol 56: 467–489.

Tandon KS, Baldwin EA, Shewfelt RL (2000) Aroma perception of individual volatile compounds in fresh tomatoes (*Lycopersicon esculentum* Mill.) as affected by the medium of evaluation. Postharvest Biol Technol 20: 261–268.

Tanksley SD (1993) Mapping polygenes. Annu Rev Genet 27: 205–233.

Tanksley SD (2004) The genetic, developmental, and molecular bases of fruit size and shape variation in tomato. Plant Cell 16: S181–S189.

Tanksley SD, Loaiza-Figueroa F (1985) Gametophytic self-incompatibility is controlled by a single major locus on chromosome 1 in *Lycopersicon peruvianum*. Proc Natl Acad Sci USA 82: 5093–5096.

Tanksley SD, McCouch SR (1997) Seed banks and molecular maps: unlocking genetic potential from the wild. Science 277: 1063–1066.

Tanksley SD, Nelson JC (1996) Advanced backcross QTL analysis: a method for the simultaneous discovery and transfer of valuable QTLs from unadapted germplasm into elite breeding lines. Theor Appl Genet 92: 191–203.

Tanksley SD, Medina-Filho H, Rick CM (1982) Use of naturally-occuring enzyme variation to detect and map genes controlling quantitative traits in an interspecific cross of tomato. Heredity 49: 11–25.

Tanksley SD, Ganal MW, Prince JP, de Vicente MC, Bonierbale MW, Broun P, Fulton TM, Giovannoni JJ, Grandillo S, Martin GB, Messeguer R, Miller JC, Paterson AH, Pineda O, Roder MS, Wing RA, Wu W, Young ND (1992) High density molecular linkage maps of the tomato and potato genomes. Genetics 132: 1141–1160.

Tanksley SD, Grandillo S, Fulton TM, Zamir D, Eshed Y, Petiard V, Lopez J, Beck-Bunn T (1996) Advanced backcross QTL analysis in a cross between an elite processing line of tomato and its wild relative *L. pimpinellifolium*. Theor Appl Genet 92: 213–224.

Termolino P, Fulton T, Perez O, Eannetta N, Xu Y, Tanksley SD, Grandillo S (2010) Advanced backcross QTL analysis of a *Solanum lycopersicum* × *Solanum chilense* cross. Proc SOL2010 7th Solanaceae Conference, Dundee, Scotland, 5–9 September 2010, p 56.

Thoday JM (1961) Location of polygenes. Nature 191: 368–370.

Thoquet P, Olivier J, Sperisen C, Rogowsky P, Laterrot H, Grimsley N (1996a) Quantitative trait loci determining resistance to bacterial wilt in tomato cultivar Hawaii 7996. Mol Plant-Microbe Interact 9: 826–836.

Thoquet P, Olivier J, Sperisen C, Rogowsky P, Prior P, Anais G, Mangin B, Bazin B, Nazer R, Grimsley N (1996b) Polygenic resistance of tomato plants to bacterial wilt in the French West Indies. Mol Plant-Microbe Interact 9: 837–842.

The Potato Genome Sequencing Consortium, Xu X, Pan S, Cheng S, Zhang B, Mu D, Ni P, Zhang G, Yang S, Li R, et al. (2011). Genome sequence and analysis of the tuber crop potato. Nature 475: 189–195.

Tieman DM, Zeigler M, Schmelz EA, Taylor MG, Bliss P, Kirst M, Klee HJ (2006) Identification of loci affecting flavour volatile emissions in tomato fruits. J Exp Bot 57: 887–896.

Tripodi P, Di Dato F, Maurer S, Al Seekh S, Brog M, Van Haaren MJJ, Frusciante L, Mohammad A, Tanksley SD, Zamir D, Gebhardt C, Grandillo S (2010) A genetic platform of tomato multi-species introgression lines: present and future. Proceedings of the 7th Solanaceae Conference, Dundee, September 5–9 2010. pp 166.

Truco MJ, Randall LB, Bloom AJ, St Clair DA (2000) Detection of QTLs associated with shoot wilting and root ammonium uptake under chilling temperatures in an interspecific backcross population from *Lycopersicon esculentum* × *L. hirsutum*. Theor Appl Genet 101: 1082–1092.

Vallejos CE, Tanksley SD (1983) Segregation of isozyme markers and cold tolerance in an interspecific backcross of tomato. Theor Appl Genet 66: 241–247.

Van der Hoeven RS, Monforte AJ, Breeden D, Tanksley SD, Steffens JC (2000) Genetic control and evolution of sesquiterpene biosynthesis in *Lycopersicon esculentum* and *L. hirsutum*. Plant Cell 12: 2283–2294.

Van der Hoeven R, Ronning C, Giovannoni J, Martin G, Tanksley S (2002) Deductions about the number, organization, and evolution of genes in the tomato genome based on analysis of a large expressed sequence tag collection and selective genomic sequencing. Plant Cell 14: 1441–1456.

van der Knaap E, Tanksley SD (2001) Identification and characterization of a novel locus controlling early fruit development in tomato. Theor Appl Genet 103: 353–358.

van der Knaap E, Tanksley SD (2003) The making of a bell pepper-shaped tomato fruit: identification of loci controlling fruit morphology in Yellow Stuffer tomato. Theor Appl Genet 107: 139–147.

van der Knaap E, Lippman ZB, Tanksley SD (2002) Extremely elongated tomato fruit controlled by four quantitative trait loci with epistatic interactions. Theor Appl Genet 104: 241–247.

van der Knaap E, Sanyal A, Jackson SA, Tanksley SD (2004) High-resolution fine mapping and fluorescence *in situ* hybridization analysis of *sun*, a locus controlling tomato fruit shape, reveals a region of the tomato genome prone to DNA rearrangements. Genetics 168: 2127–2140.

van Heusden AW, Koornneef M, Voorrips RE, Brüggemann W, Pet G, Vrielink-van Ginkel R, Chen X, Lindhout P (1999) Three QTLs from *Lycopersicon peruvianum* confer a high level of resistance to *Clavibacter michiganensis* ssp. *michiganensis*. Theor Appl Genet 99: 1068–1074.

Van Ooijen JW, Boer MP, Jansen RC, Maliepaard C (2002) MapQTL® 4.0, Software for the calculation of QTL positions on genetic maps. Plant Research International, Wageningen, The Netherlands.

Villalta I, Bernet GP, Carbonell EA, Asins MJ (2007) Comparative QTL analysis of salinity tolerance in terms of fruit yield using two Solanum populations of F7 lines. Theor Appl Genet 114: 1001–1017.

Villalta I, Reina-Sánchez A, Bolarín MC, Cuartero J, Belver A, Venema K, Carbonell EA, Asins MJ (2008) Genetic analysis of Na(+) and K(+) concentrations in leaf and stem as physiological components of salt tolerance in tomato. Theor Appl Genet 116: 869–80.

Wang J-F, Olivier J, Thoquet P, Mangin B, Sauviac L, Grimsley NH (2000) Resistance of tomato line Hawaii7996 to *Ralstonia solanacearum* Pss4 in Taiwan is controlled mainly by a major strain-specific locus. Mol Plant Microbe Interact 13: 6–13.

Wang H, Zhang YM, Li X, Masinde GL, Mohan S, Baylink DJ, Xu S (2005a) Bayesian shrinkage estimation of quantitative trait loci parameters. Genetics 170: 465–480.

Wang S, Basten CJ, Zeng ZB (2005b) Windows QTL Cartographer 2.5. Dept. of Statistics, North Carolina State University, Raleigh, NC, USA.

Wang S, Chang Y, Guo J, Chen JG (2007) Arabidopsis ovate family protein 1 is a transcriptional repressor that suppresses cell elongation. Plant J 50: 858–872.

Weller JI, Soller M, Brody T (1988) Linkage analysis of quantitative traits in an interspecific cross of tomato (*Lycopersicon esculentum x Lycopersicon pimpinellifolium*) by means of genetic markers. Genetics 118: 329–339.

Willcox JK, Catignani GL, Lazarus S (2003) Tomatoes and cardiovascular health. Crit Rev Food Sci Nutr 43: 1–18.

Whittaker JC, Thompson R, Visscher PM (1996) On the mapping of QTL by regression of phenotype on marker-type. Heredity 77: 23–32.

Wu F, Mueller LA, Crouzillat D, Petiard V, Tanksley SD (2006) Combining bioinformatics and phylogenetics to identify large sets of single-copy orthologous genes (COSII) for comparative, evolutionary and systematic studies: a test case in the euasterid plant clade. Genetics 174: 1407–1420.

Xiao H, Jiang N, Schaffner EK, Stockinger EJ, van der Knaap E (2008) A retrotransposon-mediated gene duplication underlies morphological variation of tomato fruit. Science 319: 1527–1530.

Xu X, Martin B, Comstock JP, Vision TJ, Tauer CG, Zhao B, Pausch RC, Knapp S (2008) Fine mapping a QTL for carbon isotope composition in tomato. Theor Appl Genet 117: 221–233.

Yang W, Sacks EJ, Lewis Ivey ML, Miller SA, Francis DM (2005) Resistance in *Lycopersicon esculentum* intraspecific crosses to race T1 strains of *Xanthomonas campestris* pv. *vesicatoria* causing bacterial spot of tomato. Phytopathology 95: 519–527.

Yates HE, Frary A, Doganlar S, Frampton A, Eannetta NT, Uhlig J, Tanksley SD (2004) Comparative fine mapping of fruit quality QTLs on chromosome 4 introgressions derived from two wild species. Euphytica 135: 283–296.

Yu ZH, Wang JF, Stall RE, Vallejos CE (1995) Genomic localization of tomato genes that control a hypersensitive reaction to *Xanthomonas campestris* pv. *vesicatoria* (Doidge) Dye. Genetics 141: 675–682.

Zamir D (2001) Improving plant breeding with exotic genetic libraries. Nat Rev Genet 2: 983–989.

Zamir D, Tal M (1987) Genetic analysis of sodium, potassium and chloride ion content in *Lycopersicon*. Euphytica 36: 187–191.

Zamir D, Selia Ben-David T, Rudich J, Juvik JA (1984) Frequency distributions and linkage relationships of 2-tridecanone in interspecific segregating generation of tomato. Euphytica 33: 481–488.

Zamir D, Ekstein-Michelson I, Zakay Y, Navot N, Zeidan M, Sarfatti M, Eshed Y, Harel E, Pleban T, van-Oss H, Kedar N, Rabinowitch HD, Czosnek H (1994) Mapping and introgression of a Tomato yellow leaf curl virus tolerance gene, *Ty-1*. Theor Appl Genet 88: 141–146.

Zanor MI, Rambla JL, Chaïb J, Steppa A, Medina A, Granell A, Fernie AR, Causse M (2009) Metabolic characterization of loci affecting sensory attributes in tomato allows an assessment of the influence of the levels of primary metabolites and volatile organic contents. J Exp Bot 60(7): 2139–2154.

Zeng ZB (1994) Precision mapping of quantitative trait loci. Genetics 136: 1457–1468.

Zhang L, Lin GY, Nino-Liu, DO, Foolad MR (2003a) Mapping QTLs conferring early blight (*Alternaria solani*) resistance in a *Lycopersicon esculentum* × *L. hirsutum* cross by selective genotyping. Mol Breed 12: 3–19.

Zhang LP, Lin GY, Foolad MR (2003b) QTL comparison of salt tolerance during seed germination and vegetative growth in a *Lycopersicon esculentum* × *L. pimpinellifolium* RIL population. Acta Hort 618: 59–67.

7

Molecular Breeding

Emidio Sabatini,[1],* Massimiliano Beretta,[2] Tea Sala,[2]
Nazareno Acciarri,[1] Justyna Milc[3] and Nicola Pecchioni[3]

ABSTRACT

Breeding efforts are and will be in the future critical to improve
worldwide crop production for food, feed, non-food industrial and
environmental aims. The recent development of "omics" sciences, in
particular genomics, and bioinformatics give the great opportunity
to be either integrated into traditional breeding schemes, or used to
generate new breeding schemes and paradigms. There are important
expectations from such technologies and from new genome-wide
information to enhance the effectiveness of plant breeding programs,
starting from main crops such as tomato. When genotyping tools will
not be anymore a limit, one issue crucial for future breeding will be
the availability of well characterized germplasm collections. A detailed
genetic analysis of the breeder's germplasm aided by DNA technologies
will make more and more useful good programs of pre-breeding,
to help in planning crosses and guiding the choice of inbred lines
for hybrid combinations. As a side-aspect of this, DNA tools can be
applied for DUS and hybrid purity testing. Molecular markers enable
marker-assisted selection (MAS) for gene and QTL introgression, gene
pyramiding and genetic ideotype breeding. MAS application presents
several advantages such as increased reliability and efficiency provided
by the fact that they can be scored at the seedling stage enabling indirect
selection especially for quantitative traits with low heritabilities. MAS
becomes particularly useful, or necessary when a phenotypic assay may
be influenced by the environment, is particularly costly or technically

[1]C.R.A. Agricultural Research Council, Research Unit #13 for Vegetable Crops, via Salaria 1,
63077 Monsampolo del Tronto (AP), Italy.
[2]C.R.A. Agricultural Research Council, Research Unit #12 for Vegetable Crops, via Paullese
28, 26836 Montanaso Lombardo (LO), Italy.
[3]University of Modena and Reggio Emilia, Department of Agricultural and Food Sciences,
via Amendola 2, 42122 Reggio Emilia, Italy.
*Corresponding author

difficult, MAS integration into classical breeding schemes, as well as the new possibilities are reviewed. Nowadays the cost of genotyping a plant rather than the cost of a data point, limited molecular marker polymorphism within cultivated tomato, and thus unavailability of closely-linked markers for all agronomically relevant traits, are the main limitations of MAS. However a trend towards an increased cost accessibility of genomics-based technologies is underway. On the other hand, in developing countries the unfamiliarity of many traditional plant breeders with the use of molecular markers and technologies still represents a major limitation. Eventually, for most crops MAS will be scaled up to the genomic level, consequently the breeder could practice whole genome selection. The expectations from genomics-assisted selection (GAS) are high for many species including tomato. The shifts to second and, soon, to third generation sequencing technologies will provide unique ways in which to conceive plant breeding programs. In addition, the development of phenomics and metabolomics should be taken into great consideration. Transgenic breeding, offers several advantages such as ability to overcome incompatibility barriers and the possibility to incorporate only the specific cloned gene into the recipient, thus avoiding the transfer of undesirable genes in introgressed chromosome regions from wild donors. Very strict release procedures unfortunately severely limit this opportunity, until the development of cisgenics or of other technical alternatives. Pyramiding of transgenes with either similar or complementary and/or additive effects is possible. It is here reviewed how tomato has been modified for improvement of different traits such as fruit quality, parthenocarpy, constitutive overproduction of antifungal compounds, bacterial and viral disease resistance and post-harvest and processing technology.

Keywords: MAS, genomic assisted selection, NGS, molecular breeding, tomato, marker technologies, transgenic breeding

7.1 Introduction

Breeding technologies are critical for improving crop production in our changing world with an exponentially growing population and in the face of extreme environmental changes (Tester and Langridge 2010). The development of molecular biology and later of "omics" sciences and bioinformatics has offered substantial opportunities for enhancing the effectiveness of classical plant breeding programs. Molecular and bioinformatic tools can either be integrated into traditional breeding schemes to efficiently analyze large numbers of traits and crosses during the early seedling stage, or to generate new breeding schemes and programs previously not feasible. Through these approaches and tools, both the phenotype (observable identity of an individual) and the genotype (genetic identity of an individual) of new varieties can be analyzed to predict the performance of novel introgressed traits. This new paradigm describes the intention of "Molecular Breeding", often termed "marker-assisted selection (MAS)".

The key to this silent revolution in plant breeding is to create genotype-to-phenotype trait knowledge for obtaining breeding objectives. This new knowledge can be particularly valuable for product development and subsequent deployment to resource-poor farmers. Therefore, great expectations reside in MAS technology. There is a widespread belief that its extensive adoption is inevitable, and in turn, that achieving substantial impacts on plant yields represents the greatest challenge for agricultural scientists in the short to medium term (Collard and Mackill 2008). The intensified development of "omics" sciences and associated technologies today represents a quantum jump in our molecular understanding of important plant breeding traits and, accordingly, a great promise for MAS. In particular, the shifts to second and to third generation DNA sequencing technologies will provide unique ways in which to conceive plant breeding programs.

7.2 Germplasm Characterization and Exploitation

Domestication together with classical breeding created well-documented bottlenecks in tomato allele diversity (Miller and Tanksley 1990), while breeders are relentlessly searching for new sources of variability. For this reason, deploying genetic variability is one of the promises of modern molecular breeding. The knowledge of genome sequences and the use of molecular markers will uncover biodiversity and enlarge the genetic bases of future cultivated varieties.

Cultivated tomato, similar to common bean (*Phaseolus vulgaris*) (Papa et al. 2007) and other crops, faced a reduction of allelic diversity via domestication and, within the last century, intensive breeding has led to the modern pool of elite germplasm (Bai and Lindhout 2007). It has been calculated that tomato cultivars retained only about 5% of the genetic variation of their wild relatives (Miller and Tanksley 1990). For this reason, systematic study and characterization of *Solanum* wild relatives, as well as of tomato germplasm, is of prime importance for current and future genetic improvement of the crop (see also Chapter 3, and Zamir 2001; Paran and Zamir 2003; Gur and Zamir 2004).

Such evaluation is imperative before initiating any breeding program, in order to understand the genetic background, genetic distance, and the breeding value of the available material. The available tools for molecular breeding programs provide an opportunity to the breeder for increased exploitation of the allelic diversity contained in an array of cultivated and wild germplasm collections. Morphological, agronomic and biochemical parameters have been widely used for this purpose (e.g., Rick and Holle 1990). At present, underlying functional genetic diversity needs to be better described and measured to be effectively incorporated into breeding strategies and used for the management of the world's plant genetic resources.

In order to reach this "genomically aware" method of tomato improvement, in addition to the education of a new generation of breeders and necessary laboratory investments, a key issue is an accurate association between traits and genetic determinants.

After a company or a public breeding institution has established a laboratory platform for molecular breeding and has access to a significant germplasm collection, it can begin the work of germplasm molecular characterization that can lead to effective pre-breeding, and also to more efficient ways of unequivocal germplasm identification and of hybrid purity testing.

7.2.1 Pre-breeding

"Pre-breeding" refers to a series of actions that are preparatory to breeding, that are aimed at germplasm improvement and informing the choice of parents for crosses. They are generally time consuming and costly operations that increase the value of the genetic platform of a breeder. Pre-breeding activity can, therefore, constitute one of the main tasks of public research institutions involved in plant breeding; they should preserve, evaluate and prepare suitable germplasm for national seed companies, with efforts that, once recognized, are justifiable under public funding.

The history of plant genetics includes two important observations relevant to pre-breeding. First, agronomically important phenotypic traits are often controlled by a few key genes whose mutant alleles are sometimes fixed by selection in domesticated genotypes. The fixation of alleles in this manner is particularly frequent for components of the so-called "domestication syndrome". A clear example of this is the mutant *teosinte branched* (*tb1*) locus, that determines maize (*Zea mays*) plant and inflorescence architecture (Buckler and Thornsberry 2002). The tomato plant has been shaped by key genes, often underlying quantitative trait loci (QTL), that determine growth habit (e.g., self-pruning), fruit traits (set, size, color and morphology), and essential disease resistances (Bai and Lindhout 2007). Recently it was shown that the manipulation of a single regulatory gene in tomato (*DET1*) could influence the production of different phytonutrients generated from two independent pathways, leading to a novel, complex fruit quality phenotype (Davuluri et al. 2005).

The second fact is peculiar to tomato genetics and breeding: the wild relatives of the cultivated species are valuable sources of useful, often underexploited variation (Tanksley and McCouch 1997). In addition to the domesticated *S. lycopersicum*, the section *Lycopersicon* of the *Solanum* genus includes 12 wild tomato relatives (Peralta et al. 2006). These wild species, in tomato more than in other crops, (e.g., soybean, maize, barley, have been crossed to cultivated genotypes with successful introgression of traits. As a striking example, modern commercial hybrids carry different combinations of 15 independently introgressed disease resistance genes that originated from wild species. Moreover, fruit quality genes and QTLs have been introduced from wild tomatoes during the breeding history of tomato (Zamir 2001; Fernie et al. 2006). A handful of successes clearly illustrate the potential to unlock useful allelic variation from wild *Solanum* species. QTLs for increased fruit size were introduced from the small-berried *S. pimpinellifolium* (Tanksley et al. 1996), and fruit color was increased using an introgression from *S. habrochaites* (Bernacchi et al. 1998). Three independent introgressions from *S. pennellii* into single introgression lines (ILs), on chromosomes 7, 8 and 9, once cumulated into a single genotype, were able to increase yield to 50% higher over of a leading market cultivar under both wet and dry cultivation regimes (Gur and Zamir 2004).

Molecular breeding allows an increase in the efficiency of pre-breeding activities through three different routes: the evaluation of cultivated materials for planning crosses, so-called "parent building", and the identification of valuable introgressions from wild species. Since modern tomato breeding is focused on the creation of F_1 hybrids, a common conventional process by seed companies is the crossing of proprietary and competitors' cultivated materials, evaluation, and selection of the recombinant populations obtained until generations F_4 to F_6, when crosses are initiated to produce test hybrids for agronomical progeny evaluations. It seems that recurrent selection programs to select parents on the basis of their combining abilities as a basis for inbred line development is not uncommon in tomato breeding (Bai and Lindhout 2007).

Therefore, a detailed genetic analysis of the breeder's germplasm could assist in planning crosses and guiding the choice of parents for inbred line development, in terms of their relative genetic distances and of key genes/QTLs that they possess. Using available tools, molecular breeders can estimate the breadth of genetic variation contained in their materials. In addition, but with more significant economic effort, they can estimate the extent of linkage blocks created by background selection and linkage drag. There are several studies for estimating the genetic variation in tomato breeding germplasm. Benor et al. (2008) surveyed 39 tomato inbred lines of Asiatic and US origin by means of simple sequence repeats (SSRs), while Garcia-Martinez et al. (2005) analyzed genetic variation in 48 cultivars from three Spanish tomato breeding pools by means of both dominant amplified fragment length polymorphisms (AFLPs) and codominant SSR markers. Their results were useful to classify as well as to evaluate the extent of diversity of the various geographic, growth habit (determinate and indeterminate) and breeding groups. A more extensive study was performed on a 500 genotype sample representing a large fraction of elite European elite cultivars by means of microsatellites. The study was valuable in validating "blind tests" for the identification of unknown samples (Bredemeijer et al. 2002).

Because microsatellites are generally anonymous markers, it would be interesting to increase the application of functional, gene-derived markers to germplasm evaluation. With this view, Van Deynze et al. (2007) analyzed the sequence diversity of conserved genes (COS, conserved orthologous sequences) in 12 tomato processing and fresh-market genotypes. They generated a total of 1,487 single nucleotide polymorphisms (SNPs) and 282 insertion/deletions (indels), expanding by 847 unigenes (an increase of 33%) the number of Sol Genomics Network (SGN) COSII set, for a total of 3,434 COS markers. This COSII set will be valuable in evaluating the genetic diversity of cultivated tomato genotypes, since it contains genes of metabolic and agronomic significance. Since many germplasm collections of breeders and institutions have been characterized for quantitative morpho-agronomic traits with continuous phenotypic distribution, some interesting approaches involve the joint use of molecular (marker, qualitative) and morpho-agronomic (qualitative and quantitative) data to classify and characterize breeding pools (Franco et al. 2001), including heirloom tomato varieties (Goncalves et al. 2009).

The availability of marker tools, especially if economically profitable, could stimulate their increased use in tomato, namely, to select inbred lines for hybrid combinations. The combination of MAS with phenotypic selection in accumulating favorable QTL alleles through cycles of recurrent selection has been proposed for plant molecular breeding, especially for medium to low heritability traits (Dekkers and Hospital 2002; Moose and Mumm 2008). It has been successfully applied in maize for increasing yield after recurrent selection under high nitrogen input (Coque and Gallais 2006). The former authors found, during the second cycle (C2) of recurrent selection, a majority (10) of new significant genomic regions plus 7 of 9 previously detected (at C1) positively acting genomic regions. Important trait loci such as those of root architecture and glutamine synthetase activity were consistently accurately predicted (Coque and Gallais 2006).

Parent building is the second route that is worthwhile to follow through molecular assistance, conceptually conceived as the design of parents for new crosses (Langridge and Barr 2003). Accumulating positive alleles at key trait loci (e.g., multiple pathogen resistance loci) into a single inbred line is tractable in parent design, especially if single loci are added sequentially (to reduce numbers of surveyed plants), and if the loci to be accumulated are not in repulsion linkage. This procedure can be extended to a more demanding whole-genome scan, to design a so-called "genetic ideotype" carrying an array of key loci and selected regions from different germplasm sources (Langridge et al. 2006). Such a pre-breeding strategy, while appealing and highly desirable, can be limited by the population size needed to recover the recombinant lines in which the multiple loci have been accumulated, as well as by cost. This is especially true for small to medium enterprises (SMEs), for a species such as tomato whose world seed market value was calculated at about $680 million in 2007 (Bai and Lindhout 2007).

Parent building thus mainly involves "pyramiding" of multiple loci from multiple crosses (see Section 7.3.2). The practice of parent building was proposed for the self pollinating barley (Langridge et al. 1996; Ye et al. 2007), and in this species was demonstrated to be effective in generating new genotypes with improved malting quality QTLs (Emebiri et al. 2009), as well as malting quality QTLs together with winter growth habit genes (Laidò et al. 2009). In tomato, the example of pyramiding three *S. pennellii*-derived ILs that resulted in an outperformance of modern commercial hybrids (Gur and Zamir 2004), is an example of a parent building scheme. The highly valuable pyramided genotype can be successfully used for crosses in processing tomato breeding programs to enhance yield and fruit quality.

Appropriate molecular tools are available for the aim of evaluation of breeding germplasm as well as parent building. What at present must be improved is the bioinformatics component, as has occurred in major private seed companies for applying MAS to major crops. The molecular breeder should be put in a position to use the wealth of phenotypic information obtained from multilocation field trials, as well as from proprietary greenhouse and controlled environment trials, and to filter this using bioinformatic tools in order to select appropriate materials bearing a sum of selected trait values. In addition, he must be able to handle and search the mass of public and proprietary genotypic information, related to genes, markers and map positions, together with allelic typing of the breeding material. An advancement would be a bioinformatic tool that would allow composite queries to combine phenotypic and genotypic information for a large sample of germplasm, to initiate a breeding, as well as a pre-breeding (parent building) program.

Online genomic information sources for the genetics research community include, for example, maizeGDB database for maize (Lawrence et al. 2004), or GrainGenes (Matthews et al. 2003) for wheat, barley, rye and oat. Those databases are incomplete with respect to breeders' activity, as they do not generally include both genotypic and phenotypic data. While some of them, such as GrainGenes, do store both types of information, genomic and phenotypic data are managed separately with no method for the user to correlate them. Moreover, the applicability of most of the molecular data to breeding is oftentimes limited since multiple locus genotypes are not available. Some databases have been designed to store and manage both phenotypic and genotypic data, e.g., Germinate, Panzea, AppleBreed and PlantDB (Milc et al. 2011). For example, Panzea is specific to maize and allows an advanced genomic diversity and phenotype connection (GDPC) search (Zhao et al. 2005). AppleBreed allows the management of pluri-annual data on the same individual plants, and supports apple breeders and geneticists in their genetic studies and in their exploration of germplasm collections for trait/marker associations, being sufficiently generic to allow it to be used for other perennial crops (Antofie et al. 2007).

However, in databases designed to store experimental data, the available data are generally restricted to those implemented by the developers/users with no easy method to exploit information that resides in other data sources. Recently, a database specific for molecular breeding of wheat, barley and rice (fundamental agricultural crops) CEREALAB (www. cerealab.unimore.it/) was released. The CEREALAB database aims to store genotypic and phenotypic data obtained by the homonymous project, and to integrate them with previously existing data sources in order to create a tool specific for cereal breeders. By using composite queries, such a database can help in identifying germplasm accessions of interest due to particular phenotypic trait values, markers associated to key traits, marker alleles associated to trait positive variants, and thus assist in choosing desired parentals for breeding programs (Milc et al. 2011).

The third route to pre-breeding is the prediction of value of wild tomato relatives for their use in introgression schemes. This will be of particular importance in the future, since a plateau in yield and other agronomic traits will eventually be reached if only cultivated tomato variability is used (Bai and Lindhout 2007). Nevertheless, pre-breeding is not easy and it remains costly and time consuming. A successful deployment of useful allelic variation hidden in phenotypically weak wild genotypes carries two limitations: finding an efficient way to discover loci and alleles, and the efforts needed to eliminate an undesirable wild background in the introgression schemes. Three approaches can help solve the first problem: advanced backcross-QTL (AB-QTL) mapping schemes (Tanksley and Nelson 1996), EcoTILLING to discover novel alleles (Comai et al. 2004), and next generation sequence-based resequencing of wild accessions at selected candidate genes.

AB-QTL mapping allows the discovery of positive wild alleles at newly mapped QTLs with a scheme significantly less demanding and at the same time closer to breeding, than traditional linkage mapping. It consists of a QTL mapping procedure performed in rather advanced backcross (BC) populations (BC2–BC3) derived from crosses with wild relatives, with the cultivated line as recurrent, after elimination by negative selection of unadapted genotypes. The most extensive AB-QTL experiments in plants were conducted in tomato, for which populations from five wild species (*S. pimpinellifolium* LA1589, *S. peruvianum*, LA1708, *S. hirsutum* LA1777, *S. neorickii* LA2133 and *S. pennellii* LA1657) have been tested in the laboratory and the field in a number of locations around the world. Results for LA1589, LA1708 and LA1777 have been documented by Tanksley et al. (1996), Fulton et al. (1997) and Bernacchi et al. (1998). Several years later the crosses with LA2133 (Fulton et al. 2000) and with LA1657 (Frary et al. 2004) provided results, consistently recovering QTLs with favorable alleles coming from the wild parent. After the initial proof-of-concept, the methodology demonstrated its value in uncovering positive wild QTL alleles in other crops, especially in cereals such as maize, barley, wheat and rice (Ho et al. 2002; Huang et al. 2003; Pillen et al. 2003; Rangel et al. 2005). A possible limitation of such an approach is the necessity of a single, however, challenging, project for every "wild x cultivated" combination, and the restriction of surveying only one wild accession at a time in each cross. This makes the survey of an extensive sample of the *Solanum* wild biodiversity intractable, unless other genomic strategies are adopted. Moreover, the strategy is targeted to QTLs.

A possible alternative tool to uncover useful alleles from different accessions of wild *Solanum* species is "EcoTILLING" (Comai et al. 2004). Such a method is oriented towards known (sequenced) genes and is based on a 1:1 co-amplification of cultivated and wild genomic DNAs, for a homozygous selfing species, followed by detection of mutant alleles by using the CEL1 enzyme. When the population of co-amplified fragments is denatured to single strands and re-annealed, if a mutation is present, a mismatched re-annealed product is produced (a heteroduplex). Heteroduplexes are specifically cleaved by the CEL1 nuclease and can subsequently be identified by fragment analysis. Such a strategy, even if unsuitable for detecting QTL alleles, could be directed to the growing number of cloned/annotated agronomically key genes in large samples of accessions per each wild tomato species. For example, this strategy uncovered new allelic variants of *eIF4E*, a factor that controls virus susceptibility in melon (*Cucumis*) (Calero-Nieto et al. 2007). The study characterized a new *eIF4E* allele from *Cucumis zeyheri* which, based on functional analysis, appeared to be potentially responsible for the resistance of plants to melon necrotic spot virus (MNSV). EcoTILLING was used in the tomato pathogenic fungus *Cladosporium fulvum* to uncover new avirulence gene alleles (Stergiopoulos et al. 2007).

After new useful alleles at QTLs and genes have been identified in agronomically poor wild relatives by AB-QTL and EcoTILLING, respectively, the efforts needed to introgress them into valuable genotypes and to eliminate the wild negative background may still present a substantial limitation. It should be recognized that in the molecular era this represents a far smaller technical challenge than before. High-throughput marker technologies that allow whole-genome scans can be applied to recover the cultivated background. Therefore, the limitation can more accurately be viewed in terms of costs versus benefits, including the valuable time saved, than in terms of technology. The AB-QTL approach may represent the initial step to introgression of wild genome segments into a cultivated genome. Once identified, the QTLs can be fixed at homozygosis into BC3-BC4 lines and backcrossing can be continued, aided by markers, to eliminate virtually all of the remaining wild genome. In this way valuable amounts of pre-breeding material can be generated, such as ILs bearing single wild-derived genomic regions in a cultivated background. Zamir (2001) proposed to realize such a process on a large scale, to generate a genetic infrastructure consisting of libraries of ILs, each one carrying a chromosomal segment from a wild species. Sufficient numbers of lines would be necessary so as to reconstruct with their introgressions the complete wild genome of origin, and different IL libraries could represent different accessions of the same wild species. Such a genetic infrastructure might in part provide an alternative to cisgenic breeding; however, in addition to the high cost of such an effort, it would potentially involve linkage drag of deleterious elements in cis within an introgressed segment.

7.2.2 Testing for Distinctness, Uniformity and Stability (DUS)

DUS is an essential component of the variety registration procedure. This test should provide evidence that a new candidate variety satisfies three fundamental criteria: (i) Distinctness: the new variety is clearly distinguishable for at least one descriptor from a reference collection of varieties and other candidates; (ii) Uniformity: the number of off-types does not exceed a threshold value and iii) Stability: the variety is able to maintain all properties after reproduction. Testing procedures and methods are determined by the International Union for the Protection of New Varieties of Plants (UPOV) and are species-specific. In the past such testing was performed essentially using morphological descriptors; an alternative and more cost effective system resulted in the use of molecular markers in DUS testing, particularly for the purpose of distinctiveness. An advantage to molecular over morphological markers is the ability to discriminate closely related varieties even if they are very similar in morphology; it is also known that marker assays can be used to spot heterogeneity or impurities in a seed sample (Cooke et al. 2003).

Many examples of the use of molecular markers in DUS testing are available in the literature for different horticultural and field crops. In the context of protecting new varieties of pepper (*Capsicum annuum*), it was shown that SSRs could be used to complement a DUS test of a candidate variety (Kwon et al. 2005). In pea (*Pisum sativum*) a marker system was developed as a tool for germplasm management and genetic diversity studies, as well as for rapid variety identification and DUS testing (Smykal et al. 2008). In potato (*Solanum tuberosum*) the identification of cultivars by morphological characters is a time-consuming practice. For this reason, a rapid and robust method for variety differentiation was developed; using a set of nine SSRs it was possible to differentiate over 1,000 cultivars excluding somaclonal variants and mutants. This process, from DNA extraction to cultivar identification, can be carried out in a single day, making it feasible for routine work (Reid et al. 2009).

Whereas extensive efforts in resistance breeding have led to the introgression of several resistance genes into modern varieties, it should be noted that a number of disease tests have recently become important criteria for assessing DUS. This is the case even though several environmental and genetic factors can influence the expression of resistance. When such descriptors are taken into consideration, a phenotypically unequivocal characterization of the cultivar may be difficult. In a recent study, molecular markers developed in tomato provided an alternative to pathogen reaction tests for DUS testing of candidate varieties. Results of the molecular assays were confirmed with respect to biological tests, thus enabling a possible introduction of markers as tools in tomato DUS testing (Arens et al. 2010). The results demonstrated that molecular identification could be used to assess distinctness and to complement morphological assessment, especially in cases where either long periods of time or confounding external factors played an important role.

It is therefore the duty of the UPOV to permit these technologies to acquire the required status of official descriptors in tomato DUS testing as well as in other crops. At present, although molecular markers are not accepted in DUS testing, UPOV has established a working group on the use of biochemical and molecular techniques to approve and rule on the use of molecular tools in DUS testing (http://www.cpvo.europa.eu/; Button 2006; Arens et al. 2010).

7.2.3 Hybrid Purity Testing

In tomato as well as in maize and in other important horticultural species where hybrid varieties are mainly used, hybrid purity testing is a routine task carried out by seed companies, among several tests required to deliver high quality seed to farmers. Since the pioneering efforts of tomato hybrid production of Gilbert (1912), the necessity to distinguish true hybrids from selfs has been evident. After studies of the rate of cross-pollination by the use of the recessive potato-leaf (*cc*) morphological marker in a recipient tester line (Haskell and Paterson 1966), the task was easily accomplished until the 1990s with the use of biochemical markers. Isoenzyme markers have been widely applied in hybrid purity testing not only in tomato (Tanksley and Kuehn 1985) but also in crop plants such as brussels sprouts (*Brassica oleracea* var. *gemmifera*) (Woods and Thurman 1976) and cabbage (*Brassica oleracea* var. *capitata*) (Arùs et al. 1985).

In spite of the wide adoption of DNA marker technologies, isoenzymes such as alcohol dehydrogenase are being still applied along with seed proteins, coupled with ultrathin-layer isoelectric focusing (UTLIEF) electrophoresis (van den Berg 1991), as an economically favorable method for hybrid purity testing in cereal crops such as rice (Zhao et al. 2005). Since the mid-1990s random amplified polymorphic DNA (RAPD) markers, although dominant in nature, have been applied as tools for hybrid seed testing in tomato (Paran et al. 1995). Liu et al. (2007) tested three DNA molecular marker systems, RAPD, inter-simple sequence repeat (ISSR) and SSR, to test seed genetic purity of commercial hybrid tomato and concluded that SSRs together with RAPDs were the most effective markers for this purpose. Jung et al. (2010) proposed a panel of 40 single nucleotide polymorphisms (SNPs) selected from the conserved orthologous set II (COSII) of unigenes to identify pepper hybrids, using an allele-specific (AS-PCR) method of interrogation. A similar COSII-based methodology might be proposed with the same aim for tomato.

In conclusion, although largely dependent on cost versus benefit considerations, DNA polymorphisms are destined in the medium-term to become the main tool in hybrid purity testing of tomato as well as most other hybrid variety crops.

7.3 Marker-Assisted Trait Introgression

7.3.1 Gene and QTL Introgression

MAS is based on a marker (morphological, biochemical or DNA)—trait association that enables the breeder to monitor the transmission of trait genes, using the marker genotype as a selection criterion (Stam 2003). The development of marker systems began with using mutations at the loci controlling plant morphology since the first association between seed size and coat pigmentation in *Phaseolus vulgaris* was identified by Sax (1923). However, the use of such morphological markers in breeding soon showed limitations due to pleiotropy, incomplete penetrance, or the fact that they are considerably affected by environmental factors and by the developmental stage of the plant.

Currently, tomato is not only one of several cultivated plant species whose draft genome sequence has been completed (http://solgenomics. net/), but also one of the best studied crop species with respect to agronomy, genetics, genomics and breeding. This is attested to by the large numbers of different molecular markers that in this species have been developed, mapped, and validated for association to agronomically useful traits. The availability of markers to assist tomato breeding began with the first isozyme markers developed in the 1980s, when a linkage association was reported between a tomato nematode resistance gene and an *Aps* isozyme allele (Medina-Filho 1980). Since then, the far more numerous and informative SNP, indel (Landegren et al. 1998), SSR (He et al. 2003), and COSII markers have become available (Wu et al. 2006).

At the same time, since a first molecular linkage map of tomato was published in 1986 containing 18 isozyme and 94 DNA markers (mostly cDNA clone-based restriction fragment length polymorphsisms, RFLPs) (Bernatzky and Tanksley 1986), numerous mapping efforts (see Chapter 5) have led to different map releases, mostly based on crosses between cultivated and wild tomatoes. From the Tomato-EXPEN 1992 (Tanksley et al. 1992), to the Tomato-EXPIMP 2008 (Gonzalo and van der Knaap 2008) up to 2,500 different markers had been mapped and are available for introgressing trait-associated marker alleles. Moreover, more than 1,300 putative genes related to morphological, physiological (e.g., fruit ripening, fruit abscission or male sterility), and disease resistance traits have been identified (Foolad 2007) and are being assigned to genomic sequence positions (http://solgenomics.net/, Tomato iTAG map, Tomato AGP Map).

This platform of tools is largely available to molecular breeders since it was mostly produced with publicly funded research, and can be used for both new trait-associated marker development and their application, with which we deal in the present chapter. Methods of application can be categorized into two strategies: i) the integration of markers into routine breeding programs devoted to new inbred lines and hybrid variety combinations, and ii) the use of markers to realize new breeding programs, infeasible without the genotyping tools.

The integration of markers into the routine of traditional breeding programs, can, in turn, be divided into two objectives. Firstly, the fixation of homozygosity at genes and markers associated with monogenic traits and/or QTLs in pedigree and progeny breeding schemes can be monitored. Secondly, the introgression and fixation of genes/QTLs leading to the improvement of recurrent genomic backgrounds through BC schemes can be followed. Several authors have classified this last aim of "standard" MAS as marker-assisted backcross (MABC) (Hospital 2002; Peleman and van der Voort 2003).

The use of MAS in tomato progeny breeding began during the 1980s with the development of isozyme markers and it continued into the 1990s with the use of molecular "associated" markers, together with cloned genes, i.e., "perfect" markers. Conceptually and in practice, the application to progeny breeding is quite simple. It follows the development of inbred lines in tomato, as in other selfers, by fixation at homozygosis in a certain generation, generally in F_2, the markers and genes associated to a favorable trait. It is simply a selection criterion added to the traditional breeding procedure that represents an additional cost, and that therefore should contribute to an improved system. The Southern Australian Barley Improvement Program, as a systematic application of MAS in a selfing species since the mid-1990s, reports that the greatest impact of MAS on breeding schemes lies in pedigree or other progeny schemes, thus indicating the usefulness of such a strategy, while MABC accounts for only 10–15% of the surveyed genetic materials (Barr et al. 2000). Nevertheless, the efficiency of MAS applied to progeny breeding schemes depends on some crucial issues: qualitative genes versus QTLs underlying a trait, number and distance of markers used, condition of coupling versus repulsion of favorable alleles, and traits selected by MAS. Costs and various marker platforms will be analyzed below in detail (see Sect. 8.4)

With respect to fixation at homozygosis of qualitative trait genes versus QTLs, the former is simpler if there are no pleiotropic effects and there are suitable associated markers. In this case, the availability either of a perfect marker or of a single, very closely linked marker (less than 1 cM), in complete linkage disequilibrium with the selected allele, is desirable for its wide application to different combinations of crosses. This will help to avoid retaining a 'wild', 'false-positive' allele in progeny where the introgression originates from a different wild accession that, even if of the same species, lacks the gene of interest. QTLs can be fixed at homozygosis through a pedigree scheme, together with other genes and QTLs. Most QTLs cloned in plants were observed to be located at not more than 2 cM with respect to their original mapping position (Price 2006). This implies that, if correctly performed and validated across different experiments and years, QTL mapping is accurate. As a consequence, a 5 cM region tagged by two flanking markers encompassing a central QTL peak should be more than sufficient for QTL MAS.

To avoid errors in QTL selection, namely the selection of false positives, two flanking markers per locus rather than one are necessary. Moreover, QTL MAS efficiency can carry complications from frequently observed QTL-QTL interactions, oftentimes inevident, and thus termed "background" effects. A QTL might interact with loci, often unknown, present in the QTL-carrying parent but absent in the derived selected genotype, or vice versa. This could account for a lower rate of success in QTL MAS sometimes observed in progeny- and BC-based breeding programs (Hospital 2009). As for numbers of markers, an increasing panel of markers surveyed for each breeding population could be applied by increasing the number of target traits in a complex breeding project, for improving fresh market or processing tomato.

An arrayed assay for selecting tens, hundreds, or even thousands of loci, although reducing the cost of single data points, contributes significantly to inflated costs of pedigree schemes. This issue, discussed below (see Sect. 8.4), is a limitation of MAS but holds great promise.

An analogy to describe the conservation of linkage blocks in a selfing cereal species is the term "national parks", where recombination is a rare animal. In a progeny-based breeding scheme, the presence of linkage blocks, as well as different origins of linked useful loci in *repulsion* phase in an F_1, are known limitations to exploiting tomato species' genomic potential. Although essential for recessive and incompletely dominant genes, the fixation in *coupling* in a single plant of positive alleles at tightly linked loci in a pedigree-derived F_2 population of typical size is therefore a difficult task, at present tractable only by greatly increasing the population size and by intensively applying marker genotyping. The same can be argued when the aim is to reduce linkage drag, or when trying to break a linkage block with mixed positive and negative effects.

In tomato, as in other crops, genes conferring resistance to pathogens, pests and other parasites often cluster together in complex loci (Young 1999; Grube et al. 2000). On the other hand, compiling resistances into superior inbred lines seems to be one of the imperatives of modern tomato breeding. When dealing with MAS for such traits, a frequent occurrence is the need to carry the resistance specifities in cis phase. Significantly, one recent patent deposited at the US Patent office is simply a MAS method to obtain a tomato plant carrying in coupling both resistances to tomato yellow leaf curl virus (TYLCV) (*Ty-1*) from *S. chilense* and to root-knot nematode (*Mi-1*) from *S. peruvianum* (Hoogstraten et al. 2009).

MAS becomes either appealing or imperative for traits difficult to precisely phenotype due to environmental effects and/or technical difficulties. In other cases, when selecting for traits relatively easily phenotyped, MAS can be desirable for the high number of assays that can be performed on a small number of seeds or plants, including those in earlier generations, instead of requiring a relatively large number of plants or fruits per assay as often occurs for phenotyping. The present situation of available markers or underlying genes for agronomic traits is fairly positive and expected to improve in the next few years; such a scenario makes MAS increasingly appealing for medium-sized seed companies, both for pedigree- and BC-based programs.

To date, more than 40 genes underlying both qualitative and quantitative trait loci known for conferring resistances to all major classes of pathogens, together with additional loci of resistance genes analogs (RGAs) have been mapped and/or cloned from Solanaceae species (Grube et al. 2000; Chunwongse et al. 2002; Parrella et al. 2002; Foolad et al. 2002; Zhang et al. 2002; Bai et al. 2003). Therefore, MAS has become a reality in many seed companies for improving tomato vertical resistances to diseases such as bacterial speck, corky root, Fusarium wilt, late blight, nematodes, powdery mildew, tobacco and tomato mosaic virus (TMV, ToMV), tomato spotted wilt virus (TSWV), TYLCV, and Verticillium wilt. On the other hand, in many small and medium-sized seed companies MAS is not yet a routine procedure for manipulating resistance QTLs, although some efforts are being undertaken to improve the quantitative resistance to bacterial wilt, bacterial canker, powdery mildew and TYLCV (Foolad et al 2005; Foolad 2007). MAS has also been employed to select genes for some simple morpho-physiological traits such as self-pruning (many tomato genetics and breeding programs in the USA according to Foolad 2007), and QTLs for some complex fruit quality traits (Lecomte et al. 2004; Labate et al. 2007).

In public breeding programs few breeders have used MAS for tomato improvement and it seems that MAS is not a routine approach, even if some applications have been reported. For example, Barone et al. (2005) employed PCR markers to improve vertical resistances to a few diseases. Recently, Barbieri et al. (2010) reported the fixation at homozygosis by MAS of *Ty-1* and *Ty-2* TYLCV resistance genes into F_2 segregating populations derived from a valuable fresh-market tomato to obtain resistant inbred lines. Suitable PCR markers targeting resistance genes are often selected from those publicly available (e.g., *Mi*, *Sw5* and *Tm2a* associated markers), or designed directly on the gene sequences (e.g., *I2*, *Pto* and *Ve2*), and used to select resistant genotypes in BC as well as in progeny-based selfing schemes.

Other public breeding programs that reported MAS applications are studies of resistance to bacterial canker, bacterial speck and bacterial spot (Coaker and Francis 2004; Yang and Francis 2005), and of horizontal resistance to blackmold (Robert et al. 2001) and late blight (Brouwer and St. Clair 2004). Ad-hoc sequence characterized amplified region (SCAR), cleaved amplified polymorphic sequence (CAPS) and AFLP markers were developed to be effectively used for introgression by MAS of the resistance genes *ol-1* (Huang et al. 1998) and *ol-2* (De Giovanni et al. 2004) to *Oidium lycopersicum*, causal agent of tomato powdery mildew. SSR markers linked to a late blight resistance gene in tomato have been identified and can be used in MAS (Zhu et al. 2006). Molecular markers associated with *Pyl* gene conferring resistance to the corky root causal agent *Pyrenochaeta lycopersici* were used for screening of fresh-market tomato for resistance to the pathogen (Pucci et al. 2007). Marker-assisted breeding for resistance traits is broadly diffuse, as are the respective marker tools; Table 7-1 summarizes many major reports including the associated genes, QTLs, and information on the available markers.

Because the deployment of doubled haploid (DH) production coupled with pedigree and BC schemes seems to be more prevalent in species such

Table 7-1 Bacterial, viral, fungal and nematode resistance genes in tomato (*Solanum lycopersicum*) and associated molecular genetic markers work.

Trait	Source species[a]	Gene	Chr.	Marker/locus/acc. n°	Status	Primers	Reference
BACTERIAL DISEASE							
Bacterial canker	PER		5, 7, 9	TG363, TG61, TG254	SCAR (TG61)		van Heusden et al. 1999
	HAB	Rcm2.0	2	TG537	CAPS HinfI	TG537 F: TACCCGAGGCTCAGAAACAC TG537 R: CATCAACAGGAGATCGGTTTT	Coaker and Francis 2004
			2	TG091	CAPS DraI	TG091 F: TGCAGAGCTGTAATATTTAGAC TG091R: CFFTCTCAGTTGCAACTCAA	Kabelka et al. 2002
		Rcm5.1	5	CT202	CAPS Tsp	CT202 F: TAATCCGAGAAGGTGATCCG CT202 R: GGCTTATAACCCATGCCAAAAG	Coaker and Francis 2004
			5	TG358, TG538	CAPS RsaI	TG358 F: CCAAGTGCAGAGAGTACTGGA TG358 R: TGAATGAACATGATCAAAGTATGC	Coaker and Francis 2004
Bacterial speck	LYC	Pto	5				Martin et al. 1993
	PIM	Pto					Yang and Francis 2005
		Prf					Salmeron et al. 1996
	numerous	Pto		AF220602 AF220603	cloned gene	F: GGTCACCATGGGAAGCAAGTATTC R:GGCTCTAGATTAAATAACAGACTCTTGGAG	Rose et al. 2005; Rose et al. 2007
Bacterial spot	LYC	Rx-1	1	TG236			Yu et al. 1995
		Rx-2	1	TG157			Yu et al. 1995
		Rx-3	3	TG351			Yu et al. 1995
		Rx-3	3				Yang and Francis 2005

						Reference
PIM	Rx-4	11				Robbins et al. 2009
PEN	Xv4	3	TG599	CAPS HindIII, ECoRI	TG599 F: TGTTGATCCTTGCTTGCTGT TG599 R: TTGTATGGTGCAACTTCCC	Astua-Monge et al. 2000
		3	TG134	RFLP		Astua-Monge et al. 2000
LYC	Bs4	5	TG432, CT16	RFLP		Ballvora et al. 2001; Schornack et al. 2004
LYC		11	SSR637	SSR	SSR637 F: AATGTAACAACGTGTCATGATTC SSR637 R: AAGTCACAAACTAAGTTAGG	Yang et al. 2005
		11	TOM196	SSR	TOM196 F: CCTCCAAATCCCAAAACTCT TOM196 R: TGTTTCATCCACTATCACGA	Yang et al. 2005
		11	TOM144	SSR	TOM144 F: CTGTTTACTTCAAGAAGGCTG TOM144 R: ACTTTAACTTTATTATTGCGACG	Yang et al. 2005
Bacterial wilt						
LYC		6	CT184			Danesh et al. 1994
		7	TG51b			Danesh et al. 1994
		10	CT225b			Danesh et al. 1994
LYC		4, 6, 11	TG268, TG118, GP162			Thoquet et al. 1996
LYC		6	TG118, TG240	RFLP		Thoquet et al. 1996; Mangin et al. 1999
LYC		2, 6, 8, 12	GP504, TG73, CT135, TG564	RFLP		Wang et al. 2000
LYC	Bwr-3	3	TG515	RFLP BglII		Carmeille et al. 2006

Table 7-1 contd....

Table 7-1 contd....

Trait	Source species[a]	Gene	Chr.	Marker/locus/acc. n°	Status	Primers	Reference
		Bwr-4	4	CD73	RFLP		Carmeille et al. 2006
		Bwr-6	6	TG73	RFLP		Carmeille et al. 2006
		Bwr-8	8	CD40, CT135	RFLP		Carmeille et al. 2006
		Bwr-12	12	TG564	RFLP *BamHI*		Carmeille et al. 2006
VIRAL DISEASE							
Alfalfa mosaic virus (AMV)	HAB	Am	6		AFLP *HindIII*, *MseI*	F: AGACTGCGTACCAGCTTA R: GATGAGTCCTGAGTAAC	Parrella et al. 2004
Tomato chlorotic mottle virus (TCMV)	PER	Tcm-1	NA		NA		Giordano et al. 2005
	LYC		NA		NA		Ji et al. 2007b
Tomato mottle virus (ToMoV)	CHI	Ty-3	6		RAPD, SCAR, CAPS		Griffiths and Scott 2001; Ji and Scott 2006; Ji et al. 2007a
Tomato yellow leaf curl virus (TYLCV)	CHI	Ty-1	6	TG178	CAPS *TaqI*	F: GGTACTCCTGGAAGGGTTAAGG R: CACGCTGGTTCTGTTGTATCTC	Maxwell et al. 2006

						F / R	
		Ty-1	6	TG436	CAPS StyI	F: TCTGCAAGTCGCATCGGAAGGTATCG R: GTATGGGCCACCTGGCATGCACCTCG	Maxwell et al. 2006
	HAB	Ty-1		REX	CAPS TaqI	F: TCGGAGCCTTGGTCTGAATT R: GCCAGAGATGATTCGTGAGA	Seah et al. 2004
		Ty-2	11	TG105A	CAPS TaqI	F: CTTCAGAATTCCTGTTTAGTCAGTTGAACC R: ATGTCACATTTGTTGCTTGGACCATCC	Maxwell et al. 2006
		Ty-2	11	C2_At5g25760	CAPS StyI	F: TCCTTATGATGGTGGAGTTTTCCAG R: AAAGCAATTATAGCTCGACAAACAG	Barbieri et al. 2010
		Ty-2	11	T0302	SCAR	F: TGGCTCATCCTGAAGCT- GATAGCGC R: AGTGTACATCCTTGCCATTGACT	Hanson et al. 2006
	HAB	Ty-1 Ty-3		FER FLUW25	SCAR	F: CAAGTGTGCATAT ACTTCA TA(t/g)TCACC R: CCATATATAACCTCTGTTTCTATTTCGAC	Ji et al. 2007b
	CHI	Ty-3		PG3 HinfI		F: ATGACTCCAACAAGCAAAGGCACGAG R: AAAGAGAAGCTGCAATGTGTCGCC	Ji et al. 2007a
		Ty-3		cLET-1-I13 DdeI		F: ACATTCTCTGGGTTGATTCCTCCG R: CCTCTGTTTCACTTTCTACTGC	Ji et al. 2007a
		Ty-3		cLEG-31P16 HaeIII		F: ATGGTGACTAAGGTGGATGAGCCT R: TGAGTGCCAACGATAAATGCTACC	Ji et al. 2007a
	CHI	Ty-3/ Ty-3a		FER-G8F	CAPS TaqI	F: CAT CCC GTG CAT CAT CCA AAG TGA R: CTA AGG GTG TAC CCC AAG GGA AC	Jensen et al. 2007
	CHI	Ty-4	3	C2_At4g17300 Ct_At5g60160			Ji et al. 2008; Ji et al. 2009
Cucumber mosaic virus (CMV)	CHI	Cmr	12		RFLP		Stamova and Chetelat 2000
Curly top virus (CTV) (Beet curly top virus)	PER, HAB, PIM	multigenic					Martin and Thomas 1986; Thomas and Mink 1998

Table 7-1 contd....

Table 7-1 contd....

Trait	Source species[a]	Gene	Chr.	Marker/locus/ acc. n°	Status	Primers	Reference
Potato leaf roll virus (PLRV)	PER						Hassan and Thomas 1988; Thomas and Mink 1998
Tomato yellow top virus (TYTV)	PER						Hassan and Thomas 1988
Pepino mosaic virus (PepMV)	CHI, HAB						Soler et al. 2004; Picó et al. 2002
Potato virus X (PVX)	LYC						Rashid et al. 1989
Potato virus Y (PVY)	PIM						Légnani et al. 1997; Parrella et al. 2002
Tobacco etch virus (TEV)	HAB	Pot-1	3		AFLP		Légnani et al. 1997; Parrella et al. 2002
Tomato mosaic virus (ToMV) also known as Tobacco mosaic virus (TMV)	HAB	Tm-1	2		SCAR	F: GGTGCTCCGTCGATGCAAAGTGCA R: GGTGCTCCGTAGACATAAAATCTA	Ohmori et al. 2000; Lanfermeijer et al. 2003, 2005
	PER	Tm-22	9	SNP2493/2494	Tetra primer ARMS	F1: GGGTATACTGGGAGTGTCCAATTC R1: CCGTGCACGTTACTTCAGACAA F2: CTCATCAAGCTTACTCTAGCCTACTTTAGT R2: CTGCCAGTATATAACGGTCTACCG	Lanfermeijer et al. 2005

Disease		Gene		Marker	Method	Primer	Reference
		Tm-22	9	SNP901	Tetra primer ARMS	see paper	Lanfermeijer et al. 2005
				SCN13	CAPS AccI	OPN13UF: AGCGTCACTCCATACTTGGAATAA OPN13T R: AGCGTCACTCAAAATGTACCCAAA	Sobir et al. 2000
				TG101	RFLP, STS Srd, PstI		Fazio et al. 1999
Groundnut bud necrosis virus (GBNV)	PER	Sa-5	9		PCR, cloned		Folkertsma et al. 1999
Groundnut ringspot virus (GRSV), Tomato chlorotic spot virus (TCSV)	PER	Sa-5	9		PCR, cloned		Boiteux and Giordano 1993
Tomato spotted wilt virus (TSWV)	PER	Sa-1a	9				Finlay 1953; Stevens et al. 1992
	LYC	Sa-1b	9				Stevens et al. 1996; Roselló et al. 1998
	LYC	Sa-2	9				Folkertsma et al. 1999
	LYC	Sa-3	9				Brommonschenkel et al. 2000
	LYC	Sa-4	9				Roselló et al. 2001; Scott et al. 2005

Table 7-1 contd....

Table 7-1 contd....

Trait	Source species[a]	Gene	Chr.	Marker/locus/acc. n°	Status	Primers	Reference
	PER	Sw-5	9	CT220	PCR, cloned	F: AAGCCGAATTATCTGTCAAC R: GTTCCTGACCATTACAAAAGTAC	Folkertsma et al. 1999
	PER	Sw-5	9	Sw-5b-LRR		F: TCTTATATTGTGGAGTTTTTGTCG R: TCCACCCTATCAAATCCAAC	Garland et al. 2005
	PER	Sw-5	9	Sw 5-2	SCAR	F: AATTAGGTTCTTGAAGCCCATCT R: TTCCGCATCAGCCAATAGTGT	Dianese et al. 2010
	CHI	Sw-7	9				Gordillo et al. 2008
FUNGAL DISEASE							
Alternaria stem canker	PEN	Asc	3		RFLP		van der Biezen et al. 1995; Stommel and Zhang 1998; Mesbah et al. 1999
Anthracnose ripe rot	LYC		various	numerous			Stommel and Zhang 1998
Blackmold	CHE		2, 3, 9, 12	numerous			Robert et al. 2001
Corky root	PER	Py-1	3	TG324	CAPS DraI	F:CTTCTAGTAGTCCAACAGCAACTG R: CACTTGGTTGATGGATAGTG	Doganlar et al. 1998
Early blight	HAB, LYC, PIM	QTLs	2, 7, 9				Foolad et al. 2002; Zhang et al. 2003; Foolad et al. 2005
	ARC	QTLs	2, 9		SNP, SSR, AFLP		Chaerani et al. 2007

Disease	PER	Gene	Chr	Marker	Type	Primer	Reference
Fusarium crown		*Frl*	9				Vakaloumakis et al. 1997
		Tm2	9				Vakaloumakis et al. 1997
		Frl	9	TG101, UBC 194	RAPD, SCAR	UBC194: AGG ACGTGCC	Fazio et al. 1999
Fusarium wilt	*PIM*	*I*	7, 11		RFLP		Bournival et al. 1990
	PEN	*I-1*	7, 11		RFLP		Sarfatti et al. 1991
	PIM	*I-1*	11		SCAR	At2-F3 CGAATCTGTATATTACATCCGTCGT At2-R3 GGTGAATACCGATCATAGTCGAG	Ori et al. 1997; Scott et al. 2004
		I-2	7, 11		RFLP		Tanksley and Costello 1991
					SCAR	Z1063F ATTTGAAAGCGTGGTATTGC Z1063R CTTAAACTCACCATTAAATC	Simons et al. 1998
				TAO1	CAPS RsaI, FokI	TAO1 F GGGCTCCTAATCCGTGCTTCA TAO1R GGTGGAGGATCGGGTTTGTTTC	Staniaszek et al. 2007
					SCAR	TFusF1 CTGAAACTCTCCGTATTT TFusrr2 CCTGGATGAACAGCTGAG	El Mohtar et al. 2007
		I-2C	7, 11		RFLP		Ori et al. 1997; Simons et al. 1998
	PEN	*I-3*	7, 11		RFLP		
		I-3		P7-43DF3/R1	SCAR	F: CACGGGATATGTTRTTGATAAGCATGT R: GTCTTTACCACCAGGAACTTTATCACC	Barillas et al. 2008
Gray leafspot	*PIM*	*Sm*	11		RFLP		Behare et al. 1991

Table 7-1 contd....

Table 7-1 contd....

Trait	Source species[a]	Gene	Chr.	Marker/locus/ acc. n°	Status	Primers	Reference
Late blight	PIM, HAB	Ph-1, Ph-2, Ph-3, Ph-4, QTLs	all		RFLP		Pierce 1971; Frary et al. 1998; Moreau et al. 1998; Chunwongse et al. 2002; Brouwer and St Clair 2004; Brouwer et al. 2004; Kole et al. 2006
Leaf mold	PIM	Cf-1	1,6		RFLP		Jones et al. 1993; Balint-Kurti et al. 1994
		Cf-2 Cf-4 Cf-5 Cf-9	1,6		RFLP		Lauge et al. 1998
	PIM	Cf-4 Cf-9		AJ002237.1			Parniske et al. 1997, 1999
	PIM	Cf-9				Cf 95: AAAGCAAACA TTTCTTGATTTCTT Cf 9F2: GCTTGGAATATGA CCTTCAAATCTG	Jones et al. 1994
		Cf-4 Cf-9					Wulff et al. 2009
	PIM	Cf-ECP1	1	LOXR	CAPS HincII		Soumpourou et al. 2007
	PIM	Cf-ECP4	1	TG236	CAPS SspI		Soumpourou et al. 2007
Powdery mildew	CHI	Lv	12	CT121, CT129	RFLP		Chunwongse et al. 1997
	LYC	ol2	4		AFLP, SCAR CAPS HHaI	SCARU3-2 F: AGTGGTTGGCGGATAGGTG SCARU3-2 R:TTGGCAAGGTGGGAAAACT	Ricciardi et al. 2007; Pavan et al. 2008

	HIR	Ol-1, Ol-3	6	TG25	CAPS HaeIII	F: CAACAGCTGCCACAAACACT R: AGTTTGGTGCTTCATGCAAA	Huang et al. 2000
	PAR	Ol-qtl1	6		AFLP	see paper	Bai et al. 2003
	PAR	Ol-qtl2	12	P18M51-701c			Bai et al. 2003
	PAR	Ol-qtl3	12	B432U	SCAR	F: TCAGAAAGGGAAGAATCAAG R: CCCGATCCAATGTTATGTCTGAA	Bai et al. 2003
	PER	Ol-4	6	Aps1/TaqI-Sau96I	CAPS	F: ATGGTGGGTCCAGGTTATAAG R: CAGAATGAGCTTCTGCCAATC	Bai et al. 2004
	PER	Ol-4	6	By-4/ApoI-HypCH4IV	CAPS	F: CATAGTGTAGCTTTGATTCTTGTA R: CCAATTGCCGGGAAGGAA	Bai et al. 2004
Verticillium wilt	LYC	Ve1		Ve1 allele	AS-PCR	F: GTTACATGCAATCTCTTTGG R: AATTAATGTGGACAAGCTCTG	Acciarri et al. 2007
				ve1 allele	AS-PCR	F: AGCTCATTTTGAAGGACTCTA R: AGAGTAGTCCACATAGATGG	Acciarri et al. 2007
		Ve1/Ve2			CAPS XbaI	VVF1:GAC TAC ATT GAC CCT GGG CTC TTG VVR2: TGA GAG CAC CTT AAG CTT TTC AAT	Kuklev et al. 2009
		Ve1		SNP2199	Tetra primer ARMS	Ve1_SNP2199Ft CAGGCCCTTTGGATGAATCACATT Ve1_SNP2199Ra GTTGGACAAAAGAGAGAAAGTGAAGCTAAGT	Arens et al. 2010
				SNP2199		F: CAGGCCCTTTGGATGAATCACTAT R:GTTGGACAAAAGAGAGAAAGTGAAGCTTACT	Arens et al. 2010
		Ve2		V2LeO3	CAPS XbaI	F: CGAACCATCAAAATCTATCC R: GAGGTCAAGGGGCACTGTC	Acciarri et al. 2007
	LYC		9	SNP2827	Tetra primer ARMS	Ve2_2720F: GGATCTTAGCTCACTTTATGTTTTGAAC Ve2_3040R GGTGCTGGTTTCAACTCTGAAGT	Arens et al. 2010

Table 7-1 contd....

Table 7-1 contd....

Trait	Source species[a]	Gene	Chr.	Marker/locus/acc. n°	Status	Primers	Reference
Grey mold	NEOR		3,4	QTL3, pQTL4			Finkers et al. 2008
	LYCOP		1	TG17	CAPS HindIII	TG17 F CGGCTGTGTACGTATCTGA TG17 R AAAATCAATTGAACCGGCTGT	Davis et al. 2009
			9	TG18	CAPS BamhI	TG18 F CTCAAGCTCCAGCTGTTTCC TG18 R GCTCCTTCTGCAATGGGTAA	Davis et al. 2009
			11	TG46	CAPS HaeIII	TG46 F ATCCCAACCTCTGAGCACAC TG46 R GTTCCTGGAACCGATATTGC	Davis et al. 2009
ROOT KNOT NEMATODE							
	PER	Mi1	6	Mi23	SCAR	Mi23F TGGAAAAATGTTGAATTTCTTTTG Mi23R GCATACTATATGGCTTGTTTACCC	Garcia et al. 2007
		Mi3	12	TG180	RFLP	TG180 F ATACTTCTTTGCAGGAACAGCTCAC TG180 R CACATTAGTGATCATAAAGTACCAG	Yaghoobi et al. 2005
	PER		6	REX1	CAPS, RFLP TaqI	REX F TCCGAGCCTTGGTCTGAATT REX R GCCAGAGATGATTCGTGA	Williamson et al. 1994; Ammiraju et al. 2003
	PER	Mi1-2	6		SCAR	PmiF3 GGTATGAGCATGCTTAATCAGAGCTCTC PmiR3 CCTACAAGAAATTATTGTGCGTGTGAATG	El Mehrach et al. 2005
	ARC	Mi9	6	*numerous*			Jablonska et al. 2007

[a]**ARC,** *Solanum arcanum;* **CHI,** *Solanum chilense;* **LYC,** *Solanum lycopersicum;* **HAB,** *Solanum habrochaites;* **LYCOP,** *Solanum lycopersicoides;* **NEOR,** *Solanum neorickii;* **PEN,** *Solanum pennellii;* **PER,** *Solanum peruvianum;* **PIM,** *Solanum pimpinellifolium.*

as potato, rice, maize, wheat and rapeseed (*Brassica napus*) (Forster et al. 2007) relative to tomato, the most promising developments of tomato MAS could be an early generation selection at multiple loci on large segregating populations. This might also be viewed as a genetic ideotype breeding scheme (see Section 7.3.3). An early generation selection at multiple loci, especially if performed on a population derived from complex crosses, could allow the recovery of rare positive recombinants. Such positive recombinants are frequently difficult to obtain in traditional pedigree breeding if aimed at improving low heritability traits (Hospital et al. 1997). The possibility of early generation MAS on large populations is particularly attractive in tomato compared to cereals, since in this species the practice of transplantation is well-established. This would mean, for example, performing MAS on very large numbers of F_2s at the plantlet stage, then transplanting into the field, marker pre-selected F_2s of normal size, occupying the same acreage, but already containing a large number of favorable alleles.

For introgressing monogenic traits into elite genetic materials, MABC is the most straightforward strategy (Hospital and Charcosset 1997; Barr et al. 2000; Lecomte et al. 2004). In most cases, the elite variety used for backcrossing is called the "recurrent" or "recipient" parent, with a large number of positive characters, lacking only a few desirable traits. On the other hand, the "donor" parent, oftentimes a wild-derived inbred line, possesses one or more genes controlling an important trait absent in the elite variety (e.g., disease resistance), but is agronomically quite poor. Classically, the backcrossing method aims to improve the defective trait by introgressing it from a donor while selecting against the undesirable characteristics of the unadapted parent.

In the MAS era this standard breeding procedure should be accommodated to introgression of QTLs once they are identified as single genomic regions that have real effects on phenotype, and if two flanking markers are available. In fact, before the advent of markers, only major "QTLs" with large effects on phenotype were manageable in BC introgression; moreover, they were dominant or recessive rather than additive since the BC populations to be selected for sequential crosses carry the introgressed segment in a heterozygous state. However, several authors found MABC to be difficult for introgressing QTLs, especially minor QTLs. This was mainly for two reasons: inaccurate validation of QTL effects and positions in different experiments (preintrogression), and unpredictable interactions between QTLs and different genetic backgrounds.

In tomato, Goodstal et al. (2005) successfully introgressed a single QTL with a major effect on chilling tolerance by MABC, the source of which was *S. habrochaites* chromosome 9. The developed near-isogenic line (NIL) of cultivated tomato carrying the QTL retained the effect on the trait. In the study by Lecomte et al. (2004), five major tomato QTLs involved in organoleptic quality attributes were chosen to be introgressed from cherry tomato into three different recipient lines with average flavor through MAS. In all three recipient genotypes, the introduced regions had favorable effects on all traits except for fruit weight, although some of the QTL effects were specific for only one of the recipient backgrounds. A short time later, Chaib et al. (2006) reported MABC introgression of five QTLs for fruit quality from a cherry tomato line into three modern lines with larger fruits. The authors evaluated the QTL effects on materials at different stages of the selection procedure, as BC_3S_1 families which segregated simultaneously for the five QTLs, and as three sets of QTL-NILs (at BC_3S_3 generation), which differed from the recipient parent for the single introgressed QTLs. They found that not all of the QTLs were confirmed in new backgrounds, that new QTLs were discovered in the new materials, while the large majority of the original target QTLs had consistent effects when evaluated in the same background of the initial RIL mapping population.

Lawson et al. (1997) introgressed four target chromosomal regions containing five QTLs for pest resistance (acylsugar accumulation) from wild tomato into cultivated tomato. Starting with the ILs of Eshed and Zamir (1995), each carrying one target region, they performed three backcrosses followed by one intermating generation to obtain progenies homozygous for the resistance alleles at the five QTL. Selection was based on both marker genotype and phenotype. The introgression of the four regions was successful at the genomic level. However, the level of acylsugar accumulation in the progenies introgressed for the five QTLs was lower than expected, and in particular lower than that of the interspecific F_1 hybrid, indicating that some genetic factors (QTL) of the accumulation were missing, either lost or not controlled in the program.

In other species, the results of MABC vary. Toojinda et al. (1998) introgressed two QTLs for stripe rust resistance in barley, through one BC followed by one haplo-diploidization with selection on marker genotype and phenotype, into a recipient background different from the one used to map the QTLs. Both QTLs were confirmed and additional QTLs were detected in the new background, including some resistance alleles carried by the susceptible parent. Knoll and Ejeta (2008) successfully validated in different genetic backgrounds two QTLs for early-season cold tolerance whose source was a Chinese landrace of sorghum (*Sorghum bicolor*), and found them ready to be introgressed into elite cultivars. In rice, Neeraja et al. (2007) were able to efficiently convert the submergence-sensitive cultivar Swarna into a submergent-tolerant variety by introgressing by MABC the major QTL sub1 on chromosome 9, after only three BC generations requiring two to three years.

Han et al. (1997) attempted to introgress two QTLs for malting quality into a feed barley cultivar by combined MAS and phenotypic selection with only partial success. Phenotypic analysis of the obtained lines carrying single QTLs showed that one QTL increased the malting quality of the recipient while the second QTL did not. Kandemir et al. (2000) successfully introgressed separately three yield QTLs by MAS into three barley NILs with the background of a lower yielding malting variety; this notwithstanding, no NIL was able to yield significantly more than the recipient cultivar. The authors claimed possible interaction effects, not acting in the lines, near-isogenic to the recipients apart from the QTL region. Shen et al. (2001) attempted with limited success a MABC program to introduce four QTLs for root depth into a rice cultivar, a trait that is very difficult to manage phenotypically. Starting from DH lines they produced a number of BC_3F_3 lines, each carrying one or two QTLs at most, and using only genotypic selection. When plant phenotype was analyzed for validating the introgression, they found that among the four QTLs, only one out of four gave the expected effect in the progenies. The authors hypothesized a loss due to inaccurate mapping and QTL-QTL interactions to explain the results.

These experiments highlight the necessity of accurate and validated QTL map positions together with the probable presence of unpredictable interaction effects between QTL-background and QTL-QTL. Such genetic background effects on the observed phenotypic improvement of the trait after introgression is a phenomenon also observed in animals. Koudandé et al. (2005) found a 30% reduction in the expected effects on trypanotolerance for three QTLs introgressed by MABC into a recipient mouse strain. Ribaut et al. (2002) and Ribaut and Ragot (2007) reported an interesting study of introgression for five target regions containing QTLs for drought tolerance in maize. The results showed that the success of the introgression in improving yield under drought strongly depended on the environmental conditions. Under severe drought stress the QTL-carrying progenies (F_1 hybrids) were far significantly higher yielding than controls, while they did not show any improvement in no-stress or even in mild drought stress conditions.

Zhu et al. (1999) screened DH lines of barley for the presence of several QTLs for yield, a complex and composite trait, based on selection for marker genotype alone. They phenotypically evaluated the progenies in five environments including four locations and two years; the original map positions of the QTLs were confirmed as correct in the progenies. However, the effects of the QTLs in the progenies were often different from the expectations with regards to magnitude and sign; moreover, the authors detected epistatic interactions between QTLs as well as numerous GxE interactions. The authors concluded that QTL introgression for complex traits should focus on allelic combinations based on epistatic interactions rather than on individual QTL effects.

This conclusion may apply as a general principle when attempting single QTL introgressions through MABC, that have a high risk to fail, especially when dealing with minor effect QTLs, if the extent of their interactions with environment, with background and with other QTLs are not taken into account. The alternative, as emphasized in other parts of this chapter, is to initiate the breeding program before QTL mapping and then proceed with QTL identification and validation of effects directly in the target background, i.e., following an AB-QTL introgression procedure.

Without the need to evaluate the qualitative versus quantitative genetic behavior of an introgressed segment, if we focus on the method, the use of DNA markers in backcrossing theoretically increases the efficiency of selection at three levels (Holland 2004; Collard and Mackill 2008). At the first level, namely "foreground selection", markers can aid selection of target alleles whose effects are difficult to observe phenotypically (e.g., recessive, laborious or time consuming, or expressed late in the plant growth cycle), and enable the breeder to perform selection at the seedling stage to choose the best plants for subsequent backcrossing (Holland 2004; Collard and Mackill 2008).

However, during foreground selection, "linkage drag" of negative effects carried by loci included in the introgressed region should be avoided. It has been demonstrated in the literature (Young and Tanksley 1989) as well as through personal experience (N Pecchioni, unpublished results, 2009), that in a selfing species when a BC line is fully genotyped (by whole-genome scan), the chromosome with the highest percentage of the donor genome and most persistent through BC generations is usually the chromosome carrying the targeted introgression. This occurs mostly around the introgression since this is the object of phenotypic selection for the introgressed trait, i.e., for the donor alleles.

Reports of demonstrated negative effects by linkage drag can be found in the literature. An example is the case of introgression of QTLs for late blight resistance in tomato (Brouwer and St Clair 2004), where most QTLs for horticultural traits were identified in intervals adjacent to those containing the late blight resistance QTLs. The authors concluded that only fine mapping of the resistance QTLs would allow the use of MAS for the precise introgression of *S. hirsutum* segments containing late blight resistance alleles separately from those containing deleterious alleles at horticulturally important QTLs. Similar results were obtained with QTLs for blackmold resistance from *S. cheesmaniae* (Robert et al 2001). Five QTLs were selected for introgression from *S. cheesmaniae* into cultivated tomato by using MAS. While one QTL conferred higher resistance and improved earliness, considered to be agronomically favorable, four out of the five introgressed QTLS were primarily associated with negative horticultural traits.

To avoid the negative effects of linkage drag, during the second level of selection, termed "recombinant selection", markers can be used to select the BC progenies in which recombinations reduce the size of the introgression to as small as possible. Therefore, recombinant selection can be considered as a more stringent "foreground selection". This can be achieved by using markers flanking the target locus on both sides and as near to the target as possible, e.g., less than 5 cM on either side (Holland 2004; Collard and Mackill 2008). Hospital (2002), based on computational studies, concluded that the application of very close markers is the only way to substantially reduce linkage drag.

This does not say anything about minimal population sizes and genotyping efforts necessary to obtain the desired genotypes. To obtain two relatively close flanking markers with recurrent alleles, together with the target region carrying donor alleles, double recombinant genotypes are necessary. These are very difficult to obtain through a single BC generation, so that reasonably at least two BC generations are required (Young and Tanksley 1989; Hospital 2001). Hospital and Charcosset (1997) produced a mathematical solution to calculate the optimal population sizes to obtain the desired MABC target that was then applied by Frisch et al. (1999) to compare possible alternative MABC strategies. These last authors agreed with Hospital and Charcosset (1997) in concluding that larger population sizes in advanced generations rather than in early ones substantially reduce the overall number of genotyped individuals throughout the entire BC scheme.

At the third level, namely "background selection", markers that are unlinked to the target introgression enable the selection of those progenies that contain higher percentages of background (recurrent) genome, thus accelerating the output of the MABC program. The feasibility and power of such an approach has been demonstrated by observations of real versus theoretical recurrent genome percentage in a BC plant. After two backcrosses progenies are expected to carry 87.5% (on average) recurrent parent alleles at loci unlinked to the target. However, it has been shown that this value largely varies among individual plants. Powell et al. (1996) reported a range in percentage of donor parent marker alleles of 8%–60% in BC_1 barley progenies instead of the expected 25%. Therefore, the "background markers" should be able to identify those progenies that in each BC generation are most similar to the recurrent parent. Conventional breeding methods typically require six to eight backcrosses to fully recover the recurrent parent genome (Collard et al. 2005), and several donor chromosome fragments may remain. By using markers the same result can be achieved at the BC_4, BC_3 or possibly BC_2 generation (Frisch et al. 1999; Collard and Mackill 2008).

The significantly higher cost of the marker-assisted background selection can be balanced by the time savings and by the relative economic benefits, especially if using low cost multilocus markers. Moreover, alternative MABC strategies can be planned in order to reduce the number of marker data points required to recover the recurrent genome.

In a study aimed at finding the best BC strategy for background selection to minimize the number of marker assays, Frisch et al. (1999) found that, instead of keeping the population size constant through the BC cycles, increasing it from BC_1 to BC_3 reduced the amount of required marker data by as much as 50% without affecting the percentage of recurrent parent genome recovered. Using markers to expedite the recovery of the recipient background genome was elegantly shown to be made efficient by the integration of the *Bt* transgene into different maize genetic backgrounds (Ragot et al. 1995). This confirmed the theoretical prediction that use of markers provides a gain in time of at least two BC generations. Although few other results on this matter have been published it is known that the technique is now routinely used, in particular, by private plant breeding companies.

In the above marker-based strategies of MABC, the simultaneous 'introgression of multiple targets' has not been considered. However, an oligo- or poly-genic basis may underlie some traits that are improved through introgression by backcrossing. For example, this is the case for the accumulation of resistances or for the introgression of complex traits. In general, several QTLs of small or medium effects account for the total variability of these traits. In such instances, to control several targets by markers the number of individuals that must be genotyped increases exponentially with the number of target loci. Hospital and Charcosset (1997) computed the required population sizes for different numbers of target QTLs to be introgressed, and they concluded that, in general, it is inadvisable to plan to manipulate more than three or four QTLs simultaneously in an MABC program. If the targets are known genes at single QTL directly controlling the trait (e.g., steps of a metabolic pathway) rather than QTLs indirectly selected by markers, the maximum number of targets could be slightly higher, but should not exceed five or six.

Just four QTLs for resistance to *Phytophthora capsici* from a small-fruit pepper were successfully introgressed into a single elite bell pepper line through a simple MABC scheme until BC_3 (Thabuis et al. 2004). The introgressed QTLs were able to improve resistance although a decrease of the effect for the moderate-effect QTLs was observed. Multiple introgressions would then also have the advantage to accumulate positive QTL alleles from different parents, for the same trait, looking for the exploitation of additive ("transgressive") effects. This was attempted in barley by Emebiri et al. (2009). They crossed two malting varieties, each carrying a major QTL for malting quality, with an adapted high yielding elite line carrying other resistance traits. The two F_1s were then crossed together and followed by MAS through the MABC procedure. The expected results were obtained and the best lines had malting quality values similar to the best available commercial lines in Australia.

For multiple introgressions, marker-based control of the background loci in addition to the multiple foreground MAS should not theoretically increase the size of the population required, unless we incorporate selection of recombinants at the introgressions to reduce total linkage drag. However, when several targets are controlled simultaneously in one single BC scheme, the number of individuals heterozygous at all targets is low in each BC generation, and this leaves little opportunity to further select among those individuals for the genetic background. In this scenario, introgression can be combined with a "gene pyramiding" scheme (Hospital and Charcosset 1997; Koudandé et al. 2000). For example, if the strategy is to introgress four targets, one could first perform four parallel BC schemes each introgressing one target into the genetic background or two BC schemes each introgressing two targets, and then accumulate all targets into the same background by gene pyramiding. This should capitalize on an increased efficiency of background selection in the separate BC lines (provided at least two BC generations are performed), hence, an improved efficiency with respect to a single BC scheme. However, the total duration of the program is then expected to be longer due to the additional generations for accomplishing the gene pyramiding scheme.

7.3.2 Gene Pyramiding

Gene pyramiding serves as an efficient method to accumulate genes or QTLs positively influencing various agronomic traits into the same individual (Ashikari and Matsuoka 2006), and for this reason it is a useful approach to maximize the use of existing genetic resources. MAS is an effective approach for pyramiding genes or QTLs from different sources and for different traits into elite germplasm (Ashikari and Matsuoka 2006; Foolad 2007), with the aim of enhancing overall trait performance, remedying defects, increasing the durability of disease resistance and broadening the genetic base of the released cultivars (Ye and Smith 2008). From another point of view, before the advent of MAS, gene pyramiding was not practically feasible without extensive capacities of phenotyping nurseries, extensive time and cost, and with riskier results.

In rice, five QTLs were pyramided to couple root traits with grain aroma. Four QTLs for root morphological traits such as length and thickness (i.e., drought tolerance) on different chromosomes, together with a fifth QTL for aroma were cumulated after three BC generations and two further crosses between BC$_3$ lines (Steele et al. 2006). In wheat, Kuchel et al. (2007) pyramided resistance genes to leaf and stripe rust pathogens and grain quality (dough strength) genes into an elite cultivar.

Pyramiding seems to be particularly suited to cumulate resistances (Joshi and Nayak 2010). MAS has been used for pyramiding more than one resistance gene in the same variety or breeding line in tomato, as reviewed by Barone et al. (2009). Pyramiding resistance QTLs for quantitative resistance to a given pathogen is particularly appealing in plants since there is a high probability that polygenic resistance, when not associated with hypersensitive response (HR), is more durable (Lindhout 2002; Joshi and Nayak 2010).

As discussed above for introgression breeding, pyramiding closely linked genes in *repulsion* phase can be problematic. As resistance genes tend to reside in clusters, target alleles of different resistance genes may be located in tandem but present in different accessions. Therefore, it is of paramount importance to precisely fine map the alleles of the different genes with respect to one another. This goal can be most easily achieved using DNA markers. Subsequently, the linked markers can be utilized to select for the rare recombinants that combine the favorable alleles in tandem (Peleman and van der Voort 2003).

When approaching a MAS pyramiding program an important issue to be considered is the possible tight linkage between genes to be pyramided coming from different sources. In the case of pyramiding resistance genes, another variable, important for the target phenotype as well as for the selection pressure on the pathogen's populations, is the "horizontal" rather than the "vertical" nature of the pyramiding scheme. This classification, which has nothing to do with the nature of the resistance, is defined in the former as the pyramiding of genes and/or QTLs of resistance against the same pathogen, pest or parasite into a single genotype, even if each gene can be directed against a different strain; the latter is the union of resistance specificities against different species of pathogens, pests or parasites.

Many examples for rice have been published. Katiyar et al. (2001) reported a successful cumulation of two gall midge resistance loci (*Gm-2* and *Gm-6t*) in rice, although they were linked at about 16 cM. A genotype homozygous at one locus, while segregating for the second one, was identified and selfed to recover double homozygous recombinants. In one of the first pyramiding results in plants, an example of successful "horizontal" pyramiding against the same pathogen, Huang et al. (1997) pyramided four *Xa* genes for bacterial blight resistance in rice into different combinations (2, 3 or 4 genes). The pyramided lines exhibited higher levels of resistance and/or a wider resistance spectrum than the lines with single genes. Moreover, some pyramided lines showed resistance to pathogen races to which all single-gene carrying parents were susceptible. Results were also generally successful for Hittalmani et al. (2000), who pyramided three *Pi* genes for rice blast resistance into different combinations. However, in this case some multiple-genes combinations did not perform any better than single genes, indicating that sufficient knowledge of the spectrum of gene effects is necessary prior to performing the MAS program.

The literature contains frequent reports of the cumulation of different genes and/or QTLs which confer resistance to the same pathogen, often times successful as in wheat (Liu and Anderson 2003), but sometimes unsuccessful, as reported by Tan et al. (2009) in potato. Two pyramided resistance genes to the nematode *Meloidogyne hapla* derived from wild *Solanum* stocks, when pyramided, did not reduce the number of egg masses of the nematode in comparison with a single resistance gene. Joshi and Nayak (2010) reviewed multiple studies from 1994 to 2010 that reported results of "horizontal" pyramiding projects.

Among the reports of pyramiding of multiple resistances to multiple pathogens/parasites, Yang and Francis (2005) pyramided two resistance genes to bacterial speck and bacterial spot in tomato, respectively, the cloned *Pto* gene, and *Rx3*. This presented a challenge because the two genes resided in repulsion linkage on tomato chromosome 5. The authors used MAS to select 33 pyramided homozygous plants with coupling-phase resistances out of more than 4,000 plants of F_2 and F_3 populations. In another successful example of MAS that exceeded expectations, Gebhardt et al. (2006) pyramided resistance genes to potato virus Y (PVY) and to the nematode *Globodera rostochiensis*, alternatively with resistance genes to potato virus X (PVX), or to potato wart (*Synchytrium endobioticum*), into a single potato genotype. In addition to the efficacy of the pyramided genes, the authors found new resistances to other pathotypes of *Synchytrium endobioticum* in the progenies, unexpectedly provided by the parent carrying the resistance to PVY.

Some interesting reports have described *pyramiding of transgenes*. The pyramiding MAS strategy may be more efficient and cost-effective than either multiple constructs or repeated transformation events into the same recipient line. The different strategies observed beyond such pyramiding programs involve the cumulation of different steps of a pathway, of different mechanisms of response to a stress, or the attempt to deploy an additive effect on a trait. In tomato, folate enhancement by pyramiding two transgenes and thus putting together two branches of the folate biosynthesis pathway was reported (Diaz de la Garza et al. 2007). Tomato plants overexpressing a mammalian cyclohydrolase were crossed with tomato plants overexpressing an *Arabidopsis* gene encodingan amino deoxychorismate synthase.

Maruthasalam et al. (2007) reported an interesting case in transgenic rice where two components of rice plant defense, the genes encoding chitinase and a thaumatin-like protein were pyramided with a bacterial resistance gene (R), *Xa21*, all introduced by genetic transformation, thus acquiring enhanced resistance to both a fungal and a bacterial foliar disease. Combining two different components of stress tolerance, Duan et al. (2009) pyramided a tobacco (*Nicotiana tabacum*) line overexpressing the gene *betA*, driving the synthesis of the osmoprotectant glycine betaine, with a tobacco plant overexpressing the Na+/H+ antiport *AtNHX1*. Notwithstanding the successful pyramiding and transgene expression, the gain in salt tolerance with respect to the single transgenic plants was very low, indicating that the union of different components of a cellular system of resistance does not necessarily provide the expected improvement. Samis et al. (2002) tried to join the effects of two *superoxide dismutase* (*SOD*) transgenes driven by the promoter 35S into a single alfalfa (*Medicago sativa*) plant, to improve persistence and biomass production. As in the previous example, the sum of the two transgenes did not improve performance over that of the single transgenes; in this case the effect was even lower than that of the single transgenes, leading the authors to hypothesize a negative effect due to the two 35S promoters.

Other variables that require careful consideration are the numbers of genes and/or QTLs to be pyramided, the presence versus absence of background selection, and most importantly, the optimization of the pyramiding scheme in order to keep the numbers of genotyped plants and the number of generations needed to reach the goal as low as possible.

Regarding the numbers of loci to be pyramided and the optimization of the dimensions and applied methods, a few studies have provided detailed analyses of the subject in plants (Servin et al. 2004; Ye and Smith 2008) and in animals (Koudandé et al. 2000).

According to Hospital and Charcosset (1997), when the target alleles are distributed among multiple parents one can perform marker-assisted gene pyramiding involving several initial crosses between the parents. For example, four genes (G1–G4), that are present in four different lines (L1–L4), can be combined into a single line in a two-step procedure. In the first step, two lines that are homozygous for two target loci each (G1/G2 vs. G3/G4) are developed by crossing pairs of lines (L1×L2 vs. L3×L4), followed by selection of two-locus homozygous plants among F_2s, recombinant inbred lines (RILs), or DHs (Dekkers and Hospital 2002). Then, such individuals are crossed together to produce individuals that carry at least one copy of all four target loci. This part of the pyramiding scheme is crucial for the success of the pyramiding program. It was termed by Servin et al. (2004) as the selection of the "root genotype". The second part of this scheme is aimed at the fixation of the pyramided genes into a homozygous state; this is called the "pedigree", or "fixation" (Servin et al. 2004) phase. At the end of this phase an inbred line is produced by selfing and phenotypic selection to obtain the pyramided target inbred line.

This process can be expanded to more than four genes by expanding the pyramid. For example Ye and Smith (2008) proposed a simple pyramiding procedure for six genes, and Servin et al. (2004), as proof of concept for their study, designed a pyramiding scheme for eight genes. However, such basic schemes are limited to a small number of target loci. If target loci are numerous, and in particular if some loci are linked on the same chromosome, such procedures certainly deserve a phase of project optimization, and the mathematical theory in this domain remains quite unexplored.

Based on theory developed as early as 1961 (Bailey 1961), Ye et al. (2004) put forward a theoretical prediction method and a computational tool useful to estimate the expected genotype frequencies in a three-way cross. Servin et al. (2004) accurately analyzed the theory of marker-assisted gene pyramiding. They provided an algorithm suited to compare different pyramiding schemes, namely, successions of crosses and selections through generations, in order to estimate the required population sizes and total duration (in number of generations). Based on such an algorithm, the most efficient pyramiding scheme can be chosen for every project. The authors also simulated a calculation for the above-mentioned eight gene project, and their algorithm solved it in three fewer generations than the reference method, and with less marker genotyping.

Ye and Smith (2008) proposed some practical guidelines to optimize the pyramiding programs; some of them seem to be quite obvious but were not previously classified. They were, as follows: i) founding parents with fewer target markers (genes) enter the program at earlier stages, ii) a cross involving strong repulsion linkage should be performed as early as possible (to be solved by recombination in a large population), iii) if the genotyping costs are low and large populations are not impractical, the maximum number of crosses should be performed at each generation (to reduce the number of generations required to obtain the root genotype), iv) for opposing conditions to (iii), one cross per generation is required, v) when the required population size would be intractably large, the use of backcrossing is advisable before assembling more genes. Some additional, very practical and interesting considerations were proposed for the fixation phase. For example, it was recommended to advance all satisfactory genotypes to the next generation, since more than one partially heterozygous genotype has the potential to produce the target genotype.

7.3.3 Genetic Ideotype Breeding

If the concept of gene pyramiding is taken to its extreme, one might conceive a combination of all positive allelic variants for all genes and

QTLs of a species on a genomic scale. MAS has revolutionized the concept of the "phenotypic ideotype" aimed at by the breeders of past decades, substituting the concept of "genetic ideotype" targeted by modern breeders. The idea of an allelic composite picture of the genome was born in the 90s together with the tools to create a "graphical genotype" for an individual plant, by assigning a color or a label to the alleles carried at all known loci ordered by chromosomes (van Berloo 1999). From an observed graphical genotype, breeders applying MAS started to compose desired graphic genotypes to be recovered through complex crosses and MAS, or through other MAS strategies (Barr et al. 2000). Peleman and van der Voort (2003) later defined such genetic ideotype objectives and programs with the appealing term of "breeding by design". Today such schemes are possible to perform, from a coarse to a fine scale, by assembling genomic "blocks" by MAS, each block carrying significant contributions to the improvement of the genotype.

The concept of genetic ideotype, if intuitive, attractive and simple per se, is on the other hand demanding in terms of realization, and in terms of genomic as well as of genetic knowledge required. To be realized on a fine scale, such a mosaic needs high-resolution chromosome genotyping (or preferably, haplotyping), to assay the informative ordered points that can be interrogated to create a desired genomic "picture". In addition, physically ordered informative loci are meaningless per se. An extensive series of phenotyping and mapping experiments, both by linkage and association methods, is needed to associate genes, genomic regions, and their positive allelic variants to the important agronomic traits of a species.

According to Peleman and van der Voort (2003), the breeding by design strategy requires a three-step approach. In the first step, the loci involved in all of the agronomically relevant traits must be mapped. This generally requires the summation of results of many years of research in different parts of the world. The use of ILs can help identify favorable alleles carried by single introgressed blocks that perform consistently across a spectrum of genetic backgrounds. In this case, the availability of a good and complete IL library is crucial. In the second step, the assessment of the allelic variation at the mapped loci that have an impact on phenotype must be accomplished. The third and final step is the breeding by design proper. The breeder should design the desired combination of loci with their alleles. Software tools can then be used to design an optimal scheme to obtain such a genetic ideotype, in terms of donor genotypes, crosses, and generations required.

7.4 Advantages, Limitations and Prospects of MAS

MAS has become a routine procedure in most plant breeding programs, in particular for improving resistance and quality traits. However, despite the enormous advantages, its adoption is still incomplete, particularly in small, family-operated breeding companies and public institutions. The typical issues related to MAS limits and prospects are reviewed below, after a summary of its advantages over phenotypic-based selection.

7.4.1 Advantages

MAS has been described as "precision plant breeding" (Collard and Mackill 2008). DNA markers are highly reliable selection tools as they are stable, not influenced by environmental conditions and relatively easy to score by an experienced technician. MAS can thus increase the precision, reliability and efficiency of selection and it can help breeders to overcome many problems associated with traditional phenotypic assays. Increased reliability is provided by the fact that in case of phenotypic assays, the score may be influenced by the environment, the heritability of the trait, the number of genes involved, and the effects that single genes may have on each other. Increased efficiency derives from the fact that DNA markers can be scored at the seedling stage and thus MAS can accelerate breeding for traits that are expressed late during plant development, such as flower, fruit and seed characteristics, grain or fruit quality, male sterility or photoperiod sensitivity. By selecting at the seedling stage, larger populations can be scored compared to phenotypic selection, and considerable amounts of time and space needed for experiments can be saved.

It is also possible to screen for traits that are extremely difficult, highly dependent on the environment, and expensive or time consuming to measure such as: tolerances to drought, cold and salt, to mineral deficiencies and toxicity, root morphology, resistance to quarantined pests or to specific races of disease causal agents or to insects. Moreover, selection can be practiced for several traits simultaneously on the same single plants and/or seeds, a task that is difficult or even impossible by conventional breeding (Peleman and van der Voort 2003; Foolad 2007; Collard and MacKill 2008). MAS provides an opportunity to distinguish the homozygous versus heterozygous state of many loci in a single generation without the need for progeny testing (Collard and Mackill 2008). Molecular markers are also useful to prevent genotype misclassification where many conventional BC programs fail. For example, despite carefully made backcrosses to a recurrent parent, progeny can derive exclusively from self-pollination due to failure of crossing. Breeders often work on misidentified genetic material; e.g., the misclassification of a plant for the presence of a donor gene (disease escape instead of disease resistance) and its use as a parent in further crossing could lead to failure.

These time and resource consuming errors can be avoided by MAS (Varshney et al. 2004). The use of molecular markers enables the breeder to eliminate unsuitable plants before planting and this permits further considerable money savings, avoiding the large number of lines that need to be carried through the evaluation and selection phases of a traditional plant breeding program based on phenotypic selection and progeny testing (Foolad 2007). Because many lines can be discarded by MAS early in a breeding scheme, this permits a more efficient use of limited glasshouse and/or field space because only important breeding material is maintained (Collard and Mackill 2008). Finally, the use of DNA markers for indirect selection offers the most benefits for quantitative traits with low heritabilities as these are the most difficult characters to assess in field experiments (Hospital et al. 1997; Foolad 2007).

7.4.2 Limitations and Solutions

It has been noted that throughout the last ten years MAS has not provided all of the benefits that theoretically resided, and still reside, in its premises (Young 1999; Collard and Mackill 2008). Collard and Mackill (2008) reported a careful list of ten reasons that can explain a lower than expected impact of MAS to varietal development. In our opinion, the cost of genotyping remains as the most important limitation for the implementation of MAS (Dreher et al. 2003), and not only in developing countries. At present, high-throughput marker genotyping has *elevated costs* per assayed sample or genotype, while showing drastically decreased costs of single data points. However, with advancements in technology together with availability of more DNA sequence information, cost will not remain as a permanent problem. A trend towards an increased accessibility of genomics-based technologies is underway, dependent on the fair competition between companies offering tools and platforms. This will result in a greater adoption of markers in plant breeding. So far, large-scale SNP detection based on various platforms has been made available. Detection systems such as SnaPShot, SNPlex, SNPWave, pyrosequencing, and Arrayed Genotyping Systems such as Illumina Golden Gate and Infinium genotyping have been proposed for MAS (see Barone et al. 2009).

While genomic resolution of genotyping has increased, the variable and fixed (equipment) costs have also increased, all of which influence breeder's choices (Kwok and Chen 2003). In this regard, the necessity of costly equipment, as well as the benefit from economies of scale, would recommend contracting marker assays to large commercial laboratories (Collard and Mackill 2008), even though, for example, in practical barley MAS, simple multiplexed marker assays are half as expensive relative to outsourcing (Fox et al. 2005). This last observation suggests that the cost of MAS is a complex issue that depends on marker implementation costs, the accuracy of markers, the relevance of the traits for the target environments, and importantly, on the capacity/costs for phenotypic screenings (Brennan and Martin 2007).

SNP genotyping using high-throughput systems can be applied to breeding populations, but it requires extensive preliminary work for SNP validation and selection, after discovery by (re)sequencing of large numbers of SNPs. Then, a critical issue related to cost is the number ratio, i.e., the number of markers assayed versus the number of genotypes assayed, required by a particular MAS program (Dayteg et al. 2007; Bagge and Lubberstedt 2008). For example, genomic characterization of a germplasm collection can require hundreds to thousands markers, together with hundreds to thousands of assayed genotypes; alternatively, a foreground marker-assisted BC program generally requires a few markers surveyed on hundreds of genotypes. In each case, different methodologies of SNP detection should be adopted to optimize costs. There are genotyping methods that are considered to be more elastic in terms of cost variation to changes in number ratio, and less elastic methods that vary less in cost in response to such changes (Bagge and Lubberstedt 2008).

An array-based large-scale SNP analyses can drastically reduce the cost per single data point to a few cents (Bagge and Lubberstedt 2008), providing a realistic choice for large and mid-scale breeding companies. However, the cost per genotype if many data points are collected, and consequently the cost for a MAS project, even at high-throughput, can be highly significant, ranging from about $10,000 to $140,000 per breeding program according to the number of plants to be genotyped (N Pecchioni personal communication, 2010). Therefore, the MAS cost should be interpreted within the broad context of a breeding program, that considers various possibilities of genotyping versus phenotyping intensities and phases, as was applied to an assisted breeding strategy in wheat (Kuchel et al. 2005). Moreover, streamlining the organization and automation of the entire MAS workflow, from rapid DNA extraction, sample management and storage, to polymorphic variant visualization, can impart significant impacts in decreasing the cost per sample (Brennan and Martin 2007; Dayteg et al. 2007).

Closely related to the issue of cost, another limit to MAS can be the *limited molecular marker polymorphism within cultivated tomato*. In tomato, most studies related to MAS are based on crosses between cultivated tomato and related species, while the level of genetic variation within the cultivated tomato, revealed by most of the common molecular markers, is estimated to be relatively low. To subvert this limitation efforts are being made to develop high-throughput markers with greater resolution, including SNPs and indels (Suliman-Pollatschek et al. 2002; Labate and Baldo 2005). A good example of this problem was the case of selecting for resistance genes to tomato bacterial spot (caused by *Xanthomonas campestris* pv. *Vesicatoria*) (Yang et al. 2005; Foolad 2007). The resistance genes were identified in either cultivated tomato or tomato wild relatives. In this case, MAS could not be easily employed within populations of cultivated tomato using traditional molecular markers such as RFLPs, CAPS, or AFLPs; only the development and discovery of SNPs within the cultivated species and their wide-scale application helped to alleviate the problem. Moreover, in the absence of "perfect", genic-derived markers, when using linked markers another challenge is encountered in the inconsistency of the polymorphic allele associated to the positive variant of a gene/locus through multiple breeding populations. In view of this, Yang et al. (2008) proposed the generation of many "candidate" markers linked to the trait locus from which to select the best marker to be used in MAS after a validation step of their alleles in a broad sample of germplasm.

7.4.2.1 *Unavailability of Closely-linked Markers for Many Traits*

If the biologically annotated causal gene, i.e., the "perfect" marker, is not available for a trait, genetically associated markers are used instead. Insufficient linkage between a marker and a gene or QTL is, therefore, another limitation of MAS (Collard and Mackill 2008). This points out the necessity of using several flanking markers rather than a single one that can significantly reduce the probability of selecting false positives. Crossover events between markers and the genes or QTLs of interest become a serious problem when genes or QTLs to be transferred are found within wild species. The use of markers not sufficiently close to the gene or QTL of interest makes it difficult to break negative associations between the introgressed QTL alleles and loci causing poor horticultural characteristics.

Although numerous mapping studies have identified QTLs for a wide range of traits in diverse crop species, including tomato, relatively few QTL-flanking markers have been routinely implemented in plant breeding programs (Foolad 2007; Labate et al. 2007). In seed companies, which invest in the integration of molecular markers in breeding, while the successful use of MAS for manipulating single-gene traits has been well documented, *there is still little evidence of the use of MAS for manipulating QTLs for complex traits* (Foolad 2007). The main reason for this lack of adoption is that the applied markers used have been unreliable for selection of superior phenotypes. In some cases, this can be attributed to low accuracy of QTL mapping studies, although this may be based on an assumption rather than on hypothesis testing. Price (2006) argued that accuracy of QTL mapping may realistically be in the range of 2 cM. More likely, this might be due to inadequate validation (Sharp et al. 2001). The identification of reliable QTLs is difficult unless replicated experiments are conducted at least across environments and years. In many cases, the presence of QTLxE effects limit the validity of the selection for a QTL, whose action is environment-dependent (Romagosa et al. 1999).

The effect of the genetic background, possibly due to interacting (epistatic) QTLs, contributes to limit the impact of QTL-assisted introgression (Collard and Mackill 2008). Also, the complexity of some traits such as tolerance to salt stress in tomato does not afford success to the introgression of any QTL that might influence a specific component of a highly complex tolerance mechanism (Cuartero et al. 2006). For this reason, as discussed by Barone and coworkers (2009) for tomato diseases that are under the genetic control of QTLs, and also by Cuartero et al. (2006) for salt tolerance, perhaps more complex approaches beyond linkage mapping are required to identify key candidate genes for QTLs, and thus the candidate gene-derived markers suitable for assisting the selection for polygenic traits. The potential to decipher plant transcriptome variation during plant-pathogen interactions through differential approaches, for example, should make it possible to design genic molecular markers (GMMs) able to monitor the selection of new resistant varieties.

A major limitation on the development of MAS in developing countries is the *unfamiliarity of many traditional plant breeders with the use of markers, or their limited access to molecular marker laboratories, instrumentation and reagents* (Tester and Langridge 2010). This issue is less important today than 10 years ago, but probably more often than one would imagine, identification and mapping of genes and QTLs are performed mainly by researchers who are not true breeders or do not have a direct interest in crop improvement. Conversely, breeders may have little knowledge either of molecular tools or of research results concerning QTLs and genes associated with traits. Such limits are respectively termed "application" and "knowledge" (i.e., technology transfer) gaps by Collard and Mackill (2008). They can be resolved by creating teams of experts, e.g., breeders with agronomists, molecular biologists, and other professional figures systematically communicating with each other within breeding companies.

7.4.3 Prospects for MAS

Some of the prospects and future trends for MAS were cited above in discussing MAS limitations. On the positive side, increases in throughput and in numbers of *genic, functional markers (FMs)* are expected (Andersen and Lübberstedt 2003). During the past several years, functionally characterized genes, expressed sequence tag (EST) and genome sequencing projects have facilitated the development of molecular markers from transcribed regions. Among the more important and popular molecular markers that can be developed from ESTs are SNPs, SSRs and COSs (Varshney et al. 2002, 2005). Putative functions can be deduced for markers derived from ESTs or sequenced cloned genes by indirect evidence, by using bioinformatic homology searches (BLASTX) with protein databases (e.g., NCBI NRPEP and Swiss-Prot), although they should then be experimentally validated.

An interesting approach termed "genome zipper" was proposed in barley for chromosome 1H by Mayer et al. (2009) to enhance the efficiency and accuracy of genome functional annotation within the species. To construct a validated and virtual genetically ordered gene list the authors use an integrated bioinformatics approach. The method assigns annotation and order to barley genes by deploying not just one, but multiple sources of existing information on syntenic gene order and functional annotation from reference genomes of rice and sorghum, and against wheat and barley EST datasets.

As soon as they are made available for the sequenced tomato genome (Mueller et al. 2009), FMs will allow genuine gene-assisted breeding without loss of a desirable trait due to recombination events between a marker and a gene under selection. For example, among the R-genes tagged by MAS, 16 have been cloned and functionally annotated and can be used as genic markers (Barone et al. 2009).

When compared to random markers such as AFLPs, SCARs and RAPDs, FMs possess clear advantages because they are completely linked to a desired trait allele, and because they target the functional polymorphism in the gene, thus allowing selection in different genetic backgrounds without revalidating the relationship between a marker and QTL allele. For this reason, they have been referred to as "perfect markers", even though different alleles with the same polymorphism (resulting from intragenic recombination, insertion, deletion or mutation) might produce different phenotypes. A perfect marker allows breeders to track specific alleles within pedigrees and populations, and to minimize linkage drag flanking the gene of interest. In tomato, many markers currently used in MAS were developed based on the sequence(s) of the gene(s) responsible for the observed phenotype. A possible limitation to the use of perfect markers in MAS could be their patenting, if they were identified, validated and protected by a patent through privately funded research. However, the increasing availability of genic molecular markers for tomato, as well as the enhanced information about annotated and ordered genes anchored to the tomato genome sequence, will facilitate the development of large-scale genomics-assisted breeding (Varshney et al. 2009).

7.4.3.1 From "MAS" to "GAS"

Eventually, for most crops MAS will be scaled up to the genomic level. Presently for species with a reference genome or other functional genomics resources, the biggest limitation to this development is the cost of high-throughput genotyping. In theory, the knowledge of the relative values of alleles at all segregating loci in a population could allow the breeder to design a desired genotype in silico, thus taking to a genomic level the paradigm of "breeding by design" (see Section 8.2.3). Consequently, the breeder could practice whole genome selection. The expectations from genomics-assisted selection (GAS) are high for many species including tomato. For example, in rice, it is expected that genomic selection will allow not only genomic ideotype breeding, but also a revisiting of breeding strategies such as wide hybridization (Negrao et al. 2008). In oil palm, GAS is expected to reduce the amount of time required for the release of improved germplasm, to six years versus the typical 19 years (Wong and Bernardo 2008).

In principle, it should be possible to select, based on thousands markers in early generations and on large populations, a single recombinant individual carrying an optimum combination of marker alleles, in a manner similar to what was proposed with few tens of markers by van Berloo and Stam (2001). However, the reality could be much more complex. In fact, MAS breeders may encounter obstacles in the form of linkage blocks that do not allow allelic constitutions in trans. Therefore, interactions between loci would reduce the efficiency of whole genome selection. Furthermore, complex systems of regulation of gene expression such as epigenesis could interfere with the effect of a genomic ideotype on the phenotype (Varshney et al. 2009).

GAS breeding, as well as the initiative of sequencing the tomato genome, have been facilitated by the many genomic resources now available for the species. For example, tomato has many publicly available EST collections (more than 250,000 ESTs) derived from various tissues and different developmental stages (Barone et al. 2009). In addition, these genomic resources have been widely exploited for developing new microarray-based technologies that can be used for high throughput genotyping (HTG). HTG permits custom-designed SNP assays (Bray et al. 2001; Butler and Ragoussis 2008), suitable for non-model organisms with few genomic resources. However, using custom-designed HTG SNP assays for non-model species carries a cost: the uncertainty related to genotyping errors caused by unidentified paralogs. The validity of HTG within "unknown genomic terrain" may be difficult to assess without prior estimates of sequence variation, or without time-consuming and expensive sequencing validation of genotyping results (Sandve et al. 2010). Minimizing the probability of genotyping errors caused by polymorphisms between paralogous loci is crucial. This encompasses careful experimental planning which involves selection of SNP loci to be genotyped thorough bioinformatics analysis, and post-analysis data clean-up.

Next-generation (re)sequencing of different plant genotypes, in species where a reference genome sequence is available, will represent an additional improvement in efficiency of GAS. We might envision an additional step in the methodology, due to the fact that next-generation sequencing will allow sampling the genetic (sequence) diversity within a species, thus fulfilling one of the great expectations of breeding for the near future, a deeper use of the allelic diversity (Varshney et al. 2009; Tester and Langridge 2010).

However, the advent of the GAS paradigm should not be viewed alone; other "omics" platforms are likely to contribute to the assisted selection of novel tomato varieties. In particular, the development of *phenomics* should be taken into consideration. Automated high throughput phenotyping could allow dissection of complex phenotypic traits into discrete phenotypic components such as drought and/or to salt tolerance, and the use of such components to derive models and indices to allow more efficient selection during specific growth stages and environmental conditions (Tester and Langridge 2010).

A second promising "omics" science for tomato breeding is *metabolomics*. Schauer et al. (2008) by using metabolomics in conjunction with tomato ILs, characterized 889 quantitative fruit metabolic loci involving metabolites of nutritional and organoleptic importance of tomato fruit, potentially useful for crop improvement. Fernie et al. (2008) proposed the concept of metabolomics-assisted breeding. After integrating metabolic networks with their genetic determinants, it would be theoretically possible to identify key complete metabolite sets that could be used for selection. Although this approach is in its early stages, it should be evaluated for inherent costs and validated for its efficiency.

Molecular breeding during the 21st century will likely include efforts to take advantage of the increasing knowledge of the control of *meiotic recombination*. Crossover recombination is a crucial process because it allows plant breeders to create novel allele combinations on chromosomes that can be used for creating superior F_1 hybrids. Gaining control over this process, in terms of increasing crossover incidence, altering crossover positions on chromosomes or silencing crossover activity, is essential for plant breeders to effectively engineer the allelic composition of chromosomes (Wijnker and deJong 2008). This could be particularly interesting, for example in tomato as well as in other autogamous plants, where linkage blocks are extensive, and where it would be highly desirable, e.g., to couple certain resistance gene clusters (Hoogstraten and Braun 2009).

Research is progressing along several lines: a revival of classical meiotic research combined with an increasing knowledge on the molecular control of meiosis will create new applications for plant breeding. However, in spite of the number of genes putatively involved in controlling crossover, few studies have been published on the practical applications of such genes. This is in contrast to the various patents that exist for control of crossover, indicating that such methods have attracted the attention of many researchers and are acknowledged to be of economic interest (Wijnker and deJong 2008).

Another method related to the management of recombination is the reverse breeding (RB) strategy, as described by Dirks and coworkers (2009). RB facilitates breeding of F_1-hybrids by suppression of meiotic recombination. In the final breeding steps the genes used for the genetic modification are eliminated through crossing, resulting in end-products that are free of genetically modified DNA sequences. RB generates perfectly complemented homozygous parental lines through engineered meiosis. The method is based on reducing genetic recombination in the selected heterozygote by eliminating meiotic crossing over. Male or female spores obtained from such plants contain combinations of non-recombinant parental chromosomes which can be cultured *in vitro* to generate homozygous DH plants. From these DHs, complementary parents can be selected and used to reconstitute the heterozygote in perpetuity. Since the fixation of unknown heterozygous genotypes is impossible in traditional plant breeding, RB could fundamentally change plant breeding in the future.

7.5 Transgenic Breeding

The first source of variability used in tomato breeding was intraspecific, followed by interspecific by means of introgression breeding. In cases where incompatibility was an obstacle for the cross between wild species and cultivated tomato, these barriers have sometimes been overcome either by the rescue of immature embryos and their *in vitro* cultivation, or by somatic hybridization (Guri et al. 1991). Notwithstanding well-known examples of success, the variability contained in wild relatives is neither sufficient to completely solve disease resistance problems, nor to improve some aspects of organoleptic and nutritional quality of the fruits, thus suggesting to combine transgenic with classical breeding to overcome incompatibility barriers.

Different opinions concerning the application and the future of this technique are found in recent literature on transgenic crops. According to Abdelbacki and coworkers (2010), transgenic breeding is increasingly used throughout the world to improve traits in a number of crops; however, Bai and Lindhout (2007), despite reckoning with the enormous potential for breeding applications, lay no validity in a commercial future for transgenic tomatoes. The authors claimed that transgenic breeding came to a complete halt in 2000, due to, among other reasons, expensive patent filing and consumers' concerns, especially in the European Union. The first transgenic plant variety, 'Flavr-Savr'[TM], was a commercial variety released in 1994 that did not achieve market success and that was later retired from cultivation.

However, efficient genetic transformation and generation of transgenic lines is today possible because of recent advances in recombinant DNA technology and in the development of precise and efficient gene-transfer protocols (Gosal et al. 2009, 2010). In tomato, transformation protocols based both on *Agrobacterium tumefaciens* and biolistics are available, and their efficiency is continuously increasing (Sun et al. 2006).

The transgenic approach, in contrast to classical and MAS breeding, offers several advantages. In addition to the above described ability to overcome incompatibility barriers, it allows the incorporation of only the specific cloned gene into the recipient, thus avoiding the transfer of undesirable genes from donor organisms. Through this approach, pyramiding of genes with either similar or complementary and/or additive effects is possible (Abdelbacki 2010). Transgenesis has been significantly used in breeding tomato for both organoleptic and nutritional quality traits, and tolerance to biotic and abiotic stresses, with positive results obtained in many cases (Diez and Nuez 2008). However, relatively few patent applications have been approved. Transgenic plants have also been developed for research purposes, e.g., for the restoration of fertility, establishing an *Ac/Ds* insertional mutation system, fruit parthenocarpy, and for increasing amino acid content in the seed (Diez and Nuez 2008). The transgenic approach to manipulating polygenic characters is much more complex. The genetic transformation of such traits is not yet a straightforward process, although current techniques should allow the transfer of multiple major genes that may act epistatically or additively to improve plant performance. In some cases, such as salt and drought stress tolerance, plant response is very complex and involves several genes with additive effects on different response components. Transgenic lines developed using a single gene may thus not be as productive as they would be if they had been developed using transformation of many genes (Abdelbacki 2010).

Improvement of performance of a transgenic line under natural field conditions is mandatory for stakeholders. Nevertheless, among the *weak points* of genetic engineering in tomato, it must be noted that, with the exception of a few notable cases (Table 7-2), most transgenic lines of different crops have rarely been extensively field-tested. Thus, how these transgenic lines will perform under field stress conditions is unknown, because under natural field conditions a line or cultivar faces a multitude of environmental factors (Witcombe et al. 2008; Abdelbacki 2010). This final assessment represents the most stringent test for our understanding of biological processes affected by the transgenes.

Fruit productivity and quality are both crucial to the success of any novel tomato line or hybrid. Consequently, transgenic tomatoes must meet these criteria in order to represent a product worthy of widespread

Table 7-2 GMO projects in tomato that included yield and/or quality testing under field conditions.

Citation	Phenotype	Gene(s)/species source[a]	Country	Type/cultivar	Field data collected[b]	Quality testing
Laporte et al. 1997, 2001	Improved productivity and sugar content	*Sucrose-phosphate synthase / Zea mays*	USA	Processing/inbred lines	Y	Y
Giorio et al. 2007	Increased beta-carotene content	Overexpression endogenous *Lycopene beta-cyclase (tLcy-b)—CaMV35S*	Italy	Processing/F1 hybrid	Y	Y
Rotino et al. 2005	Parthenocarpy	*DefH9-RI-iaaM/Pseudomonas syringae / Antirrhinum maius* (auxin-synthesis)	Italy	Processing/inbred cv. UC82	Y	Y
Accotto et al. 2005; Corpillo et al. 2004	Virus resistance (TSWV)	TSWV modified nucleoprotein gene	Italy	Fresh market/F1 hybrid	Y	Y
FDA 1994 (Calgene)	Fruit firmness	Transgene (antisense) for silencing endogenous *polygalacturonase* (PG)	USA	Processing/line CR3	N	Y
Porretta et al. 1998 (Zeneca)	Fruit firmness	Transgene (antisense) for silencing endogenous *PG-CaMV35S*	Italy	Processing/inbred cultivar	N	Y
Kalamaki et al. 2003; Powell et al. 2003	Fruit firmness/texture	Overexpression/silencing endogenous PG and *expansin 1*	USA	Fresh market/cv. Ailsa Craig	N	Y
Metha et al. 2002	Prolonged vine life; Polyamine content; Fruit ripening	*ySAMdc gene (yeast S-adenosylmethionine decarboxylase)/Saccharomyces cerevisiae*	USA	Commercial variety	N	Y

[a]Species source for non-endogenous genes
[b]Y = yes; N = no

cultivation. Moreover, an improved product quality of the product should not lower plant productivity.

Another important limit of the transgenic approach is the functional characterization of the genes of interest; it is imperative to sufficiently know the genes responsible for the trait to be improved before initiating any transgenic breeding project. Although gene function can be assigned either through classical forward genetics (starting with a phenotype and identifying the underlying genes), or reverse genetics (starting with a DNA sequence and identifying its biological role), rigorous functional characterization is not available for many genes. For instance, the majority of transgenes used to improve stress tolerance are transcription factors that either induce or repress the expression of other genes. Therefore, a functional analysis of many genes will be required; much of this work has been carried out in *Arabidopsis* (Witcombe et al. 2008).

Although genetic engineering offers the promise to rapidly improve a desired trait, and transgenic plant development has generally been considered to be a highly expedient route with which to develop superior cultivars, some authors correctly argue that approximately 10 years are required to create a transgenic cultivar due to an extensive safety testing and approval procedure. For example, the project of the Amflora transgenic potato of BASF® released in Europe in 2010, began in 1996 in Sweden (see news at http://www.agro.basf.com), well beyond a 10 year time-span. Such an investment in time is similar to that needed for development and testing of new lines in traditional breeding programs using exotic germplasm (Zamir 2001; Labate et al. 2007). The commercialization of transgenic tomatoes has not been very successful over time. This is due to, among other reasons, the emphasis on qualitative manipulation for altering quantitative traits, and the social issues associated with transgenic crops (Diez and Nuez 2008).

The final weak point is consumer acceptance. Particularly in Europe, for many reasons beyond the scope of this review, the idea of genetically modified tomato, "GM tomato", does not induce consumers to visualize something positive, or particularly fresh, good tasting or of high quality. This is particularly true for fresh market tomato, while it may be less important in several countries for processing tomato. If we therefore consider, i) the high development and registration costs, in a seed market whose total value has been estimated to be about $680 million (Bai and Lindhout 2007), ii) the extensive time under present rules to complete the transgenic registration procedure, and iii) unlikely consumer acceptance, a project in the primarily human consumption oriented tomato currently represents an extremely high risk.

7.5.1 Traits and Objectives

Tomato has been genetically modified for different purposes. The main objectives have included the improvement of the following features:

- Fruit quality: less cell wall degradation and fruit softening, inhibition of fruit ripening, fruit sweetness, flavor aldehydes and alcohols content, modification of carotenoid pathways.
- Parthenocarpy (seedless fruits).
- Constitutive overproduction of antifungal compounds (class I chitinase and class I β-1, 3-glucanase) so as to increase resistance to *Fusarium oxysporum* f.sp. *lycopersici*.
- Bacterial disease resistance (resistance to *Xanthomonas campestris* pv. *vesicatoria*).
- Viral disease resistance: TYLCV, Tomato leaf curl virus (TLCV), Tomato yellow leaf curl sardinia virus (TYLCSV), TSWV, Cucumber mosaic virus (CMV).
- Post-harvest and processing technology.

Based on a search of the European Commission Joint Research Centre (JRC) database (http://gmoinfo.jrc.ec.europa.eu/gmp_browse.aspx), through 2009, more than 50 transgenic tomato events were patented in Europe. In summary, i) more than 40% of them involved pathogen resistance genes or traits related to tolerance or resistance to biotic stresses, ii) 19% were for improvement of processing quality and ripening processes related to harvest time, iii) 15% involved the increase of fruit nutritional value, iv) the remaining targeted miscellaneous traits, e.g., herbicide resistance, parthenocarpy and restoration of fertility. In the following are some examples of transgenic breeding as applied to tomato.

7.5.1.1 Sugars

In plants, sucrose is the main transported form of carbohydrate. Moreover, sucrose regulates transcription and/or translation of target genes (Ciereszko et al. 2001; Leszek et al. 2004; Weise et al. 2005). Sucrose metabolism and transport might be altered by several genetic mechanisms, and sucrose-phosphate synthase activity influences whole-plant carbon allocation with consequences to both tomato productivity and fruit sugar content (Laporte et al. 2001). This conclusion is based on field trials performed with tomato lines transgenic for the maize sucrose-phosphate synthase gene (Laporte et al. 1997), driven either by the CaMV35S or by the *rbcS* promoter. Interestingly, only plant lines with a moderate (i.e., two-fold) increase in sucrose-phosphate synthase activity showed an increased fruit production (20% on average), and sometimes a slight increase of soluble solids, in genetically modified fruits versus wild type controls. Indeed, only plants with a moderate increase displayed all of the desired benefits, while a higher expression level of sucrose-phosphate synthase was not beneficial for plant growth and production (Laporte et al. 2001).

7.5.1.2 Carotenoids

Fruit color of wild tomatoes varies from green to yellow, orange and red. Various studies have demonstrated that green fruit is the ancestral character and that colored fruits arose once within the clade about one-million years ago (Peralta and Spooner 2001; Nesbitt and Tanksley 2002). Carotenoids present in tomato fruits play an important role in human nutrition: beta-carotene is well recognized as the pro-vitamin A carotenoid, and there is a positive association between lycopene and health. Carotene biosynthesis in higher plants has been reviewed extensively (Bartley and Scolnik 1995; Hirschberg 2001). The availability of cDNAs that encode for nearly all of the enzymes required for carotenoid biosynthesis in plants has stimulated considerable interest in engineering plants with altered carotenoid content (Cunningham and Gantt 1998). Numerous efforts, both successful and unsuccessful, to alter fruit carotenoid composition by manipulating expression of carotenogenesis have been reported (for a complete list see Fraser and Bramley 2009).

7.5.1.3 Virus Resistance

Whitefly vectored Tomato chlorotic mottle virus (TCMV), Tomato mottle virus (ToMoV), and TYLCV are, among at least 35 viruses, members of the *Begomovirus* genus (family Geminiviridae) (Fauquet et al. 2003). To date, five wild relatives of tomato (*S. habrochaites*, *S. chilense*, *S. pimpinellifolium*, *S. cheesmaniae*, *S. peruvianum*) were reported to bear resistance to various Begomoviruses (reviewed by Ji et al. 2007b). In turn, molecular markers linked to each of the single genes or QTLs that provide resistance to Begomoviruses have been identified (Ji et al. 2007a). TYLCV that first appeared in European tomato-growing areas in 1988, causes symptoms that include severe stunting, marked reduction in leaf size, flower abscission, and significant yield reduction. In certain regions it has caused 100% yield losses and forced netting use in greenhouses in many countries. Portuguese production has declined 48% since 1995, and in 2000, losses of 15% to 60% were reported in Greece. In the United States the virus appeared in 1989 and it reduced Florida tomato production by 20%. As in Europe, widespread insecticide use has been employed to minimize losses, while researchers from the University of Florida have transformed tomatoes to obtain resistance to TYLCV (Gianessi et al. 2003). Italian researchers have successfully transformed tomato plants by inserting a gene of the TYLC geminivirus by *Agrobacterium*. The resulting tomato plants were resistant to TYLCV when challenged either directly by the virus or by the whitefly vector. It is assumed that such types of transgenic tomato varieties should be planted on 53% of the tomato hectares of four Mediterranean countries (Spain, Italy, Greece and Portugal) to efficiently combat the disease. This would help to replace the current combined use of insecticides and insect netting that, even if widely applied in European tomato production, do not permit a complete eradication or exclusion of whiteflies, leaving a current unavoidable loss of 1% of production (Brunetti et al. 1997).

Resistance to TSWV by genetically engineering plants with the TSWV nucleoprotein (*N*) gene, first reported in 1991 in tobacco (Gielen et al. 1991), has been demonstrated in several species other than tomato. Using *Agrobacterium*-mediated transformation, Accotto and colleagues (2005) have produced tomato plants with resistance to TSWV, with one line demonstrating a complete resistance. Two field trials were conducted during spring to summer 1999 in different areas of Italy. No insecticides were applied. In the trial for resistance, TSWV infection was monitored five times throughout the duration of the experiment. At the first survey (three weeks from the beginning), 19 plants among non-transgenic hybrids showed the typical symptoms of TSWV infection, including wilting of the apex, bronzing of leaves, and reduced size. Dot blot assays confirmed TSWV infection of these plants. Infection then increased with time, reaching, at the end of the experiment, 33% and 50% of the plants. Among the transgenic hybrids and the resistant commercial hybrid, no plants were ever found to be infected in dot blot assays, and none showed symptoms of infection.

Since 1987, CMV has caused important yield losses in tomato crops, particularly in processing tomato varieties in the main production areas of Italy (Grassi and Gallitelli 1990). Transgenic plants of *S. lycopersicum* cv. UC82B were produced to express coat protein (CP) genes of CMV. CP gene cloning, construction of the vectors, tomato regeneration and transformation techniques, as well as selection of lines in a growth chamber were described by Kaniewski and Lawson (1998). Natural CMV infection of transgenic and non-transgenic UC82B plants was monitored by enzyme-linked immunosorbent assay (ELISA), and by infectivity tests on *Chenopodium* spp. In the summer of 1993, only three transformed lines were available. ELISA analysis conducted at 30 days post-inoculation revealed that UC82B was highly infected. In contrast the transgenic lines were infected to a significantly lower degree, and yield was considerably higher than in non-transformed controls.

The Campania region accounts for 6% of Italy's total tomato production. The predominant tomato grown for processing in Campania is the famous "San Marzano" variety. Tomato production in Campania has declined significantly during the last decade due to epidemics of CMV, and there is concern that production will disappear from the region. In 1992, the Italian Plant Pathology Research Institute began a project to produce transgenic tomatoes with resistance to CMV. Valanzuolo and colleagues (1993) transformed San Marzano tomato lines with a chimeric gene whose transcript is the benign satellite RNA, *Ra*. Massive production of satellite and attenuation of symptoms have both been observed in transformed plants infected with a satellite-free CMV strain. Transformed plants were tested for agronomical traits in a contained (aphid-proof) environment: the results showed that transgenic tomatoes had higher yields than control plants in the presence of infection, and qualitative determinants were unchanged following the *Agrobacterium*-mediated genetic transformation. Seven years of field tests then showed that the transgenic tomatoes were unaffected by CMV and retained all of the agronomic characteristics of the San Marzano tomato. Planting transgenic tomato varieties resistant to CMV, if it was permitted by the government, could prevent the total loss of the production of San Marzano tomatoes in Campania.

7.5.1.4 Parthenocarpy

Parthenocarpy, or fruit set without pollination and fertilization, represents a valuable trait in tomato. Parthenocarpic fruits are seedless, and thus advantageous for both the processing industry (e.g., tomato paste, tomato juice) and fresh consumption. To be a valid product a genetically modified parthenocarpic plant should have not only seedless fruits and improved plant productivity under cultivation conditions adverse for fruit set, but also fruit quality at least equal to that of its genetic background. Parthenocarpic plants transgenic for an ovule-placenta-specific auxin synthesizing gene have been tested under field trial conditions by Rotino and colleagues (2005). Seed content was significantly lower than in the wild-type background (Table 7-3), while beta-carotene increased in both parthenocarpic genetically modified lines tested.

7.5.1.5 Fruit Firmness

Fruit firmness is valuable for both post-harvest and processing technology. Several strategies have been developed to improve fruit consistency and juice viscosity. During ripening, polygalacturonase (PG) hydrolyses pectins and causes fruit softening. Consequently, PG was one of the first targets chosen for modifying fruit firmness by down-regulation of its expression. Down-regulation of PG inhibits cell wall degradation and consequently improves fruit firmness in tomato. This has been achieved by several transgenes acting most likely through a common molecular mechanism:

Table 7-3 Seed set and fruit quality traits in two tomato parthenocarpic transgenic lines (Ri4 and Ri5) in comparison with their background (UC82) and a commercial hybrid (Allflesh). L*a*b* color space: L*=brightness, a*=red to green, b*=yellow to blue; Titratable acids in meq/100 ml NaOH 0.1N. DM = dry matter. Lowercase letters indicate statistically significant differences among mean values.

Genotype	Fruit with seed (%)	Seed number	L*	a*	b*	°Brix	pH	DM (%)	Titratableacids	Firmness (Kg)
Allflesh	87a	68.8a	39.2b	36.5a	25.4c	5.3a	4.29a	6.19a	6.12a	0.43a
UC82	85a	36.5b	41.1b	35.8a	27.9a	3.8a	4.08bc	5.06bc	7.23a	0.40a
Ri4	27b	18.4c	39.9ab	34.1b	25.9bc	4.5b	3.96c	4.85c	6.12a	0.43a
Ri5	20b	11.4c	40.0ab	34.0b	27.1ab	4.2bc	4.11b	5.24b	6.70a	0.41a

silencing of the *PG* gene. Thus, although not formally proven by showing the presence of siRNA homologous to the target gene, the case studies described here most likely represent genetic strategies based on eliciting in planta RNA interference (Smith et al. 2002). The expression of an antisense *PG* gene construct silenced endogenous *PG* gene expression and consequently the production of the PG enzyme. Silencing of *PG* gene expression has also been achieved by transcription of a truncated PG mRNA introduced into tomato (FDA 1994). Transgene transcription was under the control of the CaMV35S promoter, and the *PG* gene fragment was transcribed either in the sense or in the antisense orientation. Both types of transgenic constructs silenced the expression of the endogenous *PG* gene, also most likely by eliciting an RNA interference process. The genetically modified tomato fruits had decreased PG activity and a prolonged shelf-life. The only difference between the genetically modified and the recipient variety of tomato is the molecular weight distribution of pectin molecules. No differences in tomatine, lycopene, soluble carbohydrates, fibers, minerals, malic and citric acid were reported.

7.5.1.6 Prolonged Vine Life

During tomato fruit ripening, polyamine content decreases (Galston and Sawhney 1990), while both ethylene biosynthesis and activity increase. Ethylene is a phytohormone that controls fruit ripening. Polyamines are considered to have an anti-ripening function, since it has been shown that polyamines inhibit ethylene biosynthesis and that polyamines and ethylene interact during tomato fruit ripening (Cassol and Mattoo 2002). The polyamines and ethylene biosynthetic pathways have a common substrate: S-adenosylmethionine (SAM). The enzyme SAM decarboxylase converts SAM into decarboxylated SAM; it thus controls a rate-limiting step in polyamine biosynthesis. These considerations prompted Metha and collaborators (2002) to genetically modify tomato with a transgene driving expression of the SAM decarboxylase from yeast under the control of the fruit-specific E8 promoter. The transgene was expressed only in the tomato fruit during ripening and it caused an increase in spermine and spermidine content of the fruit. Fruit ripening was significantly delayed, thus prolonging fruit vine life. Genetically modified fruits were of normal size; lycopene content was more than doubled in genetically modified fruits. This increase was considered as beneficial because of its antioxidant properties.

7.5.2 *Perspectives*

Is there a *future* for tomato transgenic breeding? The future might be called *cisgenic or intragenic* breeding (Rommens 2004; Rommens et al. 2007). The new paradigm by which genetic modifications could be realized using native plant DNA should be more acceptable by consumers, and more importantly, could help to simplify and reduce costs of the registration of cisgenics for their release to cultivation and commercialization.

This new plant breeding technique has been well described (e.g., Rommens 2007, 2010), and uses genetic modification as a direct tool to introduce new characteristics to a plant. The genetic material used for the modification originates from the same or a sexually compatible species. Two different approaches, cisgenesis and intragenesis, are possible.

Cisgenesis is the production of genetically modified plants using donor DNA from the species itself or from a cross-compatible species (Schaart and Visser 2009). The newly introduced DNA is an unchanged natural genome fragment containing a gene of interest together with its own introns and regulatory sequences, such as promoter and terminator sequences. The introduced DNA is in principle free of vector DNA, the exception being T-DNA border sequences that flank the cisgenic DNA sequences. These short sequences are naturally non-coding and are unlikely to have a phenotypic effect. *Intragenesis* is the production of genetically modified plants using donor DNA from the species itself or from a cross-compatible species. The difference from cisgenesis is that intragenesis allows the creation of new combinations of DNA fragments. In intragenesis the transformation vector may be composed of functional DNA fragments from the genome of the target crop species.

Both cisgenesis and intragenesis lead to end products that contain genetically modified DNA sequences. However, only the short T-DNA border sequences are non-native to the species itself (or a cross-compatible species). Although integration of the cisgene will most probably occur in a different position in the genome, this normally does not imply that there are inherent differences with regards to level and timing of expression. Thus, cisgenesis will almost always lead to phenotypes that could be achieved by conventional breeding.

Because in intragenesis new combinations of regulatory and coding sequences can be engineered, the expression of the intragene is expected to not always correspond to the expression of the native corresponding gene. Depending on the nature of the intragene this may have various consequences for the environment and food and feed safety, relative to the background genotype. If intragenesis is specifically aimed at silencing of a single endogenous gene, the intragenic plants may be comparable to plants with knock-out mutations obtained by mutation breeding. Such plants can be used as a baseline control.

Plants originating from either intragenesis or cisgenesis contain DNA sequences that were introduced by genetic modification. If the integration was demonstrated to be outside genes of the recipient genome and the introduced genes show an expression that corresponds to the baseline control, such plants are regarded as being essentially similar to the baseline, including in terms of environmental, food and feed safety. If it is also proven that the final product is free of Agrobacteria and free of sequences that are foreign to the species, this justifies the exemption of such cisgenic and intragenic plants from the European regulations for GMOs. In general, however, intragenesis is aimed at differential expression of genes. If for intragenesis an alternative promoter was used to alter the expression of an intragene, the intragene expression may deviate from that of the baseline. In this case additional studies are required to assess environmental and food and feed safety. Examples of the use of this technique in different crop species have recently been reviewed by Rommens (2010).

7.6 Concluding Remarks

This chapter has highlighted the achievements and the prospects of molecular breeding in tomato, including a survey of transgenic improvement. Currently, the tomato post-genomics phase holds a promise of providing comprehensive knowledge of biological processes. Thanks to the availability of a nearly completed genome sequence (Mueller et al. 2009), our understanding of the tomato genome has improved, while considerable progress is being made towards unraveling gene function and genome functionality. Moreover, next-generation genomic approaches are going to expand the knowledge of the gene pools available for crop improvement and increase the precision and efficiency with which superior individuals can be identified and selected. Nevertheless, efforts are required for full exploitation of existing bioinformatics platforms and for developing new strategies, which enable large scale data sets to be meaningfully managed. In the post-genomics era, high-throughput genotyping (HTG) approaches combined with automation, increasing amounts of sequence data in the public domain and enhanced bioinformatics techniques will contribute to genomics research for tomato improvement. Marker-assisted breeding or MAS will gradually evolve into GAS for crop improvement.

It is likely that during the next 5 to 10 years the non-GM assisted selection approaches will contribute the most to tomato improvement. However, much will depend on the availability of cisgenic constructs together with marker-free technologies, and on how these will impact the consumers' acceptance, as well as high costs and the degree of regulatory demands. Consumer acceptance together with regulatory constraints imposes the greatest limitations to the widespread diffusion of GM tomato plants.

Still, many scientific aspects remain to be fully realized, especially in relation to their importance to plant breeding. The last decade has seen the rise of the emerging field of high-throughput phenotyping (HTP), which aims to capitalize on novel high-throughput computation and informatics technology developments to derive genome-wide molecular networks of genotype–phenotype associations, or "phenomic associations." (Lussier and Liu 2007). The mechanisms of epigenetic phenomena are only starting to be understood and their potential roles in crop improvement are unknown. Similarly, tantalizing bits of information concerning the possible basis of heterosis are gradually emerging. Elucidation of the mechanism of heterosis, and regulation of genetic recombination might be two of the most important contributions of molecular genetics research to crop improvement. Finally, in tomato as in other crops, newly developed genomics tools will enhance, but not replace, the conventional breeding and evaluation process. The ultimate test of the value of a genotype is its performance in the target environment and acceptance by farmers and consumers.

Acknowledgement

The authors are thankful to Joanne Labate for providing excellent advice for this chapter, helpful discussion and critical reading of the manuscript.

References

Abdelbacki AM, Taha SH, Sitohy MZ, Abou Dawood AI, Abd-El Hamid MM, Rezk AA (2010) Inhibition of Tomato Yellow Leaf Curl Virus (TYLCV) using whey proteins. Virol J 7: 26.

Acciarri N, Rotino GL, Tamietti G, Valentino D, Voltattorni S, Sabatini E (2007) Molecular markers for *Ve1* and *Ve2* Verticillium resistance genes from Italian tomato germplasm. Plant Breed 126: 617–621.

Accotto GP, Nervo G, Acciarri N, Tavella L, Vecchiati M, Schiavi M, Mason G, Vaira AM (2005) Field evaluation of tomato hybrids engineered with Tomato spotted wilt virus sequences for virus resistance, agronomic performance, and pollen-mediated transgene flow. Phytopathology 95: 800–807.

Ammiraju JSS, Veremis JC, Huang X, Roberts PA, Kaloshian I (2003) The heat-stable root-knot nematode resistance gene *Mi-9* from *Lycopersicon peruvianum* is localized on the short arm of chromosome 6. Theor Appl Genet 106: 478–484.

Andersen JR, Lübberstedt T (2003) Functional markers in plants. Trends Plant Sci 8(11): 554–560.

Antofie A, Lateur M, Oger R, Patocchi A, Durel CE, Van de Weg WE (2007) A new versatile database created for geneticists and breeders to link molecular and phenotypic data in perennial crops: the AppleBreed DataBase. Bioinformatics 23: 882–891.

Arens P, Mansilla C, Deinum D, Cavellini L, Moretti A, Rolland S, van der Schoot H, Calvache D, Ponz F, Collonnier C, Mathis R, Smilde D, Caranta C, Vosman B (2010) Development and evaluation of robust molecular markers linked to disease resistance in tomato for distinctness, uniformity and stability testing. Theor Appl Genet 120: 655–664.

Arùs P, Shields CR, Orton TJ (1985) Application of isozyme electrophoresis for purity testing and cultivar identification of F₁ hybrids of *Brassica oleracea*. Euphytica 34: 651–657.

Ashikari M, Matsuoka M (2006) Identification, isolation and pyramiding of quantitative trait loci for rice breeding. Trends Plant Sci 11: 344–350.

Astua-Monge G, Minsavage GV, Stall RE, Vallejos CE, Davis MJ, Jones JB (2000) *Xv4-vrxv4*: A new gene-for-gene interaction identified between *Xanthomonas campestris* pv. *vesicatoria* race T3 and the wild tomato relative *Lycopersicon pennellii*. Mol Plant-Microbe Interact 13: 1346–1355.

Bagge M, Lübberstedt T (2008) Functional markers in wheat—technical and economic aspects. Mol Breed 22: 319–328.

Bai Y, Huang CC, van der Hulst R, Meijer-Dekens F, Bonnema G, Lindhout P (2003) QTLs for tomato powdery mildew resistance (*Oidium lycopersici*) in *Lycopersicon parviflorum* G1.1601 co-localize with two qualitative powdery mildew resistance genes Mol Plant-Microbe Interact 16: 169–176.

Bai Y (2004) The genetics and mechanisms of resistance to tomato powdery mildew (*Oidium lycopersici*) in *Lycopersicon* species. PhD thesis, Wageningen University, Wageningen, Netherlands.

Bai Y, Lindhout P (2007) Domestication and breeding of tomatoes: what have we gained and what can we gain in the future? Ann Bot 100: 1085–1094.

Bailey NTJ (1961) Introduction to the Mathematical Theory of Genetic Linkage. Clarendon Press, Oxford, UK.

Balint-Kurti PJ, Dixon JS, Jones DA, Norcott KA, Jones JDG (1994) RFLP linkage analysis of the *Cf-4* and *CF-9* genes for resistance to *Cladosporium fulvum* in tomato. Theor Appl Genet 88: 691–700.

Ballvora A, Pierre M, van den Ackerveken G, Schornack S, Rossier O, Ganal M, Lahaye T, Bonas U (2001) Genetic mapping and functional analysis of the tomato *Bs4* locus governing recognition of the *Xanthomonas campestris* pv. *vesicatoria* AvrBs4 protein. Mol Plant-Microbe Interact 14: 629–638.

Barbieri M, Acciarri N, Sabatini E, Sardo L, Accotto GP, Pecchioni N (2010) Introgression of resistance to two mediterranean virus species causing tomato yellow leaf curl into a valuable traditional tomato variety. J Plant Pathol 92: 485–493.

Barillas AC, Mejia L, Sanchez-Perez A, Maxwell DP (2008) CAPS and SCAR markers for detection of *I-3* gene introgression for resistance to *Fusarium oxysporum* f. sp. *lycopersici* race 3. Tomato Genet Coop Rep 58: 11–17.

Barone A, Ercolano MR, Langella R, Monti L, Frusciante L (2005) Molecular marker-assisted selection for pyramiding resistance genes in tomato. Adv Hort Sci 19: 147–152.

Barone A, Di Matteo A, Carputo D, Frusciante L (2009) High-throughput genomics enhances tomato breeding efficiency. Curr Genom 10: 1–9.

Barr AR, Jefferies SP, Warner P, Moody DB, Chalmers KJ, Langridge P (2000) Marker assisted selection in theory and practice. In: Logue S (ed) Barley Genetics VIII: Proc 8th International Barley Genetics Symposium. GRDC, Adelaide, Australia, pp 167–168.

Bartley GE, Scolnik PA (1995) Plant carotenoids: pigments for photoprotection, visual attraction, and human health. Plant Cell 7: 1027–1038.

Behare J, Laterrot H, Sarfatti M, Zamir D (1991) Restriction fragment length polymorphism mapping of the *Stemphylium* resistance gene in tomato. Mol Plant-Microbe Interact 4: 489–492.

Benor S, Zhang M, Wang Z, Zhang H (2008) Assessment of genetic variation in tomato (*Solanum lycopersicum* L.) inbred lines using SSR molecular markers. J Genet Genom 35: 373–379.

Bernacchi D, Beck-Bunn T, Emmatty D, Eshed Y, Inai S, Lopez J, Petiard V, Sayama H, Uhlig J, Zamir D, Tanksley SD (1998) Advanced backcross QTL analysis of tomato. II. Evaluation of near isogenic lines carrying single-donor introgressions for desirable wild QTL-alleles derived from *Lycopersicon hirsutum* and *L. pimpinellifolium*. Theor Appl Genet 97: 170–180.

Bernatzky R, Tanksley SD (1986) Toward a saturated linkage map in tomato based on isozymes and random cDNA sequences. Genetics 112: 887–898.

Boiteux LS, Giordano LB (1993) Heranga da resistencia a duas especies de Tospovirusem tomate. Hort Bras 11: 63.

Bournival BL, Vallejos CE, Scott JW (1990) Genetic analysis of resistances to races 1 and 2 of *Fusarium oxysporum* f. sp. *lycopersici* from the wild tomato *Lycopersicon pennellii*. Theor Appl Genet 79: 641–645.

Bray MS, Boerwinkle E, Doris PA (2001) High-throughput multiplex SNP genotyping with MALDI-TOF mass spectrometry: practice, problems and promise. Hum Mutat 17: 296–304.

Bredemeijer GMM, Cooke RJ, Ganal MW, Peeters R, Isaac P, Noordijk Y, Rendell S, Jackson J, Röder MS, Wendehake K, Dijcks M, Amelaine M, Wickaert V, Bertrand L, Vosman B (2002) Construction and testing of a microsatellite database containing more than 500 tomato varieties. Theor Appl Genet 105: 1019–1026.

Brennan JP, Martin PJ (2007) Returns to investment in new breeding technologies. Euphytica 157: 337–349.

Brommonschenkel SH, Frary A, Frary A, Tanksley SD (2000) The broad-spectrum tospovirus resistance gene *Sw-5* of tomato is a homolog of the root-knot nematode resistance gene *Mi*. Mol Plant-Microbe Interact 13: 1130–1138.

Brouwer DJ, Jones ES, St Clair DA (2004) QTL analysis of quantitative resistance to *Phytophthora infestans* (late blight) in tomato and comparisons with potato. Genome 47: 475–492.

Brouwer DJ, St Clair DA (2004) Fine mapping of three quantitative trait loci for late blight resistance in tomato using near isogenic lines (NILs) and sub-NILs. Theor Appl Genet 108: 628–638.

Brunetti A, Tavazza M, Noris E, Tavazza R, Caciagli P, Ancora G, Crespi S, Accotto GP (1997) High expression of truncated viral Rep protein confers resistance to Tomato Yellow Leaf Curl Virus in transgenic tomato plants. Mol Plant-Microbe Interact 10: 571–579.

Buckler ES, Thornsberry JM (2002) Plant moleculer diversity and applications to genomics. Curr Opin Plant Biol 5: 107–111.

Butler H, Ragoussis J (2008) BeadArray-based genotyping. Meth Mol Biol 439: 53–74.

Button P (2006) New developments in the International Union for the Protection of New Varieties of Plants. Acta Hort 714: 195–210.

Calero-Nieto F, Di Pietro A, Roncero MI, Hera C (2007) Role of the transcriptional activator XlnR of *Fusarium oxysporum* in regulation of xylanase genes and virulence. Mol Plant-Microbe Interact 20: 977–985.

Carmeille A, Prior P, Kodja H, Chiroleu F, Luisetti J, Besse P (2006) Evaluation of resistance to race 3, biovar 2 of *Ralstonia solanacearum* in tomato germplasm. J Phytopathol 154: 398–402.

Cassol T, Mattoo AK (2002) Do polyamines and ethylene interact to regulate plant growth, development and senescence? In: Nath P, Mattoo A, Ranade S, Weil J (eds) Molecular and Cellular Biology: New Trends. SMPS Publishers, Dehradun, India.

Chaerani R, Groenwold R, Stam P, Voorrips RE (2007) Assessment of early blight (*Alternaria solani*) resistance in tomato using a droplet inoculation method. J Gen Plant Pathol 73: 96–103.

Chaib J, Lecomte L, Buret M, Causse M (2006) Stability over genetic backgrounds, generations and years of quantitative trait locus (QTLs) for organoleptic quality in tomato. Theor Appl Genet 112: 934–944.

Chunwongse J, Doganlar S, Crossman C, Jiang J, Tanksley SD (1997) High-resolution genetic map of the *Lv* resistance locus in tomato. Theor Appl Genet 95: 220–223.

Chunwongse J, Chunwongse C, Black L, Hanson P (2002) Molecular mapping of the *Ph-3* gene for late blight resistance in tomato. J Hort Sci Biotechnol 77: 281–286.

Ciereszko I, Johansson H, Kleczkowski LH (2001) Sucrose and light regulation of a cold-inducible UDP-glucose pyrophosphorylase gene via a hexokinase-independent and abscisic acid-insensitive pathway in Arabidopsis. Biochem J 354: 67–72.

Coaker GL, Francis DM (2004) Mapping, genetic effects, and epistatic interaction of two bacterial canker resistance QTLs from *Lycopersicon hirsutum*. Theor Appl Genet 108: 1047–1055.

Collard BCY, Jahufer MZZ, Brouwer JB, Pang ECK (2005). An introduction to markers, quantitative trait loci (QTL) mapping and marker-assisted selection for crop improvement: The basic concepts. Euphytica 142: 169–196.

Collard BCY, Mackill DJ (2008) Marker-assisted selection: an approach for precision plant breeding in the twenty-first century. Phil Trans Roy Soc Lond B Biol Sci 363: 557–572.

Comai L, Young K, Till BJ, Reynolds SH, Greene EA, Codomo CA, Enns LC, Johnson JE, Burtner C, Odden AR, Henikoff S (2004) Efficient discovery of DNA polymorphisms in natural populations by Ecotilling. Plant J 37: 778–786.

Cooke DEL, Young V, Birch PRJ, Toth R, Gourlay F, Day JP, Carnegie SF, Duncan JM (2003) Phenotypic and genotypic diversity of *Phytophthora infestans* populations in Scotland (1995–97). Plant Pathol 52: 181–192.

Coque M, Gallais A (2006) Genomic regions involved in response to grain yield selection at high and low nitrogen fertilization in maize. Theor Appl Genet 112: 1205–1220.

Corpillo D, Gardini G, Vaira AM, Basso M, Aime S, Accotto GP, M Fasano (2004) Proteomics as a tool to improve investigation of substantial equivalence in genetically modified organism: the case of a virus-resistant tomato. Proteomics 4: 193–200.

Cuartero J, Bolari MC, Asins MJ, Moreno V (2006) Increasing salt tolerance in the tomato. Plants and Salinity Special Issue J Exp Bot 57: 1045–1058.

Cunningham FX, Gantt E (1998) Genes and enzyme in carotenoid biosynthesis in plants. Annu Rev Plant Physiol Plant Mol Biol 49: 557–583.

Danesh D, Aarons S, McGill GE, Young ND (1994) Genetic dissection of oligogenic resistance to bacterial wilt in tomato. Mol Plant-Microbe Interact 7: 464–471.

Davis J, Yu D, Evans W, Gokirmak T, Chetelat RT, Stotz HU (2009) Mapping of loci from *Solanum lycopersicoides* conferring resistance or susceptibility to *Botrytis cinerea* in tomato. Theor Appl Genet 119: 305–314.

Davuluri GR, van Tuinen A, Fraser PD, Manfredonia A, Newman R, Burgess D, Brummell DA, King SR, Palys J, Uhlig J, Bramley PM, Pennings HM, Bowler C (2005) Fruit-specific RNAi-mediated suppression of *DET1* enhances carotenoid and flavonoid content in tomatoes. Nat Biotechnol 23: 890–895.

Dayteg C, Tuvesson S, Merker A, Jahoor A, Kolodinska-Brantestam A (2007) Automation of DNA marker analysis for molecular breeding in crops: practical experience of a plant breeding company. Plant Breed 126: 410–415.

De Giovanni C, Dell'Orco P, Bruno A, Ciccarese F, Lotti C, Ricciardi L (2004) Identification of PCR-based markers (RAPD, AFLP) linked to a novel powdery mildew resistance gene (*ol-2*) in tomato. Plant Sci 166: 41–48.

Dekkers JCM, Hospital F (2002) Utilization of molecular genetics in genetic improvement of plants and animals. Nat Rev Genet 3: 22–32.

Dianese EC, Fonseca MEN, Goldbach R, Kormelink R, Inoue-Nagata AK, Resende RO, Boiteux LS (2010) Development of a locus-specific, co-dominant SCAR marker for assisted-selection of the *Sw-5* (Tospovirus resistance) gene cluster in a wide range of tomato accessions. Mol Breed 25: 133–142.

Díaz de la Garza RI, Gregory JF, Hanson AD (2007) Folate biofortification of tomato fruit. Proc Natl Acad Sci USA 104: 4218–4222.

Diez MJ, Nuez F (2008) Tomato. In: Prohens J, Nuez F (eds) Vegetables II. Handbook of Plant Breeding, vol 2. Springer, Heidelberg, Germany, pp 249–323.

Dirks R, van Dun K, de Snoo CB, van den Berg M, Lelivelt CLC, Voermans W, Woudenberg L, de Wit JPC, Reinink K, Schut JW, van der Zeeuw E, Vogelaar A, Freymark G, Gutteling EW, Keppel MN, van Drongelen P, Kieny M, Ellul P, Touraev A, Ma H, de Jong H, Wijnker

E (2009) Reverse breeding: a novel breeding approach based on engineered meiosis. Plant Biotechnol J 7: 837–845.

Doganlar S, Dodson J, Gabor B, Beck-Bunn T, Crossman C, Tanksley SD (1998) Molecular mapping of the *py-1* gene for resistance to corky root rot (*Pyrenochaeta lycopersici*) in tomato. Theor Appl Genet 97: 784–788.

Dreher K, Khairallah M, Ribaut J-M, Morris M (2003). Money matters (I): costs of field and laboratory procedures associated with conventional and marker-assisted maize breeding at CIMMYT. Mol Breed 11: 221–234.

Duan K, Yang H, Ran K, You S, Zhao H, Jiang Q (2009) Characterization of a novel stress–response member of the MAPK family in *Malus hupehensis* Rehd. Plant Mol Biol Rep 27: 69–78.

El Mehrach K, Gharsallah Chouchane S, Mejia L, Williamson VM, Vidavski F, Hatimi A, Salus MS, Martin CT, Maxwell DP (2005) PCR-based methods for tagging the *Mi-1* locus for resistance to root-knot nematode in begomovirus-resistant tomato germplasm. Acta Hort 695: 263–270.

El Mohtar CA, Atamian HS, Dagher RB, Abou-Jawdah Y, Salus MS, Maxwell DP (2007). Marker-assisted selection of tomato genotypes with the *I-2* gene for resistance to *Fusarium oxysporum* f. sp. *lycopersici* Race 2. Plant Dis 91: 758–762.

Emebiri L, Michael P, Moody DB, Ogbonnaya FC, Black C (2009) Pyramiding QTLs to improve malting quality in barley: gains in phenotype and genetic diversity. Mol Breed 23: 219–228.

Eshed Y, Zamir D (1995) An introgression line population of *Lycopersicon pennellii* in the cultivated tomato enables the identification and fine mapping of yield-associated QTL. Genetics 141: 1147–1162.

Fauquet CM, Bisaro DM, Briddon RW, Brown JK, Harrison BD, Rybicki EP, Stenger DC, Stanley J (2003) Revision of taxonomic criteria for species demarcation in the family Geminiviridae, and an updated list of begomovirus species. Arch Virol 148: 405–421.

Fazio G, Stevens MR, Scott JW (1999) Identification of RAPD markers linked to fusarium crown and root rot resistance (*Frl*) in tomato. Euphytica 105: 205–210.

FDA (1994) US Food and Drug Administration. Biotechnology consultation memorandum of conference BNF no. 000003: http://www.fda.gov/Food/Biotechnology/Submissions/ucm161132.htm. FDA, Silver Spring, MD.

Fernie AR, Tadmor Y, Zamir D (2006) Natural genetic variation for improving crop quality. Curr Opin Plant Biol 9: 196–202.

Fernie A, Schauer N, Semel Y, Kochevenko A, Fait A, Carrari F, Zamir D (2008) Integrated genomics approaches using introgression lines of tomato. Comp Biochem Physiol A-Mol Integr Physiol 150: S46–S46.

Finkers R, Bai YL, van den Berg P, van Berloo R, Meijer-Dekens F, ten Have A, van Kan J, Lindhout P, van Heusden AW (2008) Quantitative resistance to *Botrytis cinerea* from *Solanum neorickii*. Euphytica 159: 83–92.

Finlay KW (1953) Inheritance of spotted wilt virus resistance in tomato. Aust J Biol Sci 6: 153–163.

Folkertsma RT, Spassova MI, Prins M, Stevens MR, Hille J, Goldbach RW (1999) Construction of a bacterial artificial chromosome (BAC) library of *Lycopersicon esculentum* cv. Stevens and its application to physically map the *Sw-5* locus. Mol Breed 5: 197–207.

Foolad MR, Zhang LP, Khan AA, Nino-Liu D, Lin GY (2002) Identification of QTLs for early blight (*Alternaria solani*) resistance in tomato using backcross populations of a *Lycopersicon esculentum* × *L. hirsutum* cross. Theor Appl Genet 104: 945–958.

Foolad MR, Sharma A, Ashrafi H, Lin G (2005) Genetics of early blight resistance in tomato. Acta Hort 695: 397–406.

Foolad MR (2007) Molecular mapping, marker-assisted selection and map-based cloning in tomato. In: Varshney RK, Tuberosa R (eds) Genomics Assisted Crop Improvement. Vol 2: Genomics Applications in Crops. Springer, Dordrecht, The Netherlands, pp 307–356.

Forster BP, Heberle-Bors E, Kasha KJ, Touraev A (2007) The resurgence of haploids in higher plants. Trends Plant Sci 12: 368–375.

Fox RL, Hayden MJ, Mekuria G, Eglinton JK (2005) Cost analysis of molecular techniques for marker assisted selection within a barley breeding program. In: Proc 12th Australian Annual Barley Technical Symposium, Hobart, Tasmania, Australia: http://www.regional.org.au/au/#abts.

Franco J, Crossa J, Ribaut JM, Betran J, Warburton ML, Khairallah M (2001). A method for combining molecular markers and phenotypic attributes for classifying plant genotypes. Theor Appl Genet 103: 944–952.

Frary A, Graham E, Jacobs J, Chetelat RT, Tanksley SD (1998) Identification of QTL for late blight resistance from *L. pimpinellifolium* L3708. Tomato Genet Coop Rep 48: 19–2.

Frary A, Fulton TM, Zamir D, Tanksley SD (2004) Advanced backcross QTL analysis of a *Lycopersicon esculentum* x *L. pennellii* cross and identification of possible orthologs in the Solanaceae. Theor Appl Genet 108: 485–496.

Fraser PD, Bramley PM (2009) Genetic manipulation of carotenoid content and composition in crop plants. In: Britton G, Pfander H, Liaaen-Jensen S (eds) Carotenoids, Volume 5. Birkhäuser Verlag, Basel, Switzerland, pp 99–114.

Frisch M, Bohn M, Melchinger AE (1999) Comparison of selection strategies for marker-assisted backcrossing of a gene. Crop Sci 39: 1295–1301.

Fulton TM, Beck-Bunn T, Emmatty D, Eshed Y, Lopez J, Petiard V, Uhlig J, Zamir D, Tanksley SD (1997) QTL analysis of an advanced backcross of *Lycopersicon peruvianum* to the cultivated tomato and comparisons with QTLs found in other wild species. Theor Appl Genet 95: 881–894.

Fulton TM, Grandillo S, Beck-Bunn T, Fridman E, Frampton A, Lopez J, Petiard V, Uhlig J, Zamir D, Tanksley SD (2000) Advanced backcross QTL analysis of a *Lycopersicon esculentum* x *L. parviflorum* cross. Theor Appl Genet 100: 1025–1042.

Galston AW, Sawhney RK (1990) Polyamines in plant physiology. Plant Physiol 94: 406–410.

Garcia BE, Mejía L, Salus MS, Martin CT, Seah S, Williamson VM, Maxwell DP (2007) A co-dominant SCAR marker, Mi23, for detection of the *Mi-1.2* gene for resistance to root-knot nematode in tomato germplasm, pp: 1–13: http://www.plantpath.wisc.edu.

Garcia-Martinez S, Andreani L, Garcia-Gusano M, Geuna F, Ruiz JJ (2005). Evolution of amplified length polymorphism and simple sequence repeats for tomato germplasm fingerprinting: utility for grouping closely related traditional cultivars. Genome 49: 648–656.

Garland S, Sharman M, Persley D, McGrath D (2005) The development of an improved PCR-based marker system for *Sw-5*, an important TSWV resistance gene of tomato. Aust J Agri Res 56: 285–289.

Gebhardt C, Bellin D, Henselewski H, Lehmann W, Schwarzfischer J, Valkonen JPT (2006) Marker-assisted combination of major genes for pathogen resistance in potato. Theor Appl Genet 112: 1458–1464.

Gianessi L, Sankula S, Reigner N (2003). Plant biotechnology: Potential impact for improving pest management in European agriculture. A summary of nine case studies. Washington, DC: National Center for Food and Agricultural Policy: http://www.ncfap.org/reports/Europe/ExecutiveSummaryDecember.pdf.

Gielen JJL, de Haan P, Kool AJ, Peters D, van Grinsven MQJM, Goldbach RW (1991) Engineered resistance to tomato spotted wilt virus, a negative-strand RNA virus. Nat Biotechnol 9: 1363–1367.

Gilbert AW (1912) Mendalian study of tomatoes. Am Breed Assn Mag 7: 169–188.

Giordano LB, Silva-Lobo VL, Santana FM, Fonseca MEN, Boiteux LS (2005) Inheritance of resistance to the bipartite tomato chlorotic mottle begomovirus derived from *Lycopersicon esculentum* cv. 'Tyking'. Euphytica 143: 27–33.

Giorio G, AL Stigliani, C D'Ambrosio (2007) Agronomic performance and transcriptional analysis of carotenoid biosynthesis in fruits of transgenic HighCaro and control tomato lines under field conditions. Transgenic Res. 16: 15–28.

Gonçalves LSA, Rodrigues R, do Amaral Jùnior AT, Karasawa M, Sudré CP (2009) Heirloom tomato gene bank: assessing genetic divergence based on morphological, agronomic and molecular data using a Ward-modified location model. Genet Mol Res 8: 364–374.

Gonzalo MJ, van der Knaap E (2008) A comparative analysis into the genetic bases of morphology in tomato varieties exhibiting elongated fruit shape. Theor Appl Genet 116: 647–656.

Goodstal FJ, Kohler GR, Randall LB, Bloom AJ, St Clair DA (2005) A major QTL introgressed from wild *Lycopersicon hirsutum* confers chilling tolerance to cultivated tomato (*Lycopersicon esculentum*) Theor Appl Genet 111: 898–905.

Gordillo LF, Stevens MR, Millard MA, Geary B (2008) Screening two *Lycopersicon peruvianum* collections for resistance to Tomato spotted wilt virus. Plant Dis 92: 694–704.

Gosal SS, Wani SH, Kang MS (2009) Biotechnology and drought tolerance. J Crop Improv 23: 19–54.

Gosal SS, Wani SH, Kang MS (2010) Biotechnology and crop improvement. J Crop Improv 24: 153–217.

Grassi G, Gallitelli D (1990) Presenza della necrosi letale del pomodoro in Emilia Romagna. Inform Fitopat 12: 59–61.

Griffiths PD, Scott JW (2001) Inheritance and linkage of tomato mottle virus resistance genes derived from *Lycopersicon chilense* accession LA 1932. J Am Soc Hort Sci 126: 462–467.

Grube RC, Radwanski ER, Jahn M (2000) Comparative genetics of disease resistance within the Solanaceae. Genetics 155: 873–887.

Gur A, Zamir D (2004) Unused natural variation can lift yield barriers in plant breeding. PLoS Biol 2: e245.

Guri A, Dunbar LJ, Sink KC (1991) Somatic hybridization between selected *Lycopersicon* and *Solanum* species. Plant Cell Rep 10: 76–80.

Han F, Ullrich SE, Kleinhofs A, Jones BL, Hayes PM, Wesenberg DM (1997) Fine structure mapping of the barley chromosome-1 centromere region containing malting-quality QTLs. Theor Appl Genet 95: 903–910.

Hanson P, Green SK, Kuo G (2006) *Ty-2*, a gene on chromosome 11 conditioning geminivirus resistance in tomato. Tomato Genet Coop Rep 56: 17–18.

Haskell G, Paterson EB (1966) Chromosome number of a sub-antarctic *Rubus*. Nature 211: 759.

Hassan S, Thomas PE (1988) Extreme resistance to tomato yellow top virus and potato leaf roll virus in *Lycopersicon peruvianum* and some of its tomato hybrids. Phytopathology 78: 1164–1167.

He C, Poysa V, Yu K (2003). Development and characterization of simple sequence repeat (SSR) markers and their use in determining relationships among *Lycopersicon esculentum* cultivars. Theor Appl Genet 106: 363–373.

Hirschberg J (2001) Carotenoid biosynthesis in flowering plants. Curr Opin Plant Biol 4: 210–218.

Hittalmani S, Parco A, Mew TV, Zeigler RS, Huang N (2000). Fine mapping and DNA marker-assisted pyramiding of the three major genes for blast resistance in rice. Theor Appl Genet 100: 1121–1128.

Ho JC, McCouch SR, Smith ME (2002) Improvement of hybrid yield by advanced backcross QTL analysis in elite maize. Theor Appl Genet 105: 440–448.

Holland JB (2004) Implementation of molecular markers for quantitative traits in breeding programs-challenges and opportunities. Proc 4th International Crop Science Congress, Brisbane, Australia: www.cropscience.org.au.

Hoogstraten JJG, Braun CJ (2009) Methods for coupling resistance alleles in tomato. US Patent 07615689.

Hospital F, Charcosset A (1997) Marker assisted introgression of Quantitative Trait Loci. Genetics 147: 1469–1485.

Hospital F, Moreau L, Lacoudre F, Charcosset A, Gallais A (1997) More on the efficiency of marker-assisted selection. Theor Appl Genet 95: 1181–1189.

Hospital F (2001) Size of donor chromosome segments around introgressed loci and reduction of linkage drag in marker-assisted backcross programs. Genetics 158: 1363–1379.

Hospital F (2002) Marker-assisted backcross breeding: a case-study in genotype building theory. In: Kang MS (ed) Quantitative Genetics, Genomics, and Plant Breeding. CABI Publishing, Wallingford, UK.

Hospital F (2009) Challenges for effective marker-assisted selection in plants. Genetica 136: 303–310.

Huang C-C, Groot T, Meijer-Dekens F, Niks RE, Lindhout P (1998) The resistance to powdery mildew (*Oidium lycopersicum*) in *Lycopersicon* species is mainly associated with hypersensitive response. Eur J Plant Pathol 104: 399–407.

Huang, CC, Van de Putte PM, Haanstra-van der Meer JG, Meijer-Dekens F, Lindhout P (2000) Characterization and mapping of resistance to *Oidium lycopersicum* in two *Lycopersicon hirsutum* accessions: Evidence for close linkage of two Ol-genes on chromosome 6. Heredity 85: 511–520.

Huang XQ, Hsam SLK, Zeller FJ (1997) Identification of powdery mildew resistance genes in common wheat (*Triticum aestivum* L. em Thell.) IX. Cultivars, land races and breeding lines grown in China. Plant Breed 116: 233–238.

Huang XQ, Coster H, Ganal MW, Roder MS (2003) Advanced backcross QTL analysis for the identification of quantitative trait loci alleles from wild relatives of wheat (*Triticum aestivum* L.). Theor Appl Genet 106: 1379–1389.

Jablonska B, Ammiraju JSS, Bhattarai KK, Mantelin S, de Ilarduya OM, Roberts PA, Kaloshian I (2007) The *Mi-9* gene from *Solanum arcanum* conferring heat-stable resistance to root-knot nematodes is a homolog of *Mi-1*. Plant Physiol 143: 1044–1054.

Jensen KS, Martin CT, Maxwell DP (2007) A CAPS marker, FER-G8, for detection of *Ty3* and *Ty3a* alleles associated with *S. chilense* introgressions for begomovirus resistance in tomato breeding lines: http://www.plantpath.wisc.edu/.

Ji Y, Scott JW (2006) *Ty-3*, a begomovirus resistance locus linked to *Ty-1* on chromosome 6 of tomato. Tomato Genet Coop Rep 56: 22–25.

Ji Y, Schuster DJ, Scott JW (2007a) *Ty-3*, a begomovirus resistance locus near the Tomato yellow leaf curl virus resistance locus *Ty-1* on chromosome 6 of tomato. Mol Breed 20: 271–284.

Ji Y, Scott JW, Hanson P, Graham E, Maxwell DP (2007b) Sources of resistance, inheritance, and location of genetic loci conferring resistance to members of the tomato-infecting begomoviruses. In: Czosnek H (ed) Tomato Yellow Leaf Curl Virus Disease. Springer, Dordrecht, The Netherlands, pp 343–362.

Ji Y, Scott JW, Maxwell DP, Schuster DJ (2008) *Ty-4*, a tomato yellow leaf curl virus resistance gene on chromosome 3 of tomato. Tomato Genet Coop Rep 58: 29–31.

Ji Y, Scott JW, Schuster DJ (2009) Toward fine mapping of the Tomato *yellow leaf curl virus* resistance gene *Ty-2* on chromosome 11 of tomato. HortScience 44: 614–618.

Jones DA, Dickinson MJ, Balint-Kurti PJ, Dixon MS, Jones JDG (1993) Two complex resistance loci revealed in tomato by classical and RFLP mapping of the *Cf-2*, *Cf-4*, *Cf-5* and *Cf-9* genes for resistance to *Cladosporium fulvum*. Mol Plant-Microbe Interact 6: 348–357.

Jones DA, Thomas CM, Hammond-Kosack KE, Balint-Kurti PJ, Jones JD (1994) Isolation of the tomato *Cf-9* gene for resistance to *Cladosporium fulvum* by transposon tagging. Science 266: 789–793.

Joshi RK, Nayak S (2010) Gene pyramiding—A broad spectrum technique for developing durable stress resistance in crops. Biotechnol Mol Biol Rev 5: 51–60.

Jung J-K, Park S-W, Liu WY, Kang B-C (2010) Discovery of single nucleotide polymorphism in *Capsicum* and SNP markers for cultivar identification. Euphytica 175: 91–107.

Kabelka E, Franchino B, Francis DM (2002) Two loci from *Lycopersicon hirsutum* LA407 confer resistance to strains of *Clavibacter michiganensis* subsp. *michiganensis*. Phytopathology 92: 504–510.

Kalamaki MS, MH Harpster, JM Palys, JM Labavitch, DS Reid and DA Brummell (2003) Simultaneous transgenic suppression of LePG and LeExp1 influences rheological properties of juice and concentrates from a processing tomato variety. J Agric Food Chem. 51: 7456–7464.

Kandemir N, Jones BL, Wesenberg DM, Ullrich SE, Kleinhofs A (2000) Marker-assisted analysis of three grain yield QTL in barley (*Hordeum vulgare* L.) using near isogenic lines. Mol Breed 6: 157–167.

Kaniewski W, Lawson C (1998) Coat protein and replicase mediated resistance to plant viruses. In: Hadidi A, Khetarpal RK, Koganezawa H (eds) Plant Virus Disease Control (chapter 6). APS Press, St Paul, MN, USA.

Katiyar SK, Tan Y, Huang B, Chandel G, Xu Y, Zhang Y, Xie Z, Bennett J (2001) Molecular mapping of gene *Gm-6(t)* which confers resistance against four biotypes of Asian rice gall midge in China. Theor Appl Genet 103: 953–961.

Knoll J, Ejeta G (2008) Marker-assisted selection for early-season cold tolerance in sorghum: QTL validation across populations and environments. Theor Appl Genet 116: 541–553.

Kole C, Ashrafi H, Lin G, Foolad M (2006) Identification and molecular mapping of a new R gene, *Ph-4*, conferring resistance to late blight in tomato. Solanaceae Conf, Univ of Wisconsin, Madison, WI, USA, Abstr 449.

Koudandé OD, Iraqi F, Thomson PC, Teale AJ, van Arendonk JAM (2000) Strategies to optimize marker-assisted introgression of multiple unlinked QTL. Mamm Genome 11: 145–150.

Koudandé OD, van Arendonk JAM, Iraqi F (2005) Marker-assisted introgression of Trypanotolerance QTL in mice. Mamm Genome 16: 112–119.

Kuchel H, Ye G, Fox R, Jefferies S (2005) Genetic and economic analysis of a targeted marker-assisted wheat breeding strategy. Mol Breed 16: 67–78.

Kuchel H, Fox R, Reinheimer J, Mosionek L, Willey N, Bariana H, Jefferies S (2007) The successful application of a marker-assisted wheat breeding strategy. Mol Breed 20: 295–308.

Kuklev MY, Fesenko IA, Karlov GI (2009) Development of a CAPS marker for the Verticillium wilt resistance in tomatoes. Russ J Genet 45: 575–579.

Kwok P-Y, Chen X (2003) Detection of single nucleotide polymorphisms. Curr Iss Mol Biol 5: 43–60.

Kwon S-J, Park K-C, Kim JH, Lee JK, Kim N-S (2005) *Rim 2/Hipa* CACTA transposon display; A new genetic marker technique in *Oryza* species. BMC Genet 6: 15.

Labate JA, Baldo AM (2005) Tomato SNP Discovery by EST mining and resequencing. Mol Breed 16: 343–349.

Labate JA, Grandillo S, Fulton T, Muños S, Caicedo AL, Peralta I, Ji Y, Chetelat RT, et al. (2007) Tomato. In: Kole C (ed) Genome Mapping and Molecular Breeding in Plants. Vol 5: Vegetables. Springer NY, USA, pp 1–125.

Laidò G, Barabaschi D, Tondelli A, Gianinetti A, Stanca AM, Li Destri Nicosia O, Di Fonzo N, Francia E, Pecchioni N (2009) QTL alleles from a winter feed type can improve malting quality in barley. Plant Breed 128: 598–605.

Landegren U, Nilsson M, Kwok PY (1998) Reading bits of genetic information: methods for single-nucleotide polymorphism analysis. Genome Res 8: 769–776.

Lanfermeijer FC, Dijkhuis J, Sturre MJG, de Haan P, Hille J (2003) Cloning and characterization of the durable tomato mosaic virus resistance gene *Tm-22* from *Lycopersicon esculentum*. Plant Mol Biol 52: 1037–1049.

Lanfermeijer FC, Warmink J, Hille J (2005) The products of the broken *Tm-2* and the durable *Tm-22* resistance genes from tomato differ in four amino acids. J Exp Bot 56: 2925–2933.

Langridge P, Lance R, Barr A (1996) Practical application of marker assisted selection. In: Scoles G, Rossnagel B (eds) Proc 5th Int Oat Conf and 7th Int Barley Genetics Symposium. University Extension Press, Saskatoon, Saskatchewan, Canada, pp 141–149.

Langridge P, Barr AR (2003) Preface to 'Better barley faster: the role of Marker Assisted Selection'. Aust J Agri Res 54: i–iv.

Langridge P, Paltridge N, Fincher G (2006) Functional genomics of abiotic stress tolerance in cereals. Brief Funct Genom Proteom 4: 343–354.

Laporte MM, Galagan JA, Shapiro JA, Boersig MR, Shewmaker CK, Sharkey TD (1997) Sucrose-phosphate synthase activity and yield analysis of tomato plants transformed with maize sucrose-phosphate synthase. Planta 203: 253–259.

Laporte MM, Galagan JA, Prasch AL, Vanderveer PJ, Hanson PT, Shewmaker CK, Sharkey TD (2001) Promoter strength and tissue specificity effects on growth of tomato plants transformed with maize sucrose-phosphate synthase. Planta 212: 817–822.

Lauge R, Dmitriev AP, Joosten MHAJ, DeWit PJGM (1998) Additional resistance gene(s) against *Cladosporium fulvum* present on the *Cf-9* introgression segment are associated with strong PR protein accumulation. Mol Plant-Microbe Interact 11: 301–308.

Lawrence CJ, Dong Q, Polacco ML, Seigfried TE, Brendel V (2004) MaizeGDB, the community database for maize genetics and genomics. Nucl Acids Res 32 (Database issue): D393–D397.

Lawson DM, Lunde CF, Mutschler MA (1997) Marker-assisted transfer of acylsugar-mediated pest resistance from the wild tomato, *Lycopersicon pennellii*, to the cultivated tomato, *Lycopersicon esculentum*. Mol Breed 3: 307–317.

Lecomte L, Duffé P, Buret M, Servin B, Hospital F, Causse M (2004) Marker-assisted introgression of five QTLs controlling fruit quality traits into three tomato lines revealed interactions between QTLs and genetic backgrounds. Theor Appl Genet 109: 658–668.

Légnani R, Gebre-Selassie K, Marchoux G, Laterrot H (1997) Interaction between PVY pathotypes and tomato lines. Tomato Genet Coop Rep 47: 13–15.

Leszek A, Kleczkowski LA, Geisler M, Ciereszko I, Johansson H (2004) UDP-glucose pyrophosphorylase. An old protein with new tricks. Plant Physiol 134: 912–918.

Lindhout P (2002) The perspectives of polygenic resistance in breeding for durable disease resistance. Euphytica 124: 217–226.

Liu LW, Wang Y, Gong YQ, Zhao TM, Liu G, Li XY, Yu FM (2007) Assessment of genetic purity of tomato (Lycopersicon esculentum L.) hybrid using molecular markers. Sci Hort 115: 7–12.

Liu S, Anderson JA (2003) Targeted molecular mapping of a major wheat QTL for Fusarium head blight resistance using wheat ESTs and synteny with rice. Genome 46: 817–823.

Lussier Y and Y Liu (2007) Computational approaches to phenotyping: high-throughput phenomics. Proceedings of the American Thoracic Society 4, no. 1 (January): 18–25.

Mangin B, Thoquet P, Olivier J, Grimsley NH (1999) Temporal and multiple quantitative trait loci analyses of resistance to bacterial wilt in tomato permit the resolution of linked loci. Genetics 151: 1165–1172.

Martin GB, Brommonschenkel SH, Chunwongse J, Frary A, Ganal MW, Spivey R, Wu T, Earle ED, Tanksley SD (1993) Map-based cloning of a protein kinase gene conferring disease resistance in tomato. Science 262: 1432–1436.

Martin MW, Thomas PE (1986) Increased value of resistance to infection if used in integrated pest management control of tomato curly top. Phytopathology 76: 540–542.

Maruthasalam S, Kalpana K, Kumar KK, Loganathan M, Poovannan K, Raja AJ, Kokiladevi E, Samiyappan R, Sudhakar D, Balasubramanian P (2007) Pyramiding transgenic resistance in elite indica rice cultivars against the sheath blight and bacterial blight. Plant Cell Rep 26: 791–804.

Matthews DE, Carollo VL, Lazo GR, Anderson OD (2003) GrainGenes, the genome database for small-grain crops. Nucl Acids Res 31: 183–186.

Maxwell DP, Martin C, Salus MS, Montes L, Mejía L (2006) Breeding tomatoes for resistance to tomato-infecting begomoviruses: http://www.plantpath.wisc.edu.

Mayer KF, Taudien S, Martis M, Simková H, Suchánková P, Gundlach H, Wicker T, Petzold A, Felder M, Steuernagel B, Scholz U, Graner A, Platzer M, Dolezel J, Stein N (2009) Gene content and virtual gene order of barley chromosome 1H. Plant Physiol 151: 496–505.

Medina-Filho HP (1980) An electrophoretic variant as a tool for breeding tomato for nematode resistance. PhD Thesis, University of California, Davis, CA, USA.

Mesbah LA, Kneppers TJA, Takken FLW, Laurent P, Hille J, Nijkamp HJJ (1999) Genetic and physical analysis of a YAC contig spanning the fungal disease resistance locus *Asc* of tomato (*Lycopersicon esculentum*). Mol Gen Genet 261: 50–57.

Metha RA, Cassol T, Li N, Ali N, Handa AK, Mattoo AK (2002) Engineered polyamine accumulation in tomato enhances phytonutrient content, juice quality, and vine life. Nat Biotechnol 20: 613–618.

Milc J, Sala A, Bergamaschi S, Pecchioni N (2011) A genotypic and phenotypic information source for marker-assisted selection of cereals: the CEREALAB Database. Database, doi: 10.1093/database/baq038.

Miller JC, Tanksley SD (1990) RFLP analysis of phylogenetic relationships and genetic variation in the genus *Lycopersicon*. Theor Appl Genet 80: 437–448.

Moose SP, Mumm RH (2008) Molecular plant breeding as the foundation for 21st century crop improvement. Plant Physiol 147: 969–77.

Moreau P, Thoquet P, Olivier J, Laterrot H, Grimsley N (1998) Genetic mapping of *Ph-2*, a single locus controlling partial resistance to *Phytophthora infestans* in tomato. Mol Plant-Microbe Interact 11: 259–269.

Mueller LA, Lankhorst RK, Tanksley SD, Giovannoni JJ, White R, Vrebalov J, Fei Z, van Eck J, Buels R, Mills AA, et al. (2009) A snapshot of the emerging Tomato genome sequence. Plant Genome 2: 78–92.

Neeraja CN, Maghirang-Rodriguez R, Pamplona A, Heuer S, Collard BCY, Septiningsih EM, Vergara G, Sanchez D, Xu K, Ismail AM, Mackill DJ (2007) A marker-assisted backcross approach for developing submergence-tolerant rice cultivars. Theor Appl Genet 115: 767–776.

Negrao S, Oliveira MM, Jena KK, Mackill D (2008) Integration of genomic tools to assist breeding in the *japonica* subspecies of rice. Mol Breed 22: 159–168.

Nesbitt TC, Tanksley SD (2002) Comparative sequencing in the genus *Lycopersicon*: implications for the evolution of fruit size in the domestication of cultivated tomatoes. Genetics 162: 365–379.

Ohmori ST, Murata M, Motoyoshi F (2000) Molecular characterization of the SCAR markers tightly linked to the *Tm-2* locus of the genus *Lycopersicon*. Theor Appl Genet 101: 64–69.

Ori N, Eshed Y, Paran I, Presting G, Aviv D, Tanksley S, Zamir D, Fluhr R (1997) The *I2C* family from the wilt disease resistance locus *I2* belongs to the nucleotide binding, leucine-rich repeat superfamily of plant resistance genes. Plant Cell 9: 521–532.

Papa R, Bellucci E, Rossi M, Leonardi S, Rau D, Gepts P, Nanni L, Attene G (2007) Tagging the signatures of domestication in common bean (*Phaseolus vulgaris*) by means of pooled DNA samples. Ann Bot 100: 1039–1051.

Paran I, Goldman I, Tanksley SD, Zamir D (1995) Recombinant inbred lines for genetic mapping in tomato. Theor Appl Genet 90: 542–548.

Paran I, Zamir D (2003) Quantitative traits in plants: beyond the QTL. Trends Genet 19: 303–306.

Parniske M, Hammond-Kosack KE, Golstein C, Thomas CM, Jones DA, Harrison K, Wulff BBH, Jones JDG (1997) Novel disease resistance specificities result from sequence exchange between tandemly repeated genes at the *Cf-4/9* locus of tomato. Cell 91: 821–832.

Parniske M, Wulff BBH, Bonnema G, Thomas CM, Jones DA, Jones JDG (1999) Homologues of the *Cf-9* disease resistance gene (*Hcr9*s) are present at multiple loci on the short arm of tomato chromosome 1. Mol Plant-Microbe Interact 12: 93–102.

Parrella G, Ruffel S, Moretti A, Morel C, Palloix A, Caranta C (2002) Recessive resistance genes against potyviruses are localized in colinear genomic regions of the tomato (*Lycopersicon* spp.) and pepper (*Capsicum* spp.) genomes. Theor Appl Genet 105: 855–861.

Parrella G, Moretti A, Gognalons P, Lesage ML, Marchoux G, Gebre-Selassie K, Caranta C (2004) The *Am* gene controlling resistance to Alfalfa mosaic virus in tomato is located in the cluster of dominant resistance genes on chromosome 6. Phytopathology 94: 345–350.

Pavan S, Zheng Z, Borisova M, van Der Berg P, Lotti C, De Giovanni C, Lindhout P, de Jong H, Riccardi L, Visser RGF, Bay Y (2008) Map- vs. homology-based cloning for the recessive gene *ol-2* conferring resistance to tomato powdery mildew. Euphytica 162: 91–98.

Peleman JD, van der Voort JR (2003) The challenges in Marker Assisted Breeding. In: van Hintum TJL, Lebeda A, Pink D, Schut JW (eds) Eucarpia Leafy Vegetable. CGN, Netherlands, pp 125–130 What is CGN? We need the place also.

Peralta IE, Spooner DM (2001) Granule-bound starch synthase (*GBSSI*) gene phylogeny of wild tomatoes (*Solanum* L. section *Lycopersicon* [Mill.] Wettst. subsection *Lycopersicon*). Am J Bot 88: 1888–1902.

Peralta IE, Knapp S, Spooner DM (2006) Nomenclature for wild and cultivated tomatoes. Tomato Genet Coop Report 56: 6–12.

Picó B, Herraiz J, Ruiz JJ, Nuez F (2002) Widening the genetic basis of virus resistance in tomato. Sci Hort 94: 73–89.

Pierce LC (1971) Linkage tests with *Ph* conditioning resistance to race 0, *Phytophthora infestans*. Tomato Genet Coop Rep 21: 30.

Pillen K, Zacharias A, Leon J (2003) Advanced backcross QTL analysis in barley (*Hordeum vulgare* L.). Theor Appl Genet 107: 340–352.

Porretta S, G Poli, E Minuti (1998) Tomato pulp quality from transgenic fruits with reduced polygalacturonase (PG). Food Chem. 62: 283–290.

Powell W, Morgante M, Andre C, Hanafey M, Vogel J, Tingel S, Rafalski A (1996) The comparison of RFLP, RAPD, AFLP and SSR (microsatellite) markers for germplasm analysis. Mol Breed 2: 225–235.

Price AH (2006) Believe it or not, QTLs are accurate! Trends Plant Sci 11: 213–216.

Pucci N, Voltattorni S, Infantino A, Acciarri N (2007) Screening of fresh market tomato for resistance to *Pyrenochaeta lycopersici*. J Plant Pathol 89(3S): 55–56.

Ragot M, Biasiolli M, Delbut MF, Dell'Orco A, et al. (1995) Marker-assisted backcrossing: a practical example. In: Berville A, Tersac M (eds) Les Colloques, no 72, Techniques et Utilisations des Marqueurs Moleculaires. INRA, Paris, France, pp 45–56.

Rangel PHN, Brondani C, Rangel PN, Brondani RPV, Zimmermann FJP (2005) Development of rice lines with gene introgression from the wild *Oryza glumaepatula* by the AB-QTL methodology. Crop Breed Appl Biotechnol 5: 10–21.

Rashid F, Khalid S, Ahmad I, Mughal SM (1989) Potato virus X (PVX) resistance in tomato cultivars. Int J Pest Manag 35: 357–358.

Reid A, Hof L, Esselink D, Vosman B (2009) Potato cultivar genome analysis. Plant Pathol Techniq Protoc 508: 295–308.

Ribaut JM, Banziger M, Betran JA, Jiang C, Edmeades GO, Dreher K, Hoisington DA (2002). Use of molecular markers in plant breeding: drought tolerance improvement in tropical maize. In: Kang MS (ed) Quantitative Genetics, Genomics, and Plant Breeding. CABI Publishing, Wallingford, UK, pp 85–99.

Ribaut JM, Ragot M (2007) Marker-assisted selection to improve drought adaptation in maize: the backcross approach, perspectives, limitations, and alternatives. J Exp Bot 58: 351–360.

Ricciardi L, Lotti C, Pavan S, Bai Y, Lindhout P, De Giovanni C (2007) Further isolation of AFLP and LMS markers for the mapping of the *Ol-2* locus related to powdery mildew (*Oidium neolycopersici*) resistance in tomato (*Solanum lycopersicum* L.). Plant Sci 172: 746–755.

Rick CM, Holle M (1990) Andean *Lycopersicon esculentum* var. *cerasiforme*: genetic variation and its evolutionary significance. Econ Bot 44: 69–78.

Robbins MD, Darrigues A, Sim S-C, Masud MAT, Francis DM (2009) Characterization of hypersensitive resistance to bacterial spot race T3 (*Xanthomonas perforans*) from tomato accession PI 128216. Phytopathology 99: 1037–1044.

Robert VJM, West MAL, Inai S, Caines A, Arntzen, Smith JK, St Clair DA (2001) Marker-assisted introgression of blackmold resistance QTL alleles from wild *Lycopersicon cheesmanii* to cultivated tomato (*L. esculentum*) and evaluation of QTL phenotypic effects. Mol Breed 8: 217–233.

Romagosa I, Han F, Clancy JA, Ullrich SE (1999) Individual locus effects on dormancy during seed development and after ripening in barley. Crop Sci 39: 74–79.

Rommens CM (2004) All-native DNA transformation: a new approach to plant genetic engineering. Trends Plant Sci 9: 457–464.

Rommens CM (2007) Intragenic crop improvement: combining the benefits of traditional breeding and genetic engineering. J Agri Food Chem 55: 4281–4288.

Rommens CM (2010) Barriers and paths to market for genetically engineered crops. Plant Biotechnol J 8: 101–111.

Rommens CM, Haring MA, Swords K, Davies HV, Belknap WR (2007) The intragenic approach as a new extension to traditional plant breeding. Trends Plant Sci 12: 397–403.

Rose LE, Langley CH, Bernal AJ, Michelmore RW (2005) Natural variation in the *Pto* pathogen resistance gene within species of wild tomato (*Lycopersicon*). I. Functional analysis of *Pto* alleles. Genetics 171: 345–357.

Rose LE, Michelmore RW, Langley CH (2007) Natural variation in the *Pto* disease resistance gene within species of wild tomato (*Lycopersicon*). II. Population genetics of *Pto*. Genetics 175: 1307–1319.

Roselló S, Díez MJ, Nuez F (1998) Genetics of tomato spotted wilt virus resistance coming from *Lycopersicon peruvianum*. Eur J Plant Pathol 104: 499–509.

Roselló S, Ricarte B, Díez MJ, Nuez F (2001) Resistance to Tomato spotted wilt virus introgressed from *Lycopersicon peruvianum* in line UPV 1 may be allelic to *Sw-5* and can be used to enhance the resistance of hybrids cultivars. Euphytica 119: 357–367.

Rotino GL, Acciarri N, Sabatini E, Mennella G, Lo Scalzo R, Maestrelli A, Molesini B, Pandolfini T, Scalzo J, Mezzetti B, Spena A (2005) Open field trial of genetically modified parthenocarpic tomato: seedlessness and fruit quality. BMC Biotechnol 5: 32–40.

Salmeron JM, Oldroyd GE, Rommens CM, Scofield SR, Kim HS, Lavelle DT, Dahlbeck D, Staskawicz BJ (1996) Tomato *Prf* is a member of the leucine-rich repeat class of plant disease resistance genes and lies embedded within the *Pto* kinase gene cluster. Cell 86: 123–133.

Samis K, Bowley S, McKersie B (2002) Pyramiding Mn-superoxide dismutase transgenes to improve persistence and biomass production in alfalfa. J Exp Bot 53: 1343–1350.

Sandve SR, Rudi H, Dørum G, Berg PR, Rognli OA (2010) High-throughput genotyping of unknown genomic terrain in complex plant genomes: lessons from a case study. Mol Breed 26: 711–718.

Sarfatti M, Abu-Abied M, Katan J, Zamir D (1991) RFLP mapping of *I1*, a new locus in tomato conferring resistance against *Fusarium oxysporum* f. sp. *lycopersici* race 1. Theor Appl Genet 82: 22–26.

Sax K (1923) The association of size differences with seed coat pattern and pigmentation in *Phaseolus vulgaris*. Genetics 8: 552–560.

Schaart JG, Visser RGF (2009) Novel plant breeding techniques. Consequences of new genetic modification-based plant breeding techniques in comparison to conventional plant breeding: www.cogem.net.

Schauer N, Semel Y, Balbo I, Steinfath M, Repsilber D, Selbig J, Pleban T, Zamir D, Fernie AR (2008) Mode of inheritance of primary metabolic traits in tomato. Plant Cell 20: 509–523.

Schornack S, Ballvora A, Gürlebeck D, Peart J, Ganal M, Baker B, Bonas U, Lahaye T (2004) The tomato resistance protein Bs4 is a predicted non-nuclear TIR-NB-LRR protein that mediates defense responses to severely truncated derivatives of AvrBs4 and overexpressed AvrBs3. Plant J 37: 46–60.

Scott JW, Agrama HA, Jones JP (2004) RFLP-based analysis of recombination among resistance genes to fusarium wilt races 1, 2 and 3 in tomato. J Am Soc Hort Sci 129: 394–400.

Scott JW, Stevens MR, Olson SM (2005) An alternative source of resistance to Tomato spotted wilt virus. Tomato Genet Coop Rep 55: 40–41.

Seah S, Yaghoobi J, Rossi M, Gleason CA, Williamson VM (2004) The nematode-resistance gene, *Mi-1*, is associated with an inverted chromosomal segment in susceptible compared to resistant tomato. Theor Appl Genet 108: 1635–1642.

Servin B, Martin OC, Mezard M, Hospital F (2004) Toward a theory of marker-assisted gene pyramiding. Genetics 168: 513–523.

Sharp PJ, Johnston S, Brown G, McIntosh RA, Pallotta M, Carter M, Bariana HS, Khatkar S, Lagudah ES, Singh RP, Khairallah M, Potter R, Jones MGK (2001) Validation of molecular markers for wheat breeding. Aust J Agri Res 52: 1357–1366.

Shen L, Courtois B, McNally KL, Robin S, Li Z (2001) Evaluation of near-isogenic lines of rice introgressed with QTLs for root depth through marker-aided selection. Theor Appl Genet 103: 75–83.

Simons G, Groenenkijk J, Wijbrandi J, Reijans M, Groenen J, Diergaarde P, Van der Lee T, Bleeker M, Onstenk J, deBoth M, Haring M, Mes J, Cornelissen B, Zabeau M, Vos P (1998) Dissection of the fusarium *I2* gene cluster in tomato reveals six homologs and one active gene copy. Plant Cell 10: 1055–1068.

Smith CJS, Watson CF, Ray J, Bird CR, Morris PC, Schuch W, Grierson D (2002) Antisense RNA inhibition of polygalacturonase gene expression in transgenic tomato. Nature 334: 724–726.

Smykal P, Horacek J, Dostalova R, Hybl M (2008) Variety discrimination in pea (*Pisum sativum* L.) by molecular, biochemical and morphological markers. Theor Appl Genet 49: 155–166.

Sobir T, Ohmori M, Murata F, Motoyoshi F (2000) Molecular characterization of the SCAR markers tightly linked to the Tm-2 locus of the genus *Lycopersicon*. Theor Appl Genet 101: 64–69.

Soler-Aleixandre S, Cebolla Cornelio J, Nuez F (2004). Identificación de fuentes de resistencia en tomate al virus del mosaico del pepino dulce (PepMV). Actas de Horticoltura n° 41, II Congreso de Mejora Genetica de Plantas, Leon, Spain, pp 107–110.

Soumpourou E, Iakovidis M, Chartrain L, Lyall V, Thomas CM (2007) The *Solanum pimpinellifolium Cf-ECP1* and *Cf-ECP4* genes for resistance to *Cladosporium fulvum* are located at the Milky Way locus on the short arm of chromosome 1. Theor Appl Genet 115: 1127–1136.

Stam P (2003) Marker-assisted introgression: speed at any cost? In: van Hintum TJL, Lebeda A, Pink D, Schut JW (eds) Eucarpia Leafy Vegetable. CGN, Centre for Genetic Resources, Wageningen, Dordrecht, The Netherlands, pp 117–124.

Stamova BS, Chetelat RT (2000) Inheritance and genetic mapping of cucumber mosaic virus resistance introgressed from *Lycopersicon chilense* into tomato. Theor Appl Genet 101: 527–537.

Staniaszek M, Kozik EU, Marczewski W (2007) A CAPS marker TAO1902 diagnostic for the *I-2* gene conferring resistance to *Fusarium oxysporum* f. sp. *lycopersici* race 2 in tomato. Plant Breed 126: 331–333.

Steele KA, Price AH, Shashidhar HE, Witcombe JR (2006) Marker-assisted selection to introgress rice QTLs controlling root traits into an Indian upland rice variety. Theor Appl Genet 112: 208–221.

Stergiopoulos I, Groenewald M, Staats M, Lindhout P, Crous PW, de Wit PJ (2007) Mating-type genes and the genetic structure of a world-wide collection of the tomato pathogen *Cladosporium fulvum*. Fung Genet Biol 44: 415–429.

Stevens MR, Heiny DK, Griffiths PD, Scott JW, Rhoads DD (1996) Identification of co-dominant RAPD markers tightly linked to the tomato spotted wilt virus (TSWV) resistance gene *Sw-5*. Tomato Genet Coop Rep 46: 27–28.

Stevens MR, Scott SJ, Gergerich RC (1992) Inheritance of a gene for resistance to tomato spotted wilt virus (TSWV) from *Lycopersicon peruvianum* Mill. Euphytica 59: 9–17.

Stommel JR, Zhang YP (1998) Molecular markers linked to quantitative trait loci for anthracnose resistance in tomato. HortScience 33: 514.

Suliman-Pollatschek S, Kashkush K, Shats H, Hillel J, Lavi U (2002) Generation and mapping of AFLP, SSRs and SNPs in *Lycopersicon esculentum*. Cell Mol Biol Lett 7: 583–597.

Sun D, Li W, Zhang Z, Chen Q, Ning H, Qiu L, Sun G (2006) Quantitative trait loci analysis for the developmental behavior of Soybean (*Glycine max* L. Merr.). Theor Appl Genet 112: 665–673.

Tan MYA, Alles R, Hutten RCB, Visser RGF, van Eck HJ (2009) Pyramiding of *Meloidogyne hapla* resistance genes in potato does not result in an increase of resistance. Potato Res 52: 331–340.

Tanksley SD, Kuehn GD (1985) Genetics, subcellular localization, and molecular characterization of 6-phosphogluconate dehydrogenase isozymes in tomato. Biochem Genet 23: 441–454.

Tanksley SD, Costello W (1991) The size of the *L. pennellii* chromosome 7 segment containing the *I-3* gene in tomato breeding lines as measured by RFLP probing. Tomato Genet Coop Rep 41: 60–61.

Tanksley SD, Ganal MW, Prince JP, De Vicente MC, Bonierbale MW, Broun P, Fulton TM, Giovannoni JJ, Grandillo S, Martin GB, Messeguer R, Miller JC, Miller L, Paterson AH, Pineda O, Roder MS, Wing RA, Wu W, Young ND (1992) High density molecular linkage maps of the tomato and potato genomes. Genetics 132: 1141–1160.

Tanksley SD, Grandillo S, Fulton TM, Zamir D, Eshed Y, Petiard V, Lopez J, Beck-Bunn T (1996) Advanced backcross QTL analysis in a cross between an elite processing line of tomato and its wild relative *L. pimpinellifolium*. Theor Appl Genet 92: 213–224.

Tanksley SD, Nelson JC (1996) Advanced backcross QTL analysis: a method for the simultaneous discovery and transfer of valuable QTLs from unadapted germplasm into elite breeding lines. Theor Appl Genet 92: 191–203.

Tanksley SD, McCouch SR (1997) Seed banks and molecular maps: unlocking genetic potential from the wild. Science 277: 1063–1066.

Tester M, Langridge P (2010) Breeding technologies to increase crop production in a changing world. Science 327: 818–822.

Thabuis A, Lefebvre V, Bernard G, Daubeze AM, Phaly T, Pochard E, Palloix A (2004) Phenotypic and molecular evaluation of a recurrent selection program for a polygenic resistance to *Phytophthora capsici* in pepper. Theor Appl Genet 109: 342–351.

Thomas PE, Mink GI (1998) Tomato hybrids with non specific immunity to viral and mycoplasma pathogens of potato and tomato. HortScience 33: 764–765.

Thoquet P, Olivier J, Sperisen C, Rogowsky P, Laterrot H, Grimsley N (1996) Quantitative trait loci determining resistance to bacterial wilt in tomato cultivar Hawaii7996. Mol Plant-Microbe Interact 9: 826–836.

Toojinda T, Baird E, Booth A, Broers L, Hayes P, Powell W, Thomas W, Vivar H, Young G (1998) Introgression of quantitative trait loci (QTLs) determining stripe rust resistance in barley: an example of marker-assisted line development. Theor Appl Genet 96: 123–131.

Vakalounakis DJ, Laterrot H, Moretti A, Ligoxigakis K, Smardas K (1997) Linkage between *Frl* (*Fusarium oxysporum* f. sp. *radicis-lycopersici* resistance) and *Tm-2* (tobacco mosaic virus resistance-2) loci in tomato (*Lycopersicon esculentum*). Ann Appl Biol 130: 319–323.

Valanzuolo S, Catello S, Colombo M, Dani M, Monti MM, Uncini L, Petrone P, Spigno P (1993) Cucumber mosaic virus resistance in transgenic San Marzano tomatoes. Acta Hort 376.

van Berloo R (1999) GGT: software for the display of graphical genotypes. J Hered 90: 328–329.

van Berloo R, Stam P (2001) Simultaneous marker-assisted selection for multiple traits in autogamous crops. Theor Appl Genet 102: 1107–1112.

van den Berg BM (1991) A rapid and economical method for hybrid purity testing of tomato (*Lycopersicon esculentum* L.) F_1 hybrids using ultrathin-layer isoelectric focusing of alcohol dehydrogenase variants from seeds. Electrophoresis 12: 64–69.

van der Biezen EA, Glagotskaya T, Overduin B, Nijkamp HJ, Hille J (1995) Inheritance and genetic mapping of resistance to *Alternaria alternata* f. sp. *lycopersici* in *Lycopersicon pennellii*. Mol Gen Genet 247: 453–461.

Van Deynze AE, Stoffel K, Buell CR, Kozik A, Liu J, van der Knaap E, Francis D (2007) Diversity in conserved genes in tomato. BMC Genom 8: 465.

van Heusden AW, Koornneef M, Voorrips RE, Brüggemann W, Pet G, Vrielink-van Ginkel R, Chen X, Lindhout P (1999) Three QTLs from *Lycopersicon peruvianum* confer a high level of resistance to *Clavibacter michiganensis* ssp. *michiganensis*. Theor Appl Genet 99: 1068–1074.

Varshney A, Mohapatra T, Sharma RP (2004) Molecular mapping and marker assisted selection of traits for crop improvement. In: Srivastava PS, Narula A, Srivastava S (eds) Biotechnology and Molecular Markers. Amanya, New Delhi, India, pp 289–330.

Varshney RK, Graner A, Sorrells ME (2005) Genic microsatellite markers in plants: features and applications. Trends Biotechnol 23: 48–55.

Varshney RK, Thiel T, Stein N, Langridge P, Graner A (2002) In silico analysis on frequency and distribution of microsatellites in ESTs of some cereal species. Cell Mol Biol Lett 7: 537–546.

Varshney RK, Nayak SN, May GD, Jackson SA (2009) Next-generation sequencing technologies and their implications for crop genetics and breeding. Trends Biotechnol 27: 522–531.

Wang JF, Olivier J, Thoquet P, Mangin B, Sauviac L, Grimsley NH (2000) Resistance of tomato line Hawaii7996 to *Ralstonia solanacearum* Pss4 in Taiwan is controlled mainly by a major strain-specific locus. Mol Plant-Microbe Interact 13: 6–13.

Weise SE, Kim KS, Stewart RP, Sharkey TD (2005) β-maltose is the metabolically active anomer of maltose during transitory starch degradation. Plant Physiol 137: 756–761.

Wijnker E, de Jong H (2008) Managing meiotic recombination in plant breeding. Trends Plant Sci 13: 640–646.

Williamson VM, Lambert KN, Kaloshian I (1994). Molecular biology of nematode resistance in tomato. In: Lamberti F, de Giorgi C, Bird DM (eds) Advances in Molecular Plant Nematology. Plenum Press, New York, USA, pp 211–219.

Witcombe JR, Hollington PA, Howarth CJ, Reader S, Steele KA (2008) Breeding for abiotic stresses for sustainable agriculture. Phil Trans Roy Soc Lond B Biol Sci 363: 703–716.

Wong CK, Bernardo R (2008) Genomewide selection in oil palm: increasing selection gain per unit time and cost with small populations. Theor Appl Genet 116: 815–824.

Woods S, Thurman DA (1976) The use of seed acid phosphatases in the determination of the purity of F1 hybrid brussels sprout seed. Euphytica 25: 707–712.

Wu F, Mueller LA, Crouzillat D, Petiard V, Tanksley SD (2006) Combining bioinformatics and phylogenetics to identify large sets of single-copy orthologous genes (COSII) for comparative, evolutionary and systematic studies: A test case in the euasterid plant clade. Genetics 174: 1407–1420.

Wulff BBH, Heese A, Tomlinson-Buhot L, Jones DA, de la Peña M, Jones JDG (2009) The major specificity-determining amino acids of the tomato *Cf-9* disease resistance protein are at hypervariable solvent-exposed positions in the central leucine-rich repeats. Mol Plant-Microbe Interact 22: 1203–1213.

Yaghoobi J, Yates JL, Williamson VM (2005) Fine mapping of the nematode resistance gene *Mi-3* in *Solanum peruvianum* and construction of a *S. lycopersicum* DNA contig spanning the locus. Mol Genet Genom 274: 60–69.

Yang H, Renshaw D, Thomas G, Buirchell B, Sweetingham M (2008) A strategy to develop molecular markers applicable to a wide range of crosses for marker assisted selection in plant breeding: a case study on anthracnose disease resistance in lupin (*Lupinus angustifolius* L.). Mol Breed 21: 473–483.

Yang W, Francis DM (2005) Marker assisted selection for combining resistance to bacterial spot and bacterial speck in tomato. J Am Soc Hort Sci 130: 716–721.

Yang W, Sacks EJ, Lewis Ivey ML, Miller SA, Francis DM (2005) Resistance in *Lycopersicon esculentum* intraspecific crosses to race T1 strains of *Xanthomonas campestris* pv. *vesicatoria* causing bacterial spot of tomato. Phytopathology 95: 519–527.

Ye G, Eagles HA, Dieters MJ (2004) Parental selection using known genes for inbred line development. In: Black CK, Panozzo JF and Rebetzke GJ (eds) Cereals 2004. In: Proc

54th Australian Cereal Chemistry Conference and 11th Wheat Breeders Assembly, pp 245–248.

Ye G, Moody D, Emebiri L, van Ginkel M (2007) Designing an optimal marker-based pedigree selection strategy for parent building in barley in the presence of repulsion linkage, using computer simulation. Aust J Agri Res 58: 243–251.

Ye G, Smith KF (2008) Marker-assisted gene pyramiding for inbred line development: basic principles and practical guidelines. Int J Plant Breed 2: 1–10.

Young ND (1999) A cautiously optimistic vision for marker-assisted breeding. Mol Breed 5: 505–510.

Young ND, Tanksley SD (1989) RFLP analysis of the size of chromosomal segments retained around the *Tm-2* locus of tomato during backcross breeding. Theor Appl Genet 77: 353–359.

Yu ZH, Wang JF, Stall RE, Vallejos CE (1995) Genomic localization of tomato genes that control a hypersensitive reaction to *Xanthomonas campestris* pv. *vesicatoria* (Doidge) dye. Genetics 141: 675–682.

Zamir D (2001) Improving plant breeding with exotic genetic libraries. Nat Rev Genet 2: 983–989.

Zhang GF, Maudens KE, Storozhenko S, Mortier KA, Van Der Straeten D, Lambert WE (2003) Determination of total folate in plant material by chemical conversion into para-aminobenzoic acid followed by high performance liquid chromatography combined with on-line postcolumn derivatization and fluorescence detection. J Agri Food Chem 51: 7872–7878.

Zhang LP, Khan A, Nino-Liu D, Foolad MR (2002) A molecular linkage map of tomato displaying chromosomal locations of resistance gene analogs based on a *Lycopersicon esculentum* x *Lycopersicon hirsutum* cross. Genome 45: 133–146.

Zhao W, Canaran P, Jurkuta R, Fulton T, Glaubitz J, Buckler E, Doebley J, Gaut B, Goodman M, Holland J, Kresovich S, McMullen M, Stein L, Ware D (2005) Panzea: a database and resource for molecular and functional diversity in the maize genome. Nucl Acids Res 34 (Database issue): D752–D757.

Zhu H, Briceno G, Dovel R, Hayes PM, Liu BH, Liu CT, Ullrich SE (1999) Molecular breeding for grain yield in barley: an evaluation of QTL effects in a spring barley cross. Theor Appl Genet 98: 772–779.

Zhu HS, Wu T, Zhang ZX (2006) Inheritance analysis and identification of SSR markers linked to Late Blight resistant gene in tomato. Agri Sci Chin 5: 517–521.

Positional Cloning Strategies in Tomato

Cornelius S. Barry

ABSTRACT

Information gained through the study of genetic variants provides valuable insight into the function of biological systems. In particular, the availability of large collections of monogenic mutants in model organisms that disrupt aspects of the growth, development and physiology are valuable resources for geneticists and biologists, facilitating discoveries across the sub-disciplines of biological sciences. In order to take full advantage of these genetic resources, tools and approaches must be available that allow for the identification of the molecular basis underlying mutations. Positional cloning utilizes a combination of genetic populations derived from polymorphic parents along with molecular markers and genetic and physical maps to identify genetic loci conferring mutant phenotypes. This approach has been particularly successful in plant biology when coupled with the ability, especially in *Arabidopsis thaliana*, to perform large scale genetic screens to identify mutants of interest. As genetic and genomic tools and resources have become more widely available for other plant species, including important model crop plant species, positional cloning has become an increasingly feasible approach for the functional analysis of genes underlying mutant phenotypes.

Keywords: mapping populations, molecular markers, genetic maps, physical maps, genome sequence, candidate gene analysis

Department of Horticulture, Michigan State University, East Lansing, Michigan 48824, USA.

8.1 Introduction

Mutant variants in traits of interest are fundamental to understanding biological phenomena and in many species collections of mutants have been a cornerstone that has driven scientific discovery. In plants, the process of identifying a gene of interest from a mutant phenotype has traditionally been cumbersome, taking several years to identify and confirm candidate genes. The widespread adoption of *Arabidopsis* as a model organism, the associated development of tools for molecular genetic analysis, and ultimately a genome sequence revolutionized the process of gene cloning and facilitated the use of genetic screens to isolate loci of interest. Indeed at the turn of the millennium, prior to the completion of the *Arabidopsis* genome, it was suggested that following the development of a mapping population it was possible to progress from mutant phenotype to candidate gene in as little as six to eight weeks (Somerville 2000). Although positional cloning strategies have become more widespread for crop plants, there is still a considerable time investment required to conduct these experiments and the timeframe proposed for *Arabidopsis* is untenable in the majority of plant systems. However, it is the author's estimate that at the time of writing this review, approximately 50 loci have been isolated using positional cloning or a combination of genetic mapping and candidate gene analysis in tomato. The emerging genome sequence of tomato, together with the associated tool development including bacterial artificial chromosome (BAC) libraries, genetic markers, physical contig development, and BAC end sequencing have reduced and will continue to reduce the effort required to isolate genes using forward genetics based approaches (http://solgenomics. net/organism/Solanum_lycopersicum/genome). Highlighting examples from the literature, this review describes the factors that must be taken in to account when designing positional cloning strategies in tomato, together with the options that are available to ensure that these strategies are successful. In addition, current limitations are discussed, together with prospects for future technology and tool development.

There are four important criteria that must be met in order to ensure successful outcome of a positional cloning experiment:

1) A robust mutant phenotype that can be readily scored in the background of an interspecific cross.
2) A high-density molecular map with a sufficient number of molecular markers in the region of interest that display polymorphisms between the parents used in the interspecific cross.
3) A moderate to high level recombination rate surrounding the locus of interest to facilitate the construction of a genetic map.

4) Tools for generating a physical map of the locus including large insert genomic clones.

A flow diagram outlining the suggested experimental steps for a successful positional cloning strategy in tomato is described in Fig. 8-1. These steps will be described in detail in the subsequent sections of this review.

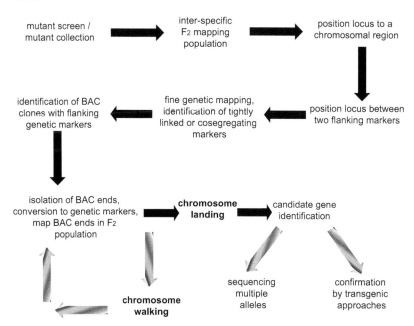

Figure 8-1 Principle steps involved in a positional cloning strategy. Solid arrows represent required steps and shaded arrows represent optional steps that are dependent upon experimental outcomes. For example chromosome walking is only necessary if chromosome landing is unsuccessful following an initial round of BAC library screening. Confirmation of candidate genes can be achieved through sequencing of multiple alleles if available or by transgenic approaches to complement or recreate mutant phenotypes.

8.2 Mutant Collections

The availability of a large mutant collection is a prerequisite for the genetic dissection of biological phenomena. In tomato, a number of mutants are available that arose spontaneously during cultivation and have subsequently been saved by breeders or growers and been made available to the scientific community. In addition, there is a long history of utilizing chemical and physical mutagens to induce mutations in different tomato varieties (Stubbe 1964). A number of these mutant stocks are curated and distributed by the CM Rick Tomato Genetics Resource Center (TGRC), hosted in the

Department of Plant Sciences at UC Davis (http://tgrc.ucdavis.edu/). The TGRC collection includes monogenic mutants, wild species germplasm, several cultivars, and a series of *S. pennellii* and *S. habrochaites* introgression lines. The current collection includes approximately 1,000 monogenic stocks at 622 loci. A significant number of these loci have been assigned to a chromosomal region through classical linkage or molecular mapping and many have been isolated by positional cloning or candidate gene approaches. The mutant collection held at TGRC represents a community effort to generate and preserve tomato germplasm that has spanned the last six decades. Recently, a large systematic mutant screen has been performed by the Zamir group at the Hebrew University of Jerusalem. Using the M82 cultivar, 6,000 EMS and 7,000 fast-neutron M_2 families were screened under field conditions for a set of 15 primary characters that span all stages of plant development, together with a set of 48 secondary phenotypic characters that can be used to refine mutant classification (Menda et al. 2004). This analysis led to the discovery of approximately 3,500 mutants that have been cataloged and made available through a searchable online database (http://zamir.sgn.cornell.edu/mutants/). This collection represented a significant research effort but has greatly expanded the numbers of tomato mutants available in the public domain including the identification of a number of confirmed additional alleles at loci previously identified in other mutant collections (Ori et al. 2007; Kimura et al. 2008).

The principle barrier to generating and screening large mutant populations of tomato is the logistics associated with prohibitive size of the standard cultivated tomato. One potential way to avoid this issue of plant size is to utilize the Micro-Tom variety developed at the University of Florida (Scott and Harbaugh 1989). The Micro-Tom variety stands just a few inches tall and can be grown in large numbers in a relatively small space. Micro-Tom has been promoted as a model system for genetic analysis in tomato and has been subsequently adopted by a significant number of research groups as their primary cultivar and many resources have now been developed for this variety including expressed sequence tag (EST) collections and efficient transformation protocols (Meissner et al. 1997; Yamamoto et al. 2005; Dan et al. 2006; Sun et al. 2006). Mutant collections have also been developed using chemical, physical and genetic mutagens including transposon tagged and promoter trap lines (Meissner et al. 1997, 2000). Micro-Tom mutant populations have been utilized for discovering genes involved in carotenoid biosynthesis, phosphate transport, wax biosynthesis and floral organ identity (Isaacson et al. 2002; Vogg et al. 2004; Nagy et al. 2005; de Martino et al. 2006).

In contrast to *Arabidopsis*, very few targeted genetic screens of mutagenized populations have been performed in tomato due to logistical problems associated with growing and screening large numbers of plants, particularly for traits that require plant growth beyond the seedling stage. The adoption of Micro-Tom may alleviate some of the space constraints in performing such screens. In addition, the M82 cultivar can be readily grown to maturity in a seed tray format and may also be suitable for specific genetic screens (Menda et al. 2004). Nevertheless despite the problems associated with plant size, a few targeted mutant screens have been performed including those for identifying loci involved in wound signaling and disease resistance in tomato (Salmeron et al. 1994; Howe and Ryan 1999; Hu et al. 2005).

8.3 Strategies for Generating Mapping Populations

In order to facilitate molecular mapping, an interspecific cross must be generated that segregates for the mutant phenotype of interest and polymorphic loci throughout the genome. The vast majority of monogenic mutants of tomato have arisen spontaneously or been induced by various mutagens in the cultivated *S. lycopersicum* species. Therefore, a wild tomato species must typically be used as the second mapping parent. In theory, any of the wild tomato species that are sexually compatible with *S. lycopersicum* can be utilized for generating mapping populations, but in practice the species that are typically chosen include *Solanum pennellii*, *Solanum pimpinellifolium* and *Solanum cheesmaniae*. Each of these species has advantages and disadvantages as a mapping parent and ultimately the choice may be dictated by the trait of interest and whether the mutant has already been assigned to a chromosomal location using classical linkage mapping. F_2 mapping populations are generally preferred for mapping monogenic traits but F_2 progeny of crosses between *S. lycopersicum* and *S. pennellii* have greatly reduced fertility and are therefore of limited use for mapping traits associated with fruit development or ripening. A backcross strategy can be utilized to overcome fertility issues between interspecific crosses through the generation of a backcross (BC) population. However, this population structure results in reduced mapping resolution as only one of the parental alleles can be resolved in the homozygous state. Alternatively, for traits associated with fruit development and ripening in a fertile F_2 population can be generated using either *S. cheesmaniae* or *S. pimpinellifolium* as a mapping parent together with *S. lycopersicum*. For example, several loci that inhibit fruit ripening have been mapped using F_2 populations generated from crosses between *S. lycopersicum* and *S. cheesmaniae* including *ripening inhibitor*, *non-ripening*, *Colorless non-ripening*, and *Green-ripe* (Giovannoni et al. 1995; Barry et al. 2005; Manning et al. 2006).

In choosing a wild species as a parent for a mapping population, the level of genetic diversity between species also needs to be considered, together with the availability of markers with previously defined polymorphisms. For example, the phylogenetic distance between *S. lycopersicum* and *S. pennellii* is greater than that between *S. lycopersicum* and *S. pimpinellifolium* and therefore the levels of polymorphism at a given locus is generally far higher between the former (Miller and Tanksley 1990). In practical terms, this higher level of diversity translates into more readily identifiable markers that display polymorphisms between the parents of the mapping population. Indeed, the high polymorphism rate between *S. lycopersicum* and *S. pennellii* alleles has been exploited for the generation of the most widely used and marker-rich maps of tomato including the EXPEN2000 F_2 map, which serves as the reference map for the tomato genome sequencing initiative (http://solgenomics.net/). In contrast, fewer polymorphic markers have been generated and mapped for *S. cheesmaniae* and *S. pimpinellifolium* alleles. A genetic map is available for a population generated between *S. lycopersicum* and *S. pimpinellifolium* and this has recently been populated with additional PCR-based markers (http://solgenomics.net/) (Doganlar et al. 2002). However, genome-wide polymorphic markers are not available for distinguishing *S. lycopersicum* and *S. cheesmaniae* alleles and positional cloning experiments that have utilized mapping populations generated between these parents have been empirically determined for targeted loci as required.

In cases where the map position of a mutant is already known, for example if the mutant has been placed on the classical genetic map by linkage mapping or on the molecular map at low resolution, it may be possible to use an *S. pennellii* introgression line (IL) as a mapping parent. The *S. pennellii* ILs represent a collection of 76 individual lines that have overlapping, marker defined segments of the *S. pennellii* genome introgressed within an *S. lycopersicum* background (Eshed and Zamir 1995; Liu et al. 2003). A fully fertile F_2 population segregating for the gene of interest and *S. pennellii* alleles spanning the locus can be generated. This population structure, therefore, takes advantage of the higher polymorphism rate of *S. pennellii* and the associated increased marker availability without the fertility problems. However, recombination rates within IL generated F_2 populations can be lower, hence more F_2 individuals may need to be screened in order to achieve a mapping resolution comparable to a typical cross between *S. lycopersicum* and a wild tomato species. However, several mutant loci have been mapped using this approach including those corresponding to the *green-flesh, high-pigment 1* and *fasciated* mutants (Liu et al. 2004; Barry et al. 2008; Cong et al. 2008).

8.4 Strategies for Extracting DNA for Genetic Mapping

As the approaches to molecular mapping have evolved there have been gradual changes in the methods that have been employed for DNA extraction from plant material. Molecular mapping utilizing restriction fragment length polymorphism (RFLP) markers requires relatively large quantities of DNA to be extracted, particularly if the sample is to be genotyped with several markers. This may well be the case in plants identified as informative recombinants within the mapping interval. Initially, in order to facilitate mapping with RFLP markers, large scale DNA isolations were performed. However, this scale limits the number of samples that can be handled by an individual researcher. A scaled down version of a DNA extraction protocol, that allows the use of micro-centrifuge tubes, provides an increase in efficiency in terms of the numbers of samples that can be isolated by an individual researcher (Fulton et al. 1995). The yield from these scaled down extractions is reduced but sufficient to enable two separate digests to be obtained from one sample. This scale of extraction is therefore suitable for screening large numbers of plants for recombinants between two flanking markers during experiments to fine map genetic loci. This method has been adapted for use with a bead beater machine for pulverizing the leaf tissue and, together with the use of repeater pipettes a practiced researcher can isolate three hundred samples per day (Barry et al. 2005).

The development of PCR-based genetic markers has greatly reduced the labor and cost associated with positional cloning. As PCR-based markers require less DNA per marker assay than RFLP markers, efficiency has been gained in the time and resources required to extract DNA. For example, a 96-well format DNA extraction protocol has been developed for tomato that yields sufficient DNA for screening several PCR-based markers. This has the benefit of increasing throughput to several hundred samples per day (Fridman et al. 2003). Therefore, two or three rounds of extractions using a 96-well plate based protocol would typically be sufficient for fine mapping a locus of interest. These in-house methods are still fairly labor intensive and time consuming and commercial kits are now available from several vendors that allow the isolation of genomic DNA from multiple samples in as little as thirty minutes. These kits together with liquid handling equipment can semi-automate the DNA extraction process and therefore provide increased efficiency in terms of throughput and labor savings. However, for the vast majority of academic laboratories adoption of these kits and technologies may not be a cost-effective proposition. An alternative to the immediate chemical based extraction protocol is offered by the FTA® system developed by Whatman®. This system allows the collection and storage of leaf material on a card format at room temperature. Lysis of the material occurs through chemicals on the card and the nucleic acids are released and bind to the card matrix. Tissue punches can subsequently be taken and the DNA eluted for use in PCR genotyping. This method may be particularly useful when laboratory facilities are not immediately available or when working with collaborators at distinct locations.

8.5 Molecular Markers

A molecular map saturated with genetic markers is a prerequisite for a successful positional cloning strategy and the greater the number of markers that are mapped the more chance is there of landing within or immediately adjacent to a locus of interest. The tomato-EXPEN 2000 reference map currently contains over 2,500 markers that are largely evenly distributed across the genome (http://solgenomics.net/). A large number of the tomato molecular markers are derived from gene coding regions and indeed on very rare occasions a molecular marker has been found to correspond to the target gene. This was the case for the *high pigment 2* (*hp-2*) locus, which corresponds to the CT151 marker (Mustilli et al. 1999).

RFLPs were the first molecular markers developed for genetic mapping and the tomato-EXPEN 2000 reference map currently contains almost 1,350 RFLP markers. Some of these markers were the originally mapped on the tomato-EXPEN 1992 map and the remainder, consist largely of a set of conserved ortholog set (COS) markers developed from the tomato EST collection (Tanksley et al. 1992; Fulton et al. 2002). RFLP markers were the mainstay of positional cloning for the first 15 years but are typically cumbersome and expensive to use as they require large quantities of DNA, restriction enzymes, agarose, hybridization membrane, and either radioactive or non-radioactive detection methods. To screen 2,000 F_2 individuals for recombinants between two flanking RFLP markers at a locus of interest could take an individual researcher several months to complete.

The sequences or partial sequences for a large number of the RFLP markers are now available through SGN and a significant number of RFLP markers have been converted into cleaved amplified polymorphic sequences (CAPS) markers (Frary et al. 2005). In addition, a new set of PCR-based markers termed the COSII markers have recently been developed (Wu et al. 2006). COSII markers are designed from single copy *Arabidopsis* genes and can be utilized across Solanaceae species with a common set of primers to facilitate comparative mapping approaches. Together, the COSII and other CAPS markers, derived from RFLP markers, comprise approximately 1,070 of the mapped markers on the tomato-EXPEN 2000 map. The development of PCR-based marker systems for tomato and other plants has greatly improved the efficiency of molecular mapping. However, the majority of markers only have defined polymorphisms for *S. lycopersicum* and *S. pennellii* parents. Therefore, researchers mapping using populations with different wild species as parents need to define polymorphisms experimentally as mapping progresses and additional markers are required.

The EST collection of tomato has also been computationally mined for the presence of regions that can be utilized for simple sequence repeat

(SSR) and single nucleotide polymorphism (SNP)-based markers, allowing the detection of marker polymorphisms through direct amplification of DNA without the need for subsequent digestion with restriction enzymes (Labate and Baldo 2005; Yamamoto et al. 2005; Van Deynze et al. 2007). These approaches are particularly powerful for identifying marker polymorphisms for mapping in intraspecific breeding populations but are also readily adaptable to polymorphism discovery between different species. For example, sequence analysis of 967 loci from *S. lycopersicum* breeding lines with a *S. pimpinellifolium* accession identified 275 loci with SNPs and 98 loci containing either insertions or deletions representing polymorphism discovery rates of 28.4 and 10% respectively (Van Deynze et al. 2007). These polymorphisms have the potential to be extremely useful in positional cloning experiments that use *S. lycopersicum* and *S. pimpinellifolium* as mapping parents. The tomato EST collections can also be mined for polymorphisms between tomato species. Currently within the SGN database 7,812 and 8,255 ESTs are available for *S. pennellii* and *S. habrochaites*, respectively. ESTs from these species are assembled into unigenes along with ESTs derived from *S. lycopersicum*. Comparison of orthologous ESTs from these divergent tomato species can reveal SNPs that can subsequently be utilized as genetic markers. The development of massively parallel sequencing technology also holds great promise for additional marker development in tomato for both breeding strategies and positional cloning. This technology has been utilized for gene expression profiling in *Arabidopsis* and could also potentially be utilized for sequencing ESTs from diverse tomato germplasm to search for polymorphic loci (Weber et al. 2007).

8.6 Assigning Loci to Chromosomal Regions

The first step toward positional cloning of a locus of interest is assignment to a chromosomal location. This is typically achieved using a small mapping population of approximately 100 F_2 individuals. The population can be screened for the mutant phenotype and then linkage of the phenotype to molecular markers can be determined. To accomplish this task, a set of markers evenly spaced throughout the 12 chromosomes at 10–15 cM intervals are utilized. Markers that are unlinked to the mutant phenotype will display a recombination frequency of 50% whereas markers that are linked to the mutant locus will display a recombination frequency of less than 50%, that will be lower the more tightly linked the genetic marker is to the locus. Following assignment of linkage to a chromosomal region, additional markers from the region can be utilized to place the locus between two flanking markers. A population of approximately 100 F_2 individuals usually provides satisfactory recombination rates to refine the map position to an interval of between 5 and 10 cM, providing a foundation for fine mapping of the locus.

The effort of assigning a locus to a chromosomal region using molecular markers can be reduced substantially if the mutant has already been assigned to the classical genetic map by linkage analysis. Approximately 290 mutant loci have been assigned to the 12 chromosomes of the classical genetic map of tomato and these assignments can potentially help significantly in molecular mapping experiments (Rick and Yoder 1988). However there are a few examples where following molecular mapping, a mutant locus was assigned to a different chromosome than had originally been determined by linkage mapping. For example the *jointless-2* (*j-2*) locus was placed on chromosome 11 by classical mapping but molecular mapping revealed that it was located on chromosome 12 (Zhang et al. 2000). Similarly, the *high pigment 1* (*hp-1*) locus was originally mapped to chromosome 12 by linkage mapping but was subsequently found to be localized on chromosome 2 by molecular mapping (Yen et al. 1997). Therefore, a degree of caution must be exercised when designing crosses for developing mapping populations. In particular, it is prudent not to utilize the *S. pennellii* IL lines as a mapping parent unless the chromosomal location of the locus has been clearly identified.

8.7 Fine Genetic Mapping

Following assignment of a locus to a chromosome and positioning between two flanking markers, it is usually necessary to increase the mapping resolution by increasing the population size to identify additional recombination events close to the locus. This is most effectively achieved by screening plants at the seedling stage for recombinants between the two closest flanking markers and subsequently screening these new recombinants for the corresponding mutant phenotype as they mature. To facilitate this screening process, it is preferable if the flanking genetic markers are PCR-based or are converted to this format for the purpose. As new recombinants are discovered and additional molecular markers from the interval are mapped, the map position of the locus is refined. Therefore, this approach becomes a process of elimination in which genetic markers and recombinant plants are eliminated as more tightly linked markers are identified. Typically the process of refinement continues until all additional markers have been exhausted or a given marker co-segregates with the mutant phenotype, i.e., no additional recombinant individuals are available. The population size required to fine map a locus at sufficient resolution to attempt a chromosome landing or walking experiment varies dramatically depending on the genome location and the ratio of the genetic to physical distance within the chromosomal region. An average value of 700 kb per cM has been described for the tomato genome but some mapping experiments have noted ratios as little as 43 kb per cM to as high as 25 Mb per cM for the *jointless-2* (*j-2*) locus that resides close to the centromere of chromosome 12 (Segal et al. 1992; Tanksley et al. 1992; Budiman et al. 2004). Typically, loci that reside within the euchromatic region of the tomato genome can be mapped to a resolution that facilitates gene cloning in populations ranging in size from 800–2,500 F_2 individuals. Although loci that lie within, or are adjacent to the heterochromatin or other areas of the genome where recombination is suppressed, typically require larger mapping populations to be screened in order to get sufficient resolution for gene identification. For example, a population of 5,000 F_2 plants were screened in order to map the *j-2* locus whereas 7,850 F_2 plants were screened to fine map the *hp-1* locus, which is located adjacent to the 45S ribosomal repeat region of chromosome 2 (Budiman et al. 2004; Liu et al. 2004). Therefore, the success of a positional cloning experiment and the amount of effort required to fine map the locus is determined somewhat by its physical location in the genome.

8.8 Resources for Generating Physical Maps

Following fine genetic mapping, it is necessary to identify and isolate a physical piece of DNA spanning the locus of interest so that ultimately potential candidate genes can be identified. If the closest flanking genetic markers do not co-segregate with the locus of interest in a fairly large mapping population, it may be advantageous to obtain an estimate of the ratio of genetic to physical distance across the mapping interval prior to attempting to initiate a chromosome walk. Earlier studies utilized pulse field gel electrophoresis of high molecular weight DNA fragments followed by hybridization to markers that flank the locus of interest to gain an estimate of this ratio (Ganal et al. 1989; Giovannoni et al. 1995). However, this approach has been largely replaced by fluorescence *in situ* hybridization (FISH) on pachytene chromosomes using low copy DNA sequences as probes and has in some instances been utilized to aid the construction of a physical map of a target locus (Peterson et al. 1999; Tor et al. 2002; Budiman et al. 2004).

The construction of physical maps spanning a locus was initially accomplished using yeast artificial chromosome (YAC) libraries. A three-fold genome equivalent YAC library was constructed from tomato and a PCR pooling strategy was subsequently developed to facilitate rapid screening (Martin et al. 1992; Pillen et al. 1996). This YAC library has been utilized successfully for several positional cloning experiments to isolate genes involved in disease resistance, plant development, nutrient uptake, wound responses and fruit ripening (Martin et al. 1993; Pnueli et al. 1998; Ling et al. 1999; Ling et al. 2002; Vrebalov et al. 2002; Li et al. 2003). Despite their utility for generating physical maps, YAC clones can be unstable and rearrangements are not uncommon, which can lead to problems in contig assembly. As such YAC libraries have now been superseded by BAC libraries, which are usually more stable and are easily propagated in *E. coli* vectors. BAC libraries are often constructed to contain between 5 and 10× genome equivalents and individual clones have insert sizes that typically range from 80–150 kb in size. Several BAC libraries have been made from both *S. lycopersicum* and the wild species *S. pennellii* and *S. cheesmaniae* and have been successfully used in many positional cloning experiments and are being utilized for the community effort to sequence the tomato genome (Budiman et al. 2000; Vrebalov et al. 2002; Li et al. 2005; Mueller et al. 2005; Barry and Giovannoni 2006; Manning et al. 2006; Chen et al. 2007; Cong et al. 2008).

BAC libraries can be screened with flanking genetic markers by direct colony hybridization on macro-array filters or by PCR, if a pooling strategy has been utilized to order the BAC library. Screening a BAC library is a matter of chance as a target sequence may not be represented in a given library or can be over-represented relative to the overall coverage of the library (Budiman et al. 2000). This phenomenon may be due in part to the choice of restriction enzyme and the size selection criteria used to make the library. For example, genomic regions that contain a low or high incidence of a particular restriction enzyme recognition sequence may be excluded during the cloning step on the basis of size. The problem of library coverage can be reduced if multiple libraries are available that were constructed using different restriction enzymes or through random shearing of genomic DNA. In *S. lycopersicum* three BAC libraries constructed with *Hind*III, *Mbo*I, and *Eco*RI are available together with a fosmid library constructed with randomly sheared DNA (http://solgenomics.net/organism/Solanum_lycopersicum/genome).

Following putative identification of a BAC clone that contains a flanking genetic marker, it is necessary to confirm the clone as a true positive. This can be accomplished by a combination of hybridization to digested BAC DNA, PCR, or direct DNA sequencing using primers designed to the original genetic marker. Furthermore, it is necessary to orient the BAC relative to the locus of interest and also to additional BAC clones, if several clones were identified during screening. Finally, it is also necessary to determine whether the identified BAC clone contains the gene of interest. To accomplish these aims it is necessary to isolate the BAC ends and convert them into probes and genetic markers. BAC ends can be isolated by direct sequencing and then primers designed to amplify each end by PCR. As part of the tomato genome sequencing project, just over 400,000 BAC end sequences from three *S. lycopersicum* BAC libraries are available (http://solgenomics.net/organism/Solanum_lycopersicum/genome). Therefore, if these libraries are utilized for screening there is a high probability that BAC end sequences are already available, eliminating the need to directly sequence the BAC end prior to primer design. Following mapping of BAC end derived genetic markers, if the mapping population contains sufficient resolution, informative recombinants may be available to establish that the gene of interest resides within the BAC clone. This scenario has been termed chromosome landing as no additional steps are required to construct a physical map of the region (Tanksley et al. 1995).

If the initial YAC or BAC clone(s) identified using the closest flanking markers do not contain the target locus, it is necessary to initiate a chromosome walk to identify overlapping clones. To achieve this goal the YAC or BAC end closest to the locus must be utilized to rescreen the library and repeat the process of characterization, clone-end isolation, genetic marker generation and mapping to move physically closer to the locus. Chromosome walking to a locus can add substantially to the effort involved to identify a gene of interest and in some cases has been required to accomplish gene isolation (Ling et al. 1999; Ling et al. 2002; Manning et al. 2006). However, as more markers and BAC libraries have become available for tomato the probability of chromosome landing has greatly increased. However, even with the availability of several BAC libraries it is sometimes possible that library coverage may still be lacking, rendering development of a physical contig spanning the locus of interest challenging. Under these circumstances cosmid libraries containing genomic inserts of approximately 40 kb have been utilized successfully to try and cover gaps in BAC contigs or to refine locus positions and, long range PCR has been used to amplify genomic DNA from regions where no clone coverage is available (Barry and Giovannoni 2006; Manning et al. 2006). Problems may also arise in mapping BAC ends or using them as probes for additional chromosome walking if they are comprised of repetitive sequences. This is quite a common occurrence even within euchromatic BACs, but can be particularly challenging if the locus of interest resides in the heterochromatin, as is the case with *j-2*. Although, these problems can be overcome by directly sequencing the region of interest (Budiman et al. 2004; Yang et al. 2005).

8.9 Identification and Confirmation of Candidate Genes

High resolution genetic mapping generally leads to the locus of interest being placed between two adjacent flanking markers that are typically separated by several kilo bases of intervening DNA, which has the potential to contain many genes. Rarely does fine mapping alone result in the identification of a single candidate gene. Therefore, the candidate gene typically needs to be identified from a group of several genes. Some of the first map based cloning experiments in tomato utilized YACs to construct a physical contig spanning the locus of interest and then YAC DNA or subclones were used to directly screen cDNA libraries to identify candidates that could then be confirmed by mapping or genetic complementation (Martin et al. 1993; Ling et al. 1999; Vrebalov et al. 2002). This strategy was not without risk and required that cDNA libraries be available or constructed and that the target gene be sufficiently expressed in the tissue type to be represented in the library. The development of BACs, automated DNA sequencing technology, and EST collections has eliminated the need to adopt this approach.

Once a BAC clone identifying the target locus has been determined, sequenced and annotated a single candidate gene must be identified from a group of several genes. In some cases, the choice of a candidate gene may be obvious from the mutant phenotype. For example, if a mutant phenotype exhibits chloroplast defects, a strong candidate may be a protein that is targeted to the plastid. Alternatively, if a mutant has a tissue-specific phenotype, experimental determination of expression patterns or in silico analysis of EST data may provide information suggestive of tissue-specific localization. However, it may be necessary to sequence the mutant and wild type alleles of several genes from the mapping interval in order to identify a candidate with a mutant variant, particularly if the mapping interval is around 100 kb. In these cases, if the mutant phenotype is readily identifiable at the seedling stage and the rate of recombination in the region is fairly high it may be more efficient to perform additional fine mapping to try and obtain additional informative recombinants from within the mapping interval to narrow down the physical interval. If this is deemed to be an ineffective strategy, sequence analysis of wild type and mutant alleles from the mapping interval must be performed. As the majority of mutant alleles reside within the coding regions of genes, it is often sufficient to determine the sequence of cDNA clones amplified from mutant and wild type RNA samples. If multiple mutant alleles are available at a given locus, determining the sequence of the additional alleles can provide enhanced confidence that the correct candidate gene has been identified and may alleviate the need to perform complementation analysis by transgenic approaches. However, some mutations at certain loci may also occur outside of the protein coding region and it may also be pertinent as part of the routine candidate gene screening process to include a survey of gene expression in wild type and mutant tissue samples to check for altered expression. For example, the *Green-ripe* (*Gr*) mutant that causes an inhibition cf fruit ripening, due to reduced ethylene responsiveness, is the result of a deletion immediately upstream of the protein coding region that leads to enhanced expression levels in tomato fruit (Barry and Giovannoni 2006). Similarly, recent characterization of the semi-dominant *Petroselinum* (*Pts*) locus, that controls leaf shape revealed the presence of a 1 bp deletion in the promoter of a novel *KNOX* gene homolog termed *PTS/TKD1* (Kimura et al. 2008). This deletion was sufficient to cause elevated levels of *PTS/TKD1* expression in the wild tomato species, *S. galapagense* resulting in altered leaf shape. These examples illustrate that it can be insufficient to look for sequence alterations solely within the coding region of candidate genes.

In rare cases caused by stably inherited epigenetic mutations, it is possible that sequence variation between wild type and mutant alleles is not present. For example, the *Colorless non-ripening* (*Cnr*) locus was mapped to a 13 kb interval on the long arm of chromosome 2 but sequence analysis failed to reveal any nucleotide differences between wild type and mutant alleles. Similarly, no changes in DNA sequence could be detected between wild type and *Cnr* in 95 kb of flanking DNA. However, a single gene encoding a tomato homolog of a *SQUAMOSA* promoter binding protein (SBP-box), designated *LeSPL-CNR*, from within the mapping interval displayed reduced expression in mutant fruit compared to wild type. Virus-induced gene silencing (VIGS) of the *LeSPL-CNR* gene in tomato fruit resulted in sectors that failed to fully ripen, therefore mimicking the *Cnr* mutant phenotype. In addition, analysis of the promoter region of *LeSPL-CNR* revealed higher levels of cytosine methylation within a 286 bp interval, which is consistent with altered levels of gene expression and can be stably inherited (Manning et al. 2006).

Following identification of a candidate gene, transgenic approaches are often required as final proof that the correct gene has been identified, particularly if only a single mutant allele is available. In the case of recessive mutations complementation of the mutant with the wild type allele in stable transgenic lines is a standard approach using either a constitutive promoter such as the *CaMV35S* or the native gene promoter and has been utilized extensively for confirming gene identity (Martin et al. 1993; Vrebalov et al. 2002; Barry et al. 2008). The choice of native promoter may be necessary when detrimental effects are observed with high activity, constitutive promoters. Alternatively if the mutation exhibits dominance, recreation of the mutant phenotype via expression of the mutant allele in a wild type background may be required (Barry and Giovannoni 2006; Kimura et al. 2008). A few studies have taken advantage of the presence of multiple alleles at a given locus to add a degree of confidence that the correct candidate gene has been identified eliminating the need for confirmation via transgenic approaches. For example, five alleles have been reported at the *Lanceolate* (*La*) locus, three at the *bipinatta* (*bip*) locus, and two separate alleles were sequenced at the *hp-1*, and *tangerine* (*t*) loci (Isaacson et al. 2002; Liu et al. 2004; Ori et al. 2007; Kimura et al. 2008). However, in the case of the *CRTISO* gene at the *t* locus, functional analysis was confirmed via heterologous expression in *E. coli* followed by carotenoid profiling (Isaacson et al. 2002). The sequence analysis of multiple alleles can save a considerable amount of time over the generation of transgenic plants to confirm the function of candidate genes as this latter approach can take between 12 and 18 months to accomplish, particularly if segregating T_1 progeny are analyzed.

Functional analysis of candidate genes using transient expression approaches may help to overcome some of the effort required to produce and analyze stably transformed tomato. For example, virus-induced gene silencing (VIGS) was used to silence the expression of *LeSPL-CNR* in wild type fruits to recreate the mutant phenotype (Manning et al. 2006). Rather than utilizing complementation assays to restore wild type function, the use of VIGS to recreate the mutant phenotype has the potential to serve as final confirmation of candidate gene identity and has been utilized in tomato for silencing in both vegetative and reproductive tissues (Liu et al. 2002; Fu et al. 2005). However, there is the potential for silencing approaches to yield different phenotypes to those of the original mutant allele, particularly if the latter alters or reduces protein activity rather than conferring a null mutant phenotype. Therefore, caution may be required in interpreting the results of such approaches. Similarly, it may be possible to utilize transient expression of the wild type allele to complement a mutant phenotype. Transient expression through *Agrobacterium*-mediated infiltration has worked successfully in tomato fruits (Orzaez et al. 2006). However, expression levels peak at three to four days post-infiltration and decline thereafter indicating that the timing of infiltration may be critical for the observation of the desired phenotype.

8.10 The Impact of an Emerging Genome Sequence

The euchromatic region of tomato is currently being sequenced by an international consortium using a combination of BAC by BAC and whole genome shotgun approaches (http://solgenomics.net/organism/Solanum_lycopersicum/genome). During the process of writing and editing of this article, a high quality draft genome sequence has become available. Clearly, the genome resource will greatly facilitate positional cloning strategies as the need to screen BAC libraries, and generate a physical map spanning a gene of interest should be alleviated. Similarly, the sequence resources will generate the capacity to produce an unlimited source of genetic markers. Together, these factors will greatly reduce the effort involved in isolating a gene of interest using a map based approach.

Even during the period when the tomato genome resources were being developed, early release of various data sets provided tools and information that had a positive impact on positional cloning strategies. For example, a number of BACs were anchored to genetic markers to create seed BACs for the genome sequencing effort. Greater than 400,000 BAC and 100,000 fosmid end sequences were made available through the Solanaceae Genome Network (SGN) website. Additionally, the *Hind*III and *Mbo*I BAC libraries were genetically fingerprinted and assembled into contigs with the aim of generating a physical map of the genome and creating a minimal tiling path for increased sequencing efficiency (http://solgenomics.net/organism/ Solanum_lycopersicum/genome). These resources increased the likelihood that a BAC clone containing a gene of interest could be readily identified using computational approaches without the need for library screening or chromosome walking.

The recent identification of the *green-flesh* (*gf*) locus represents a good example of how these resources were utilized (Barry et al. 2008). The *gf* locus was mapped to a chromosomal interval on the long arm of chromosome 8 between the flanking markers CT148 and CT265. A BAC clone, HBa0165B06 that contains the CT148 marker was fully sequenced as part of the chromosome 8 sequencing project. Furthermore, fingerprinted contig (FPC) analysis had placed this BAC clone within a contig containing three additional BAC clones. The BAC end sequences of these additional clones were analyzed with the aim of identifying more tightly-linked genetic markers residing between CT265 and CT148. This analysis revealed that the end sequence of one of these over-lapping BACs, HBa0020k14, corresponded to a tomato unigene that had homology to the *STAY GREEN* (*SGR*) gene of rice (Park et al. 2007). The *SGR* gene is required for chlorophyll degradation and mutations at this locus in several species fail to degrade chlorophyll (Armstead et al. 2007; Ren et al. 2007; Sato et al. 2007). Sequence analysis of wild type and mutant alleles of this gene from tomato revealed the presence of a single nucleotide polymorphism that resulted in the substitution of an invariant arginine residue to a serine. Subsequent confirmation that the correct candidate had been identified was achieved through complementation analysis of transgenic plants.

The tomato genome sequence was directly utilized, without the use of genetic mapping, to identify a candidate gene for the *bip* mutant that increases leaf shape complexity (Kimura et al. 2008). The *SAWTOOTH 1* and *SAWTOOTH 2* genes control leaf shape in *Arabidopsis* and double homozygous mutant lines at these loci display enhanced leaf serration (Kumar et al. 2007). A homolog of these genes was identified within a sequenced BAC clone from tomato chromosome 2 and this BAC clone was located in a region of the chromosome close to the map position of the *bip* mutant on the classical genetic map. Sequence analysis of the *SAW* gene homolog from *bip* and two additional alleles, *bip2* and *bip3* confirmed the identity of the mutant locus. These examples, illustrate how the emerging genome sequence resources for tomato can be used to facilitate gene identification with limited or no genetic mapping and without the need for chromosome walking.

8.11 Alternative Strategies for Identification of Mutant Loci

The current number of phenotyped mutant loci available for tomato is in the region of approximately 5,000, whereas the total number of predicted genes has been estimated to be in the region of 35,000 (Van der Hoeven et al. 2002). If the function of all of the genes in tomato is to be determined, additional phenotypic screens must be performed or alternative methods of mutant identification must be utilized. The majority of the available mutant loci thus far discovered in tomato, have been identified in general phenotypic screens that typically affect whole plant or organ morphology. Biochemical screens or those examining responses to compounds such as hormones, or screens for abiotic or biotic stresses have been underutilized. Similarly, forward genetic screens are limited in their ability to reveal altered phenotypes within gene families that act redundantly. Alternative platforms exist for large-scale creation of mutants within genes of interest and some of these have been applied to tomato. For example, insertional mutagenesis using various T-DNAs has been highly successful for creating knockout mutants in *Arabidopsis*. A transposon tagging system has also been developed for insertional mutagenesis of Micro-Tom but has yet to be converted to a high-throughput format and it has been calculated that it could take as many as 480,000 insertion lines to generate a saturated knockout mutant collection (Meissner et al. 2000; Emmanuel and Levy 2002). This would represent a huge undertaking for the tomato community even with the adoption of high-throughput transformation protocols for Micro-Tom (Dan et al. 2006; Sun et al. 2006).

Several EMS mutagenized populations of tomato have recently been generated, including populations within the Micro-Tom variety to be utilized together with targeted induced local lesions in genomes (TILLING) (McCallum et al. 2000). Whereas traditional mutant screening of populations requires the identification of a mutant phenotype, TILLING can be utilized to identify mutant alleles within genes without a priori knowledge of a mutant phenotype. Following identification of mutations within a gene of interest, screening can be performed to search for phenotypes attributable to the mutated gene. TILLING also offers the increased potential for identifying an allelic series of mutations in a target gene, including null mutations and amino acid substitutions that may be particularly useful for performing structure-function analysis. TILLING holds great promise for the tomato community as a high-throughput functional genomics tool and TILLING projects have been initiated in the US, France, Italy, Japan, and India (http://tilling.ucdavis.edu; http://www.eu-sol.net/science). Eventually these projects may lead to service facilities that can be utilized by researchers for identifying mutations in genes of interest.

8.12 Future Perspectives

The sequence of the tomato genome will overcome many of the problems associated with positional cloning strategies that have been outlined in this review, particularly in respect to the generation of physical contigs spanning loci of interest, the identification of candidate genes and the availability of genetic markers. Therefore, the process of progressing from mutant phenotype to candidate gene will become more efficient. This may drive more genetic screening of mutated populations for interesting phenotypes. While genetic screening of tomato for phenotypes of interest is unlikely to be performed to the same depth as the multitude of screens that have been performed in *Arabidopsis*, tomato remains an important model system for genetic studies of fruit development and ripening, leaf shape, plant architecture, plant-insect interactions, plant-microbe interactions, wound signaling, and specialized metabolism. The improved tools to facilitate positional cloning strategies may stimulate researchers in some of these areas to perform genetic screens for mutant phenotypes, particularly those that can be identified in small plants. In addition, suppressor and enhancer screens to identify modifiers and regulators of loci of interest may also become more feasible in tomato. While forward genetic strategies will continue to play a role as part of the functional genomics tool box, this will become balanced by targeted reverse genetics approaches such as gene silencing and TILLING that will potentially directly lead to the identification of multiple mutant alleles in genes of interest. During the last 20 years, positional cloning in crop plants has become a viable proposition and extensive resources have been made to facilitate this process. The next 20 years will likely see similar dramatic changes in how plant biologists investigate gene function. One key area which looks set to be revolutionized is the development of new sequencing technologies (Bentley 2006). Advances in this area may ultimately lead to the ability to simply re-sequence whole genomes of individuals from within a mutant collection or natural population to identify mutant variants or natural variation in genes of interest.

Acknowledgements

Research in the author's laboratory is supported through start up funds from Michigan State University and Michigan AgBioResearch.

References

Armstead I, Donnison S, Aubry J, Harper J, Hortensteiner S, James C, Mani J, Moffet M, Ougham H, Roberts L, Thomas A, Weeden N, Thomas H, King I (2007) Cross-species identification of Mendel's I locus. Science 315: 73–73.

Barry CS, Giovannoni JJ (2006) Ripening in the tomato Green-ripe mutant is inhibited by ectopic expression of a protein that disrupts ethylene signaling. Proc Natl Acad Sci USA 103: 7923–7928.

Barry CS, McQuinn RP, Chung M-Y, Besuden A, Giovannoni JJ (2008) Amino acid substitutions in homologs of the STAY-GREEN protein are responsible for the green-flesh and chlorophyll retainer mutations of tomato and pepper. Plant Physiol 147: 179–187.

Barry CS, McQuinn RP, Thompson AJ, Seymour GB, Grierson D, Giovannoni JJ (2005) Ethylene insensitivity conferred by the Green-ripe and Never-ripe 2 ripening mutants of tomato. Plant Physiol 138: 267–275.

Bentley DR (2006) Whole-genome re-sequencing. Curr Opin Genet Dev 16: 545–552.

Budiman MA, Chang SB, Lee S, Yang TJ, Zhang HB, de Jong H, Wing RA (2004) Localization of jointless-2 gene in the centromeric region of tomato chromosome 12 based on high resolution genetic and physical mapping. Theor Appl Genet 108: 190–196.

Budiman MA, Mao L, Wood TC, Wing RA (2000) A deep-coverage tomato BAC library and prospects toward development of an STC framework for genome sequencing. Genome Res 10: 129–136.

Chen KY, Cong B, Wing R, Vrebalov J, Tanksley SD (2007) Changes in regulation of a transcription factor lead to autogamy in cultivated tomatoes. Science 318: 643–645.

Cong BL, Barrero S, Tanksley SD (2008) Regulatory change in YABBY-like transcription factor led to evolution of extreme fruit size during tomato domestication. Nat Genet 40: 800–804.

Dan YH, Yan H, Munyikwa T, Dong J, Zhang YL, Armstrong CL (2006) MicroTom-a high-throughput model transformation system for functional genomics. Plant Cell Rep 25: 432–441.

de Martino G, Pan I, Emmanuel E, Levy A, Irish VF (2006) Functional analyses of two tomato APETALA3 genes demonstrate diversification in their roles in regulating floral development. Plant Cell 18: 1833–1845.

Doganlar S, Frary A, Ku H-M, Tanksley SD (2002) Mapping quantitative trait loci in inbred backcross lines of *Lycopersicon pimpinellifolium* (LA1589). Genome 45: 1189–1202.

Emmanuel E, Levy AA (2002) Tomato mutants as tools for functional genomics. Curr Opin Plant Biol 5: 112–117.

Eshed Y, Zamir D (1995) An introgression line population of *Lycopersicon pennellii* in the cultivated tomato enables the identification and fine mapping of yield-associated QTL. Genetics 141: 1147–1162.

Frary A, Xu YM, Liu JP, Mitchell S, Tedeschi E, Tanksley S (2005) Development of a set of PCR-based anchor markers encompassing the tomato genome and evaluation of their usefulness for genetics and breeding experiments. Theor Appl Genet 111: 291–312.

Fridman E, Pleban T, Zamir D (2003) A method for DNA extraction from leaves in a 96-well format. Rep Tomato Genet Coop 53: 19–21.

Fu DQ, Zhu BZ, Zhu HL, Jiang WB, Luo YB (2005) Virus-induced gene silencing in tomato fruit. Plant J 43: 299–308.

Fulton TM, Chunwongse J, Tanksley SD (1995) Microprep protocol for extraction of DNA from tomato and other herbaceous plants. Plant Mol Biol Rep 13: 207–209.

Fulton TM, Van-der-Hoeven R, Eannetta NT, Tanksley SD (2002) Identification analysis and utilization of conserved ortholog set markers for comparative genomics in higher plants. Plant Cell 14: 1457–1467.

Ganal MW, Young ND, Tanksley SD (1989) Pulsed field gel-electrophoresisand physical mapping of large DNA fragments in the TM-2A region of chromosome-9 in tomato. Mol Gen Genet 215: 395–400.

Giovannoni JJ, Noensie EN, Ruezinsky DM, Lu X, Tracy SL, Ganal MW, Martin GB, Pillen K, Alpert K, Tanksley SD (1995) Molecular genetic analysis of the ripening-inhibitor and non-ripening loci of tomato: A first step in genetic map-based cloning of fruit ripening genes. Mol Gen Genet 248: 195–206.

Howe GA, Ryan CA (1999) Suppressors of systemin signaling identify genes in the tomato wound response pathway. Genetics 153: 1411–1421.

Hu GS, deHart AKA, Li YS, Ustach C, Handley V, Navarre R, Hwang CF, Aegerter BJ, Williamson VM, Baker B (2005) EDS1 in tomato is required for resistance mediated by TIR-class R genes and the receptor-like R gene *Ve*. Plant J 42: 376–391.

Isaacson T, Ronen G, Zamir D, Hirschberg J (2002) Cloning of tangerine from tomato reveals a carotenoid isomerase essential for the production of beta-carotene and xanthophylls in plants. Plant Cell 14: 333–342.

Kimura S, Koenig D, Kang J, Yoong FY, Sinha N (2008) Natural variation in leaf morphology results from mutation of a novel KNOX gene. Curr Biol 18: 672–677.

Kumar R, Kushalappa K, Godt D, Pidkowich MS, Pastorelli S, Hepworth SR, Haughn GW (2007) The Arabidopsis BEL1-LIKE HOMEODOMAIN proteins SAW1 and SAW2 act redundantly to regulate KNOX expression spatially in leaf margins. Plant Cell 19: 2719–2735.

Labate JA, Baldo AM (2005) Tomato SNP discovery by EST mining and resequencing. Mol Breed 16: 343–349.

Li C, Schilmiller AL, Liu G, Lee GI, Jayanty S, Sageman C, Vrebalov J, Giovannoni JJ, Yagi K, Kobayashi Y, Howe GA (2005) Role of {beta}-oxidation in jasmonate biosynthesis and systemic wound signaling in tomato. Plant Cell 17: 971–986.

Li CY, Liu GH, Xu CC, Lee GI, Bauer P, Ling HQ, Ganal MW, Howe GA (2003) The tomato Suppressor of prosystemin-mediated responses2 gene encodes a fatty acid desaturase required for the biosynthesis of jasmonic acid and the production of a systemic wound signal for defense gene expression. Plant Cell 15: 1646–1661.

Ling HQ, Bauer P, Bereczky Z, Keller B, Ganal M (2002) The tomato fer gene encoding a bHLH protein controls iron-uptake responses in roots. Proc Natl Acad Sci USA 99: 13938–13943.

Ling HQ, Koch G, Baumlein H, Ganal MW (1999) Map-based cloning of chloronerva a gene involved in iron uptake of higher plants encoding nicotianamine synthase. Proc Natl Acad Sci USA 96: 7098–7103.

Liu Y, Schiff M, Dinesh-Kumar SP (2002) Virus-induced gene silencing in tomato. Plant J 31: 777–786.

Liu Y, Gur SA, Ronen G, Causse M, Damidaux R, Buret M, Hirschberg J, Zamir D (2003) There is more to tomato fruit colour than candidate carotenoid genes. Plant Biotechnol J 1: 195–207.

Liu YS, Roof S, Ye ZB, Barry C, van Tuinen A, Vrebalov J, Bowler C, Giovannoni J (2004) Manipulation of light signal transduction as a means of modifying fruit nutritional quality in tomato. Proc Natl Acad Sci USA 101: 9897–9902.

Manning K, Tor M, Poole M, Hong Y, Thompson AJ, King GJ, Giovannoni JJ, Seymour GB (2006) A naturally occurring epigenetic mutation in a gene encoding an SBP-box transcription factor inhibits tomato fruit ripening. Nat Genet 38: 948–952.

Martin GB, Brommonschenkel SH, Chunwongse J, Frary A, Ganal MW, Spivey R, Wu T, Earle ED, Tanksley SD (1993) Map-based cloning of a protein kinase gene conferring disease resistance in tomato. Science 262: 1432–1436.

Martin GB, Ganal MW, Tanksley SD (1992) Construction of a yeast artificial chromosome library of tomato and identification of cloned segments linked to two disease resistance loci. Mol Gen Genet 233: 25–32.

McCallum CM, Comai L, Greene EA, Henikoff S (2000) Targeting induced local lesions in genomes (TILLING) for plant functional genomics. Plant Physiol 123: 439–442.

Meissner R, Chague V, Zhu QH, Emmanuel E, Elkind Y, Levy AA (2000) A high throughput system for transposon tagging and promoter trapping in tomato. Plant J 22: 265–274.

Meissner R, Jacobson Y, Melamed S, Levyatuv S, Shalev G, Ashri A, Elkind Y, Levy A (1997) A new model system for tomato genetics. Plant J 12: 1465–1472.

Menda N, Semel Y, Peled D, Eshed Y, Zamir D (2004) In silico screening of a saturated mutation library of tomato. Plant J 38: 861–872.

Miller JC, Tanksley SD (1990) RFLP analysis of phylogenetic relationships and genetic variation in the genus Lycopersicon. Theor Appl Genet 80: 437–448.

Mueller LA, Tanksley SD, Giovannoni JJ, van Eck J, Stack S, Choi D, Kim BD, Chen MS, Cheng ZK, Li CY, Ling HQ, Xue YB, Seymour G, Bishop G, Bryan G, Sharma R, Khurana J, Tyagi A, Chattopadhyay D, Singh NK, Stiekema W, Lindhout P, Jesse T, Lankhorst RK, Bouzayen M, Shibata D, Tabata S, Granell A, Botella MA, Giullano G, Frusciante L, Causse M, Zamir D (2005) The Tomato Sequencing Project, the first cornerstone of the International Solanaceae Project (SOL). Compar Funct Genom 6: 153–158.

Mustilli A-C, Fenzi F, Ciliento R, Alfano F, Bowler C (1999) Phenotype of the tomato high pigment-2 mutant is caused by a mutation in the tomato homolog of DEETIOLATED1. Plant Cell 11: 145–157.

Nagy R, Karandashov V, Chague W, Kalinkevich K, Tamasloukht M, Xu GH, Jakobsen I, Levy AA, Amrhein N, Bucher M (2005) The characterization of novel mycorrhiza-specific phosphate transporters from *Lycopersicon esculentum* and *Solanum tuberosum* uncovers functional redundancy in symbiotic phosphate transport in solanaceous species. Plant J 42: 236–250.

Ori N, Cohen AR, Etzioni A, Brand A, Yanai O, Shleizer S, Menda N, Amsellem Z, Efroni I, Pekker I, Alvarez JP, Blum E, Zamir D, Eshed Y (2007) Regulation of LANCEOLATE by miR319 is required for compound-leaf development in tomato. Nat Genet 39: 787–791.

Orzaez D, Mirabel S, Wieland WH, Granell A (2006) Agroinjection of tomato fruits, A tool for rapid functional analysis of transgenes directly in fruit. Plant Physiol 140: 3–11.

Park SY, Yu JW, Park JS, Li J, Yoo SC, Lee NY, Lee SK, Jeong SW, Seo HS, Koh HJ, Jeon JS, Park YI, Paek NC (2007) The senescence-induced staygreen protein regulates chlorophyll degradation. Plant Cell 19: 1649–1664.

Peterson DG, Lapitan NLV, Stack SM (1999) Localization of single- and low-copy sequences on tomato synaptonemal complex spreads using fluorescence in situ hybridization (FISH). Genetics 152: 427–439.

Pillen K, Alpert KB, Giovannoni JJ, Ganal MW, Tanksley SD (1996) Rapid and reliable screening of a tomato YAC library exclusively based on PCR. Plant Mol Biol Rep 14: 58–67.

Pnueli L, Carmel-Goren L, Hareven D, Gutfinger T, Alvarez J, Ganal M, Zamir D, Lifschitz E (1998) The SELF-PRUNING gene of tomato regulates vegetative to reproductive switching of sympodial meristems and is the ortholog of CEN and TFL1. Development 125: 1979–1989.

Ren GD, An K, Liao Y, Zhou X, Cao YJ, Zhao HF, Ge XC, Kuai BK (2007) Identification of a novel chloroplast protein AtNYE1 regulating chlorophyll degradation during leaf senescence in *Arabidopsis*. Plant Physiol 144: 1429–1441.

Rick CM, Yoder JI (1988) Classical and molecular genetics of tomato—highlights and perspectives. Annu Rev Genet 22: 281–300.

Salmeron JM, Barker SJ, Carland FM, Mehta AY, Staskawicz BJ (1994) Tomato mutants altered in bacterial disease resistance provide evidence for a new locus controlling pathogen recognition. Plant Cell 6: 511–520.

Sato Y, Morita R, Nishimura M, Yamaguchi H, Kusaba M (2007) Mendel's green cotyledon gene encodes a positive regulator of the chlorophyll-degrading pathway. Proc Natl Acad Sci USA 104: 14169–14174.

Scott JW, Harbaugh BK (1989) Micro-Tom—a minature dwarf tomato. FL Agri Expt Sta Circ 370: 1–6.

Segal G, Sarfatti M, Schaffer MA, Ori N, Zamir D and Fluhr R (1992) Correlation of genetic and physical structure in the region surrounding the I2 Fusarium-oxysporum resistance locus in tomato. Mol Gen Genet 231: 179–185.

Somerville C (2000) The twentieth century trajectory of plant biology. Cell 100: 13–25.

Stubbe H (1964) Mutanten der Kulturtomate *Lycopersico esculentum* Miller V Genet Resour Crop Evol 12: 121–152.

Sun HJ, Uchii S, Watanabe S, Ezura H (2006) A highly efficient transformation protocol for Micro-Tom a model cultivar for tomato functional genomics. Plant Cell Physiol 47: 426–431.

Tanksley SD, Ganal MW, Martin GB (1995) Chromosome landing—a paradigm for map-based gene cloning in plants with large genomes. Trends Genet 11: 63–68.

Tanksley SD, Ganal MW, Prince JP, De-Vicente MC, Bonierbale MW, Broun P, Fulton T-M, Giovannoni JJ, Grandillo S (1992) High density molecular linkage maps of the tomato and potato genomes. Genetics 132: 1141–1160.

Tor M, Manning K, King GJ, Thompson AJ, Jones GH, Seymour GB, Armstrong SJ (2002) Genetic analysis and FISH mapping of the Colourless non-ripening locus of tomato. Theor Appl Genet 104: 165–170.

Van der Hoeven R, Ronning C, Giovannoni J, Martin G, Tanksley S (2002) Deductions about the number organization and evolution of genes in the tomato genome based on analysis of a large expressed sequence tag collection and selective genomic sequencing. Plant Cell 14: 1441–1456.

Van Deynze A, Stoffel K, Buell CR, Kozik A, Liu J, van der Knaap E, Francis D (2007) Diversity in conserved genes in tomato. BMC Genom 8.

Vogg G, Fischer S, Leide J, Emmanuel E, Jetter R, Levy AA, Riederer M (2004) Tomato fruit cuticular waxes and their effects on transpiration barrier properties: functional characterization of a mutant deficient in a very-long-chain fatty acid beta-ketoacyl-CoA synthase. J Exp Bot 55: 1401–1410.

Vrebalov J, Ruezinsky D, Padmanabhan V, White R, Medrano D, Drake R, Schuch W, Giovannoni J (2002) A MADS-box gene necessary for fruit ripening at the tomato ripening-inhibitor (rin) locus. Science 296: 343–346.

Weber APM, Weber KL, Carr K, Wilkerson C, Ohlrogge JB (2007) Sampling the *Arabidopsis* transcriptome with massively parallel pyrosequencing. Plant Physiol 144: 32–42.

Wu FN, Mueller LA, Crouzillat D, Petiard V, Tanksley SD (2006) Combining bioinformatics and phylogenetics to identify large sets of single-copy orthologous genes (COSII) for comparative evolutionary and systematic studies: A test case in the euasterid plant clade. Genetics 174: 1407–1420.

Yamamoto N, Tsugane T, Watanabe M, Yano K, Maeda F, Kuwata C, Torki M, Ban Y, Nishimura S, Shibata D (2005) Expressed sequence tags from the laboratory-grown miniature tomato (*Lycopersicon esculentum*) cultivar Micro-Tom and mining for single nucleotide polymorphisms and insertions/deletions in tomato cultivars. Gene 356: 127–134.

Yang TJ, Lee S, Chang SB, Yu Y, Jong H, Wing RA (2005) In-depth sequence analysis of the tomato chromosome 12 centromeric region: identification of a large CAA block and characterization of pericentromere retrotranposons. Chromosoma 114: 103–117.

Yen HC, Shelton BA, Howard LR, Lee S, Vrebalov J, Giovannoni JJ (1997) The tomato high-pigment (hp) locus maps to chromosome 2 and influences plastome copy number and fruit quality. Theor Appl Genet 95: 1069–1079.

Zhang HB, Budiman MA, Wing RA (2000) Genetic mapping of jointless-2 to tomato chromosome 12 using RFLP and RAPD markers. Theor Appl Genet 100: 1183–1189.

9

Structural Genomics: Studying Variation in Gene Sequences

Allen E. Van Deynze,* Theresa Hill and *Hamid Ashrafi*

ABSTRACT

Structural genomics, the study of how DNA content contributes to the expression of genes and phenotypes (traits) in an organism can be leveraged effectively for crop improvement. Although it is understood that the coding portion of genes (exons) is important to its function, it is becoming clear that non-coding regulatory regions such as promoters, 3′ UTRs and even introns and intergenic regions play an important role in gene regulation. With the availability of expressed sequence tags, scientists now focus on studying diversity related to genes rather than random genomic DNA. In developing DNA markers, objectives must be clearly defined to determine the appropriate genomic targets, (coding, gene space or intergenic DNA) or gene sets, to focus on. Equally important, is that the populations used to develop markers must be relevant to those that they are being applied in. The rapid advancement in technologies is now allowing researchers to study the whole genomes of populations in similar time that it took to study single genes only a couple decades ago.

Keywords: Gene structure, tomato, next generation sequencing, DNA markers

9.1 Introduction

Structural genomics can be broadly defined as understanding how DNA content contributes to the expression of genes and phenotypes (traits) in an organism. Genomic research initially focused on coding regions of genes

Seed Biotechnology Center, University of California, 1 Shields Ave, Davis, CA, 95616 USA.
*Corresponding author

such as expressed sequence tags (ESTs), but as we learn more about the complexity of DNA, the blueprint for life, it is becoming clear that even regions of genes that are not transcribed into mRNA such as promoters, 3′ untranslated regions (UTR), introns, as well as intergenic sequences may play a significant role in an organism's phenotype. In this chapter we will focus on gene related sequences in tomato (*Solanum lycopersicum*) and demonstrate how to tap into these features to measure and track variation in breeding and genetic programs.

9.2 Sources of Sequence

9.2.1 Expressed Sequence Tags

Until recently, ESTs were the primary source of gene sequence immediately available in tomato to develop DNA markers, but that is rapidly changing with progress made by the International Tomato Sequencing Project (See Chapter 10) and next generation sequencing. ESTs represent the transcribed portion of genes including 5′ and 3′ UTRs and exons. Because they represent expressed genes, researchers have historically focused on developing DNA markers from these sequences. Markers developed from such sequences can be closely linked or even be the causative mutation defining a change in a phenotype or trait. This is in contrast to anonymous markers derived from genomic sequences that may or may not be associated with genes (see below and Chapter 10). Furthermore, ESTs can be annotated by comparing homology at the DNA or protein level within and across organisms where gene functions have been empirically defined. Because ESTs are gene coding sequence, they tend to be more conserved than other sources of genomic DNA allowing for some markers to be used across species and thus enabling studies in synteny within (paralogous regions) and among (orthologous regions) species (Doganlar et al. 2002; Fulton et al. 2002; Ku et al. 2001; 1992; Wu et al. 2006). Accordingly, they tend to have less polymorphism than non-coding DNA.

9.2.1.1 Sources of ESTs

There are numerous sources of domesticated and wild tomato ESTs contributing to a total of 330,396 in *S. lycopersicum* (Dana Faber Cancer Institute, 2010a), 8,346 in *S. pennellii,* and 8,000 in *S. habrochaites* (National Institute of Health, 2012). In this chapter, we will consider only those from *S. lycopersicum*, the domesticated tomato. Of these, at least 57,222 are full-length cDNA clones of the Micro-Tom miniature tomato (Yamamoto et al. 2005; Yano 2006).

There are several online databases for tomato ESTs including: the Solanaceae Genomics Network (www.sgn.cornell.edu), an integrated genomics database for ESTs, molecular maps and genomic sequence; the Tomato Stress EST Database (abrc.sinica.edu.tw/tsed), ESTs from stress-induced subtractive libraries; PlantGDB (www.plantgdb.org), a summary of NCBI dbEST; MiBAse (www.kazusa.or.jp/jsol/microtom), a database of ESTs from Micro-Tom miniature tomatoes (Yamamoto et al. 2005); and Dana Farber Cancer Institute (DFCI) (compbio.dfci.harvard.edu/tgi), which now hosts the TIGR database gene indices (Dana Faber Cancer Institute, 2010a). In addition, there are at least two secondary databases to analyze the relative expression and function of ESTs and contigs, namely, TomatEST (D'Agostino et al. 2007) and SolGene (http://210.218.199.240/SOL/index.php). Of the above databases, DFCI is the most up-to-date (Release 13.0, April 13, 2010). It provides several tables to summarize the data including source tissues. From 299,800 ESTs, of which 28,167 are tentative consensus sequences (TCS) from 153 libraries, 52,502 unique sequences were derived. An analysis of the source tissues of these ESTs indicates that fruit-derived sequences make up the majority of ESTs followed by leaf sequences. Seed, shoot, root and flower are also represented and the miscellaneous category includes ESTs from all classes (Fig. 9-1).

Categorization of the ESTs by molecular, biological processes and cellular compartments indicates that the majority of ESTs are involved in metabolic and cellular processes found in the cell (Dana Faber Cancer Institute, 2010b). Five varieties make up the majority of the ESTs in the tomato database including TA496 (derived from E6203), Rio Grande, Micro-Tom and R11-13 and R11-12, derived from a Rio Grande x Moneymaker cross with the first three making up the vast majority of ESTs (Barone et al. 2008; Yamamoto et al. 2005; Yang et al. 2004)

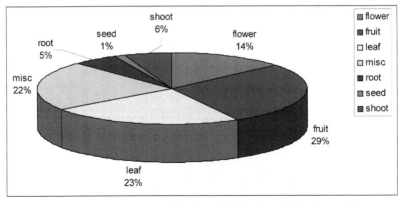

Figure 9-1 Analysis of source tissues of the 299,800 ESTs from DFCI release 13.0.

Color image of this figure appears in the color plate section at the end of the book.

9.2.2 *Gene Space*

In tomato, research is rapidly developing to take advantage of genomics beyond ESTs (See Chapter 11) to enhance our knowledge about gene regulation and interactions. As a first step to sequencing the tomato genome, an accurate estimate of gene space, the low copy fraction of the genome associated with genes, was necessary. Several approaches resulted in amazingly similar estimates. Based on Feulgen-stained pachytene chromosome analysis, Petersen (1996) estimated that the tomato genome was made up of 23% euchromatin, and gene space with pericentric heterochromatin.

This was independently substantiated by using methyl filtration of genomic DNA, where bacteria containing the enzyme McrBC digests hypermethylated DNA associated with high copy heterochromatic sequences (Rabinowicz et al. 2003). Subsequently, cloning and sample sequencing the remaining fraction resulted in an estimate of 27% of the genome being low copy (Fig. 9-2, Nathan Lakey, Orion Genomics, St Louis, USA). Using homology to sequenced bacterial artificial chromosomes (BACs) and ESTs, 35,000 genes were estimated to be in tomato, 31,000 in the euchromatin (Van der Hoeven et al. 2002; Wang et al. 2006). Furthermore 30% of the unigenes were not homologous to *Arabidopsis* proteins. It is not expected that ESTs will represent the complete gene complement of tomato. For example, a unigene set of 27,000 from 120,000 ESTs matched to only 47% of 76 genes sequenced in BACs (Van der Hoeven et al. 2002). Estimates in other crops have shown that ESTs may only cover 60% of the genes (Bedell et al. 2005) as a result of specificity of gene expression in tissues, time points and growing conditions sampled for EST libraries.

It is clear that gene regulation in promoter regions, alternative splicing microRNA encoding sequences and intergenic regions also play a major role in regulation of genes (Wang et al. 1999). For example, a mutation in the promoter of the fruit weight gene in tomato, *fw2.2*, is associated with

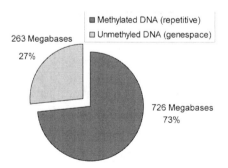

Figure 9-2 Proportion of tomato genome filtered using methylation sensitive cloning.

reduced expression during cell division in early fruit development (Cong et al. 2002; Tanksley, 2004). Although the gene space can be accessed using genome complexity reduction techniques such as methyl filtration or HiCot (Peterson et al. 2002), these techniques have not been used extensively in tomato research in lieu of genome sequencing of the euchromatic regions. The noncoding gene space will be important in marker-assisted breeding and functional genomics as a source of variation and basis of traits.

9.3 Targeted Marker Discovery

9.3.1 The Basis of DNA Markers

Not all DNA markers are equal as discussed above. The choice of type and source of DNA markers can depend largely on your goals, however all DNA markers are used to define variation in a plant genome (in this case). One can ask, are we interested in variation in a particular gene or non-genic sequence? Does it matter? Are we interested in markers associated with regulatory sequences, coding sequences, only mutations that are the cause of allelic differences in phenotype? The answers to the above questions determine how one might target or design marker discovery and applications.

Furthermore the populations that are being studied will also determine the set of markers one should use. The set of markers used must be relevant to the goals of the study. For example, Yamasaki et al. (2005), sequenced 1,095 genes in 14 maize inbred lines and defined 35 that showed no diversity. Eight were identified to have diversity in a broader set of landraces and teosinte, a progenitor of modern maize. The use of the 35 genes as markers in breeding germplasm would have little value in breeding programs, but using the subset of eight with diversity in landraces would be interesting to study trait associations and introgression from with wide crosses.

A similar case can be made for tomato where initial mapping of markers were based on cDNA RFLPs (restriction fragment length polymorphisms) mapped in interspecific crosses. This has led to the discovery of over 2,200 markers (Mueller et al. 2005; Tanksley et al. 1992) and introgression of simply inherited and complex loci for fruit and disease resistance genes (Frary et al. 2000; Fridman et al. 2000; Kabelka et al. 2004). A close look at allelic diversity of two of these traits, fruit shape and fruit size, indicated that a mutation in the second exon of the *ovate* gene causes a premature stop codon and tomato fruit to become elongated vs. round. This gene segregates in *S. lycopersicum* breeding germplasm (Liu et al. 2002). Conversely, the functional mutation in the fruit weight gene, *fw2.2*, was found in the promoter region and not the coding region (Cong et al. 2002). Although with significant effect, *fw2.2* is expressed only for a short time and at low levels and no transcripts were found in preliminary EST databases of 150,000 in tomato. Analogous to maize scenario, the mutation was only found in tomato wild relatives and showed no variation in tomato breeding lines.

The two above examples, underscore the need to understand and define clear goals for marker discovery and application. Just as the number of markers developed from interspecific crosses may be limiting in breeding programs (Van Deynze et al. 2007b), markers developed from a narrow set of breeding germplasm will have limited applicability for studies on domestication and phylogeny (Barone et al. 2008; Van Deynze et al. 2007b; Yamasaki et al. 2005).

9.3.2 Annotated Sequence

There is ample genomic or EST sequence to be used as template for marker development in tomato, although annotation of the genomic or gene space is currently limiting, i.e., gene models are just being developed to accurately predict promoter regions and the gene space per se in tomato from the developing genomic sequence (see Chapter 10 and Barone et al. 2008). Perhaps the first annotated sequence in tomato is a result of sequencing 272 mapped RFLP clones, of which 155 had a gene match (Ganal et al. 1998). Although these sequences represent cDNA clones (ESTs), it can be assumed that the majority of variation mapped is likely outside of the coding sequence of genes based on fragment sizes found in RFLPs, i.e., in the gene space. As a result, polymorphism rate from RFLPs (even from RFLPs derived from cDNA clones) overestimates that found in coding regions or ESTs per se.

The availability of EST sequences in tomato has dramatically changed plant geneticists' and plant breeders' philosophies on development and use of DNA markers. Annotated markers or markers associated with genes are now the priority. The initial EST sequence provides the template to develop markers, but developing sequence- (EST-sequence) based markers still remains challenging and expensive for genomic-scale studies and breeding. Release 1.0 of the TIGR gene index represented 4,361 unique sequences from 9,024 ESTs (Dana Faber Cancer Institute 2010a) in 1999.

Simple sequence repeat (SSR) markers can be applied directly without exact knowledge of sequence polymorphism per se. A single set of unique primers for amplification of specific SSR loci using the polymerase chain reaction (PCR, Mullis et al. 1986) can be applied across any tomato lines to assess genotypes. Therefore, one of the first exercises was to mine SSRs from the ESTs. Smulders et al. (1997) was the first to mine 2,000 EST sequences for SSRs and identify 220 of which 87% were polymorphic across tomato species and 28% within *S. lycopersicum*. Several other groups (He et al. 2003; Mueller et al. 2005) have used EST resources to mine and map SSR markers. Recently, Tang et al. (2008) reported an extensive analysis of the current EST database (249,794 ESTs). These were assembled into 28,728 unigenes of which 13% (7,163) harbored SSRs. The program developed, PolySSR, accurately identified 265 SSRs with polymorphism within the lines represented in the EST database. Unlike previous reports (Smulders et al. 1997), the authors did not report differences in polymorphism rate between long (>8–10 repeats) and short repeats. Tang et al. (2008) also characterized the position of SSRs within ESTs. The highest density (number of SSRs/ 100 kb) was found in the 5' UTR (266 SSRs/100kb) with 67–71 SSRs/100 kb in coding and 3'UTRs. Some SSRs also encompassed start and stop codons. Relatively few SSRs (44 SSRs out of 967 loci) were identified in introns of tomato genes (Van Deynze et al. 2007b). The above distribution allows for regulation of genes by variation in SSRs (Kashi and King 2006).

9.3.3 Categorizing ESTs for Marker Discovery

Annotation of ESTs allows one to select specific genes, pathways, chromosome and genetic map locations, and copy number of ESTs to be targeted for marker discovery regardless of marker type. Genomic resources in sequenced genomes can be leveraged to identify single-copy genes in related organisms. The genomic sequence of *Arabidopsis* was leveraged to identify single-copy genes in tomato and define a conserved ortholog set (COS) between species (Fulton et al. 2002). A COS is defined as a set of genes that are conserved throughout plant evolution in both sequence and copy number (Fulton et al. 2002; Lyons et al. 1997). Consequently this set can be used to bridge relationships across genes identified through ESTs in crops that can be applied to marker development for crop improvement and assist in comparative genomic analyses. A COS of 1,025 tomato unigenes (SGN COS I) was identified by comparing the set of unique gene sequences in tomato (*S. lycopersicum* only) to all translated proteins in the fully sequenced genome of *Arabidopsis* (Fulton et al. 2002). Approximately 10% of the 27,000 tomato unigene sequences met the defined COS criteria. This set was further refined to include pepper, potato, eggplant and coffee for a combined COS (SGN COS II) of 2,869 unigenes (Wu et al. 2006) across five species. Southern hybridization revealed that 85% of the COS tested are single-copy sequences in tomato and 13 other plant species from the Solanaceae, Asteraceae, Leguminosae, Cucurbitaceae, Rosaceae, Brassicaceae, Poaceae, and Malvaceae families (Fulton et al. 2002).

A third COS (UCD COS) was identified with 2,185 single-copy sequences common between *Arabidopsis* and tomato, lettuce, sunflower, soybean and maize, of which 1,704 are represented in tomato. The parameters for developing this dataset are described at Compositae Genome Project Database (Kozik and Michelmore 2003). Van Deynze et al. (2007b) compared the UCD COS and the SGN COS II datasets for *Solanum* species. The 1,704 UCD COS are homologous to 1,611 SGN Unigenes (Bombarely et al. 2011), and represent 857 tomato, 729 potato, 418 pepper and 442 coffee COS II sequences (see Table 9-1). The UCD COS represents 847 unigenes that are not in SGN COS II, which expanded the total COS by 33% from 2,587 (SGN COS II) to 3,434 unigenes (Van Deynze et al. 2007b).

As these resources represent orthologous single-copy genes, they can be used for comparative mapping purposes in Solanaceae. For example, a set of primers conserved across species has been used to amplify loci in tomato, pepper, potato and coffee. Fifteen loci were sequenced across these species for phylogenetic analysis (Wu et al. 2006). Primers designed from the UCD COS tomato sequences were also used to amplify loci in eggplant and pepper (Van Deynze et al. 2007b).

Table 9-1 Number of UCD COS in common with SGN COS II. Modified from Table 1 of Van Deynze et al. (2007b).

Species	Total[1]
Tomato	857
Tomato map	122
Potato	729
Pepper	418
Coffee	442

[1]Homology based on single hits of UCD COS at e-20 with matching over 80% of the length of the SGN COS II

9.3.4 Diversity of Tomato ESTs

9.3.4.1 Diversity in the Gene Space of Tomato

Initial estimates of sequence variation in tomato came from studies of specific genes (Nesbitt and Tanksley 2002; Ramakrishna et al. 2002; Stephan and Langley 1998) using RFLPs. Although RFLPs can be derived from the hybridization of cDNA clones, the majority access sequences outside the coding sequence of genes. Based on the size of RFLP fragments (0.5–20 kb), a large proportion of the RFLPs are the result of variation in restriction sites located outside the cDNA region homologous to the clone or probe being used (Botstein et al. 1980). Consequently, RFLPs derived from cDNA clones measure variation in the gene space. Table 9-2 summarizes reported polymorphisms in different regions of genes of tomato. Although the data cannot be directly compared due to the different lines, number of lines and loci being examined, the single nucleotide polymorphism (SNP) diversity is consistent with phylogenetic trees between *S. lycopersicum* and related species (Table 9-2; Marshall et al. 2001; Miller and Tanksley 1990;

Table 9-2 Summary of polymorphisms in tomato.

Species	Number of lines	Region	Loci	SNP density (bp/SNP)	Reference
S. cheesmaniae	8	Gene space	cDNA RFLP	538	(Stephan and Langley 1998)
S. chmielewskii	16	Gene space	cDNA RFLP	233	(Stephan and Langley 1998)
S. habrochaites	20	Gene space	cDNA RFLP	56	(Stephan and Langley 1998)
S. neorickii	4	Gene space	cDNA RFLP	3,500	(Stephan and Langley 1998)
S. pennellii	20	Gene space	cDNA RFLP	44	(Stephan and Langley 1998)
S. peruvianum	15	Gene space	cDNA RFLP	33	(Stephan and Langley 1998)
S. pimpinellifolium	20	Gene space	cDNA RFLP	82	(Stephan and Langley 1998)
S. lycopersicum/S. cheesmaniae	2	coding	FW2.2, orf44, Adh2	298	(Nesbitt and Tanksley 2002)
	2	noncoding	FW2.2, orf44, Adh2	152	(Nesbitt and Tanksley 2002)
S. lycopersicum/S. habrochaites	2	coding	FW2.2, orf44, Adh2	76	(Nesbitt and Tanksley 2002)
	2	noncoding	FW2.2, orf44, Adh2	23	(Nesbitt and Tanksley 2002)
S. lycopersicum/S. neorickii	2	coding	FW2.2, orf44, Adh2	142	(Nesbitt and Tanksley 2002)
	2	noncoding	FW2.2, orf44, Adh2	39	(Nesbitt and Tanksley 2002)
S. lycopersicum/S. pennellii	2	coding	FW2.2, orf44, Adh2	124	(Nesbitt and Tanksley 2002)
	2	noncoding	FW2.2, orf44, Adh2	24	(Nesbitt and Tanksley 2002)
	2	5' UTR, coding, 3' UTR	EST	111	(Yamamoto et al. 2005)
S. lycopersicum/S. peruvianum	2	coding	FW2.2, orf44, Adh2	248	(Nesbitt and Tanksley 2002)
	2	noncoding	FW2.2, orf44, Adh2	25	(Nesbitt and Tanksley 2002)
	5	coding and introns	sucr and Sod-2	74–263	(Stephan and Langley 1998)
S. lycopersicum/S. pimpinellifolium	3	noncoding	FW2.2, orf44, Adh2	125	(Nesbitt and Tanksley 2002)
	3	coding	FW2.2, orf44, Adh2	435	(Nesbitt and Tanksley 2002)
	12	introns	ESTs	851	(Van Deynze et al. 2007b)

Table 9-2 contd....

Table 9-2 contd....

Species	Number of lines	Region	Loci	SNP density (bp/SNP)	Reference
S. lycopersicum	30	exon and intron	ESTs	107	(Labate et al. 2006)
	30	introns	COS II ESTs	230	(Labate et al. 2006)
	30	noncoding	Arbitrary genes	179	(Labate et al. 2006)
	15	5' UTR, coding, 3' UTR	EST	9,452	(Labate and Baldo 2005)
	2	introns	*vis1*	22-49	(Ramakrishna et al. 2002)
	46	Gene space	cDNA RFLP	135	(Stephan and Langley 1998)
	5	5' UTR, coding, 3' UTR	EST	1,313	(Yamamoto et al. 2005)
	2	5' UTR, coding, 3' UTR	EST	8,500	(Yang et al. 2004)
S. lycopersicum breeding	10	introns	ESTs	1,647	(Van Deynze et al. 2007b)
S. lycopersicum Processing	8	introns	ESTs	2,372	(Van Deynze et al. 2007b)
S. lycopersicum US Fresh Market	2	introns	ESTs	5,675	(Van Deynze et al. 2007b)
S. lycopersicum/S. lycopersicum var. cerasiformae	1	introns	ESTs	1,627	(Van Deynze et al. 2007b)
Indels					
S. lycopersicum/S. pimpinellifolium	12	introns	ESTs indels	6,266	(Van Deynze et al. 2007b)
S. lycopersicum	15	5' UTR, coding, 3" UTR	EST indels	883	(Labate and Baldo 2005)
S. lycopersicum breeding	10	introns	ESTs indels	4,624	(Van Deynze et al. 2007b)
S. lycopersicum Processing	7	introns	ESTs indels	4,935	(Van Deynze et al. 2007b)
S. lycopersicum US Fresh Market	2	introns	ESTs indels	6,028	(Van Deynze et al. 2007b)
S. lycopersicum/S. lycopersicum var. cerasiformae	1	introns	ESTs indels	6,185	(Van Deynze et al. 2007b)

Stephan and Langley 1998; Wu et al. 2006). Differences in estimated genetic relationships are dependent on the diversity of loci as well. For example, the phylogenetic tree obtained from *Adh2* or *orf44* have limited resolution compared to that obtained from other loci tested (Nesbitt and Tanksley 2002). Conversely, phylogenies within tomato species are highly similar when obtained from the internal transcribed spacer region of nuclear ribosomal RNA and COS II intron or exon loci (compare Marshall et al. 2001; Wu et al. 2006). Furthermore, using a large number of polymorphic loci will increase resolution. For example, introns were shown to be 2.7–5.3 fold more polymorphic than exons (Van Deynze et al. 2007b; Wu et al. 2006). As a result, branch lengths of phylogenetic trees created with intronic data were longer than with exonic ones.

There is large difference in the level of divergence within a species with *S. neorickii* (*Lycopersicon parviflorum*) showing little divergence (1 SNP per 3,500 bp) across 36 genomic and genic loci compared to *S. peruvianum* (1 SNP per 33 loci), although more lines were sampled in the latter (Nesbitt and Tanksley 2002; Table 9-2).

9.3.4.2 Mining for Variation

At least three groups have electronically mined tomato EST databases for SNPs and insertion/deletions (indels) with limited results. In 2004, 148,373 ESTs in GenBank were examined for polymorphism electronically (Yang et al. 2004). The authors chose the two most prevalent lines in the database, Rio Grande and TA496 and filtered the data to include contigs that had at least three sequences from each genotype as criteria to call a SNP. From 1,245 contigs meeting these conditions, 101 SNPs were identified, 83 of which were confirmed empirically. The result is 1 SNP in 8,500 bp. Of the 83 validated SNPs, 53% proved useful within cultivated germplasm (Yang et al. 2004). A similar study using an algorithm that considered sequence context, but not sequence redundancy, identified 2,527 putative SNPs from 764 clusters from a similar dataset. In this case, 15 lines in the database were considered from which only 27% (28/103) of the putative SNPs tested could be validated experimentally. The validated frequency was therefore 1 SNP in 9,542 bp (Labate and Baldo 2005). In a third study, Micro-Tom tomato ESTs were compared to current EST databases including *S. pennellii*. From these, 2,039 putative SNPs and 121 putative indels were mined from a total of 186,405 ESTs reduced to 26,363 unigenes. A 69% validation rate suggests that up to 1,490 SNPs exist in this dataset. The average frequency of SNPs relative to Micro-Tom tomato was 1 SNP in 1,313 bp for *S. lycopersicum* lines and 1 SNP in 111 bp for *S. pennellii*. Micro-Tom is clearly divergent from the other lines sequenced in GenBank (Yamamoto et al. 2005). The paucity of SNPs within the ESTs can be attributed to low rate of polymorphism within gene coding regions of tomato; a limited sampling of genotypes in the databases, and shallow sampling within each of these genotypes. The current resources are thus insufficient for breeding studies.

9.3.4.3 ESTs as Templates

To assess and address the paucity of SNPs and indels discovered in tomato EST databases, several projects used known sequence to amplify and sequence loci in several varieties. Unlike RFLPs, PCR amplifies genomic sequence between two locus-specific oligos or primers. The same primer set can be used within species and also across species. The use of conserved primers across species or primers derived from conserved genes across species (such as COS, Fulton et al. 2002; Van Deynze et al. 2007b; Wu et al. 2006) allows one to study variation across species for comparative analysis. Wu et al. (2006) used COS II primer sets to demonstrate this for phylogenetic studies among Solanaceae relatives.

Within *S. lycopersicum* there is significant variation in the level of diversity, again depending on the number of lines, number of loci, breadth of germplasm used and EST region being sampled. A report showed that by randomly designing primers across 435 ESTs (including exons, introns and untranslated regions) and sequencing amplicons across eight inbred lines, polymorphism ranged from 3% to 24% (Ganal et al. 2006). The SNP frequency was not reported. Labate et al. (2006) sequenced 48 EST loci including exons and introns of 30 landraces from primary and secondary centers of diversity. SNP frequency ranged from 1 SNP in 107–230 bases (Table 9-2) indicating a large pool of alleles available in these sets of germplasm. Conversely, tenfold less variation (1 SNP in 1,647 bp) was observed when sequencing through introns of a 967 COS loci of a narrower panel (10 lines) representing US breeding germplasm. The germplasm represented processing and fresh market lines with pedigrees originating from the United States, Europe and Brazil (Van Deynze et al. 2007b).

Assuming 35,000 genes (Van der Hoeven et al. 2002) and 1,000 bp/gene, the current literature suggests that there are from 3,703–26,656 SNPs in ESTs in *S. lycopersicum* (Labate and Baldo 2005; Yamamoto et al. 2005; Yang et al. 2004). If introns are included, the estimate would be 21,250 in breeding germplasm to as high as 327,103 SNPs if landraces are included (Labate et al. 2006; Van Deynze et al. 2007b). Although, the estimates appear high, practically the number of SNPs for a given breeding program tends to be at the lower end of the range. An estimate of 3.0 to 7.9% of all loci differs in elite–by-elite crosses with the set of lines sampled by (Ganal et al. 2006; Van Deynze et al. 2007b).

9.3.5 The Next Generation of Discovery

Due to its relatively low rate of polymorphism, tomato is one of the crops that has the most to gain from new sequencing and genotyping technologies. A technology to rapidly discover and assay polymorphisms is short oligo arrays such as Nimblegen (Nimblegen Madison, USA) or Affymetrix (Affymetrix, St Clara, USA) arrays. Oligo arrays can detect polymorphisms based on differential hybridization of genotypes on individual oligos, known as single feature polymorphisms (SFPs). SFPs are not routinely used for breeding to labor and cost, but can be very efficient when interested in assaying thousands of genes simultaneously. Consequently they are an efficient tool for mapping and possibly identifying gene candidates for quantitative trait loci (QTLs) and simple traits (Dietrich et al. 2006). For example Sim et al. (2007) used a Nimblegen array with 15,270 unigenes represented to assay variation in tomato germplasm. A publicly available 10,000 unigene expression Affymetrix array exists. Affymetrix arrays with 23,000 unigenes in tomato have been used to physically map genes using genomic DNA from 76 tomato introgression lines (Eshed and Zamir 1995). Over 6,000 genes were located to unique bins and an additional 14,000 genes assigned based on BLAST hits. This information is being used to anchor the physical map for genome sequencing (see Chapter 10). The above examples used 11–15 oligos per gene, limiting the coverage to uncover potential SNPs. To address this limitation Caldwell et al. (2006) and Van Deynze et al. (2007a) designed 6.6 M oligo Affymetrix arrays in lettuce and pepper, respectively. Each of the 31,000 to 35,000 unigenes is represented by an average of 173–210 oligos, in a 2 bp overlapping tile. This afforded technical reproducibility as well as up to 20 fold coverage over traditional oligos arrays. In lettuce, over 15,000 unigenes have been mapped thus far (van Leeuwen et al. 2009). Hill et al. (2008) showed that the pepper array can be used to assay polymorphisms in tomato (Fig. 9-3). Current new technologies such as 454 (Roche biosciences, Palo Alto, USA) can deliver 800 Mb of sequence with 250–700 bp sequence reads in a single run. This allows rapid generation of *de novo* sequencing either from genomic, cDNA or gene-enriched DNA samples. The read length allows for contig assembly and annotation. The sequence, in turn, can be directly used as template or reference sequence to compare sequences from other tomato lines or varieties to define SNPs and indels.

Higher throughput technologies such as the Illumina Genome analyzer (Illumina Inc, San Diego, CA, USA) and SOLiD (Applied Biosystems, Foster City, CA, USA) technology use sequencing by synthesis to deliver 5 to 450 GB in a single run. The read length ranges from 50 to 150 bp. The short read length makes assembly challenging in complex plant genomes such as tomato. Usually a reference sequence such as an EST database or genomic

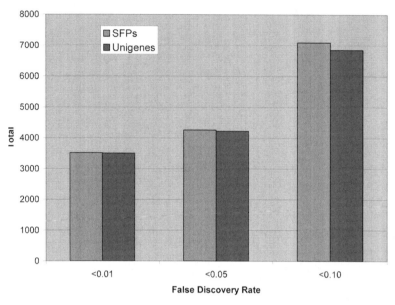

Figure 9-3 Single feature polymorphisms (SFPs) and unigenes detected in *S. pennellii* x *S. lycopersicum* using an Affymetrix pepper array with 31,000 unigenes.

sequence is required to differentiate paralogs and accurately call SNPs. The unprecedented depth of sequence compensates for the higher error rates found in these technologies relative to Sanger sequencing. At a cost of about $US 10,000–40,000/run ($40,000 for 1 run or 16 lanes on Hiseq 2000) including template preparations, these new technologies allow sequencing of several representative varieties for detection of sequence variation and subsequently designing high-throughput genotype assays for thousands of individuals including mapping populations. A new project entitled Solanaceae Coordinated Agricultural Project or SolCAP based at Michigan State University, Lansing, US is leveraging the Illumina Genome Analyzer and normalized cDNA libraries from six tomato lines to uncover SNPs and indels relevant to US breeding germplasm. From this, 7600 SNPs were then assayed in mapping populations and 480 lines to determine population structure, allele frequencies and linkage disequilibrium. These genotypes may in turn serve as a basis for linkage association studies (SolCAP) for key traits in tomato. These are a few examples of application of ultra-high throughput sequencing technologies. These technologies are advancing rapidly in throughput and decreasing costs. It is conceivable to be able to use sequencing across 100's of lines in a single breeding program as the main source of genotyping and marker discovery.

9.4 Conclusion

Although tomato was one of the first crops to incorporate genetic DNA marker technology into research, the limited diversity found in breeding populations and focus on wide crosses has limited its use in breeding populations. However, the rapid expansion of EST sequencing in tomato has provided excellent resources to develop markers associated with genes for breeding. The application of new sequencing and genotyping technologies coupled with the upcoming genomic sequence will inevitably make polymorphic DNA markers accessible and affordable for public and private breeders. The end result will be integration of DNA markers in tomato breeding programs; a better understanding of the genetic control of traits; more precise selection of traits, quality control and more rapid release of improved varieties.

References

Barone A, Chiusano ML, Ercolano MR, Giuliano G, Grandillo S, Frusciante L (2008) Structural and functional genomics of tomato. Int J Plant Genomics 2008: 820274.

Bedell JA, Budiman MA, Nunberg A, Citek RW, Robbins D, Jones J, Flick E, Rholfing T, Fries J, Bradford K, McMenamy J, Smith M, Holeman H, Roe BA, Wiley G, Korf IF, Rabinowicz PD, Lakey N, McCombie WR, Jeddeloh JA, Martienssen RA, (2005) Sorghum genome sequencing by methylation filtration. PLoS Biol 3(1): e13.

Bombarely A, Menda N, Tecle IY, Buels RM, Strickler S, Fischer-York T, Pujar A, Leto J, Gosselin J, Mueller LA (2011) The Sol Genomics Network (solgenomics.net): growing tomatoes using Perl. Nucl Acids Res 39: D1149–D1155.

Caldwell D, Kozik A, Chen F, Truco MJ, Michelmore R, Van Deynze AE (2006) Development Of A Lettuce (*Lactuca sativa*) 6.6 Million Feature Affymetrix Genechip For Massively Parallel Genotyping And Gene Expression Analysis. In: Plant and Animal Genome XVI, Scherago International, San Diego, CA, pp P183.

Cong B, Liu J, Tanksley SD (2002) Natural alleles at a tomato fruit size quantitative trait locus differ by heterochronic regulatory mutations. Proc Natl Acad Sci USA 99(21): 13606–11.

D'Agostino N, Aversano M, Frusciante L, Chiusano ML (2007) TomatEST database: in silico exploitation of EST data to explore expression patterns in tomato species. Nucleic Acids Res 35 (Database issue): D901–D905.

Dana Farber Cancer Institute (2010a) The Gene Index Project: DFCI Tomato Gene Index. http://compbio.dfci.harvard.edu/tgi/cgi-bin/tgi/gimain.pl?gudb=tomato (verified August 29, 2008). Accessed 2 January 2012.

Dana Farber Cancer Institute (2010b) The Gene Index Project: DFCI Tomato Gene Ontologies. http://compbio.dfci.harvard.edu/cgi-bin/tgi/GO_browser.pl?species=Tomato&gi_dir=lgi. Accessed 2 January 2012.

Dietrich B, Grivet L, Zhu T, Chilcott C, Dunn M, Xia Y, Zhou N, Long N, Puddephat I, Bontems S, Bonnet G (2006) Affymetrix microarrays as tools for generating maps and markers in tomato. In: Spooner D (ed) Solanaceae Genomics Conference, Madison, WI, p 466.

Doganlar S, Frary A, Daunay MC, Lester RN, Tanksley SD (2002) A comparative genetic linkage map of eggplant (*Solanum melongena*) and its implications for genome evolution in the solanaceae. Genetics 161(4): 1697–711.

Eshed Y, Zamir D (1995) An introgression line population of *Lycopersicon pennellii* in the cultivated tomato enables the identification and fine mapping of yield-associated QTL. Genetics 141: 1147–1162.

Frary A, Nesbitt TC, Frary A, Grandillo S, van der Knaap E, Cong B, Liu JP, Meller J, Elber R, Alpert KB, Tanksley SD (2000) fw2.2: A quantitative trait locus key to the evolution of tomato fruit size. Science 289(5476): 85–88.

Fridman E, Pleban T, Zamir D (2000) A recombination hotspot delimits a wild-species quantitative trait locus for tomato sugar content to 484 bp within an invertase gene. Proc Nat Acad Sci USA 97: 4718–4723.

Fulton TM, Van der Hoeven R, Eannetta NT, Tanksley SD (2002) Identification, analysis, and utilization of conserved ortholog set markers for comparative genomics in higher plants. Plant Cell 14(7): 1457–67.

Ganal MW, Czihal R, Hannappel U, Kloos DU, Polley A, Ling HQ (1998) Sequencing of cDNA clones from the genetic map of tomato (*Lycopersicon esculentum*). Genome Res 8(8): 842–7.

Ganal MW, Luerssen H, Polley A, Wolfe M (2006) High-throughput SNP identification and validation in tomato and pepper. In: Spooner D (ed) *Solanaceae Genomics Conference*, Madison, WI, p 159.

He C, Poysa V, Yu K (2003) Development and characterization of simple sequence repeat (SSR) markers and their use in determining relationships among *Lycopersicon esculentum* cultivars. Theoretical and Applied Genetics 106(2): 363–373.

Hill TA, Ashrafi H, Yao J, De Jong W, Francis D, Kozik A, Van Deynze A (2008) The application of a whole genome pepper array to Solanaceae crops. In: 5th Solanaceae Genome Workshop, Cologne, Germany, 12–16 October 2008.

Kabelka E, Yang WC, Francis DM (2004) Improved tomato fruit color within an inbred backcross line derived from *Lycopersicon esculentum* and *L. hirsutum* involves the interaction of loci. J Amer Soc Hort Sci 129(2): 250–257.

Kashi Y, King DG (2006) Simple sequence repeats as advantageous mutators in evolution, Trends in Genet 22(5): 253–259.

Kozik A, Michelmore R (2003) Compositae Genome Project Database: Conserved Orthologs in Plants. http://cgpdb.ucdavis.edu/COS_Arabidopsis/ Accessed 2 January 2012.

Ku HM, Liu J, Doganlar S, Tanksley SD (2001) Exploitation of *Arabidopsis*-tomato synteny to construct a high-resolution map of the ovate containing region in tomato chromosome 2, Genome 44(3): 470–5.

Labate JA, Baldo AM (2005) Tomato SNP discovery by EST mining and resequencing. Molecular Breeding 16(4): 343–349.

Labate JA, Robertson L, Sheffer SM, Wu F, Tanksley S, Baldo AM (2006) DNA polymorphism estimates within domesticated tomato. In: Potato Association of America-Solanaceae 2006: Genomics Meets Diversity, Madison, WI, USA, 23–27 July 2006.

Liu JP, Van Eck J, Cong B, Tanksley SD (2002) A new class of regulatory genes underlying the cause of pear-shaped tomato fruit. Proc Natl Acad Sci USA 99(20): 13302–13306.

Lyons LA, Laughlin TF, Copeland NG, Jenkins NA, Womack JE, OBrien SJ (1997) Comparative anchor tagged sequences (CATS) for integrative mapping of mammalian genomes. Nature Genetics 15(1): 47–56.

Marshall JA, Knapp S, Davey MR, Power JB, Cocking EC, Bennett MD, Cox AV (2001) Molecular systematics of *Solanum* section *Lycopersicum* (*Lycopersicon*) using the nuclear ITS rDNA region. Theor Appl Genet 103(8): 1216–1222.

Miller JC, Tanksley SD (1990) RFLP analysis of phylogenetic-relationships and genetic-variation in the genus *Lycopersicon*. Theor Appl Genet 80(4): 437–448.

Mueller LA, Solow TH, Taylor N, Skwarecki B, Buels R, Binns J, Lin C, Wright MH, Ahrens R, Wang Y, Herbst EV, Keyder ER, Menda N, Zamir D, Tanksley SD (2005) The SOL genomics network: a comparative resource for Solanaceae biology and beyond. Plant Physiol 138(3): 1310–7.

Mullis K, Faloona F, Scharf S, Saiki R, Horn G, Erlich H (1986) Specific enzymatic amplification of DNA *in vitro*: the polymerase chain reaction. Cold Spring Harb Symp Quant Biol 51 Pt 1: 263–73.

National Institutes of Health (2012) National Center for Biotechnology Information http://www.ncbi.nlm.nih.gov/ Accessed 2 January 2012.

Nesbitt TC, Tanksley SD (2002) Comparative sequencing in the genus *Lycopersicon*: Implications for the evolution of fruit size in the domestication of cultivated tomatoes. Genetics 162: 365–379.

Peterson DG, Stack SM, Price HJ, Johnston JS (1996) DNA content of heterochromatin and euchromatin in tomato (*Lycopersicon esculentum*) pachytene chromosomes. Genome 39(1): 77–82.

Peterson DG, Wessler SR, Paterson AH (2002) Efficient capture of unique sequences from eukaryotic genomes. Trends Genet 18(11): 547–50.

Rabinowicz PD, McCombie WR, Martienssen RA (2003) Gene enrichment in plant genomic shotgun libraries. Curr Opin Plant Biol 6(2): 150–6.

Ramakrishna W, Dubcovsky J, Park YJ, Busso C, Emberton J, SanMiguel P, Bennetzen JL (2002) Different types and rates of genome evolution detected by comparative sequence analysis of orthologous segments from four cereal genomes. Genetics 162(3): 1389–400.

Sim SC, Yang W, van der Knaap E, Hogenhout S, Xiao H, Francis D (2007) Microarray-based SNP discovery for tomato genetics and breeding. In: Proceedings of the XV International Plant and Animal Genome, San Diego, CA, USA, 13–17 January 2007. http://www.intl-pag.org/15/abstracts/.

Smulders MJM, Bredemeijer G, RusKortekaas W, Arens P, Vosman B (1997) Use of short microsatellites from database sequences to generate polymorphisms among L*ycopersicon esculentum* cultivars and accessions of other *Lycopersicon* species. Theoretical and Applied Genetics 94(2): 264–272.

Stephan W, Langley CH (1998) DNA polymorphism in lycopersicon and crossing-over per physical length. Genetics 150(4): 1585–93.

Tang J, Baldwin SJ, Jacobs JME, van der Linden CG, Voorrips RE, Leunissen JAM, van Eck H, Vosman B (2008) Large-scale identification of polymorphic microsatellites using an in silico approach. BMC Bioinformatics 9: doi: 10.1186/1471-2105-9-3744.

Tanksley SD (2004) The genetic, developmental, and molecular bases of fruit size and shape variation in tomato. Plant Cell 16 Suppl: S181–9.

Tanksley SD, Ganal MW, Prince JP, de Vicente MC, Bonierbale MW, Broun P, Fulton TM, Giovanonni JJ, Grandillo S, Martin GB, Messeguer R, Miller JC, Miller L, Paterson AH, Pineda O, Roder M, Wing RA, Wu W, Young ND (1992) High density molecular linkage maps of the tomato and potato genomes. Genetics 132: 1141–1160.

Van der Hoeven R, Ronning C, Giovannoni J, Martin G, Tanksley S (2002) Deductions about the number, organization, and evolution of genes in the tomato genome based on analysis of a large expressed sequence tag collection and selective genomic sequencing. Plant Cell 14(7): 1441–56.

Van Deynze A, Yao J, Choi D, Prince J, Kozik A (2007a) A whole genome microarray designed for marker discovery, genotyping and mapping in pepper. In: 4th Solanaceae Genome Workshop, Jeju, Korea, 9–13 September 2007.

Van Deynze AE, Stoffel K, Buell RC, Kozik A, Liu J, van der Knaap E, Francis D (2007b) Diversity in conserved genes in tomato, *BMC Genomics* 8(1): 465.

van Leeuwen H, Stoffel K, Kozik A, Cui X, Ashrafi H, McHale L, Lavelle D, Wong G, Chen H, Van Deynze A, Michelmore RW (2009) High-density mapping of the lettuce genome with SFP markers in over 15,000 unigenes. In: Proceedings of the XVII International Plant and Animal Genome, San Diego, CA, USA, 10–14 January 2009. http://www.intl-pag.org/17/abstracts/.

Wang RL, Stec A, Hey J, Lukens L, Doebley J (1999) The limits of selection during maize domestication. Nature 398(6724): 236–9.

Wang Y, Tang X, Cheng Z, Mueller L, Giovannoni J, Tanksley SD (2006) Euchromatin and pericentromeric heterochromatin: comparative composition in the tomato genome, Genetics 172(4): 2529–2540.

Wu FN, Mueller LA, Crouzillat D, Petiard V, Tanksley SD (2006) Combining bioinformatics and phylogenetics to identify large sets of single-copy orthologous genes (COSII) for comparative, evolutionary and systematic studies: A test case in the euasterid plant clade. Genetics 174(3): 1407–1420.

Yamamoto N, Tsugane T, Watanabe M, Yano K, Maeda F, Kuwata C, Torki M, Ban Y, Nishimura S, Shibata D (2005) Expressed sequence tags from the laboratory-grown miniature tomato (*Lycopersicon esculentum*) cultivar Micro-Tom and mining for single nucleotide polymorphisms and insertions/deletions in tomato cultivars. Gene 356: 127–34.

Yamasaki M, Tenaillon MI, Bi IV, Schroeder SG, Sanchez-Villeda H, Doebley JF, Gaut BS, McMullen MD (2005) A large-scale screen for artificial selection in maize identifies candidate agronomic loci for domestication and crop improvement. Plant Cell 17(11): 2859–72.

Yang W, Bai X, Kabelka E, Eaton C, Kamoun S, van der Knaap E, Francis D (2004) Discovery of single nucleotide polymorphisms in *Lycopersicon esculentum* by computer aided analysis of expressed sequence tags. Mol Breed14: 21–24.

Yano K (2006) KaFTom: Kasusa full-length tomato cDNA database. http://www.pgb.kazusa.or.jp/kaftom/Accessed 31 December 2011.

10

The Tomato Genome Sequencing Project

Lukas A. Mueller

ABSTRACT

The tomato genome is being sequenced by an international consortium that includes teams from Korea, China, UK, India, Netherlands, France, Japan, Spain, Italy and the US. The initial sequencing strategy was based on the sequencing of bacterial artificial chromosome clones ("BAC by BAC approach"), which was supplanted by a whole genome shotgun sequencing (WGS) approach in late 2009. Tomato is a widely used model plant, as well as a high value agricultural crop, and the best characterized plant in the Asterid clade. The availability of the tomato genome sequence will expedite research in areas such as fruit biology, plant defense, development, and enable new genome-based approaches will ultimately contribute to more efficient breeding strategies. The genome will shed light on the evolution of the Solanaceae and the wider plant phylogeny.

Keywords: bacterial artificial chromosome (BAC), genome annotation, tomato genome, Solanaceae, whole genome shotgun sequencing

10.1 Introduction

The Tomato Genome Sequencing Project has a quite long and interesting history, during the course of which tomato even changed its scientific name. The first grants to sequence the tomato genome were submitted to funding agencies in the early 2000s, but the project gained traction only after a meeting in 2003 in Washington, DC, where over a hundred tomato researchers from all over the world gathered to push for an international

Boyce Thompson Institute for Plant Research, Tower Road, Ithaca, NY 14853, USA.

effort. As a result, each of the 12 chromosomes was attributed to nine countries willing to participate in the project. Initially, these countries included Korea, China, UK, Netherlands, France, Japan, Spain and Italy, who sequenced one chromosome each, with the US responsible for the four remaining chromosomes. However, in 2004, India took on chromosome 5, and in 2007, another group in China took on chromosome 11, leaving the US and China with two chromosomes each, with the US responsible for chromosome 1 and 10. The chromosomes to be sequenced were selected by the participating countries on the basis of a number of criteria, such as the size or the presence of certain genes of interest. In the Netherlands, a project to sequence the top arm of chromosome 6 was already underway, making it the natural choice for its sequencing contribution. In early 2009, the US project was awarded the grant by the National Science Foundation (NSF) to sequence its two chromosomes.

Why sequence tomato? Tomato (*Solanum lycopersicum*, formerly called *Lycopersicum esculentum*), is the central model system of the Solanaceae (also called the nightshades) family and indeed of the entire Asterid clade. The Solanaceae are a medium-sized flowering plant family of more than 9,000 species that includes many economically important species (tomato, potato, pepper, eggplant, tobacco, petunia) (Knapp et al. 2004). Many Solanaceae species have been studied for various traits such as fruit development (Gray et al. 1992; Fray and Grierson 1993; Brummell and Harpster 2001; Alexander and Grierson 2002; Adams-Phillips et al. 2004; Tanksley 2004; Giovannoni 2004; Seymour et al. 2008), tuber development (Prat et al. 1990; Bachem et al. 1996; Fernie and Willmitzer 2001), biosynthesis of anthocyanin and carotenoid pigments (Gerats et al. 1985; Giuliano et al. 1993; Mueller et al. 2000; Spelt et al. 2002; De Jong et al. 2004; Quattrocchio et al. 2006) plant defense (Bogdanove and Martin 2000; van der Vossen et al. 2000; Gebhardt and Valkonen 2001; Kessler and Baldwin 2001; Li et al. 2001; Bai et al. 2003; Hui et al. 2003; Pedley and Martin 2003; Sacco et al. 2007) and secondary metabolites, some of which have medicinal properties (Schijlen et al. 2006; Oksman-Caldentey 2007). The tomato genome sequence is a natural basis for comparative genomics approaches within the Solanaceae, which exhibit remarkably conserved gene content and to a high degree gene order between the different species all the while exhibiting highly varied phenotypes (Tanksley et al. 1992; Knapp et al. 2004; Mueller et al. 2009). These attributes make the Solanaceae an excellent model for the study of plant adaptation to natural and agricultural environments (Knapp et al. 2004). Many species of the Solanaceae are diploid and share a basic set of 12 chromosomes (Olmstead et al. 1999). In line with these observations is the fact that recent polyploidizations during the evolutionary history of the family are limited to a few clades such as the potatoes and tobaccos (Clarkson et al. 2005; Mueller et al. 2009).

The tomato genome, in conjunction with other sequences that will soon be available from other Solanaceae and Asterids, will help answer a number of questions. First, how do genomes code for extensive phenotypic differences using relatively conserved sets of genes, and second, how can phenotypic diversity be harnessed for the improvement of agricultural products? Sequence data from related Solanaceae species, such as expressed sequence tags (ESTs; Adams et al. 1991), methylation (Palmer et al. 2003; Whitelaw et al. 2003; Fu et al. 2004) and Cot filtered sequence (Peterson et al. 2002; Yuan et al. 2003) together with sequencing by novel very high-throughput approaches such as 454 sequencing (Margulies et al. 2005) or Solexa sequencing (Shendure et al. 2005) in combination with good comparative maps (Tanksley et al. 1992; Doganlar et al. 2002; Fulton et al. 2002) between many Solanaceae plants (Hoeven et al. 2002; D'Agostino et al. 2007;) will enable insights into evolution, domestication, development, response and signal transduction pathways (Mueller et al. 2009).

A full genome sequence remains incomplete without the associated organellar genomes in plastids and mitochondria. The tomato chloroplast genome was recently completed by a European consortium (Kahlau et al. 2006) and the mitochondrial genome is being sequenced by the Instituto Nacional de Tecnología Agropecuaria in Argentina within the framework of the EU-SOL project (http://www.eu-sol.net).

To enable researchers to immediately make use of the emerging tomato sequence, a high quality annotation is essential. A genome annotation project has been initiated by the International Tomato Annotation Group (ITAG) that annotates bacterial artificial chromosome (BAC) contigs updates them at regular intervals. Annotations are available on the SGN website (http://solgenomics.net/), where a number of web-based tools have been deployed that allow researchers to download, query and visualize the annotations.

In this chapter, we review the progress, prospects and significance of the tomato genome sequencing project. Most of the observations are taken from a number of recent publications on the subject (Mueller et al. 2009; Peters et al. 2009) including a detailed account on the status of the sequencing of chromosome 6 that was recently published (Peters et al. 2009).

10.2 Sequencing Strategy

As a result of the rapid progress in sequencing technology in recent years the strategies chosen initially for a genome sequencing project should be re-evaluated from time to time and possibly revised, complemented or replaced wholesale by newer technology, as appropriate. Currently, the thinking in the tomato genome project is that the BAC-based approach still gives the most accurate reference sequence. However, certain modifications are necessary to the initial approach to exploit next generation technologies. For example,

BACs should be sequenced much more efficiently using multiplexed approaches, based on next generation sequencing, that considerably reduce the cost of initial sequencing a BAC to a certain coverage; although finishing costs remain high; overall costs should be somewhat lower. In addition, complementing the BAC-based sequencing with a whole genome shotgun approach will cover hard to fill gaps in the BAC contigs, as well as provide sequencing coverage in heterochromatic regions. The heterochromatic regions will be difficult to assemble, but genic regions that are suspected to exist in heterochromatin will also be covered by this approach. The final product will thus be a hybrid of sequencing approaches based on a very high quality BAC tiling path, in which recalcitrant gaps, which have occurred on a number of the more complete chromosomes, and which would be very expensive or impossible to fill with the BAC by BAC approach, will likely be filled with shotgun sequencing.

10.2.1 BAC by BAC Approach

Some plant genomes have a high degree of interspersion of heterchromatic sequences within the euchromatin (Gill et al. 2008), whereas other plants, including tomato, have genomes that are more distinctly structured into pericentromeric heterochromatin that is gene-poor, and distal euchromatin, that is gene-rich and comparatively low in repeat sequence. Since the vast majority of researchers are interested in the gene containing fraction of the genome, focusing initially on the euchromatic sequence reduced the complexity of the tomato project by almost a factor of four, to about 220 Mb (Peterson et al. 1996), while including an estimated minimum of 90% of all genes (Wang et al. 2006). However, a recent report on chromosome 6 estimates the fraction of genes located in the heterochromatin somewhat higher (Peters et al. 2009). A method that lends itself well for the targeted sequencing of euchromatic subregions is the BAC by BAC approach.

Initially, the tomato genome project set out solely based on a BAC by BAC approach to target sequencing at the euchormatic regions, as opposed to a whole genome shotgun strategy, which more recently has become popular even with higher complexity genomes, including plants such as poplar and grapevine (Tuskan et al. 2006; Jaillon et al. 2007). For tomato, it was felt that the BAC by BAC approach generated the highest quality "gold standard" sequence, which would be essential for use as a reference genome (International Rice Genome Sequencing Project 2005) and serving as a scaffold for other closely related genomes.

The BAC by BAC strategy is based on the anchoring of BACs (or contigs of BACs) to a reference genetic map, for example through the overgo process (Cai et al. 1998). The anchored BACs, or "seed" BACs, are then sequenced and the sequence information used to identify overlapping BACs or BAC contigs ("BAC walking"), which in turn are sequenced. Gaps between BAC contigs are closed by targeting novel markers or BACs to these gaps, which is then followed by successive rounds of BAC walking (Mueller et al. 2005, 2009; Peters et al. 2009).

For tomato, the high density F_2-2000 map is used as a reference genetic map (Fulton et al. 2002). It is based on 80 F_2 individuals from the cross *Solanum lycopersicum* LA925 x *S. pennellii* LA716 and contains a subset of restriction fragment length polymorphism (RFLP) markers from the Tomato-EXPEN 1992 map (Tanksley et al. 1992). Most of the markers are conserved ortholog set (COS) markers (Fulton et al. 2002; Wu et al. 2006) derived from a comparison of Solanaceae ESTs against the entire *Arabidopsis* genome. The COS markers were selected to be single/low copy, and having a highly significant match with a putative orthologous locus in *Arabidopsis*. Maps constructed using COS markers can readily be compared and analyzed for chromosome inversions, duplications, and other large-scale genome rearrangements. Using the comparative maps enables the transfer of knowledge from tomato to other species.

Several BAC libraries were constructed from the Heinz 1706 tomato line. The *Hind*III library consists of 129,024 clones and was available before the project started (Budiman et al. 2000). During the course of the project, two additional restriction-based BAC libraries were generated, an *Eco*RI library of 72,264 clones, and an *Mbo*I library of 52,992 clones. Together, these libraries provide more than 25x genome coverage, with BAC sizes averaging about 120 kb in the libraries. In addition, a sheared BAC library is being constructed that will be used in a new fingerprinted contig (FPC) build. A sheared library may contain regions of the genome that are not contained in restriction-based libraries due to lack of restriction sites in certain regions. The *Hind*III, *Eco*RI and *Mbo*I BAC libraries were end-sequenced by the US project, yielding over 340,000 high quality reads equivalent to 20% of the entire genome sequence. These BAC libraries have been complemented by a fosmid library that was also generated using shearing. Currently, over 180,000 high quality fosmid end-sequences from the Wellcome Trust Sanger Institute and the University of Padua are available, equivalent to 15% of the entire genome sequence. Fosmid libraries are crucial in a genome sequencing project because their narrowly defined insert length can be used as an analytical tool to detect potential mis-assemblies of BACs, and their generally shorter insert length is ideal for filling smaller gaps and thereby reducing redundant sequencing (Kim et al. 1995).

All BACs from the *Hind*III library and from the *Mbo*I library were fingerprinted and contigs of overlapping BACs were generated using the FPC tool (Soderlund et al. 2000). However, the fingerprinting, based on the older agarose-gel method, was not comparable to the quality one can achieve with the newer capillary-based approaches. Therefore, the fingerprinting is currently being re-run with the more precise capillary-based methods (Nelson et al. 2005).

An important validation of the mapped BACs is carried out using fluorescence *in situ* hybridization (FISH) on pachytene complements with entire BAC clones as probes (Chang et al. 2007; Szinay et al. 2008; see also FISH map on SGN, http://solgenomics.net/cview/map.pl?map_id=13). Another relatively fast and convenient validation method is to use genetic mapping of anchored BACs using panels of tomato introgression line populations (Eshed and Zamir 1995). These methods assure that the BACs are sequenced on the correct chromosome in the right location.

Since the current BAC by BAC sequencing effort focuses on the tomato euchromatin, determining the chromosomal borders between euchromatin and heterochromatin is essential. FISH is being used to identify BAC inserts from euchromatin/heterochromatin boundaries based on linkage map information and on the specific staining by FISH of the repetitive fraction of the tomato genome (Szinay et al. 2008).

The Tomato Genome Project developed common standards early on to ensure that the genome is completed with uniform quality across all chromosomes. The full quality standards are described in the Tomato Sequencing Guidelines document available online at: http://docs.google.com/View.aspx?docid=dggs4r6k_1dd5p56 (Mueller et al. 2009).

To summarize, the BACs are being sequenced to the following quality standards:

- The BAC sequence submitted in High Throughput Genome Sequence (HTGS) consisting of **a single contig (phase 3).**
- All bases of the HTGS Phase 3 consensus sequence must have a phred **quality score of at least 30.**
- As a result of the shotgun process the **bulk of sequence will be derived from multiple subclones sequenced from both strands**. Any regions of uni-directional sequence coverage with a single sequencing chemistry must pass manual inspection for sequence problems, but need not be annotated. Regions covered by only a single subclone must be attempted from an alternate sub clone or by direct walking on BAC DNA or by BAC polymerase chain reaction (PCR). These regions must concur with a restriction digest analysis of the clone. In addition these regions must be annotated.
- At least **99% of the sequence must have less than one error in 10,000 bp** as reported by Phrap or other sequence assembly consensus scores. Exceptions must be manually checked and pass inspection for possible problems. Any areas not meeting this standard must be annotated as such.

To date, nearly 1,200 BACs representing 120 Mb (including overlaps) have been sequenced and reported in the SGN BAC registry database (either HTGS phase 2 or phase 3) (Fig. 10-1). All these sequences are available from both Genbank (http://www.ncbi.nlm.nih.gov/) and the SGN database (http://solgenomics.net/). Of these 1,200 BACs, 762 are reported finished, and are included in the SGN Accessioned Golden Path (AGP) map representation. In total they represent close to half of the tomato euchromatin.

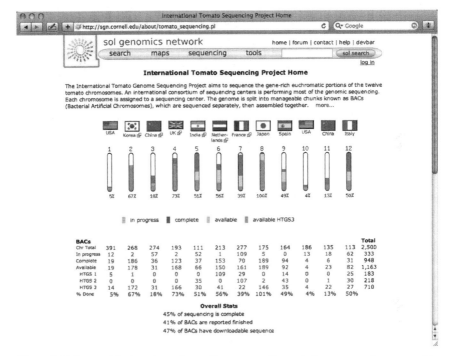

Figure 10-1 Status of the tomato euchromatin sequence. For each chromosome the responsible country is shown. Progress in the sequencing of each chromosome is given as well as the status and the availability of the BACs.

Color image of this figure appears in the color plate section at the end of the book.

10.2.2 Whole Genome Shotgun

As a complement to the BAC by BAC sequencing effort, a whole genome shotgun approach was initiated in 2008. This effort is centered around next-generation sequencing technologies, using Roche's 454 technology (Margulies et al. 2005), and the Applied Biosystem SOLiD system (Morozova et al. 2008), and also includes some whole genome shotgun sequencing based on Sanger (Sanger et al. 1977). In this approach, the entire genome is sheared and sequenced, and no distinctions are made between the different chromosomes or sequence regions (hetero- or euchromatin). As such, the effort will contribute to all existing chromosome projects. As with the BAC by BAC project, a number of different countries are contributing, including the Netherlands, who also played a major role in initiating this effort, as well as India, Italy, Japan, France, and the US.

Based on 454 technology, the Netherlands and India will provide a 15× coverage, France 10×; Italy, 10×; and the USA, 10× (all based on 3 kb and 5 kb paired ends, although libraries of 8 kb and 40 kb are still being discussed). Total coverage from 454 technology will thus be 45×, exceeding the updated specifications for coverage from Roche. Other genomes sequenced using shotgun approaches generally have had somewhat lower coverage; for example, *Vitis vinifera* (grapevine) was shotgun sequenced, using Sanger technology, to a coverage of less than 20x (Jaillon et al. 2007), and *Sorghum bicolor* was sequenced to a coverage of less than 10x, also using Sanger technology (Paterson et al. 2009).

Using the SOLiD system and paired end libraries of 2, 5 and 10 kb, the Netherlands, Spain, the UK, and Italy will each provide 15x coverage, for a total of 65x. These approaches will be complemented by a shotgun sequence produced earlier in Japan, based on the Sanger sequencing method, at 3x coverage. Of course, the BAC and fosmid ends will also be of great interest when assembling the whole genome shotgun sequence, because they provide longer range paired ends.

Producing the assembly is undoubtedly the most challenging part of the whole genome shotgun sequencing effort. However, timelines can be much more aggressive on the sequencing part, because no clone tiling paths need to be constructed. Most of the shotgun sequence is scheduled to be available before the end of 2009 and the assembly, including annotation by the International Tomato Annotation Group (ITAG) consortium, should be available shortly thereafter. The status of the sequence in 2011 is that the first draft of the genome integrating both BACs and shotgun sequence (version SL2.4) is complete and awaits publication.

10.3 Genome Annotation by ITAG

Producing a genome sequence is of very little interest in itself; a relatively high quality annotation needs to be available that researchers can use to identify candidate genes and formulate other hypotheses. To render the sequence immediately useful to the community, the International Tomato Annotation Group has formed with the goal of producing a high quality automated annotation of the tomato genome. Like the sequencing, the annotation is a collaborative effort, which involves groups from Europe,

Asia and the US. The centerpiece of the structural annotation is the Eugene gene prediction platform (Foissac et al. 2008), a powerful predictor capable of integrating both evidence-based alignments and ab initio predictions. For the functional annotation, InterPro domains are determined using InterProScan and homology searches are performed. Where possible, other sequence features (i.e., non-coding RNAs) are predicted. An important initial activity of the ITAG group was to generate a training and test set of gene sequences to train gene finding programs for tomato. Gene finding programs that are trained or have been trained include Eugene (Foissac et al. 2008), GeneMark (Isono et al. 1994), TwinScan (Korf et al. 2001), and Augustus (Stanke et al. 2008). Results of predicted gene models and their functional annotations are available on the SGN website. In the current genome build (SL2.4) has been annotated by ITAG (ITAG version 2.4), predicting almost 35,000 genes.

The characterization of repeat content is important because repeats represent a significant part of the genome, and as such define much of the genome structure and evolution. In practice, repeats interfere with genome assembly and gene annotation. Therefore, several different approaches were taken to define the repeat content of tomato. Initially, de novo repeat analysis was performed on the available BAC end sequences, and the resulting repeats were used to analyze both the BAC end sequences as well as the complete BAC sequences. The de novo repeat set masked 57% of BAC ends and 24% of full BAC sequence, indicating that the BACs selected from the euchromatin contain fewer repeats than the genome as a whole. These results support the recently described distribution of tomato repetitive sequences as determined by FISH (Chang et al. 2008). The fraction of long terminal repeat (LTR) elements was much higher in BAC ends (30%) than in the full BAC sequences (12.6%), indicating that there are large differences in the nature of repeats occurring in different genome regions (Mueller et al. 2009). This result has been confirmed with FISH using CoT-100 DNA, which represents a fraction of repetitive DNA (Szinay et al. 2008). A tomato transposon repeat dataset has been generated from finished BAC sequence, which has been used to study the transposon content of chromosome 6 (Peters et al. 2009) and which is used to mask transposons in the ITAG annotation pipeline. Based on that dataset, clear differences in transposon content were seen in heterochromatic versus euchromatic regions (Peters et al. 2009).

The emerging sequence has been analyzed for functional classification of the genes (Mueller et al. 2009). In particular, transcription factors, which play key roles in regulation of gene expression in various biological processes, have been analyzed (Mueller et al. 2009) and 66 transcription factor families have been identified in tomato (from a known total of 71 families occurring in plants). The three major transcription factor gene families in tomato include AP2-EREBP (APETALA2- ethylene responsive element binding protein), MYB and bHLH (basic helix-loop-helix) families (Mueller et al. 2009). Based on these preliminary results, it was estimated that more than 4% of all genes in tomato could code for transcription factors (Mueller et al. 2009).

The Solanaceae have served as model systems for the study of plant disease resistance. Sequence analysis of cloned plant disease resistance (R) genes conferring resistance to viral, bacterial, and fungal pathogens has shown that the majority of them possess common sequences and structural motifs. These R-genes can be grouped into three major classes (NBS-LRR type, LZ-NBS-LRR type or LRR-Tm type) on the basis of their encoded protein motifs such as leucine zippers (LZ), nucleotide binding site (NBS), leucine rich repeat (LRR), protein kinase (PK) domains, trans-membrane (Tm) domains and Toll-IL-IR homology regions (TIR). It was found that a total of 155 annotations similar to resistance-like genes and 83 annotations showed homology to the defense response like genes (Mueller et al. 2009).

These R-gene and defense response gene homologues were mapped in silico on to the sequenced BACs of the different chromosomes to find their physical locations, resulting in the localization of 59 R-gene homologues and of 21 defense response gene homologues. The mapped resistance-like and defense response genes represented about one-third of all expressed putative unique transcripts (PUTs) assembled from the tomato EST database.

10.4 Conclusion

The Tomato Genome Sequencing Project is a large, international collaboration to sequence the 220 Mb of euchromatin space of the tomato genome, which has been predicted to contain the majority of tomato genes. Although progress has been, for various reasons, somewhat slower than expected, the tomato genome project has spawned numerous international collaborations. The International SOL meetings fostered a spirit of community in Solanaceae research and so far produced over 1,200 BAC sequences, representing almost half of the targeted genome space. Sequences are being deposited at Genbank and the SGN database and are being annotated using a pipeline established by an international group (ITAG) of bioinformatics centers. Tools have been created that allow both researchers and tomato breeders to work with the emerging sequence. Through the extensive comparative maps that are available, much of the information from the tomato sequence should be readily transferable to other Solanaceae and related Asterids (Mueller et al. 2009).

A BAC by BAC sequencing approach was initially chosen to sequence the tomato genome because it was regarded to provide the highest possible sequence quality. However, since the project was started, novel "next generation" sequencing technologies have become available, which are now being applied to whole genome shotgun sequencing (WGS) for complex genomes. The BAC by BAC approach has inherent advantages, and yields insights beyond sequence space as the approach is based on careful evaluation of BAC positions by genetic mapping and by FISH. For example, several inversions could be identified between the cultivated tomato and its wild relative parent used in the reference map (Tang et al. 2008). The main drawback of the BAC by BAC approach is that it is significantly more expensive and slower than the WGS approach. Recently, the grape genome was sequenced using a shotgun approach, resulting in over 2,000 un-ordered contigs. However, it was estimated that over 95% of grape gene sequences were recovered in the sequence (Velasco et al. 2007). Thus, in the future, a hybrid approach for sequencing the tomato genome will be pursued by using whole genome shotgun (WGS) as an additional resource for finishing the euchromatic part of the genome, and for obtaining sequence for the heterochromatic part of the genome.

It is important to note that potato is also being sequenced, and that considerable synergies will be derived from this very similar genome (http://www.potatogenome.net/), from which tomato differs, on a chromosomal level, by only five pericentromeric inversions (Tanksley et al. 1992). Initially, potato was also sequenced by a BAC by BAC approach, but is now also complemented by different shotgun approaches. Because of their phylogenetic proximity, we expect that it will be possible to close sequence gaps in the tomato genome based on potato data and vice versa.

All data related to the tomato genome sequencing project can be found on SGN (http://solgenomics.net) and BAC sequences are deposited to Genbank (http://www.ncbi.nlm.nih.gov/). We expect that the euchromatin sequence, including the shotgun sequence, will be finished in 2011.

Acknowledgments

I would like to thank all the people who are involved in the tomato genome sequencing project and all the funding agencies that have made this project possible.

References

Adams MD, Kelley JM, Gocayne JD, Dubnick M, Polymeropoulos MH, Xiao H, Merril CR, Wu A, Olde B, Moreno RF (1991) Complementary DNA sequencing: expressed sequence tags and human genome project. Science 252: 1651–1656.

Adams-Phillips L, Barry C, Giovannoni J (2004) Signal transduction systems regulating fruit ripening. Trends Plant Sci 9: 331–8.

AGI (2000) Analysis of the genome sequence of the flowering plant *Arabidopsis thaliana*. Nature 408: 796–815.

Alexander L, Grierson D (2002) Ethylene biosynthesis and action in tomato: a model for climacteric fruit ripening. J Exp Bot 53: 2039–55.

Bachem CW, van der Hoeven RS, de Bruijn SM, Vreugdenhil D, Zabeau M, Visser RG (1996) Visualization of differential gene expression using a novel method of RNA fingerprinting based on AFLP: analysis of gene expression during potato tuber development. Plant J 9: 745–753.

Bai Y, Huang CC, van der Hulst R, Meijer-Dekens F, Bonnema G, Lindhout P (2003) QTLs for tomato powdery mildew resistance (*Oidium lycopersici*) in *Lycopersicon parviflorum* G1.1601 co-localize with two qualitative powdery mildew resistance genes. Mol Plant-Microbe Interact 16: 169–176.

Bogdanove AJ, Martin GB (2000) AvrPto-dependent Pto-interacting proteins and AvrPto-interacting proteins in tomato. Proc Natl Acad Sci USA 97: 8836–40.

Brummell DA, Harpster MH (2001) Cell wall metabolism in fruit softening and quality and its manipulation in transgenic plants. Plant Mol Biol 47: 311–40.

Budiman MA, Mao L, Wood TC, Wing RA (2000) A deep-coverage tomato BAC library and prospects toward development of an STC framework for genome sequencing. Genome Res 10: 129–36.

Cai WW, Reneker J, Chow CW, Vaishnav M, Bradley A (1998) An anchored framework BAC map of mouse chromosome 11 assembled using multiplex oligonucleotide hybridization. Genomics 54: 387–97.

Chang SB, Yang TJ, Datema E, van Vugt J, Vosman B, Kuipers A, Meznikova M, Szinay D, Klein Lankhorst R, Jacobsen E, de Jong H (2008) FISH mapping and molecular organization of the major repetitive sequences of tomato. Chrom Res: in press.

Chang SB, Anderson LK, Sherman JD, Royer SM, Stack SM (2007) Predicting and testing physical locations of genetically mapped Loci on tomato pachytene chromosome 1. Genetics 176: 2131–2138.

Clarkson JJ, Lim KY, Kovarik A, Chase MW, Knapp S, Leitch AR (2005) Long-term genome diploidization in allopolyploid *Nicotiana* section *Repandae* (Solanaceae). New Phytol 168: 241–252.

D'Agostino N, Traini A, Frusciante L, Chiusano ML (2007) Gene models from ESTs (GeneModelEST): an application on the Solanum lycopersicum genome. BMC Bioinformat 8 (Suppl 1): S9.

De Jong WS, Eannetta NT, De Jong DM, Bodis M (2004) Candidate gene analysis of anthocyanin pigmentation loci in the Solanaceae. Theor Appl Genet 108: 423–32.

Doganlar S, Frary A, Daunay MC, Lester RN, Tanksley SD (2002) Conservation of gene function in the solanaceae as revealed by comparative mapping of domestication traits in eggplant. Genetics 161: 1713–26.

Eshed Y, Zamir D (1995) An introgression line population of Lycopersicon pennellii in the cultivated tomato enables the identification and fine mapping of yield-associated QTL. Genetics 141: 1147–62.

Fernie AR, Willmitzer L (2001) Molecular and biochemical triggers of potato tuber development. Plant Physiol 127: 1459–65.

Foissac S, Gouzy JP, Rombauts S, Mathé C, Amselem J, Sterck L, Van de Peer Y, Rouzé P, Schiex T (2008) Genome Annotation in Plants and Fungi: EuGene as a model platform. Curr Bioinformat 3: 87–97.

Fray RG, Grierson D (1993) Molecular genetics of tomato fruit ripening. Trends Genet 9: 438–43.

Fu Y, Hsia AP, Guo L, Schnable PS (2004) Types and frequencies of sequencing errors in methyl-filtered and high cot maize genome survey sequences. Plant Physiol 135: 2040–5.

Fulton TM, Van der Hoeven R, Eannetta NT, Tanksley SD (2002) Identification, analysis, and utilization of conserved ortholog set markers for comparative genomics in higher plants. Plant Cell 14: 1457–67.

Gebhardt C, Valkonen JP (2001) Organization of genes controlling disease resistance in the potato genome. Annu Rev Phytopathol 39: 79–102.

Gerats AG, Vrijlandt E, Wallroth M, Schram AW (1985) The influence of the genes *An1, An2,* and *An4* on the activity of the enzyme UDP-glucose: flavonoid 3-O-glucosyltransferase in flowers of *Petunia hybrida*. Biochem Genet 23: 591–598.

Gill N, Hans CS, Jackson S (2008) An overview of plant chromosome structure. Cytogenet Genome Res 120: 194–201.

Giovannoni JJ (2004) Genetic regulation of fruit development and ripening. Plant Cell 16 (Suppl): S170–80.

Giuliano G, Bartley GE, Scolnik PA (1993) Regulation of carotenoid biosynthesis during tomato development. Plant Cell 5: 379–387.

Gray J, Picton S, Shabbeer J, Schuch W, Grierson D (1992) Molecular biology of fruit ripening and its manipulation with antisense genes. Plant Mol Biol 19: 69–87.

Hoeven RVd, Ronning C, Giovannoni J, Matin G, Tanksley S (2002) Deductions about the number, organization and evolution of genes in the tomato genome based on analysis of a large expressed sequence tag collection and selective genomic sequencing. Plant Cell 14: 1441–1456.

Hui D, Iqbal J, Lehmann K, Gase K, Saluz HP, Baldwin IT (2003) Molecular interactions between the specialist herbivore *Manduca sexta* (lepidoptera, sphingidae) and its natural host *Nicotiana attenuata*: V Microarray analysis and further characterization of large-scale changes in herbivore-induced mRNAs. Plant Physiol 131: 1877–93.

International Rice Genome Sequencing Project (2005) The map-based sequence of the rice genome. Nature 436: 793–800.

Isono K, McIninch JD, Borodovsky M (1994) Characteristic features of the nucleotide sequences of yeast mitochondrial ribosomal protein genes as analyzed by computer program GeneMark. DNA Res 1: 263–269.

Jaillon O, Aury JM, Noel B, Policriti A, Clepet C, Casagrande A, Choisne N, Aubourg S, Vitulo N, Jubin C, Vezzi A, Legeai F, Hugueney P, Dasilva C, Horner D, Mica E, Jublot D, Poulain J, Bruyere C, Billault A, Segurens B, Gouyvenoux M, Ugarte E, Cattonaro F, Anthouard V, Vico V, Del Fabbro C, Alaux M, Di Gaspero G, Dumas V, Felice N, Paillard S, Juman I, Moroldo M, Scalabrin S, Canaguier A, Le Clainche I, Malacrida G, Durand E, Pesole G, Laucou V, Chatelet P, Merdinoglu D, Delledonne M, Pezzotti M, Lecharny A, Scarpelli C, Artiguenave F, Pe ME, Valle G, Morgante M, Caboche M, Adam-Blondon AF, Weissenbach J, Quetier F, Wincker P (2007) French-Italian Public Consortium for Grapevine Genome Characterization. The grapevine genome sequence suggests ancestral hexaploidization in major angiosperm phyla. Nature 449: 463–467.

Kahlau S, Aspinall S, Gray JC, Bock R (2006) Sequence of the tomato chloroplast DNA and evolutionary comparison of solanaceous plastid genomes. J Mol Evol 63: 194–207.

Kessler A, Baldwin IT (2001) Defensive function of herbivore-induced plant volatile emissions in nature. Science 291: 2141–2144.

Kim UJ, Shizuya H, Sainz J, Garnes J, Pulst SM, de Jong P, Simon MI (1995) Construction and utility of a human chromosome 22-specific Fosmid library. Genet Anal 12: 81–84.

Knapp S, Bohs L, Nee M, Spooner DM (2004) Solanaceae—a model for linking genomics with biodiversity. Comp Funct Genom 5: 285–291.

Korf I, Flicek P, Duan D, Brent MR (2001) Integrating genomic homology into gene structure prediction. Bioinformatics 17 (Suppl 1): S140–8.

Li L, Li C, Howe GA (2001) Genetic analysis of wound signaling in tomato. Evidence for a dual role of jasmonic acid in defense and female fertility. Plant Physiol 127: 1414–1417.

Margulies M, Egholm M, Altman WE, Attiya S, Bader JS, Bemben LA, Berka J, Braverman MS, Chen YJ, Chen Z, Dewell SB, Du L, Fierro JM, Gomes XV, Godwin BC, He W, Helgesen S, Ho CH, Irzyk GP, Jando SC, Alenquer ML, Jarvie TP, Jirage KB, Kim JB, Knight JR, Lanza JR, Leamon JH, Lefkowitz SM, Lei M, Li J, Lohman KL, Lu H, Makhijani VB, McDade KE, McKenna MP, Myers EW, Nickerson E, Nobile JR, Plant R, Puc BP, Ronan MT, Roth GT, Sarkis GJ, Simons JF, Simpson JW, Srinivasan M, Tartaro KR, Tomasz A, Vogt KA, Volkmer GA, Wang SH, Wang Y, Weiner MP, Yu P, Begley RF, Rothberg JM (2005) Genome sequencing in microfabricated high-density picolitre reactors. Nature 437: 376–380.

Morozova O, Marra MA (2008) Applications of next-generation sequencing technologies in functional genomics. Genomics 92(5): 255–64.

Mueller LA, Klein Lankhorst R, Tanksley SD, Giovannoni JJ, White R, Vrebalov J, Fei Z, van Eck J, Buels R, Mills AA, Menda N, Tecle IY, Bombarely A, Stack S, Royer SM, Chang S, Shearer LA, Kim BD, Jo S, Hur C, Choi D, Li C, Zhao J, Jiang H, Geng Y, Dai Y, Fan H, Chen J, Lu F, Shi J, Sun S, Chen J, Yang X, Lu C, Chen M, Cheng Z, Li C, Ling H, Xue Y, Wang Y, Seymour GB, Bishop GJ, Bryan G, Rogers J, Sims S, Butcher S, Buchan D, Abbott J, Beasley H, Nicholson C, Riddle C, Humphray S, McLaren K, Mathur S, Vyas S, Solanke AU, Kumar R, Gupta V, Sharma AK, Khurana P, Khurana JP, Tyagi A, Bhutty S, Chowdhury P, Shridhar S, Chattopadhyay D, Pandit A, Singh P, Kumar A, Dixit R, Singh A, Praveen S, Dalal V, Yadav M, Ghazi IA, Gaikwad K, Sharma TR, Mohapatra T, Singh NK, Szinay D, de Jong H, Peters S, van Staveren M, Datema E, Fiers MWEJ, van Ham RCHJ, Lindhout P, Philippot M, Frasse P, Regad F, Zouine M, Bouzayen M, Asamizu E, Sato S, Fukuoka H, Tabata S, Shibata D, Botella MA, Perez-Alonso M, Fernandez-Pedrosa V, Osorio S, Mico A, Granell A, Zhang A, He J, Huang S, Du Y, Qu D, Liu L, Liu D, Wang J, Ye Z, Yang W, Wang G, Vezzi A, Todesco S, Valle G, Falcone G, Pietrella M, Giuliano G, Grandillo S, Traini A, D'Agostino N, Chiusano ML, Ercolano M, Barone A, Frusciante L, Schoof H, Jöcker A, Bruggmann R, Spannagl M, Mayer KXF, Guigó R, Camara F, Rombauts S, Fawcett JA, Van de Peer Y, Knapp S, Zamir D, Stiekema W (2009) A snapshot of the emerging tomato genome sequence. Plant Genome 2: 78–92.

Mueller LA, Goodman CD, Silady RA, Walbot V (2000) AN9, a petunia glutathione S-transferase required for anthocyanin sequestration, is a flavonoid-binding protein. Plant Physiol 123: 1561–1570.

Mueller LA, Tanksley SD, Giovannoni J, van Eck J, Stack S, Choi D, Kim BD, Chen MS, Cheng ZK, Li CYea (2005) The Tomato Sequencing Project, the first cornerstone of the International Solanaceae Project (SOL). Compar Funct Genom 6: 153–158.

Nelson WM, Bharti AK, Butler E, Wei F, Fuks G, Kim H, Wing RA, Messing J, Soderlund C (2005). Whole-genome validation of high-information-content fingerprinting. Plant Physiol 139(1): 27–38.

Oksman-Caldentey KM (2007) Tropane and nicotine alkaloid biosynthesis-novel approaches towards biotechnological production of plant-derived pharmaceuticals. Curr Pharm Biotechnol 8: 203–210.

Olmstead RG, Sweere JA, Spangler RE, Bohs L, Palmer JD (1999) Phylogeny and provisional classification of the Solanaceae based on chloroplast DNA. In: Nee M, Symon DE, Lester RN, Jessop JP (edn) Solanaceae IV, Advances in Biology and Utilization. Royal Botanic Gardens, Kew, UK, pp 111–137.

Palmer LE, Rabinowicz PD, O'Shaughnessy AL, Balija VS, Nascimento LU, Dike S, de la Bastide M, Martienssen RA, McCombie WR (2003) Maize genome sequencing by methylation filtration. Science 302: 2115–7.

Paterson AH, Bowers JE, Bruggmann R, Dubchak I, Grimwood J, Gundlach H, Haberer G, Hellsten U, Mitros T, Poliakov A, Schmutz J, Spannagl M, Tang H, Wang X, Wicker T,

Bharti AK, Chapman J, Feltus FA, Gowik U, Grigoriev IV, Lyons E, Maher CA, Martis M, Narechania A, Otillar RP, Penning BW, Salamov AA, Wang Y, Zhang L, Carpita NC, Freeling M, Gingle AR, Hash CT, Keller B, Klein P, Kresovich S, McCann MC, Ming R, Peterson DG, Mehboob-ur-Rahman, Ware D, Westhoff P, Mayer KF, Messing J, Rokhsar DS (2009) The *Sorghum bicolor* genome and the diversification of grasses. Nature 457: 551–556.

Pedley KF, Martin GB (2003) Molecular basis of Pto-mediated resistance to bacterial speck disease in tomato. Annu Rev Phytopathol 41: 215–43.

Peters SA, Datema E, Szinay D, van Staveren MJ, Schijlen EG, van Haarst JC, Hesselink T, Abma-Henkens MH, Bai Y, de Jong H, Stiekema WJ, Klein Lankhorst RM, van Ham RC (2009) *Solanum lycopersicum* cv. Heinz 1706 chromosome 6: distribution and abundance of genes and retrotransposable elements. Plant J 55: 857–869.

Peterson DG, Schulze SR, Sciara EB, Lee SA, Bowers JE, Nagel A, Jiang N, Tibbitts DC, Wessler SR, Paterson AH (2002) Integration of Cot analysis, DNA cloning, and high-throughput sequencing facilitates genome characterization and gene discovery. Genome Res 12: 795–807.

Peterson DG, Stack SM, Price HJ, Johnston JS (1996) DNA content of heterochromatin and euchromatin in tomato (Lycopersicon esculentum) pachytene chromosomes. Genome 39: 77–82.

Prat S, Frommer WB, Hofgen R, Keil M, Kossmann J, Koster-Topfer M, Liu XJ, Muller B, Pena-Cortes H, Rocha-Sosa M (1990) Gene expression during tuber development in potato plants. FEBS Lett 268: 334–8.

Quattrocchio F, Verweij W, Kroon A, Spelt C, Mol J, Koes R (2006) pH4 of Petunia is an R2R3 MYB protein that activates vacuolar acidification through interactions with basic-helix-loop-helix transcription factors of the anthocyanin pathway. Plant Cell 18: 1274–1291.

Sacco MA, Mansoor S, Moffett P (2007) A RanGAP protein physically interacts with the NB-LRR protein Rx, and is required for Rx-mediated viral resistance. Plant J 52: 82–93.

Sanger F, Nicklen S, Coulson AR (1977) DNA sequencing with chain-terminating inhibitors. Proc Natl Acad Sci USA 74: 5463–5467.

Schijlen E, Ric de Vos CH, Jonker H, van den Broeck H, Molthoff J, van Tunen A, Martens S, Bovy A (2006) Pathway engineering for healthy phytochemicals leading to the production of novel flavonoids in tomato fruit. Plant Biotechnol J 4: 433–444.

Seymour G, Poole M, Manning K, King GJ (2008) Genetics and epigenetics of fruit development and ripening. Curr Opin Plant Biol 11: 58–63.

Shendure J, Porreca GJ, Reppas NB, Lin X, McCutcheon JP, Rosenbaum AM, Wang MD, Zhang K, Mitra RD, Church GM (2005) Accurate multiplex polony sequencing of an evolved bacterial genome. Science 309: 1728–1732.

Soderlund C, Humphray S, Dunham A, French L (2000) Contigs built with fingerprints, markers, and FPC V4.7. Genome Res 10: 1772–1787.

Spelt C, Quattrocchio F, Mol J, Koes R (2002) ANTHOCYANIN1 of petunia controls pigment synthesis, vacuolar pH, and seed coat development by genetically distinct mechanisms. Plant Cell 14: 2121–2135.

Stanke M, Diekhans M, Baertsch R, Haussler D (2008) Using native and syntenically mapped cDNA alignments to improve de novo gene finding. Bioinformatics 24: 637–644.

Szinay D, Chang SB, Khrustaleva L, Peters S, Schijlen E, Bai Y, Stiekema WJ, van Ham RC, de Jong H, Klein Lankhorst RM (2008) High-resolution chromosome mapping of BACs using multi-colour FISH and pooled-BAC FISH as a backbone for sequencing tomato chromosome 6. Plant J 56: 627–637.

Tang X, Szinay D, Lang C, Ramanna MS, van der Vossen EA, Datema E, Klein Lankhorst R, de Boer J, Peters SA, Bachem C, Stiekema W, Visser RG, de Jong H, Bai Y (2008) Cross-species BAC-FISH painting of the tomato and potato chromosome 6 reveals undescribed chromosomal rearrangements. Genetics 180: 1319–28.

Tanksley SD (2004) The genetic, developmental, and molecular bases of fruit size and shape variation in tomato. Plant Cell 16 (Suppl): S181–189.

Tanksley SD, Ganal MW, Prince JP, de Vicente MC, Bonierbale MW, Broun P, Fulton TM, Giovannoni JJ, Grandillo S, Martin GB (1992) High density molecular linkage maps of the tomato and potato genomes. Genetics 132: 1141–60.

Tuskan GA, Difazio S, Jansson S, Bohlmann J, Grigoriev I, Hellsten U, Putnam N, Ralph S, Rombauts S, Salamov A, Schein J, Sterck L, Aerts A, Bhalerao RR, Bhalerao RP, Blaudez D, Boerjan W, Brun A, Brunner A, Busov V, Campbell M, Carlson J, Chalot M, Chapman J, Chen GL, Cooper D, Coutinho PM, Couturier J, Covert S, Cronk Q, Cunningham R, Davis J, Degroeve S, Dejardin A, Depamphilis C, Detter J, Dirks B, Dubchak I, Duplessis S, Ehlting J, Ellis B, Gendler K, Goodstein D, Gribskov M, Grimwood J, Groover A, Gunter L, Hamberger B, Heinze B, Helariutta Y, Henrissat B, Holligan D, Holt R, Huang W, Islam-Faridi N, Jones S, Jones-Rhoades M, Jorgensen R, Joshi C, Kangasjarvi J, Karlsson J, Kelleher C, Kirkpatrick R, Kirst M, Kohler A, Kalluri U, Larimer F, Leebens-Mack J, Leple JC, Locascio P, Lou Y, Lucas S, Martin F, Montanini B, Napoli C, Nelson DR, Nelson C, Nieminen K, Nilsson O, Pereda V, Peter G, Philippe R, Pilate G, Poliakov A, Razumovskaya J, Richardson P, Rinaldi C, Ritland K, Rouze P, Ryaboy D, Schmutz J, Schrader J, Segerman B, Shin H, Siddiqui A, Sterky F, Terry A, Tsai CJ, Uberbacher E, Unneberg P, Vahala J, Wall K, Wessler S, Yang G, Yin T, Douglas C, Marra M, Sandberg G, Van de Peer Y, Rokhsar D (2006) The genome of black cottonwood, *Populus trichocarpa* (Torr. & Gray). Science 313: 1596–1604.

van der Vossen EA, van der Voort JN, Kanyuka K, Bendahmane A, Sandbrink H, Baulcombe DC, Bakker J, Stiekema WJ, Klein-Lankhorst RM (2000) Homologues of a single resistance-gene cluster in potato confer resistance to distinct pathogens: a virus and a nematode. Plant J 23: 567–576.

Velasco R, Zharkikh A, Troggio M, Cartwright DA, Cestaro A, Pruss D, Pindo M, Fitzgerald LM, Vezzulli S, Reid J, Malacarne G, Iliev D, Coppola G, Wardell B, Micheletti D, Macalma T, Facci M, Mitchell JT, Perazzolli M, Eldredge G, Gatto P, Oyzerski R, Moretto M, Gutin N, Stefanini M, Chen Y, Segala C, Davenport C, Dematte L, Mraz A, Battilana J, Stormo K, Costa F, Tao Q, Si-Ammour A, Harkins T, Lackey A, Perbost C, Taillon B, Stella A, Solovyev V, Fawcett JA, Sterck L, Vandepoele K, Grando SM, Toppo S, Moser C, Lanchbury J, Bogden R, Skolnick M, Sgaramella V, Bhatnagar SK, Fontana P, Gutin A, Van de Peer Y, Salamini F, Viola R (2007) A high quality draft consensus sequence of the genome of a heterozygous grapevine variety. PLoS ONE 2: e1326.

Wang Y, Tang X, Cheng Z, Mueller L, Giovannoni J, Tanksley SD (2006) Euchromatin and pericentromeric heterochromatin: comparative composition in the tomato genome. Genetics 172: 2529–2540.

Whitelaw CA, Barbazuk WB, Pertea G, Chan AP, Cheung F, Lee Y, Zheng L, van Heeringen S, Karamycheva S, Bennetzen JL, SanMiguel P, Lakey N, Bedell J, Yuan Y, Budiman MA, Resnick A, Van Aken S, Utterback T, Riedmuller S, Williams M, Feldblyum T, Schubert K, Beachy R, Fraser CM, Quackenbush J (2003) Enrichment of gene-coding sequences in maize by genome filtration. Science 302: 2118–20.

Wu F, Mueller LA, Crouzillat D, Petiard V, Tanksley SD (2006) Combining bioinformatics and phylogenetics to identify large sets of single-copy orthologous genes (COSII) for comparative, evolutionary and systematic studies: A test case in the Euasterid plant clade. Genetics 174: 1407–20.

Yuan Y, SanMiguel PJ, Bennetzen JL (2003) High-Cot sequence analysis of the maize genome. Plant J 34: 249–55.

11

Comparative Genome Sequencing of Tomato and Potato: Methods and Analysis

Sanwen Huang* and Zhong-hua Zhang

ABSTRACT

The direct comparison of genomic sequences of two closely related species can serve as a powerful tool as part of a whole genome sequencing strategy. The internationally sponsored genome sequencing projects of tomato (*Solanum lycopersicum*) and potato (*S. tuberosum*) have benefitted from each other by applying comparative genome sequencing techniques. The BAC-by-BAC and shot-gun sequencing strategy utilized by both projects involves several challenges. We describe a step-by-step approach to help overcome these challenges that includes i) reliably identifying and obtaining sufficient numbers of seed BACs, ii) informing the construction of an accurate physical map and iii) BAC finishing. A sequencing and bioinformatics pipeline is presented using chromosome 11 as an example. After BAC finishing, compared sequences were subjected to bioinformatics analysis to confirm synteny, estimate conservation and divergence, and to gain insights into biological and evolutionary processes. Gene order and orientation, extent of duplication and deletion, and abundance of repetitive elements such as retrotransposons were studied. Comparative sequencing methods improved the efficiency and accuracy of sequencing and assembly of potato and tomato chromosome 11 and will continue to be beneficial for these and other species, especially in euchromatic regions.

Keywords: synteny, colinearity, bacterial artificial chromosome, physical maps, Solanaceae

Institute of Vegetables and Flowers, Chinese Academy of Agricultural Sciences, No. 12, Zhong Guan Cun Nan Da Jie, Beijing, 100081, China.
*Corresponding author

11.1 Introduction

The genomes of tomato (*Solanum lycopersicum*) and potato (*S. tuberosum*) are being sequenced by the Solanaceae Genome Project and the Potato Genome Sequencing Consortium, respectively (Mueller et al. 2005). Both of the projects employ the BAC-by-BAC strategy, which has been widely applied in genome sequencing projects for other organisms such as *Arabidopsis thaliana*, human (*Homo sapiens*), and rice (*Oryza sativa*) (*Arabidopsis* Genome Initiative 2000; International Human Genome Sequencing Consortium 2001; International Rice Genome Sequencing Project 2005). Although highly accurate genome sequences can be generated with this strategy, the progress of genome sequencing is limited in several aspects. Firstly, anchoring seed bacterial artificial chromosomes (BACs) on chromosomes using genetic markers is labor-intensive. Currently there are too few seed BACs for comprehensive tomato and potato genome sequencing. Secondly, constructing a physical map towards compiling a minimal tiling path (MTP) is a long and complex process. Third, BAC finishing requires many PCR reactions and sequencing, and can become cost-prohibitive. Thus, new methods must be explored to overcome these disadvantages. Because tomato and potato genomes were determined to be macro-colinear to a high degree (Tanksley et al. 1992), comparative sequencing promises to accelerate both the genome projects. This sequencing method allows identifying additional seed BACs, improving physical maps, and accelerating the finishing of BAC sequences.

This chapter presents a detailed example of comparative sequencing techniques and analysis of tomato and potato chromosome 11. A glossary of technical terms is provided in Fig. 11-1. Here we describe how to facilitate the two genome sequencing projects through comparative sequencing, and review what we have learned through analyzing syntenic BAC clones. In these comparative analyses, several insights have emerged. The synteny between tomato and potato has now been confirmed at the sequence level, ensuring the utility of comparative sequencing. In addition, the analysis of the content and distribution of repetitive elements in these BACs can explain why the new sequencing method is not greatly affected by repetitive sequences in the euchromatic region.

11.2 A Pipeline for the Comparative Sequencing of Tomato and Potato Genomes

The pipeline for comparative sequencing consists of five steps (Fig. 11-2).

> **Anchored BAC:** Localization of a BAC sequence on a genetic linkage map based on its match to a known genetic marker sequence.
>
> **Bacterial Artificial Chromosome (BAC):** A DNA construct used for cloning, often used to sequence the genome of an organism. A short piece (~150-350 kb) of the organism's DNA is amplified as a BAC insert, and then sequenced.
>
> **BAC End Sequence (BES):** Several-hundred bp of BAC sequence from its 5' or 3' end.
>
> **Fingerprinted Contig (Contig^{FPC}):** In genome sequencing, thousands of clones are fingerprinted and assembled into contigs. To determine the order of clones, clones are digested with one or more restriction enzymes and resulting fragments are analyzed. The probability of two clones overlapping is based on the similarity of their restriction digest fragments. A contig contains two or more overlapping clones; a minimal tiling path of clones is selected to be sequenced.
>
> **Phase III:** Production of a 'finished' version of a genome sequence, with gaps filled in and errors resolved.
>
> **Physical Map:** A physical map positions sequence-tagged sequences (STSs) by binning to chromosomal fragments or by determining ordered STS positions across large insert clones such as BACs.
>
> **Scaffold Sequence:** A set of contiguous sequences within a genome. A draft genomic sequence generated by a shotgun approach covers a large fraction of the genome that is distributed over a number of 'scaffolds' of varying lengths (many more than there are chromosomes).
>
> **Seed BAC:** An initiation BAC clone, against which minimum overlapping (tiling path) 5' and 3' BAC clones are computationally identified and sequenced. Seed BACs are spaced as evenly as possible across the genome, ideally every 1–2 Mb.
>
> **Sequence Map:** A sequence map is determined by identifying a minimum overlapping (tiling) path of BAC clones across each chromosome and then sequencing the individual BAC clones. A shotgun sequencing approach randomly cleaves each BAC insert into 1- to 4-kb fragments, subclones them into plasmid or phage vectors, and sequences one or both ends of these small inserts to ~10-fold sequence coverage. Resultant fragments are computationally assembled into a consensus sequence. Directed sequencing of incomplete or low-quality areas then permits 'BAC finishing'—finishing the sequence to a high degree of accuracy.
>
> **Synteny:** The physical co-localization of genetic loci on the same chromosome.

Figure 11-1 Glossary of technical terms used in this chapter.

Step 1

Scaffold sequences of tomato (or potato) BACs were aligned against potato (or tomato) non-repeat BAC end sequences (BESs) by BLASTN at an E-value threshold of 1e-40. If more than 75% of a BES was covered by an alignment, it was considered as one BES hit. According to the positions and orientations of BES hits on scaffold sequences, the orders and orientations of their matching BACs were determined.

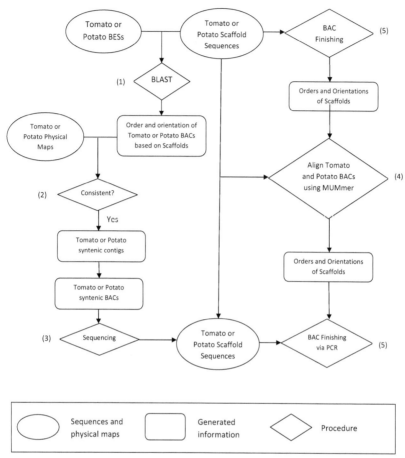

Figure 11-2 Flow Diagram of comparative sequencing of tomato and potato chromosome 11.
(1). Scaffold sequences were aligned against BESs using BLAST; (2). Orders and orientations of BACs based on scaffolds were checked for consistency with physical maps; (3). Sequencing new selected seed BACs; (4). Aligning tomato and potato scaffolds using MUMmer software program; (5). Based on the orders and orientations of scaffolds, new primers were designed for BAC finishing.

Step 2

Based on their physical maps, all matching BACs of BES hits were grouped into different FingerPrinted Contigs (FPContigs) using using an in-house pipeline. The FPContigs containing the maximum numbers of matching BACs were selected as candidate syntenic FPContigs. Then, the order and orientations of the BACs determined by scaffolds were compared with that in FPContigs. If they were consistent, the FPContigs were considered as syntenic FPContigs. When the number of syntenic FPContigs was greater than one, the merging of these FPContigs was attempted using FingerPrinted Contigs (FPC) software (Soderlund et al. 1997). For potato FPContigs, all merging was done by Jan de Bore (Wageningen University). From syntenic FPContigs one BAC, which covered the maximum homologous regions of the scaffolds, was selected as a syntenic BAC.

Step 3

The selected syntenic BACs were sequenced and assembled into scaffolds, thus the homologous scaffold sequences of tomato and potato BACs were obtained.

Step 4

The homologous sequences of tomato and potato BACs were aligned using the MUMmer program (Kurtz et al. 2004). Based on the alignments, some scaffold sequences were ordered and oriented.

Step 5

If the orders and orientations of scaffolds could be determined, the specific primers for PCR were designed to finish BAC sequencing.

11.3 Advantages of Comparative Genome Sequencing

11.3.1 Improving the Physical Maps

In both the tomato and potato genome sequencing projects, when a few BACs are sequenced, they can be used not only for BAC extension based within-species BESs and physical maps, but also to develop the physical map of the other genome as shown in "step 1" of the pipeline (Fig. 11-1). The following is an example of how to use available sequences for improving a physical map.

During the initial stages of sequencing chromosome 11 of tomato and potato, 10 tomato and 15 potato BACs (Table 11-1; Table 11-2) were shot-gun sequenced and one to six scaffolds was/were obtained for each BAC. These sequences, with approximately 6 fold average depth of coverage, were very incomplete. This became the starting point for comparative sequencing.

Five FPContigs in potato were identified as syntenic to four tomato BACs (Table 11-1) through attempts to align scaffold sequences of the 10 tomato seed BACs against potato BESs. Physical maps were referenced to ensure synteny, and conversely, the syntenic relationship was considered in constructing the physical map. As shown in Fig. 11-2A, the potato Ctg6488 and Ctg6847 contained many BACs, whose BESs were homologous to tomato BAC LE_HBa0089M02, and also the positions of most BESs on BAC sequences were consistent with the physical map. This implied that the two

Table 11-1 Summary of tomato chromosome 11 seed BACs and identified syntenic BACs in potato.

	tomato seed BACs				identified syntenic BACs in potato			
BAC name	Marker	cM	Contigs[FPC]	# of Scaffolds[a]	Syntenic BAC	Marker[b]	Bin	Syntenic Contigs[FPC]
LE_HBa0168B23	P47	0	Ctg6062	2	→ RH083G18	EACTMCGA_115.9	B031	Ctg6409
LE_HBa0054I23	cTOA-5-G7	3	Ctg6889	1				
LE_HBa0096D22	T1012	12	Ctg5821	2				
LE_HBa0107K14	J1	32	Singleton	2				
LE_HBa0062I24	T1161	39	Ctg6575	1				
LE_HBa0245M17	CT55	51	Singleton	3				
LE_HBa0029C01	T1046	54	Ctg5723	1				
LE_HBa0139J14	T1014	67	Ctg6600	1	→ RH186L12			Ctg6506
LE_HBa0161D01	T1205	77	Singleton	3	→ RH121D01			Ctg6479
LE_HBa0089M02	T1949	91	Ctg8884	2	→ RH195J04[c]	EAGAMCTG_453.1	B084	Ctg6488 Ctg6847

[a]Number of scaffolds assembled using shot-gun reads representing ~6X coverage for each BAC;
[b]The markers were identified on BAC sequences *in silico* using GenEST program;
[c]BAC RH195J04 was selected from Ctg6488. For Ctg6847, RH092I19 was sequenced by Wageningen University.

Table 11-2 Summary of potato chromosome 11 seed BACs and identified syntenic BACs in tomato.

		potato seed BACs			identified syntenic BACs in tomato			
BAC name	Marker	Bin	Contigs[FPC]	# of Scaffolds	Syntenic BAC	Marker[a]	cM	Syntenic Contigs[FPC]
RH130I13	EAACMCAC_118.3H	B002	Ctg2072	2				
RH204G21	EACAMCTT_252.5	B007	Ctg410	1				
RH179O17	EACTMAGC_7H	B019-B020	Ctg1765	6				
RH097I18	EAAGMACC_206.5H	B025-B027	Ctg1903	4				
RH067L08	EAGAMCCA_162.1	B035	Ctg6624	1	→ LE_HBa0034I10	TG147	45	Ctg6029
		B035	Ctg6624	3	→ SL_MboI0052K14[b]			Ctg7031
RH162O21	EAACMCAT_1	B041	Ctg4687	2				
RH128H14	EAACMCTC_376.8H	B041-B042	Ctg1550	2				
RH018F07	EACAMACC_393.3H	B043	Ctg4944	1				
RH095N07	EACAMCTG_262.8H	B049-B057	Ctg1082	3				
RH201N04	EAAGMCTA_382H	B052-B058	Ctg1065	1				
RH059C07	EAAGMAGC_455	B060	Ctg1673	3				
RH021O18	EACAMCCA_484.3	B061	Ctg3578	1	→ LE_HBa0245M17[c]	CT55	51	Ctg6326
RH058F17	EACAMCTC_157.4rh	B068	Ctg1150	5				
RH042C12	EACAMCGA_262.5H	B082-B083	Ctg6593	2				

[a]The markers were identified on BAC sequences *in silico* using BLAST;

[b]This BAC was localized on tomato chromosome 11 by FISH, and is close to LE_HBa0034I10;

[c]BAC LE_HBa0245M17 is a singleton and was selected as a seed BAC as indicated in Table 1. Marker CT55 was identified on this BAC, and Ctg6326 was mapped using this marker.

FPContigs were both syntenic FPContigs for BAC LE_HBa0089M02 and might be merged into one. When we adjusted the E-value to 1e-10 using FPC software, they did merge into one FPContig, but the alignments of DNA fragments within overlapping regions were poor. Through checking the alignments manually, it was found that BAC RH0149P13, for which the BES position on BAC LE_HBa0089M02 was inconsistent with its position on the physical map, produced the poor alignments. Therefore, we tried to merge the two FPContigs again after removing RH0149P13. As a result, they were merged into one FPContig with good alignments (Fig. 11-2B). In addition, it was found that the position of BAC RH090P19 in the new FPContig was altered, making it consistent with the BES alignments. It was concluded that comparative sequencing improves the physical map to some extent.

11.3.2 Selecting Additional Seed BACs

It is very laborious to anchor many BACs on given chromosomes using genetic markers. Usually it is not easy to obtain ample numbers of seed BACs from anchored BACs. Fortunately, for the tomato and potato genome projects, additional seed BACs can be obtained by aligning available BAC sequences from one species to the BESs from the other species. The following example demonstrates how new seed BACs were obtained in the sequencing of chromosomes 11 of tomato and potato.

After several syntenic FPContigs to tomato BACs were identified and reconstructed for potato chromosome 11, new seed BACs were chosen from these contigs for sequencing. Thus, four additional seed BACs were selected for potato chromosome 11 (Table 11-2). Similarly, three tomato FPContigs were obtained by aligning available sequences of 15 potato seed BACs against tomato BESs. Based on this information, two seed BACs were selected from each of tomato contigs Ctg6029 and Ctg7031. LE_HBa0245M17, which had been chosen as a seed BAC carrying the molecular marker CT55, was confirmed to represent the same tomato genomic regions as Ctg6326. Therfore, it was determined to be the syntenic tomato BAC to potato BAC RH021O18.

The reliability of selecting seed BACs using comparative sequencing was validated through mapping the selected seed BACs *in silico* or by fluorescence in situ hybridization (FISH). For example, two potato and two tomato BACs were mapped on chromosome 11 by identifying markers on BAC sequences (Table 11-1, Table 11-2). The tomato BACs, SL_MboI0052K14, LE_HBa0034I10 and LE_HBa0245M17, were localized on chromosome 11 by fluorescent in situ hybridization (FISH) analysis. The FISH pattern for BAC SL_MboI0052K14 is shown in Fig. 11-3 and the data for the latter two BACS are available on the Solanaceae Genomics Network (SGN) (www.

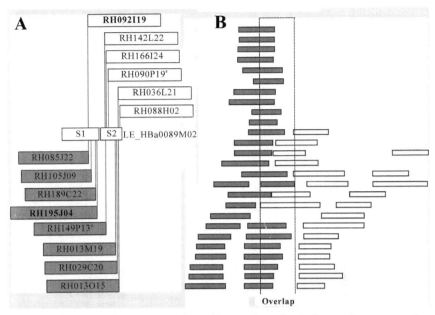

Figure 11-3 Schematic representation of improving physical maps by comparative sequencing method. Gray BACs are from Contig6488, and white are from Contig6847. A. Positions of potato BESs on scaffolds of tomato BAC LE_HBa0089M02. S1: Scaffold 1, S2: Scaffold 2. *The position of BAC on Scaffolds was inconsistent with the physical map. B. Merging of Contig6488 and Contig6847 to build the potato physical map. RH149P13 was removed in order to produce a better alignment. The position of RH090P19 was adjusted, thus it was consistent with that on scaffold sequences.

sgn.cornell.edu). This evidence demonstrated that comparative sequencing was effective and reliable for selecting additional seed BACs for the BAC-by-BAC strategy.

11.3.3 Benefits to BAC Finishing

One of the shortcomings of the BAC-by-BAC sequencing strategy is that BAC finishing requires many PCR reactions and sequencing, which makes it laborious, expensive and time-consuming. However, when tomato and potato syntenic BACs are sequenced simultaneously, comparative sequencing becomes a powerful tool to accelerate BAC finishing. The example described below is from sequencing chromosome 11 of tomato and potato.

As mentioned above, tomato BAC LE_HBa0161D01 and potato BAC RH121D91 are syntenic. Using shot-gun reads with 4–5 fold coverage, the tomato BAC was assembled into four scaffolds, and the potato BAC was assembled into six scaffolds (Fig. 11-4A). To determine the order and orientations of these scaffolds, a number of PCR reactions and additional

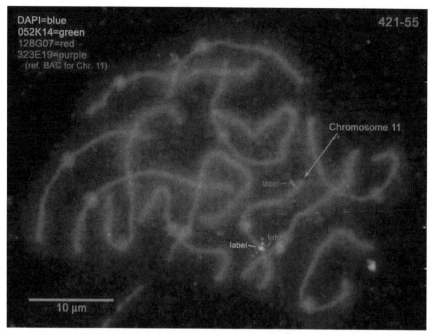

Figure 11-4 Validation of the location of BAC SL_MboI0052K14 on tomato chromosome 11 by FISH. 052K14: SL_MboI0052K14; 128G07: LE_HBa0128G07; 323E19: LE_HBa0323E19, reference BAC for chromosome 11.

Color image of this figure appears in the color plate section at the end of the book.

shot-gun reads are usually performed. However, good alignments among the tomato and potato scaffold sequences were obtained using the MUMmer program (Fig. 11-4B). All scaffolds were ordered and orientated based on the alignments (Fig. 11-4C). We validated these orders and orientations by finishing the BACs. In this way, many PCR and sequencing reactions were avoided, thus accelerating BAC finishing to a significant extent.

In addition, BAC scaffolds can be ordered and oriented using BESs and physical maps. For example, two scaffolds were obtained for tomato BAC LE_HBa0089M02. Combining the potato physical map and the alignments between potato BESs and the scaffolds, two scaffolds were ordered and oriented as illustrated in Fig. 11-2A.

11.4 Comparative Analysis of Syntenic Clones Obtained from Comparative Sequencing

Sequences obtained through comparative sequencing methods are put into informatics analysis in order to confirm the synteny and to obtain insights into the evolution of the tomato and potato genomes.

11.4.1 Characterization of Syntenic Sequences

Nine pairs of syntenic regions were generated on chromosome 11 between tomato and potato in the comparative sequencing initiative of tomato and potato, four on the short arm and five on the long arm (Table 11-3). All tomato and eight potato BACs on loci 3–9 were finished in Phase III. Three potato BACs on locus 1 and locus 2 contained five gaps. The gaps and the flanking sequences were absent in the corresponding tomato BACs, mainly due to inverted repeats, high GC content sequences, and retrotransposons. The total length of syntenic regions was 912 and 1,120 Kb for tomato and potato, respectively. Within syntenic regions, about 62% of the sequences were matched with an average identity of 88%.

The nine syntenic regions were, in general, randomly distributed on chromosome 11 (Fig. 11-5). Thus, it was possible to deduce the conservation and divergence of chromosome 11 between tomato and potato based on these sequences. Locus 4 and locus 5 were located very close to heterochromatic regions according to BAC fluorescence *in situ* hybridization (FISH) results available at SGN, while the remainder of the loci were located in the euchromatic regions. In addition to syntenic BACs, we sequenced two potato BACs (RH130I13 and RH097I18) that were anchored on potato chromosome 11 and could be mapped on tomato in silico (Fig. 11-5). Combining the

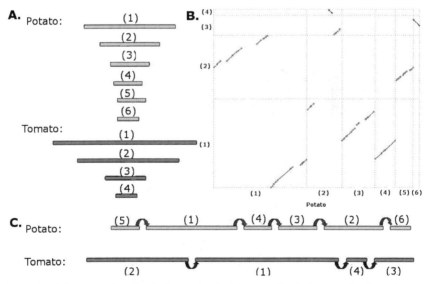

Figure 11-5 An example of ordering and orientating scaffolds by aligning homologous tomato (LE_HBa0161D01) and potato (RH121D01) BAC sequences. A. Scaffolds from homologous potato and tomato BACs; B. Sequences alignments of scaffolds; C. Deduced order and orientation of scaffolds based on alignments.

Table 11-3 Summary of syntenic regions on chromosome 11 of tomato and potato.

Locus	Chr.	Tomato			Potato			Match Length (bp)	Coverage (%)	Average Identity (%)	No. of gene pairs
		BAC	Synteny Length (bp)	No.of genes	BAC	Synteny Length (bp)	No. of genes				
1	Chr11S	a0168B23	68008	9	RH083G18	158487	13	50205	74	88	9
2	Chr11S	a0143O06	132611	17	RH179O17 RH089H13	221204	19	100312	76	89	16
3	Chr11S	a0034I10	86209	5	RH067L08	97131	8	46585	54	91	5
4	Chr11S	m0052K14	96789	10	RH052D21	84461	8	48897	51	88	8
5	Chr11L	a0245M17	134179	4	RH021O18	77663	4	44080	33	88	4
6	Chr11L	a0072I13	130777	11	RH193D01 RH058F17	182052	14	67659	52	88	11
7	Chr11L	a0139J14	101940	13	RH186L12	109587	13	75477	74	89	13
8	Chr11L	a0161D01	95022	11	RH121D01	112399	11	76770	81	87	11
9	Chr11L	a0089M02	66274	10	RH195J04	77942	10	58610	88	88	10
Total			911809	90		1120926	100	568595	62	88	87

Chr11S: short arm of chromosome 11; Chr11L: long arm of chromosome 11; Match Length: the length of regions in tomato which matched corresponding potato BACs; Coverage: percentage of the match length against synteny length in tomato locations of locus 1 and locus 2, a chromosome inversion on the short arm was confirmed, although the inversion did not involve the entire arm.

11.4.2 Colinearity between Tomato and Potato

Eighty-seven orthologous gene pairs were identified within the nine syntenic regions (Table 11-3). Among them, 86 showed conservation in their orders and orientations between tomato and potato, with one inverted gene at locus 2 (Fig. 11-6). This provided sequence evidence for the high level of colinearity between tomato and potato previously reported (Tanksley et al. 1992; Zhu et al. 2008).

Interestingly, we found different patterns for tandem duplicated genes between tomato and potato. Potato had more copies of duplicated genes than tomato within the investigated region. There were six duplicated genes in potato versus three at locus 1 in tomato, and five in potato versus two in tomato at locus 6 (Fig. 11-6). Therefore, six additional genes were identified in potato.

In addition to orthologous genes, there were a few inserted or deleted (indel) genes. In tomato, there were one and two inserted genes at locus 2 and locus 4, respectively, relative to potato (Fig. 11-6). In potato, a larger number of inserted genes were observed, one at locus 1, three at locus 2, and three at locus 3 (Fig. 11-6).

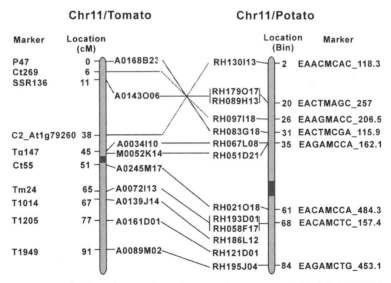

Figure 11-6 Distribution of syntenic regions on chromosome 11. SL_MboI0052K14 was localized on chromosome by FISH (see SGN). Dotted lines represent the regions which were sequenced for potato, and on which the markers in tomato could be identified. The dark regions on chromosomes denote the centromeres.

It was remarkable that all indel genes occurred at loci 1–4, which were from the short arm of chromosome 11, and only one different tandem duplication was observed on the long arm at locus 6 (Fig. 11-6). Furthermore, a higher level of sequence identity was observed on the long arm. These results implied that the level of colinearity on the long arm was higher than that on the short arm. This might be associated with the chromosomal inversion of the short arm.

To investigate the conservation and divergence of gene structures between tomato and potato, a high resolution comparison at Locus 1 was performed. As shown in Fig. 11-7, gene structures were highly conserved between tomato and potato. However, several large insertions or deletions (~2 kb) were observed within intronic regions, suggesting greater divergence in intronic sequences between tomato and potato.

11.4.3 Content and Distribution of Repetitive Elements

Applying homology and structure-based approaches, 12.47% and 11.26 % of the syntenic sequences were identified as repetitive elements in tomato and potato, respectively (Table 11-4). These percentages were clearly lower than those derived from BAC end sequences (BESs) (Datema et al. 2008; Zhu et al. 2008). This is attributable to the fact that these syntenic sequences are mainly from euchromatin regions which contain fewer repetitive elements (Wang et al. 2006). As expected, retrotransposons were the most abundant repeats within these regions in both tomato and potato (10.22% and 9.22%, respectively). Surprisingly, there were more Ty1-copia retrotransposons than Ty3-gypsy in both tomato (5.12% vs. 1.85%) and potato (3.61% vs. 0.74%). This is contrary to the distribution found from the comparison of BESs (Datema et al. 2008; Zhu et al. 2008), but is consistent with the distribution from tomato BACs reported previously (Zhu et al. 2008).

For matched regions, the proportion of repetitive elements was only 2.19% (Table 11-4). For unmatched regions, however, the proportions were dramatically increased to 29.50% and 20.75% in tomato and potato, respectively. This suggested that repetitive elements were a major contribution to the differences between the tomato and potato genomes.

11.4.4 Inserted or Deleted (indel) Genes within Syntenic Regions

A total of ten putative genes were identified as indel genes in unmatched regions of syntenic sequences at loci 1–4 (Fig. 11-6). These genes showed homology to known proteins, but no corresponding ESTs were found in the current EST collections. To help understand the evolution of the tomato and potato genomes, it is important to know how these indel genes originated. Eight of the ten genes had multiple exons, thus this ruled out the possibility

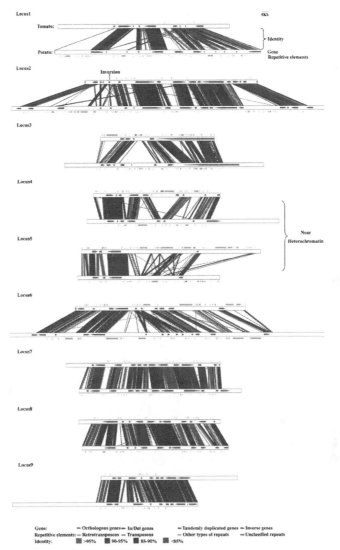

Figure 11-7 Graphic representation of the sequence alignments between syntenic tomato and potato regions, and the distribution of predicted genes and repetitive elements on them. The tomato BAC is located on top, and potato BAC is located on the bottom of each pair. The ribbons connecting two BACs represent the matched regions, and different color denotes the identity. Genes are indicated with pentagons, red indicates orthologous genes, blue indicates Inserted or Deleted (indel) genes, sienna indicates genes which have been tandemly duplicated, and green indicates genes which are inverted between tomato and potato. Transposable elements are showed by colored rectangles. Locus 1–4 are on the short arm of chromosome 11, and Locus 5–9 on the long arm as listed in Table 11-3 Locus 4 and 5 are very close to heterochromatin.

Color image of this figure appears in the color plate section at the end of the book.

Table 11-4 Statistics on proportions of repetitive elements within different regions.

Classification	Syntenic regions (%)		Non-matched regions (%)		Matched regions (%)
	Tomato	potato	Tomato	Potato	
Retrotransposons	10.22	9.22	24.54	16.97	1.58
Ty1-copia	5.12	3.61	13.49	7.83	0.07
Ty3-gypsy	1.85	0.74	4.87	1.57	0.02
LINE	0.78	0.83	0.77	0.67	0.79
SINE	0.05	0.15	0.11	0.27	0.01
Unclassified Retrotransposons	2.42	3.89	5.29	6.63	0.69
Transposons	1.00	1.38	2.16	2.70	0.30
CACTA, En/Spm	0.08	0.11	0.11	0.20	0.06
Mutator (MULE)	0.00	0.01	0.00	0.00	0.00
Unclassified Transposons	0.92	1.26	2.05	2.50	0.24
Ribosomal RNA genes	0.10	0.22	0.07	0.29	0.12
5S rDNA	0.06	0.15	0.05	0.20	0.06
45S rDNA	0.04	0.07	0.02	0.09	0.06
MITE	0.10	0.05	0.14	0.05	0.07
Unclassified	1.05	0.39	2.60	0.75	0.12
Total	12.47	11.26	29.50	20.75	2.19

that these genes were integrated as processed pseudogenes. The remaining two single-exon genes were located in the second indel gene at locus 3 and the first indel gene at locus 4. The single-exon gene at locus 3 was similar to an F-box-like gene named *ATPP2-B10* in *A. thaliana* which is a multi-exon gene, suggesting that it might be a processed pseudogene. The single-exon gene at locus 4 was similar to only a segment of one gene (Galactose mutarotase), so we could not conclude whether it was a pseudogene. In addition, frame-shifts or pre-stop codons within these sequences were not observed, thus most indel genes were not predicted to be pseudogenes. In vertebrates and plants, it has been reported that DNA sequences flanking transposable elements can be inserted into other positions within a genome by transposable element-mediated transposition (Pickeral et al. 2000; Elrouby and Bureau 2001; Wang et al. 2006; Xing et al. 2006). We observed large numbers of transposable elements adjacent to these indel genes (Fig. 11-6). Therefore, it is hypothesized that most indel genes originated from translocation events mediated by transposable elements.

11.4.5 Divergence of Synteny Length between Tomato and Potato

The haploid genome size of tomato was reported to be ~950 Mb, and that of potato was reported to be ~840 Mb (Arumuganathan and Earle 1991). Surprisingly, we found the length of seven out of nine syntenic regions in potato to be longer than those in tomato (Fig. 11-8). The two that were not longer were locus 4 and locus 5, which were very close to heterochromatin. This indicated that the length of potato euchromatic regions was longer than that of the corresponding tomato regions, while the converse was observed for heterochromatic regions.

Euchromatic regions were gene-rich, and heterochromatic regions were gene-poor in the tomato genome (Wang et al. 2006). To further validate this observation, we calculated the proportion of genic regions for each Locus. We found that the loci with relatively higher proportions (>45%) of genic regions were mostly those with longer length in potato (loci 1, 2, 3, 7, 8 and 9). Only locus 6, in which the proportion of genic region was about 30%, was an exception. For locus 4 and locus 5, the proportions were significantly lower (~12% and 28%, respectively). These results supported the contention that the lengths of euchromatic regions of potato were longer than those of tomato.

11.5 Limitations and Opportunities

This chapter described an example of the comparative sequencing approach taken for chromosome 11 of tomato and potato. When the two genome projects operate synergistically, accelerated progress can be achieved for each project. We, therefore, encourage laboratories working on other chromosomes or other genome projects to apply similar approaches. However, the utility of comparative sequencing is limited by certain factors. First of all, an accurate physical map is the precondition to utilizing the comparative sequencing method. Secondly, comparative sequencing theoretically works well within regions with good conservation between tomato and potato containing few repetitive sequences. Inversions, transpositions, duplications, insertions, and deletions interfere with the comparisons between the positions of BES hits on BAC scaffolds and physical maps. It was reported that euchromatic regions were largely devoid of transposons or other repetitive sequences in the tomato genome, implying high conservation between tomato and potato within these regions (Wang et al. 2006). This suggests that the comparative sequencing method can play an important role in sequencing euchromatic regions of tomato and potato genomes.

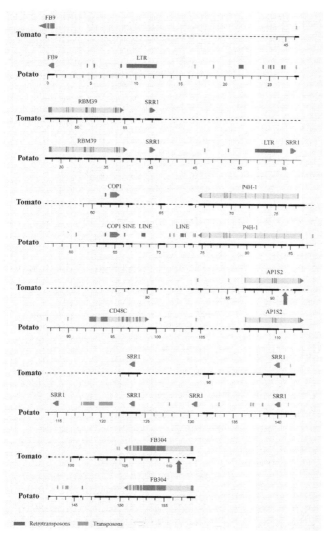

Figure 11-8 Organization of genes and transposable elements on Locus 1 syntenic regions. Thick and thin lines represent matched and unmatched regions respectively, and dashed lines are used to show the lack of sequences. Pentagons denote genes, exons are in chocolate, and introns are in yellow. Types of transposable elements are indicated using differently colored rectangles. Only those transposable elements assigned to specific types are shown. Red arrows point to insertions or deletions (~2 kb) which were observed in the intronic regions of gene AP1S2 and FB304. FB9: F-box protein At2g17036; RBM39: RNA-binding motif protein 39; SRR1: sensitivity to red light reduced 1 protein; COP1: Constitutive photomorphogenesis protein 1; P4H-1: P4H isoform 1 (procollagen-proline 4-dioxygenase); CD48C: Cell division control protein 48 homolog C; AP1S2: AP-1 complex subunit sigma-2; FB304: F-box protein At3g54455.

Color image of this figure appears in the color plate section at the end of the book.

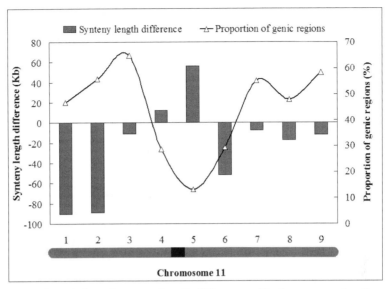

Figure 11-9 Association of synteny length difference with chromosome location and proportion of genic regions. Synteny length difference equals the length of tomato syntenic regions minus the length of the corresponding potato regions. The proportion of genic regions is generated from tomato syntenic regions. Rounded rectangle denotes chromosome 11, and the black rectangle within it is the centromere. The numbers above rounded rectangle denote different locus listed in Table 11-3.

Bioinformatics plays a great role in both sequencing and analysis. During the process of sequencing, analyzing the scaffolds helps facilitate BAC finishing. After obtaining sequences, bioinformatics is a powerful tool to investigate the biological questions that emerge from the data. In sequencing tomato and potato chromosome 11, comparative analysis of the syntenic sequences not only confirmed the correctness of the seed BAC collection but also provided useful information for subsequent sequencing and biological studies.

References

Arabidopsis Genome Initiative (2000) Analysis of the genome sequence of the flowering plant *Arabidopsis thaliana*. Nature 408: 796–815.

Arumuganathan K, Earle ED (1991) Nuclear DNA content of some important plant species. Plant Mol Biol Rep 9: 208–218.

Datema E, Mueller LA, Buels R, Giovannoni JJ, Visser RG, Stiekema WJ, van Ham RC (2008) Comparative BAC end sequence analysis of tomato and potato reveals overrepresentation of specific gene families in potato. BMC Plant Biol 8: 34.

Elrouby N, Bureau TE (2001) A novel hybrid open reading frame formed by multiple cellular gene transductions by a plant long terminal repeat retroelement. J Biol Chem 276: 41963–41968.

International Human Genome Sequencing Consortium (2001) Initial sequencing and analysis of the human genome. Nature 409: 860–921.

International Rice Genome Sequencing Project (2005) The map-based sequence of the rice genome. Nature 436: 793–800.

Kurtz S, Phillippy A, Delcher A, Smoot M, Shumway M, Antonescu C, Salzberg S (2004) Versatile and open software for comparing large genomes. Genome Biol 5(2):R12

Mueller LA, Tanksley SD, Giovannoni JJ, van Eck J, Stack S, Choi D, Kim BD, Chen M, Cheng Z, Li C, Ling H, Xue Y, Seymour G, Bishop G, Bryan G, Sharma R, Khurana J, Tyagi A, Chattopadhyay D, Singh NK, Stiekema W, Lindhout P, Jesse T, Lankhorst RK, Bouzayen M, Shibata D, Tabata S, Granell A, Botella MA, Giuliano G, Frusciante L, Causse M, Zamir D (2005) The Tomato Sequencing Project, the first cornerstone of the International Solanaceae Project SOL. Compar Funct Genom 6: 153–158.

Pickeral OK, Makalowski W, Boguski MS, Boeke JD (2000) Frequent human genomic DNA transduction driven by LINE-1 retrotransposition. Genome Res 10: 411–415.

Soderlund C, Longden I, Mott R (1997) FPC: a system for building contigs from restriction fingerprinted clones. Comput Appl Biosci 13: 523–535.

Tanksley SD, Ganal MW, Prince JP, de Vicente MC, Bonierbale MW, Broun P, Fulton TM, Giovannoni JJ, Grandillo S, Martin GB, Messeguer R, Miller JC, Miller L, Paterson AH, Pineda O, Roder MS, Wing RA, Wu W, Young ND (1992) High density molecular linkage maps of the tomato and potato genomes. Genetics 132: 1141–60.

Wang Y, Tang X, Cheng Z, Mueller L, Giovannoni J, Tanksley SD (2006) Euchromatin and pericentromeric heterochromatin: comparative composition in the tomato genome. Genetics 172: 2529–2540.

Xing J, Wang H, Belancio VP, Cordaux R, Deininger PL, Batzer MA (2006) Emergence of primate genes by retrotransposon-mediated sequence transduction. Proc Natl Acad Sci USA 103: 17608–17613.

Zhu Q, Song B, Zhang C, Ou Y, Xie C, Liu J (2008) Construction and functional characteristics of tuber-specific and cold-inducible chimeric promoters in potato. Plant Cell Rep 27: 47–55.

12

Developments in Tomato Transcriptomics

*Antonio J. Matas,[1] Zhangjun Fei,[2,] James J. Giovannoni[2]
and Jocelyn K.C. Rose[1,]**

ABSTRACT

Advances over the last decade in tomato "omics" have consolidated
the species as an experimental model for the study of fleshy fruit
biology. This chapter reviews existing tools and methods that have been
applied to characterize the tomato transcriptome, and also the changes
that the field is experiencing as a consequence of new sequencing
technologies and resulting sequence-based resources. The enhanced
rate and accuracy of transcript identification and expression profiling,
together with plummeting sequencing costs, have displaced the research
bottleneck from data acquisition to data processing and integration.
This constitutes the next big challenge in the field of genomics but also
represents numerous exciting opportunities for tomato research. The
potential practical applications, together with some of the associated
challenges are discussed in the context of an unprecedented and ever-
increasing quantity of transcript expression data.

Keywords: Transcriptome, gene expression, microarray, high
throughput sequencing, RNA-seq, data analysis

[1]Department of Plant Biology, Cornell University, Ithaca, NY 14853, USA.
[2]Boyce Thompson Institute for Plant Research, Tower Rd., Cornell University campus, Ithaca,
NY 14853 USA and USDA-ARS Robert W. Holley Center for Agriculture and Health, Tower
Rd., Cornell University campus, Ithaca, NY 14853 USA.
*Corresponding author

12.1 Introduction

The study of the transcriptome, that is, the complete set of RNAs that are present in an organism, organ, tissue or cell type, has emerged as a highly productive experimental strategy to dissect many aspects of tomato (*Solanum lycopersicum* L.) biology, including vegetative development, fruit ripening and responses to biotic/abiotic stresses. Indeed, advances in tomato genomics, and transcriptomics in particular, have helped propel the species to its prominence as an experimental model for the study of fleshy fruit biology, pathogen responses and more general areas of crop breeding and improvement. In this section, we review existing tools and methods that have been applied to characterize the tomato transcriptome, and also the linked and revolutionary changes that the field is experiencing as a consequence of new sequencing technologies and resulting sequence-based resources. The remarkable increase in the rate and accuracy of transcript identification and expression profiling, together with plummeting sequencing costs and emerging genome sequence of tomato and several of its relatives, which are presenting numerous exciting opportunities for tomato research. We also discuss some potential practical applications, together with some of the challenges associated with the unprecedented and ever-increasing quantity of transcript expression data.

12.2 Transcriptomics Tools Developed

The methodologies developed to study transcriptomics can be classified into two groups: those with a 'closed-architecture', such as microarray profiling, which depends on previous knowledge of the genetic composition of the target organism but which has until now been generally least expensive, and 'open-architecture' strategies, such as serial analysis of gene expression (SAGE) and RNA sequencing (RNA-seq), which have previously been more expensive but that allow new gene discovery as they are not limited by existing knowledge of gene sequence.

12.2.1 Microarrays

Microarrays have generally been the basis of the most widely exploited methodology in tomato transcriptomics research. A single microarray typically consists of thousands of microscopic spots (termed 'features', or 'targets') of amplified cDNAs, or cDNA oligonucleotides, of specific DNA sequences arrayed on a solid surface (Schena et al. 1995; Shalon et al. 1996; Chen et al. 1998; Churchill 2002; Alba et al. 2004). A labeled sample of cDNA(s) or cRNA(s), termed 'probes', is then hybridized with the microarray and the probe-target hybridization is detected and quantified by means of fluorescence or chemiluminescence, depending on the nature of label. Protocols to prepare the sample and perform the hybridization can be modified depending on the sample and the platform used, but currently there are well established methodologies that are available in the literature or in the microarray data repositories that are compliant with the "Minimum Information About a Microarray Experiment" (MIAME) guidelines (Brazma et al. 2001) employed as a standard by many peer-reviewed journals. Briefly, the typical steps include labeling of the probes (cDNA or RNA) with one of two cyanine fluorochromes (Cy3 or Cy5) that fluoresce at distinct wave-lengths, followed by hybridization of the labeled probes with the targets on the array chip. Several washing steps are then used to reduce non-specific hybridization and finally, a digital scanning process records the signal intensity of each spot, resulting in an image of the entire array. This image provides the basis for quantification as the first step of estimating relative transcript abundance after background subtraction and normalization steps.

Many different microarray platforms have been specifically designed to study tomato transcriptomics based on the extensive collection of tomato expressed sequence tags (ESTs) (Mueller et al. 2005). Two of the most widely found in publications, TOM1 and TOM2, use the two-color method, involving simultaneous hybridization of probes used in a direct comparison with Cy3 and Cy5 labels, respectively. TOM1 is a spotted cDNA array that represents over 8,000 independent genes, while TOM2 is an oligonucleotide array of 11,860 different 70-mers specifically designed to target unique regions of each gene with reference to the entire cDNA collection available at the time of synthesis (Alba et al. 2004, 2005). TOM2 arrays are currently available from the Tomato Functional Genomics Database (http://ted.bti. cornell.edu/cgi-bin/TFGD/order/home.cgi) and through the EU-SOL project (http://www.eu-sol.net). A third public platform, which uses the one-color method where only one sample is hybridized and quantified with each hybridization, is the GeneChip® Tomato Genome Array that has been commercialized by Affymetrix (www.affymetrix.com) and consists of approximately 10,000 tomato probe sets (11 different probes of 25-mers) that can be used to interrogate over 9,200 distinct transcripts. It should be noted that a more comprehensive Affymetrix tomato array is under development.

Since 2004, many research groups have used tomato microarrays to study gene expression and many of these experimental results have been collated in the Tomato Functional Genomics Database (TFGD) (Fei et al. 2004, 2006), including the array datasets that have been deposited in GEO (Barret et al. 2007) and ArrayExpress (Brazma et al. 2003). The list of experiments using each of the three platforms is shown in Table 12-1 and includes transcript expression profiling analyses of different tomato organs

Table 12-1 Experiment list of hybridization experiments using the tomato array platforms hosted at the Tomato Functional Genomics Database (http://ted.bti.cornell.edu/cgi-bin/TFGD/miame/home.cgi).

TOM1 cDNA array

ID	Title	Organism	Tissue	Keywords	PUBMED ID
E011	Wild type tomato fruit development (set 1)	Tomato AC	Pericarp	Time course, ripening, quality	Alba et al. 2005
E025	Wild type tomato fruit development (set 2)	Tomato AC	Pericarp	Time course, ripening, quality	Alba et al. 2005
E016	tomato *Never-ripe* mutant fruit development	Tomato AC, *Nr*	Pericarp	Time course, ripening, mutant	Alba et al. 2004
E026	tomato *rin* mutant fruit development	Tomato AC, *rin*	Pericarp	Time course, ripening, mutant	
E027	tomato *nor* mutant fruit development	Tomato AC, *nor*	Pericarp	Time course, ripening, mutant	
E028	tomato *hp1* mutant fruit development	Tomato AC, *hp1*	Pericarp	Time course, ripening, mutant	
E020	Wild type pepper fruit development	Pepper, tomato	Pericarp	Time course, ripening, regulation, comparative transcriptomics	
E012	Tomato (breaker-stage) vs. Pepper (breaker-stage)	Pepper, tomato	Pericarp	Tissue and inter-specie comparison, ripening	Alba et al. 2004
E018	Solanaceae fruit comparison	Tomato, pepper, eggplant	Fruit	Maturation, inter-specie comparison	Moore et al. 2005
E019	IL carotenoid	Tomato, introgression line	Fruit	Carotenoid, ripe fruit, regulation	
E017	Tomato introgression line transcript analysis	Tomato, introgression line	Fruit	Biochemical trait, 20daa fruit, regulation	Baxter et al. 2005
E013	Transcriptional changes determined by expression of the *Xanthomonas* Avr/Rxv effector in resistant tomato plants	Tomato	Leaf	Pathogen resistance, AvrRxv effector protein, gene modulation	Bonshtien et al. 2005
E014	Time course of *Pseudomonas syringae* DC3000 vs. delta hrp on tomato Prf-3	Tomato	Leaf	*Pseudomonas syringae* pv. *tomato* type III effectors, DC3000 type III secretion system, inoculation, gene regulation.	Cohn et al. 2005

ID	Title	Species	Tissue	Description	Reference
E015	*Pseudomonas syringae* DC3000 effector mutants	Tomato	Leaf	*Pseudomonas syringae* pv. *tomato* type III effectors, inoculation, AvrPto, AvrPtoB, gene regulation.	Cohn et al. 2005
E021	Matt's wild cherry tomato vs. *Phytophthora infestans*	Tomato		Time course, *Phytophthora infestans*, compatible interaction, inoculation, gene regulation	
E022	Gene expression profiling of infection of tomato by *Phytophthora infestans* in the field	Tomato		Field essay, *Phytophthora infestans*, gene suppression	
E039	Transcriptional profiling of the tomato fruit pericarp tissue: fruits over-expressing yeast *ySAMdc* gene and controls	Tomato	Pericarp	Transgenic plants, Spd/Spm-accumulation, transcriptome pattern alterations, ripening.	
E040	Tomato root culture and Al treatment	Tomato	Root	Abiotic stress, aluminum stress, transcript accumulation	Zhou et al. 2008
E042	Microarray Analysis on Salt Stress Affected Genes in Tomato Seedlings	Tomato	Seedlings	Abiotic stress, salt stress, gene expression	Zhou et al. 2007
E043	Transcriptional Profiling of high pigment-2dg Tomato Mutant Fruit	Tomato	Fruit	Ripening, carotenoid mutant, high pigment-2dg, transcriptional alterations	Kolotilin et al. 2007
TOM2 oligo array					
E029	Transcriptome profiling of *S. pennellii* ILs	Tomato, introgression line	Fruit	Ripe fruit, gene regulation	
E038	Impact of sucrose retrieval upon *ex vitro* acclimatization	Tomato	Plantlets	Expression profiling, acclimatization time scale, short-term sugar deprivation	
E041	Transcriptional profiles of drought-responsive genes in modulating transcription, signal transduction and biochemical pathways in tomato	Tomato, *S. pennellii*, introgression lines	Plant	Drought response, gene expression profile, drought-tolerant lines	

Table 12-1 contd....

Table 12-1 contd....

TOM2 oligo array

ID	Title	Organism	Tissue	Keywords	PUBMED ID
E044	Transcription profiling of tomato strains during fruit set	Tomato	Ovary, young fruits	Transcription profiling, development, co-expression	Wang et al 2009
E045	Transcription profiling of tomato plants at three time points in the light:dark cycle	Tomato	Seedlings	Time course, circadian clock, gene expression, co-expression	Facella et al. 2008
E046	Transcription profiling of tomato shoots and roots from plants systemically infected by TSWV reveals organ-specific transcriptional responses	Tomato	Leaf, root	TSWV infection, gene response	Catoni et al. 2009
E047	Transcription profiling of roots and leaves of tomato plants colonized by the arbuscular mycorrhizal fungus *Glomus mosseae*	Tomato	Leaf, root	*Glomus mosseae*, arbuscular mycorrhizal, gene response, co-expression	Fiorilli et al. 2009
E049	Ectopic expression of a MYB transcription factor defines a set of co-regulated genes involved in phenylalanine and phenylpropanoid synthesis in tomato fruit	Tomato	Fruit	Ectopic expression, MYB transcription factor, gene expression, gene response	

Affymetrix Tomato Genome Array

ID	Title	Organism	Tissue	Keywords	PUBMED ID
E023	Expression profiles of the tomato line M82 in response to the fungal elicitor EIX	Tomato	Leaf	HR response, Ethylene inducing xylanase (EIX) elicitor, gene response.	
E024	Transcriptome analysis of the tomato-*Clavibacter michiganensis* subsp. michiganensis compatible interaction	Tomato	Stem	*Clavibacter michiganensis* subsp. Michiganensis, inoculated tissue, gene response, co-expression	Balaji et al. 2008
E030	Gene regulation in parthenocarpic tomato fruit	Tomato	Ovule, carpel	Transgenic plants, auxin, parthenocarpy, tissue development, co-expression	Martinelli et al. 2009

E031	Expression data of tomato fruit responses to *Botrytis cinerea*	Tomato	Fruit	*Botrytis cinerea*, susceptibility, ripening, gene expression profiling	Cantu et al. 2009
E032	Expression data from PSARK::IPT Nicotiana and wild type plants	Tobacco	Leaf	Transgenic plants, PSARK::IPT, drought-responsive, senescence, transcriptional response	
E033	Trafficking of mRNA from tomato to dodder	Tomato, lespedeza dodder	Leaf, stem	Parasitic plant, mRNA trafficking, phloem transport.	Roney et al. 2007
E034	Transcription profiling of tomato seedlings Condine wild type vs. Divaricata a-DOX2 mutant	Tomato	Hypocotyl, cotyledons	*Divaricarta* mutant, alpha-dioxygenases, fatty acids biosynthesis, complementation	Bannenberg et al. 2009
E035	Transcription profiling of tomato strains RP75/59 and UC82 carpel development	Tomato	Fruit, ovary, flower, petal	Parthenocarpic fruit, ethylene, gibberellins, anthesis, fruit set, time course	Pascual et al. 2009
E036	Tomato wild type, vector control and transgenic lines expressing InsP 5-ptase: mature fruits, root tips, and leaves	Tomato	Fruit, root, leaf	Transgenic plant, inositol-1,4,5-triphosphate, Micro-Tom, ripe fruit	
E037	Expression data from various tomato plant tissues	Tomato	24 diff tissues	Tissue comparison, gene expression	Ozaki et al. 2010
E048	Effects of Potato Spindle Tuber Viroid on Tomato Gene Expression	Tomato		RNA silencing, Potato Spindle Tuber Viroid, infection, gene expression, co-expression	

Abbreviations: AC, Ailsa Craig; IL, introgression line; *Nr*, *Never-ripe* mutant; *rin*, *ripening-inhibitor* mutant; *nor*, *non-ripening* mutant; *hp1*, high-pigment mutant; HR hypersensitive response; TSWV, Tomato Spotted Wilt Virus.

and tissues, comparisons between tomato and other related Solanaceous species and the effects of abiotic and biotic stresses on gene expression.

It is important to bear in mind that microarrays are subject to a number of important limitations (Bier and Kleinjung 2001; Draghici et al. 2006). For example, they represent a closed architecture analytical platform, so the expression levels of only those genes represented on the array are analyzed. This was illustrated by the identification of several putative genes that regulate fruit quality by suppression subtraction hybridization that are not present on the TOM2 array (Page et al. 2008). In addition, the number of probes that can be used simultaneously is limited and low level transcripts are usually not well represented. Other limitations are the potential for cross-hybridization of closely related paralogous sequences, which can confound accurate estimate of transcript abundance, background signal (nonspecific hybridization) and limited dynamic range of hybridization. The primary use of microarrays is for gene expression profiling by relative quantification of the amount of each gene transcript present in a sample, or by the ratio comparison of the same target in two samples. However, they can also be used to detect single nucleotide polymorphisms (SNPs) or alternative transcript splice variants (Hacia et al. 1999; Cutler et al. 2001; Syvanen 2001; Lin et al. 2004), although this can be technically difficult and great care is needed with controlling the hybridization conditions. Based on the literature, this later use has not been widely employed in tomato to date, presumably in part due to the high rate of inbreeding and resulting low degree of polymorphism within cultivated tomato.

12.2.2 SAGE and Related Technologies

Serial analysis of gene expression (SAGE) is an open architecture methodology that is used for transcriptomics (Velculescu et al. 1995). The technique is based on the assumption that short sequences (10-14 base pairs) contain sufficient information to uniquely identify a transcript. These so-called 'sequence tags' are created and then ligated together to form long serial molecules that can be vectorized and introduced into bacteria for replication and subsequent sequencing. Quantification of the number of times a particular tag is observed in a large collection of sequenced insets provides an estimation of the relative expression level of the corresponding transcript as compared to other sequence tags comprising the total tag population. Obviously these observations are not based on hybridization, rather on direct, quantitative, digital values corresponding to the relative number of transcripts, absent the background cross-hybridization noise that can occur with microarrays. One of the main advantages of SAGE is that it allows for new genes and gene variants to be identified (granted as very short nucleotide sequences), since a priori knowledge of the mRNA sequences is not necessary. SAGE-based methodologies have generated high quality cDNA sequence sets and gene expression profiling data that relate to regulatory pathways and host-pathogen interactions in tomato (Irian et al. 2007; Uehara et al. 2007) and other related species (Fukuoka et al. 2010; Yamaguchi et al. 2010).

Even though SAGE experiments are not as common as microarrays, in part due to the relatively high cost, the basic methodology has been modified and improved (Powell 1998), resulting in enhanced tools such as SuperSAGE (Matsumura et al. 2003) and Robust-LongSAGE (RL-SAGE) (Gowda et al. 2004). While SAGE technologies have improved and it seems likely that this technology will remain as a minor methodology for quantitative expression analysis as it is primarily represented as a means of reducing sequencing costs via creation of multimeric tags that could be sequenced in a single reaction. The development and application of next generation sequencing technologies, as described below, will in many cases surpass the efficiencies of SAGE thus limiting application of this technology to specialized applications.

12.2.3 Massive Parallel Signature Sequencing

Massive parallel signature sequencing (MPSS) is a potentially powerful, but relatively infrequently adopted, open-architecture approach to study transcript expression (Brenner et al. 2000; Reinartz et al. 2002). cDNA samples are tagged and amplified in a PCR reaction so that each mRNA could give rise to ~100,000 unique tagged products. The products are attached through the tag to a micro-bead, and a combination of the restriction enzyme *BbvI* and several rounds of ligation-based sequence determination create a sequence signature, typically comprising 17 base pairs of high quality sequence. When performed in parallel, approximately one million MPSS tag sequence signatures are generated from each experiment and each is analyzed and compared with all other signatures. Those that are identical will be counted and used to estimate the level of expression of mRNA species in the sample, normalized by the total number of signatures of all mRNAs. The sensitivity of MPSS is at the level of a few molecules of mRNA per cell and the information is directly stored in digital format for management and analysis. This technology has been applied to study several plant species (Nakano et al. 2006), but to date no MPSS study of tomato has been reported.

12.3 Advances in Sequencing Technologies

The last few years have witnessed remarkable innovations in DNA sequencing technologies that are already fundamentally changing the ways in which life science experimentation is designed and implemented. These open-architecture techniques, which are variously called 'second generation', 'next-generation', 'high-throughput' or 'deep' sequencing, allow the sequencing of hundreds of thousands, to millions of small fragments of DNA in a short time at a far lower cost (per base) than any previous technology (Wall et al. 2009; Metzker 2010). However, the assembly of short sequences is technically challenging in the absence of an existing DNA sequence scaffold, or template, such as a high quality genome sequence: thus, data storage and analysis have become some of the new limitations to overcome. Here we describe the three most commonly used second-generation sequencing platforms (Mardis 2008) and discuss their advantages and limitations when applied to the study of the transcriptome: an approach that is commonly termed 'RNA-seq'.

12.3.1 454 Genome Sequencer FLX

When working with an 'orphan' organism, for which there is not currently a complete genome sequence, the general consensus is that 454 sequencing technology (produced by 454 Life Sciences, a Roche Company; http:// www.454.com) is currently the most useful second generation sequencing platform for RNA-seq studies primarily due to the high quality and relatively long read lengths generated by this technology. The Genome Sequencer FLX (GS FLX) from 454 can currently generate more than 1 million high-quality reads in a 10 hour run, with read lengths of 400 bases when using the proprietary Titanium Chemistry. The method is based on sequencing by synthesis, or pyrosequencing (Ronaghi et al. 1996, 1998; Nyrén 2007), where the activity of a DNA polymerase is detected by chemiluminescence. For transcriptome analysis purposes, a cDNA library is constructed from an RNA sample such that the resulting cDNA molecules include two adaptors that link the cDNA to a microbead. A PCR reaction is then performed in an emulsion to increase the number of copies of the modified cDNA attached to each bead and the beads are deposited in the wells of a plate, where the second strand of each cDNA is synthesized one base at a time. When a new dNTP is incorporated during each base synthesis, a chemiluminiscent signal is generated that is detected by a sensitive camera and associated with the specific base of dNTP that was present in the reaction cycle. Individual overlapping sequences can then be assembled into overlapping 'contigs' resulting in an EST collection against which these individual and future transcript reads can be mapped.

At present, the main advantage of 454 technology over other existing methods is the relatively long reads, which substantially increase the ease of sequence assembly for transcripts, or even genomes, in the absence of an additional reference sequence. The main disadvantage is the number of reads obtained per run, which is typically several times to orders of magnitude lower than competing technologies, such as Illumina sequencing, with a similar run cost. However, it is currently the preferred approach when no genome sequence is available or as a sequencing complement to SAGE-based or other shorter read next generation sequencing approaches (Nielsen et al. 2006; Molina et al. 2008; Gilardoni et al. 2010).

12.3.2 Illumina

The next generation sequencing technology from Illumina (http://www.illumina.com) uses a sequencing-by-reaction approach that occurs in parallel on the surface of a 'flow cell', which provides a large surface for thousands of chemical reactions. The basic steps of the process start with a DNA sample (for transcriptomics, RNA-derived cDNA) that is sheared to a targeted size. The ends of the DNA fragments are modified and ligated to two unique adapters before size fractionation and PCR amplification. Single-stranded, adapter-ligated fragments are linked to the surface of the flow cell and exposed to reagents for DNA polymerase-based extension. The priming of the reaction initiates when the distal end of a ligated fragment bridges to a complementary oligonucleotide on the surface. Cycles of denaturation and extension result in localized amplification of single molecules in millions of unique locations across the flow cell surface. The flow cell with millions of clusters is then subjected to cycles of extension and imaging (Bentley et al. 2008). Total throughput on a single Illumina run (8 channels) currently exceeds 150 million reads.

The main advantage of Illumina sequencing is the large number of reads that allow deep sequence coverage, which allows for biological replicates at a reasonable cost. On the other hand, the relatively short length of the reads, as compared to 454, increases the risk of inefficient transcript assembly. However, with the rich collection of tomato ESTs that are now available, in addition to the recently released tomato genome sequence (http://solgenomics.net), the Illumina sequencing platform is fast becoming the preferable approach for tomato transcriptome studies.

12.3.3 ABI Solid Sequencing

SOLiD (Sequencing by Oligonucleotide Ligation and Detection; Valouev et al. 2008) follows a similar approach of cDNA library preparation to 454 sequencing; linking the cDNA fragments to magnetic beads through a universal P1 adapter and performing an emulsion PCR step in micro-reactors. The resulting PCR products attached to the beads are covalently bound to a glass slide where primers hybridize to the P1 adapter within the library template. Four fluorescent labeled di-base probes compete for ligation to the primer and their specificity is achieved by interrogating every 1st and 2nd base in each ligation reaction. The read length is determined by multiple cycles of ligation, detection and cleavage. After a series of ligation cycles, the extension product is removed and the template is reset with a primer complementary to the n-1 position for a second round of ligation cycles. Five rounds of primer reset are completed for each sequence tag.

At the time of writing, the SOLiD platform currently generates by far the greatest number of reads but it is also the most expensive per run with the effective cost per base-pair being similar to Illumina. Perhaps the main disadvantage is that a high quality DNA scaffold is required to allow high fidelity base calling, due to the potentially ambiguous two-bases detection system, and so it has proved most useful with organisms that have a complete and well curated genome sequence.

12.3.4 Third Generation and other Sequencing Technologies

Several other sequencing technologies are being launched that promise further improvements in sequencing efficiency and quality. One such system is Polonator (http://www.polonator.org), an open architecture platform that is also based on a flow cell, but for a fraction of the cost of other technologies and with open source software for data acquisition and analysis. Perhaps the most intriguing approaches focus on single molecule sequencing, which eliminates many of the disadvantages of technologies dependent PCR amplification. Three examples are the HeliScope Single Molecule Sequencer from Helicos Bioscience (http://www.helicosbio.com), a single molecule real time sequencing platform from Pacific Bioscience (http://www.pacificbiosciences.com, Korlach et al. 2008) and the Nanopore sequencing system from Oxford, Nanopore Technologies (http://www. nanoporetech.com). Doubtless additional new technologies will be on the horizon by the time this article is published as this is a highly dynamic field. We note that is the intent of this review to avoid endorsing any one technology as all have unique attributes and advantages.

12.3.5 RNA-Seq

RNA-seq analyses are at present typically based on 454, Illumina or SOLiD sequencing platforms. These generate not only quantitatively accurate measures of individual gene expression, but also promote the discovery and quantitative expression analysis of novel transcribed sequences, alternatively spliced variants of the same genes and SNPs between different alleles, all of which can be difficult to differentiate on all but the most robust microarray platforms. An additional advantage over traditional microarrays is that limits of the dynamic range are determined only by the number of sequences obtained (Wang et al. 2009; Wilhelm and Landry 2009; Nagalakshmi et al. 2010). It is important, however, to consider the new challenges that these approaches present and the additional steps that are needed, such as normalization of the RNA-Seq datasets to improve accuracy of expression analysis (Robinson and Oshlack 2010).

RNA-seq has already been applied to characterize the transcriptional complexity of fruits (Zenoni et al. 2010), to compare gene profiling of different genotypes and their dynamics through fruit development (Alagna et al. 2009), and to study allele-specific expression (Main et al. 2009). It also allows for complex system analysis where transcript populations from different species are present allowing for individualized analysis and relationship elucidation (Ruzicka et al. 2010). There remain inherent technical biases in the system, such as those associated with the amplification and ligation steps of their protocols, which necessitates the development of bioinformatic tools for bias discovery and compensatory data processing.

12.4 Bioinformatic Challenges and Data Analysis

A common feature among the different expression profiling technologies available to study the tomato transcriptome is their ability to produce vast amounts of data. In a single microarray experiment over 10,000 genes can be monitored simultaneously and this is typically scaled up in the context of a project that involves multiple comparisons, such as examining a time course or comparison of expression in multiple tissues or treatments. This capacity to generate data has undergone a quantum leap with the new generation of sequencers and experiments can now generate hundreds of thousands to millions of ESTs, which occupy terabytes of hard drive space and require ever increasing amounts of processing time and complex algorithms (Matukumalli and Schroeder 2009). The bottleneck in current genomics scale biological research has changed from being one of data acquisition to data processing, and while the cost of DNA sequencing is constantly decreasing, this is not the case for computational hardware. This limitation, together with a shortage in trained bioinformaticians with the requisite expertise, is an increasingly common limitation for many research laboratories involved in transcriptome analysis (Fuller et al. 2009; McPherson 2009).

In response to this highly dynamic situation, community oriented initiatives are becoming a valuable/essential, means to share knowledge, to set standards and solve problems. One example is SEQanswers (http://seqanswers.com), a forum where people from computer science, bioinformatics, sequencing facilities and industry discuss the various facets of the new era of sequencing-related technologies, including transcriptomics. The interest in, and challenges of, these new sequencing technologies has furthered the development of the Bioperl project (http://www.bioperl.org/wiki/Main_Page), which focuses on providing tools and solutions for sequence and data analysis. Many of these tools can be integrated, as is the case with Gbrowse (http://gmod.org/wiki/Gbrowse, Stein et al. 2002), an open source free application that can display both genome and transcript data as a graphical output through a web-browser. This allows the display of transcript reads from RNA-seq experiments at single base resolution aligned with the corresponding region of the genome, to elucidate gene arrangements, alternative splice variants, or reveal bias in the sequence that could otherwise be difficult to visualize.

In addition to the challenges of data analysis and presentation, another major issue is data storage and curation. This is not only a problem of space due to the large amounts of raw data generated, but also the way in which data is stored, which should be in a form that is readily accessible for the research community. The Sol Genomics Network (http://solgenomics.net) is the central hub of the International Tomato Genome Sequence project and centralized bioinformatics resource for tomato, as well as other Solanaceous species. However, at present (mid 2010) infrastructure for gene expression profiling, including both microarray and second generation sequencing data, are principally housed in the Tomato Functional Genomics Database (http://ted.bti.cornell.edu; Fei et al. 2006). This database uses the collected microarray data from GEO and ArrayExpress together with some custom datasets to create a powerful tool for gene co-expression analysis, which allows the characterization of a specific gene present in any of the tomato microarray platforms across all the gene expression experiments ever performed and deposited. The database also hosts the experiment protocols and hybridization conditions, metabolite information and several tools for data correlation and analysis, and provides access information about tomato small RNAs and an analysis of digital expression data from non-normalized EST libraries (Fei et al. 2004). Other resources for tomato include comparatively smaller datasets such as full-length cDNA collections from the tomato cultivar Micro-Tom (Yano et al. 2006; Barone et al. 2008; Aoki et al. 2010).

The final challenge is how to process the resulting transcriptomics data. New tools are constantly being developed that allow better control of the assembly parameters and reduce both time and computer power needs (Huttenhower and Hofmann 2010). At the time of writing, no fewer than 43 different assembly tools are available for the various platforms and over 35 relate to RNA-seq, including those specifically design for gene expression analysis (SEQanser wiki bioinformatics tools for assembly). When choosing a given tool, several factors should be considered, including the length of the reads, whether SNPs and alternative splicing should be considered and the threshold to set for rejecting mismatches.

12.5 Use of Related Species Resources

Comparative genomic studies have shown that different members of the Solanaceae family share a high degree of gene synteny, with sufficient regions of similarity that most non-coding regions can be aligned, but with enough divergent areas to elucidate evolutionary dynamics and selective pressures (Wang et al. 2008). Identification of large sets of single-copy conserved ortholog set II (COSII) markers has accelerated comparative mapping studies among major Solanaceous species, including tomato, potato, eggplant, pepper and diploid *Nicotiana* species, allowing for the reconstruction of a hypothetical genome configuration for the ancestors of those species and explaining changes in chromosomal distribution that could be related with species diversification and domestication (Wu and Tanksley 2010).

At the gene expression level, studies have also revealed a high degree of transcript sequence similarity, for example allowing the use of tomato microarray platform to study gene expression in pepper and eggplant, where 34% of the transcripts were detected in the three species simultaneously, while another 29% was present in at least two of them (Moore et al. 2005). This study also revealed another third of the genes present in the microarray that could account for some the differences between the Solanaceous species studied (Moore et al. 2005). It is clear that both commonalities and differences (Slocombe et al. 2008) that are seen among the Solanaceae can be used to discover important genes in tomato biology with the corollary of course also true.

12.6 Application of Functional Genomics in Genomics-Assisted Breeding

Recent progress in assorted tomato 'omics' platforms, including those applicable to transcriptomics, has enabled the discovery and functional analysis of large numbers of genes associated with growth and development, yield, tolerance to biotic and abiotic stresses and improved fruit quality. The use of high-throughput functional genomics and systems approaches can allow more accurate predictions of regulatory processes and the genes that underlie them (Phillips 2008). Applying such methodologies to mapping of quantitative trait loci (QTL) in tomato has already resulted in the detection of epistatic interactions among genes and their effect in fruit quality parameters of tomato populations (Baxter et al. 2005; Causse et al. 2007; Stevens et al. 2007). Epigenetic modifications and small RNAs also play an important role in gene regulation and have been shown in some cases to be tissue and developmentally specific in their activities (Moxon et al. 2008; Elling and Deng 2010). Thus, transcript profiling can be combined with proteomic and metabolomic studies to provide a systems approach to complex biological phenomena, as has already been demonstrated in studies of fruit development and ripening (Carrari et al. 2006; Faurobert et al. 2008; Schilmiller et al. 2010), and to identify key regulatory genes that control host-pathogen interactions at the cellular level (Fiorilli et al. 2009).

Coordinated efforts from genome sequences to advanced mapping populations and development of corresponding molecular marker resources has facilitated rapid acceleration in the rate of isolating agronomically important QTLs, creating new opportunities for crop improvement. At present, an important challenge remains concerning how to convert enormous amounts of data into knowledge that can be readily applied to practical crop breeding programs (Edwards and Batley 2010). Any effort should consider a comprehensive approach that will define the performance of a crop grown under real production conditions, from the genome-scale level to the summed array of final complex traits of the organism (currently known in genomics terminology as the 'phenome'). In this regard, comprehensive transcriptome profiling can play a key role in revealing critical regulatory processes, as well as having an inherent value in enhancing or enabling other levels of knowledge, such as defining expression attributes as an aspect of genome annotation and providing an essential reference for proteomic analyses (Mochida and Shinozaki 2010).

In this context, ongoing initiatives like the US Department of Agriculture SolCAP (http://solcap.msu.edu), which aims to provide infrastructure to link allelic variation in genes to valuable traits in cultivated germplasm (specifically of potato and tomato), is of great interest. SolCAP focuses on elite breeding material that may increase the probability that these Solanaceous crops will benefit from genotype-based selection. Another integrative platform is Plant MetGenMAP (Joung et al. 2009), which should facilitate correlation analyses of public/published data on metabolic pathways with gene expression values derived from tomato microarray or second generation sequencing projects. Emerging bioinformatic tools will also advance the analysis of pathways and biological processes shown to be in flux, and thus potentially target process related, in the dataset of interest. Other initiatives, such as ISOL@ (Chiusano et al. 2007), iPlant (http://www.iplantcollaborative.org) and GMOD (http://www.gmod.org), will certainly provide additional useful tools and methodologies for establishing a common framework for data acquisition and analysis, with special emphasis on phenotypic description and its relationship to transcriptome data.

With rapidly advancing genomics data and associated bioinformatics tools that facilitate analysis and discovery, the future of tomato breeding will likely lie in a much more target oriented 'breeding by design' approach than has been previously possibly. Such strategies will be genomics enabled and will take advantage of the extensive biodiversity that exists in the thousands of tomato lines and wild species stored in tomato germplasm collections. This process and the next generation of genomics-enabled plant improvement will be driven by the discovery, characterization and exploitation of genes that are unveiled through functional genomics based high-throughput technologies (Bai and Lindhout 2007; Gupta et al. 2009), resulting in greatly improved efficiency in the development of enhanced quality tomato lines (Meli et al. 2010).

12.7 Summary

The last decade has seen the birth and rapid expansion of global transcriptomics studies in many species, including tomato, providing researches and applied breeders with a body of knowledge that has played a large part in advancing tomato as both a model system for plant research and an important vegetable crop the world-over. To date, microarrays have played a central role in tomato transcriptome analysis; however, the last two years have witnessed a revolution in the functional genomics paradigm. The rapid advancement of high-throughput next generation sequencing technologies has advanced the research bottleneck from data acquisition to data processing and integration. The development of new tools for the analysis of RNA-seq experiments and the formats in which these data are presented to the research community, for both basic and applied purposes, represents the next challenge. Given the rapid and successful deployment of genomics tools and technologies to tomato and the related Solanaceae in recent years, the outlook for continued advancement and prospects for meeting future challenges seems quite promising.

Acknowledgements

Work in all participating labs was supported by NSF Plant Genome grant DBI-0606595 and USDA National Research Initiative grant 07-02773. Additional support in the Rose lab is from the NSF Plant Genome grant DBI-0605200, the U.S.-Israel Binational Science Foundation (BSF) and Binational Agricultural Research and Development (BARD) Fund, and in the Giovannoni and Fei labs from NSF Plant Genome grants DBI-0923312 and DBI-0820612.

References

Alagna F, D'Agostino N, Torchia L, Servili M, Rao R, Pietrella M, Giuliano G, Chiusano ML, Baldoni L, Perrotta G (2009) Comparative 454 pyrosequencing of transcripts from two olive genotypes during fruit development. BMC Genom 10: 399.

Alba R, Payton P, Fei Z, McQuinn R, Debbie P, Martin GB, Tanksley T, Giovannoni JJ (2005) Transcriptome and selected metabolite analyses reveal multiple points of ethylene control during tomato fruit development. Plant Cell 17: 2954–65.

Alba R, Fei ZJ, Payton P, Liu Y, Moore SL, Debbie P, Cohn J, D'Ascenzo M, Gordon JS, Rose JKC, Martin G, Tanksley SD, Bouzayen M, Jahn MM, Giovannoni J (2004) ESTs, cDNA microarrays, and gene expression profiling: Tools for dissecting Plant Physiol and development. Plant J 39: 697–714.

Aoki K, Yano K, Suzuki A, Kawamura S, Sakurai N, Suda K, Kurabayashi A, Suzuki T, Tsugane T, Watanabe M, Ooga K, Torii M, Narita T, Shin-I T, Kohara Y, Yamamoto N, Takahashi H, Watanabe Y, Egusa M, Kodama M, Ichinose Y, Kikuchi M, Fukushima S, Okabe A, Arie T, Sato Y, Yazawa K, Satoh S, Omura T, Ezura H, Shibata D (2010) Large-scale analysis of full-length cDNAs from the tomato (*Solanum lycopersicum*) cultivar Micro-Tom, a reference system for the Solanaceae genomics. BMC Genom 11: 210.

Bai YL, Lindhout P (2007). Domestication and breeding of tomatoes: What have we gained and what can we gain in the future? Ann Bot 100: 1085–94.

Balaji V, Mayrose M, Sherf O, Jacob-Hirsch J, Eichenlaub R, Iraki N, Manulis-Sasson S, Rechavi G, Barash I, Sessa G (2008) Tomato transcriptional changes in response to *Clavibacter michiganensis* subsp. *michiganensis* reveal a role for ethylene in disease development. Plant Physiol 146: 1797–809.

Bannenberg G, Martínez M, Rodríguez M, López M, Ponce de León I, Hamberg M, Castresana C (2009) Functional analysis of alpha-DOX2, an active alpha-dioxygenase critical for normal development in tomato plants. Plant Physiol 151: 1421–32.

Barone A, Chiusano ML, Ercolano MR, Giuliano G, Grandillo S, Frusciante L (2008) Structural and functional genomics of tomato. Int J Plant Genom 2008: 820274.

Barrett T, Troup DB, Wilhite SE, Ledoux P, Rudnev D, Evangelista C, Kim IF, Soboleva A, Tomashevsky M, Edgar R (2007) NCBI GEO: Mining tens of millions of expression profiles —database and tools update. Nucl Acids Res 35: D760–5.

Baxter CJ, Sabar M, Quick WP, Sweetlove LJ (2005) Comparison of changes in fruit gene expression in tomato introgression lines provides evidence of genome-wide transcriptional changes and reveals links to mapped QTLs and described traits. J Exp Bot 56: 1591–604.

Bentley DR, Balasubramanian S, Swerdlow HP, Smith GP, Milton J, Brown CG, Hall KP, Evers DJ, Barnes CL, Bignell HR, et al. (2008) Accurate whole human genome sequencing using reversible terminator chemistry. Nature 456: 53–9.

Bier FF, Kleinjung F (2001) Feature-size limitations of microarray technology—a critical review. Fresenius' J Anal Chem 371: 151–6.

Bonshtien A, Lev A, Gibly A, Debbie P, Avni A, Sessa G (2005) Molecular properties of the *Xanthomonas* AvrRxv effector and global transcriptional changes determined by its expression in resistant tomato plants. Mol Plant-Microbe Interact 18: 300–10.

Brazma A, Parkinson H, Sarkans U, Shojatalab M, Vilo J, Abeygunawardena N, Holloway E, Kapushesky M, Kemmeren P, Lara GG, Oezcimen A, Rocca-Serra P, Sansone SA (2003) ArrayExpress—a public repository for microarray gene expression data at the EBI. Nucl Acids Res 31: 68–71.

Brazma A, Hingamp P, Quackenbush J, Sherlock G, Spellman P, Stoeckert C, Aach J, Ansorge W, Ball CA, Causton HC, Gaasterland T, Glenisson P, Holstege FCP, Kim IF, Markowitz V, Matese JC, Parkinson H, Robinson A, Sarkans U, Schulze-Kremer S, Stewart J, Taylor R, Vilo J, Vingron M (2001) Minimum information about a microarray experiment (MIAME) —toward standards for microarray data. Nat Genet 29: 365–71.

Brenner S, Johnson M, Bridgham J, Golda G, Lloyd DH, Johnson D, Luo SJ, McCurdy S, Foy M, Ewan M, Roth R, George D, Eletr S, Albrecht G, Vermaas E, Williams SR, Moon K, Burcham T, Pallas M, DuBridge RB, Kirchner J, Fearon K, Mao J, Corcoran K (2000) Gene expression analysis by massively parallel signature sequencing (MPSS) on microbead arrays. Nat Biotechnol 18: 630–4.

Cantu D, Blanco-Ulate B, Yang L, Labavitch JM, Bennett AB, Powell AL (2009) Ripening-regulated susceptibility of tomato fruit to Botrytis cinerea requires NOR but not RIN or ethylene. Plant Physiol 150: 1434–49.

Carrari F, Baxter C, Usadel B, Urbanczyk-Wochniak E, Zanor MI, Nunes-Nesi A, Nikiforova V, Centero D, Ratzka A, Pauly M, Sweetlove LJ, Fernie AR (2006) Integrated analysis of metabolite and transcript levels reveals the metabolic shifts that underlie tomato fruit development and highlight regulatory aspects of metabolic network behavior. Plant Physiol 142: 1380–96.

Catoni M, Miozzi L, Fiorilli V, Lanfranco L, Accotto GP (2009) Comparative analysis of expression profiles in shoots and roots of tomato systemically infected by tomato spotted wilt virus reveals organ-specific transcriptional responses. Mol Plant-Microbe Interact 22: 1504–13.

Causse M, Chaib J, Lecomte L, Buret M, Hospital F (2007) Both additivity and epistasis control the genetic variation for fruit quality traits in tomato. Theor Appl Genet 115: 429–42.

Chiusano ML, D'Agostino N, Traini A, Licciardello C, Raimondo E, Aversano M, Frusciante L, Monti L (2008) ISOL@: An italian SOLAnaceae genomics resource. BMC Bioinformat 9: Suppl 2–S7.

Churchill GA (2002) Fundamentals of experimental design for cDNA microarrays. Nat Genet 32: 490–5.

Cohn JR, Martin GB (2005) *Pseudomonas syringae* pv. *tomato* type III effectors AvrPto and AvrPtoB promote ethylene-dependent cell death in tomato. Plant J 44: 139–54.

Cutler DJ, Zwick ME, Carrasquillo MM, Yohn CT, Tobin KP, Kashuk C, Mathews DJ, Shah NA, Eichler EE, Warrington JA, Chakravarti S (2001) High-throughput variation detection and genotyping using microarrays. Genome Res 11: 1913–25.

Draghici S, Khatri P, Eklund AC, Szallasi Z (2006) Reliability and reproducibility issues in DNA microarray measurements. Trends Genet 22: 101–9.

Edwards D, Batley J (2010) Plant genome sequencing: Applications for crop improvement. Plant Biotechnol J 8: 2–9.

Elling AA, Deng XW (2009) Next-generation sequencing reveals complex relationships between the epigenome and transcriptome in maize. Plant Signal Behav 4: 760–2.

Facella P, Lopez L, Carbone F, Galbraith DW, Giuliano G, Perrotta G (2008) Diurnal and circadian rhythms in the tomato transcriptome and their modulation by cryptochrome photoreceptors. PLoS One 3: e2798.

Faurobert M, Chaib J, Barre M, Tricon D, Munos S, Causse M (2009) Genetic and proteomic approach of tomato fruit quality. Acta Hort 817: 119–26.

Fei Z, Tang X, Alba R, Giovannoni J (2006) Tomato expression database (TED): A suite of data presentation and analysis tools. Nucl Acids Res 34: D766–70.

Fei ZJ, Tang X, Alba RM, White JA, Ronning CM, Martin GB, Tanksley SD, Giovannoni JJ (2004) Comprehensive EST analysis of tomato and comparative genomics of fruit ripening. Plant J 40: 47–59.

Fiorilli V, Catoni M, Miozzi L, Novero M, Accotto GP, Lanfranco L (2009) Global and cell-type gene expression profiles in tomato plants colonized by an arbuscular mycorrhizal fungus. New Phytol 184: 975–87.

Fukuoka H, Yamaguchi H, Nunome T, Negoro S, Miyatake K, Ohyama A (2010) Accumulation, functional annotation, and comparative analysis of expressed sequence tags in eggplant (*Solanum melongena* L.), the third pole of the genus *Solanum* species after tomato and potato. Gene 450: 76–84.

Fuller CW, Middendorf LR, Benner SA, Church GM, Harris T, Huang XH, Jovanovich SB, Nelson JR, Schloss JA, Schwartz DC, Vezenov DV (2009) The challenges of sequencing by synthesis. Nat Biotechnol 27: 1013–23.

Gilardoni PA, Schuck S, Jungling R, Rotter B, Baldwin IT, Bonaventure G (2010) SuperSAGE analysis of the *Nicotiana attenuata* transcriptome after fatty acid-amino acid elicitation (FAC): Identification of early mediators of insect responses. BMC Plant Biol 10: 66.

Gowda M, Jantasuriyarat C, Dean RA, Wang GL (2004) Robust-LongSAGE (RL-SAGE): A substantially improved LongSAGE method for gene discovery and transcriptome analysis. Plant Physiol 134: 890–7.

Gupta V, Mathur S, Solanke AU, Sharma MK, Kumar R, Vyas S, Khurana P, Khurana JP, Tyagi AK, Sharma AK (2009) Genome analysis and genetic enhancement of tomato. Crit Rev Biotechnol 29: 152–81.

Hacia JG, Fan JB, Ryder O, Jin L, Edgemon K, Ghandour G, Mayer RA, Sun B, Hsie L, Robbins CM, Brody LC, Wang D, Lander ES, Lipshutz R, Fodor SPA, Collins FS (1999) Determination of ancestral alleles for human single-nucleotide polymorphisms using high-density oligonucleotide arrays. Nat Genet 22: 164–7.

Huttenhower C, Hofmann O (2010) A quick guide to large-scale genomic data mining. PLoS Comput Biol 6: e1000779.

Irian S, Xu P, Dai XB, Zhao PX, Roossinck MJ (2007) Regulation of a virus-induced lethal disease in tomato revealed by LongSAGE analysis. Mol Plant-Microbe Interact 20: 1477–88.

Joung JG, Corbett AM, Fellman SM, Tieman DM, Klee HJ, Giovannoni JJ, Fei ZJ (2009) Plant MetGenMAP: An integrative analysis system for plant systems biology. Plant Physiol 151: 1758–68.

Kolotilin I, Koltai H, Tadmor Y, Bar-Or C, Reuveni M, Meir A, Nahon S, Shlomo H, Chen L, Levin I (2007) Transcriptional profiling of *high pigment-2dg* tomato mutant links early fruit plastid biogenesis with its overproduction of phytonutrients. Plant Physiol 145: 389–401.

Korlach J, Marks PJ, Cicero RL, Gray JJ, Murphy DL, Roitman DB, Pham TT, Otto GA, Foquet M, Turner SW (2008). Selective aluminum passivation for targeted immobilization of single DNA polymerase molecules in zero-mode waveguide nanostructures. Proc Natl Acad Sci USA 105: 1176–81.

Lin M, Wei LJ, Sellers WR, Lieberfarb M, Wong WH, Li C (2004) dChipSNP: Significance curve and clustering of SNP-array-based loss-of-heterozygosity data. Bioinformatics 20: 1233–40.

Main BJ, Bickel RD, McIntyre LM, Graze RM, Calabrese PP, Nuzhdin SV (2009) Allele-specific expression assays using Solexa. BMC Genom 10: 422.

Mardis ER (2008) The impact of next-generation sequencing technology on genetics. Trends Genet 24: 142–149.

Martinelli F, Uratsu SL, Reagan RL, Chen Y, Tricoli D, Fiehn O, Rocke DM, Gasser CS, Dandekar AM (2009) Gene regulation in parthenocarpic tomato fruit. J Exp Bot 60: 3873–90.

Matsumura H, Reich S, Ito A, Saitoh H, Kamoun S, Winter P, Kahl G, Reuter M, Kruger DH, Terauchi R (2003) Gene expression analysis of plant host-pathogen interactions by SuperSAGE. Proc Natl Acad Sci USA 100: 15718–23.

Matukumalli LK, Schroeder SG (2009) Sequence based gene expression analysis. 191–207. In: Edwards, D., J. Stajich and D. Hansen [eds] Bioinformatics: Tools and Applications Springer, New York, USA.

McPherson JD (2009) Next-generation gap. Nat Meth 6: S2–5.

Meli VS, Ghosh S, Prabha TN, Chakraborty N, Chakraborty S, Datta A (2010) Enhancement of fruit shelf life by suppressing N-glycan processing enzymes. Proc Natl Acad Sci USA 107: 2413–8.

Metzker ML (2010) Sequencing technologies—the next generation. Nat Rev Genet 11: 31–46.

Mochida K, Shinozaki K (2010) Genomics and bioinformatics resources for crop improvement. Plant Cell Physiol 51: 497–523.

Molina C, Rotter B, Horres R, Udupa SM, Besser B, Bellarmino L, Baum M, Matsumura H, Terauchi R, Kahl G, Winter P (2008) SuperSAGE: The drought stress-responsive transcriptome of chickpea roots. BMC Genom 9: 553.

Moore S, Payton P, Wright M, Tanksley S, Giovannoni J (2005) Utilization of tomato microarrays for comparative gene expression analysis in the Solanaceae. J Exp Bot 56: 2885–95.

Moxon S, Jing RC, Szittya G, Schwach F, Pilcher RLR, Moulton V, Dalmay T (2008) Deep sequencing of tomato short RNAs identifies microRNAs targeting genes involved in fruit ripening. Genome Res 18: 1602–9.

Mueller LA, Solow TH, Taylor N, Skwarecki B, Buels R, Binns J, Lin CW, Wright MH, Ahrens R, Wang Y, Herbst EV, Keyder ER, Menda N, Zamir D, Tanksley SD (2005) The SOL genomics network. A comparative resource for Solanaceae biology and beyond. Plant Physiol 138: 1310–7.

Nagalakshmi U, Waern K, Snyder M (2010) RNA-seq: A method for comprehensive transcriptome analysis. Curr Protoc Mol Biol Chapter 4: Unit 4.11.1–13.

Nakano M, Nobuta K, Vemaraju K, Tej SS, Skogen JW, Meyers BC (2006) Plant MPSS databases: Signature-based transcriptional resources for analyses of mRNA and small RNA. Nucleic Acids Res 34: D731–5.

Nielsen KL, Hogh AL, Emmersen J (2006) DeepSAGE—digital transcriptomics with high sensitivity, simple experimental protocol and multiplexing of samples. Nucl Acids Res 34: e133.

Nyrén P (2007) The history of pyrosequencing. In: Walker JM, Marsh S (eds) Pyrosequencing Protocols. Methods in Molecular Biology, vol 373. Humana Press, Totowa, NJ, USA, pp 1–14.

Ozaki S, Ogata Y, Suda K, Kurabayashi A, Suzuki T, Yamamoto N, Iijima Y, Tsugane T, Fujii T, Konishi C, Inai S, Bunsupa S, Yamazaki M, Shibata D, Aoki K (2010) Coexpression analysis of tomato genes and experimental verification of coordinated expression of genes found in a functionally enriched coexpression module. DNA Res 17: 105–16.

Page D, Marty I, Bouchet JP, Gouble B, Causse M (2008) Isolation of genes potentially related to fruit quality by subtractive selective hybridization in tomato. Postharvest Biol Technol 50: 117–24.

Pascual L, Blanca JM, Cañizares J, Nuez F (2009) Transcriptomic analysis of tomato carpel development reveals alterations in ethylene and gibberellin synthesis during *pat3/pat4* parthenocarpic fruit set. BMC Plant Biol 9: 67.

Phillips PC (2008) Epistasis—the essential role of gene interactions in the structure and evolution of genetic systems. Nat Rev Genet 9: 855–67.

Powell J (1998) Enhanced concatemer cloning—a modification to the SAGE (serial analysis of gene expression) technique. Nucleic Acids Res 26: 3445–6.

Reinartz J, Bruyns E, Lin JZ, Burcham T, Brenner S, Bowen B, Kramer M, Woychik R (2002) Massively parallel signature sequencing (MPSS) as a tool for in-depth quantitative gene expression profiling in all organisms. Brief Funct. Genom Proteom 1: 95–104.

Robinson MD, Oshlack A (2010) A scaling normalization method for differential expression analysis of RNA-seq data. Genome Biol 11: R25.

Ronaghi M, Karamohamed S, Pettersson B, Uhlén M, Nyrén P (1996) Real-time DNA sequencing using detection of pyrophosphate release. Anal Biochem 242: 84–9.

Roney JK, Khatibi PA, Westwood JH (2007) Cross-species translocation of mRNA from host plants into the parasitic plant dodder. Plant Physiol 143: 1037–43.

Ruzicka D, Barrios-Masias F, Hausmann N, Jackson L, Schachtman D (2010) Tomato root transcriptome response to a nitrogen-enriched soil patch. BMC Plant Biol 10: 75.

Schena M, Shalon D, Davis RW, Brown PO (1995) Quantitative monitoring of gene-expression patterns with a complementary-DNA microarray. Science 270: 467–70.

Schilmiller AL, Miner DP, Larson M, McDowell E, Gang DR, Wilkerson C, Last RL (2010) Studies of a biochemical factory: Tomato trichome deep EST sequencing and proteomics. Plant Physiol: DOI 10.1104/pp.110.157214.

Shalon D, Smith SJ, Brown PO (1996) A DNA microarray system for analyzing complex DNA samples using two-color fluorescent probe hybridization. Genome Res 6: 639–45.

Slocombe SP, Schauvinhold I, McQuinn RP, Besser K, Welsby NA, Harper A, Aziz N, Li Y, Larson TR, Giovannoni J, Dixon RA, Broun P (2008) Transcriptomic and reverse genetic analyses of branched-chain fatty acid and acyl sugar production in *Solanum pennellii* and *Nicotiana benthamiana*. Plant Physiol 148: 1830–46.

Stein LD, Mungall C, Shu SQ, Caudy M, Mangone M, Day A, Nickerson E, Stajich JE, Harris TW, Arva A, Lewis S (2002) The generic genome browser: A building block for a model organism system database. Genome Res 12: 1599–610.

Stevens R, Buret M, Duffe P, Garchery C, Baldet P, Rothan C, Causse M (2007) Candidate genes and quantitative trait loci affecting fruit ascorbic acid content in three tomato populations. Plant Physiol 143: 1943–53.

Syvanen AC (2001) Accessing genetic variation: Genotyping single nucleotide polymorphisms. Nat Rev Genet 2: 930–42.

Uehara T, Sugiyama S, Masuta C (2007) Comparative serial analysis of gene expression of transcript profiles of tomato roots infected with cyst nematode. Plant Mol Biol 63: 185–94.

Valouev A, Ichikawa J, Tonthat T, Stuart J, Ranade S, Peckham H, Zeng K, Malek JA, Costa G, McKernan K, Sidow A, Fire A, Johnson SM (2008) A high-resolution, nucleosome position map of *C. elegans* reveals a lack of universal sequence-dictated positioning. Genome Res 18: 1051–63.

Velculescu VE, Zhang L, Vogelstein B, Kinzler KW (1995) Serial analysis of gene-expression. Science 270: 484–487.

Wall PK, Leebens-Mack J, Chanderbali AS, Barakat A, Wolcott E, Liang HY, Landherr L, Tomsho LP, Hu Y, Carlson JE, Ma H, Schuster SC, Soltis DE, Soltis PS, Altman N, dePamphilis CW (2009) Comparison of next generation sequencing technologies for transcriptome characterization. BMC Genom 10: 347.

Wang H, Schauer N, Usadel B, Frasse P, Zouine M, Hernould M, Latché A, Pech AC, Fernie AR, Bouzayen M (2009) Regulatory features underlying pollination-dependent and -independent tomato fruit set revealed by transcript and primary metabolite profiling. Plant Cell 21: 1428–52.

Wang Y, Diehl A, Wu FN, Vrebalov J, Giovannoni J, Siepel A, Tanksley SD (2008) Sequencing and comparative analysis of a conserved syntenic segment in the Solanaceae. Genetics 180: 391–408.

Wang Z, Gerstein M, Snyder M (2009) RNA-seq: A revolutionary tool for transcriptomics. Nat Rev Genet 10: 57–63.

Wilhelm BT, Landry JR (2009) RNA-seq-quantitative measurement of expression through massively parallel RNA-sequencing. Methods 48: 249–57.

Wu FN, Tanksley SD (2010) Chromosomal evolution in the plant family Solanaceae. BMC Genom 11: 182.

Yamaguchi H, Fukuoka H, Arao T, Ohyama A, Nunome T, Miyatake K, Negoro S (2010) Gene expression analysis in cadmium-stressed roots of a low cadmium-accumulating solanaceous plant, *Solanum torvum*. J Exp Bot 61: 423–37.

Yano K, Watanabe M, Yamamoto N, Tsugane T, Aoki K, Sakurai N, Shibata D (2006) MiBASE: A database of a miniature tomato cultivar Micro-Tom. Plant Biotechnol 23: 195–8.

Zenoni S, Ferrarini A, Giacomelli E, Xumerle L, Fasoli M, Malerba G, Bellin D, Pezzotti M, Delledonne M (2010) Characterization of transcriptional complexity during berry development in *Vitis vinifera* using RNA-seq. Plant Physiol 152: 1787–95.

Zhou S, Sauve R, Boone B, Levy S (2008) Identification of genes associated with aluminium toxicity in tomato roots using cDNA microarrays. Plant Stress 2: 113–120.

Zhou S, Wei S, Boone B, Levy S (2007) Microarray analysis of genes affected by salt stress in tomato. Afr J Environ Sci Technol 1: 014–026.

13

Proteomics and Metabolomics

Mireille Faurobert,[1] Yoko Iijima[2] and Koh Aoki[3],*

ABSTRACT

The profile of proteins and metabolites can be key determinants of productivity and quality of tomato fruit. Recent advances in analytical technologies, particularly, in mass spectrometry, have facilitated comprehensive, high-throughput profiling of proteins and metabolites. In the first half of this chapter, we describe analytical methods for comprehensive profiling of proteins, application of the methods to elucidate developmental and environmental changes of the protein profiles in response to various stresses, and link of proteomic-phenotyping to QTL. For analytical methods, in addition to the conventional two-dimensional electrophoresis-based method, overview of so-called shotgun proteomics approach using liquid chromatography-coupled mass spectrometry (LC-MS) is provided. Recent progresses in this LC-MS-based method achieve quantitative analysis of component proteins by isotopic or chemical modification of peptides, and by MS/MS spectral count. Post-translational modifications are also detected quantitatively by using affinity selection methods. These high throughput methods were employed for detecting differentially accumulated proteins in response to developmental cues and environmental stimuli. This approach is then extended to detect proteins whose accumulation is closely associated with specific genotype, or, the presence of specific QTL. In the second half of this chapter, we describe the methods and application to the phenotyping studies of metabolomics. The chemical diversity of metabolite is larger than that of protein or peptides.

[1]INRA, Unite Genetique et Amelioration des Fruits et Legumes, UR1052, 84000 Avignon, France.
[2]Department of Nutrition and Life Science, Faculty of Applied Bioscience, 1030 Shimo-ogino, Atsugi, Kanagawa, 243-0292 Japan.
[3]Graduate School of Life and Environmental Sciences, Osaka Prefecture University, B4-215, 1-1 Gakuen-cho, Naka-ku, Sakai, 599-8531 Japan.
*Corresponding author

Thus, comprehensive metabolite profiling requires several analytical platforms including gas chromatography-coupled mass spectrometry, capillary electrophoresis-coupled mass spectrometry, and LC-MS. In addition to MS-based methods, nuclear magnetic resonance (NMR) instrument has potential advantage in quantitative metabolite analysis. Metabolite peaks detected by above mentioned approach contains large numbers of unknown or unidentified metabolites. Progress in analytical methods of mass spectral signals enables us to compare peak profiles between multiple samples, even without knowing chemical identity of the unidentified peaks. This facilitate efficient search of metabolites accumulated in a specific conditions or in specific genotypes. Finally, pivotal role of proteome and metabolome databases in molecular breeding is emphasized.

Keywords: database, mass spectrometry, metabolic profiling, metabolite annotation, post-translational modification, quantitative trait loci (QTL), shotgun proteomics

13.1 Introduction

Tomato genetic engineering primarily relies not only on genome sequence information but also requires comprehensive understanding of the function of tomato genes. In a 'functional genomics' approach, the targets of research are the key molecules that determine the biological activity of cells: transcripts, proteins and metabolites. Transcripts represent the first step of gene expression. However, their presence and quantity in a cell is not necessarily linearly correlated with those of proteins or metabolites (Gibon et al. 2004). Thus, transcript profiling is not sufficient to understand the plasticity of phenotype, which is undoubtedly driven by altered levels of proteins and metabolites (Weckwerth 2008). This leads us to the understanding that proteomics and metabolomics are valuable components of functional genomics. The profile of proteins and metabolites can be key determinants of productivity and quality of tomato fruit.

A review of tomato transcriptomics is presented in Chapter 12; here we present an overview of the current status of tomato proteomics and metabolomics. We review tomato proteomics and metabolomics separately to avoid unnecessary confusion with respect to the technical terms of mass spectrometry (MS). Technical advances, physiological and molecular biological studies, and quantitative trait loci (QTL) mapping studies using proteomic approaches are first presented, followed by those using metabolomics approaches. Finally, application of the proteomic- and metabolomic-phenotyping to phenomics-assisted breeding is discussed. Abbreviations for proteomics and metabolomics technologies are summarized (Table 13-1).

Table 13-1 Abbreviations for proteomics and metabolomics technologies.

Abbreviation	Description
Chromatography	
CE	capillary electrophoresis
GC	gas chromatography
LC	liquid chromatography
PDA	photodiode array
RP	reverse phase
SCX	strong cation exchange
Electrophoresis	
2DE	two-dimensional electrophoresis
2D-DIGE	two-dimensional differential in-gel electrophoresis
BN-PAGE	blue native polyacrylamide gel electrophoresis
SDS-PAGE	sodium dodecyl sulfate polyacrylamide gel electrophoresis
Ionization	
APCI	atmospheric pressure chemical ionization
APPI	atmospheric pressure photo ionization
EI	electron ionization
ESI	electrospray ionization
MALDI	matrix-assisted laser desorption ionization
Mass Detection	
FTICR	Fourier transform ion cyclotron resonance
MS	mass spectrometry
NMR	nuclear magnetic resonance
TOF	time of flight
Q	quadrupole
Technology	
FD	fluorescence detection
ICAT	isotope coded affinity tag
IMAC	immobilized metal affinity chromatography
iTRAQ	isobaric tag for relative and absolute quantification
MuDPIT	multi-dimensional protein identification technology
PMF	peptides mass fingerprinting
TAP	tandem affinity purification
General	
MIAMET	minimum information about a metabolomics experiment
MIAPE	minimum information about a proteomics experiment
PIN	protein interaction network
PSN	protein signaling network
PTM	post-translational modification

13.2 Tomato Proteomics: Techniques, Applications, and QTL Analyses

13.2.1 Proteomics Methods

13.2.1.1 Protein Extraction

The nature of proteins is very diverse according to their primary amino acid sequence, state of modification and association with other molecules. In addition, the abundance of proteins ranges from 10 to 10^7 molecules per cell. This diversity in proteins' structure and abundance often leads to poor detection of less abundant proteins. The presence of high amounts of proteases and interfering compounds such as pigments, polysaccharides, lipids, or phenolic compounds, also leads to low efficiency in protein extraction (Rose et al. 2004; Chen and Harmon 2006). In practice, the choice of an extraction procedure largely depends on the downstream analysis to be performed (Rose et al. 2004; Faurobert et al. 2007b). When the proteome of a subcellular compartment (plastid, mitochondria, nucleus, membrane, cell wall, etc.) is of interest, isolation of subcellular compartments is performed prior to protein extraction (Yates et al. 2005).

13.2.1.2 Traditional 2DE-based Methods

Despite the emergence of 'gel free' separation methods, two-dimensional electrophoresis (2DE) is still the most widely used method in tomato proteomics. Proteins are separated by isoelectric focusing followed by sodium dodecyl sulfate polyacrylamide gel electrophoresis (SDS-PAGE); the proteins are then visualized by non-specific staining using silver nitrate, colloidal Coomassie brilliant blue or fluorescent dye (e.g., Sypro Ruby) (Chevalier et al. 2004; Chevalier et al. 2007). A single 2DE gel generally contains approximately 1,000–2,000 distinguishable protein spots. Although dedicated software packages are available, the image analysis process is a major bottleneck in comparative proteomics projects (Rose et al. 2004; Zivy 2007). Two-dimensional differential in-gel electrophoresis (2D-DIGE) technology offers an accurate quantitative method for comparative 2DE (Ünlü et al. 1997). Proteins from different samples are first labeled with high-sensitivity cyanine fluorescent dyes, mixed together, and then run on the same 2D gel. Gels are scanned for each fluorophore, the images are merged and differences between them are determined by image analysis (Tonge et al. 2001). The dyes, compatible with MS analysis, are designed to exhibit a linear response to varied protein concentration over five orders of magnitude, and sensitivity that detects sub-nanogram protein quantities.

13.2.1.3 Mass Spectrometry

Mass spectrometry (MS) has established itself as an indispensable technology for proteomic analysis with improved accuracy, resolution, sensitivity and ease of use (Cravatt et al. 2007). Availability of full genome sequences of many organisms lays the foundation on which high-throughput protein identification with MS can be performed.

The basic protocol starts with in-gel digestion of the protein spot with a site-specific protease, which generates a set of peptides. The accurate masses of the peptides are measured by MS and matched to the peptide pattern predicted from deduced amino acid sequences (Mann et al. 2001). This type of analysis, called 'peptides mass fingerprinting' (PMF), can be achieved with matrix-assisted laser desorption ionization (MALDI), which is a robust tool used to analyze relatively simple peptide mixtures. However, liquid chromatography electrospray ionization (LC-ESI)-MS/MS systems are preferred for the analysis of complex samples (Aebersold and Mann 2003). In this approach, two mass spectrometers are linked in tandem. Individual peptide ions are separated in the first mass spectrometer, and then further fragmented by collision with an inert gas in the second mass spectrometer to give a series of ion fragments. For details regarding principles of the method see Steen and Mann (2004).

13.2.1.4 Shotgun Proteomics

The 2DE-based methods have drawbacks such as difficulty in automating image analysis and under-representation of hydrophobic proteins. Tandem MS combined with liquid chromatography (LC) facilitates the direct analysis of a complex mixture of peptides without 2DE. LC separation using a reverse phase (RP) column, or a chromatographic approach (multi-dimensional protein identification technology, MuDPIT) consisting of a strong cation exchange (SCX) column prior to an RP column, has been developed (Wolters et al. 2001). The so-called 'shotgun proteomics' approach efficiently detects peptides derived from a complex mixture of proteins but is qualitative rather than quantitative. Isotope dilution methods, by which peptides are labeled with isotopes either chemically (isotope coded affinity tag, ICAT™ ; isobaric tag for relative and absolute quantification, iTRAQ™) or metabolically, have been developed to quantify peptides (Brunner et al. 2007). Unfortunately, labeling methods are expensive, time-consuming, and not suitable for large-scale studies. A variety of label-free methods have been developed, including methods based on mass spectral peak intensities and MS/MS spectral count (Wiener et al. 2004; Higgs et al. 2005; Fu et al. 2008). A recent paper described the use of the 'shotgun proteomics' and quantitative analysis of tomato fruit proteins (America et al. 2006). The sequential combination of SCX and RP chromatography allowed the detection of 6,360 peptides. In triplicate analyses, very good reproducibility of peak detection (70% of peaks were reproducible) and of the peak quantification (median CV <10%) were obtained. A plant proteomics database ProMEX, which consists of mass spectral libraries of experimentally derived tryptic peptide ions, allows matching of peptides derived from unknown protein to those from known proteins, thus accelerating the identification of unknown proteins (Hummel et al. 2007).

13.2.1.5 Analysis of Post-translational Modifications

Post-translational modifications (PTMs) of proteins are of major importance as they influence protein properties and cellular localization. Proteomics provides the best available tools to investigate PTMs (Jensen 2004). There are two ways of systematically identifying PTMs. First, proteins having different modification patterns can be separated by gel electrophoresis. However, acrylamide radicals in the gel sometimes induce artificial protein modifications such as oxidation of methionine residue and modification of cysteine residue, which masks *in vivo* PTMs. Second, PTMs can be systematically detected by MS based on the detection of mass addition to the peptide mass.

Protein phosphorylation is the most intensively studied modification. It can be investigated by 2DE by using ^{32}P-labeling, antibodies and phospho-specific stains. To improve sensitivity of the method, phosphorylated peptides are enriched either by immobilized metal ion chromatography (IMAC) (Andersson and Porath 1986) or by other chemical methods (Gevaert et al. 2007). MS-based methods for the detection and quantification of phosphorylated peptides have been developed and reviewed elsewhere (Collins et al. 2007; Sugiyama et al. 2008).

N-glycosylation is formed by the covalent attachment of carbohydrates to the parent protein at asparagine residues. In a recent study, MALDI-Quadrupole- quadrupole time-of-flight MS (MALDI-Qq-TOF-MS) was employed to elucidate the structure of glycan attached to tomato pathogenesis-induced subtilisin-like P69B protease (Bykova et al. 2006).

PTMs such as carbonylation, oxidation of thiols to sulphenic acid, tyrosine nitrosylation and formation of new disulfide bonds are detrimental to protein structure and function (Davies 2005). These PTMs can be discerned by separation in 2DE and immunoblotting (Sheehan 2006). Affinity-selection approaches based on biotin and derivatized beads are suitable to purify the protein sample that can be further studied by shotgun proteomic approaches (Azarkan et al. 2007).

13.2.1.6 Protein Complexes

Proteins are engaged in a complex array of physical and functional interactions with other proteins and macromolecules. Experimental procedures aiming at characterizing pairwise protein-protein interactions such as the yeast two-hybrid system, the tandem affinity purification method (TAP) and the protein/peptides chip have been reviewed (Pieroni et al. 2008). The review also presented the current status of the prediction of protein interaction networks (PINs) and protein-signaling networks (PSNs). Blue native polyacrylamide gel electrophoresis (BN-PAGE) is a useful method dedicated to the analysis of protein complexes (Schagger and von Jagow 1991). It was successfully applied to the investigation of subunits of alternative oxidase (AOX) within the tomato mitochondrial respiratory chain (Navet et al. 2004).

13.2.2 Application of Proteomics to Functional Studies

13.2.2.1 Fruit Development and Ripening

Rocco et al. (2006) provided the first proteome data on tomato fruit ripening, comparing two different large-fruited ecotypes, Ailsa Craig and San Marzano. Functions of the 83 identified spots were assigned to important physiological processes underlying fruit ripening: redox status control, stress and defense, carbon metabolism and energy production. Some interesting discrepancies were detected between the two ecotypes. For example, malate dehydrogenase was down-regulated upon ripening in Ailsa Craig, while up-regulated in San Marzano, which exhibits a unique sweet and non-acid taste. This line of work was recently completed by the scrupulous analysis of the major pericarp proteome variations of cherry tomato during fruit development and ripening (Faurobert et al. 2007a). Proteins were extracted at six developmental stages and 1,791 protein spots were detected. Identification of proteins and hierarchical clustering analysis demonstrated that proteins related to amino acid metabolism or protein synthesis were mainly expressed during cell division and down-regulated later, while proteins related to photosynthesis and cell wall formation transiently increased during the cell expansion phase. In contrast, the proteins related to carbohydrate metabolism and oxidative processes showed maximal abundance in mature fruit.

13.2.2.2 Abiotic Stress

Influence of a variety of stresses such as high light, high temperature, waterlogging, salinity and nutritional deficiencies has been investigated at the proteomics level. The density at which tomato plants can be grown in a greenhouse is limited due to the shade-avoidance response. A recent study compared protein profiles in leaves of plants grown in direct sunlight to those of plants grown under shade by using protein DIGE technology (Hattrup et al. 2007). In shade-grown plants, 12 proteins related to the light-harvesting function were up-regulated, suggesting that the shaded-plants were boosting their ability to convert light energy into chemical energy. On the other hand, the plants growing in direct sunlight had much higher levels of Ribulose-1,5-bisphosphate carboxylase/oxygenase (Rubisco), ATP synthase and stress-related proteins.

High temperature is known to induce a delay in tomato fruit ripening. An analysis of the pericarp proteome of fruits delayed in ripening after one day of high temperature treatment was reported (Iwahashi and Hosoda 2000). Among 1,200 proteins resolved by 2DE, mitochondrial small heat shock proteins increased by heat treatment, whereas proteins related to photosynthesis, sucrose hydrolysis (acid invertase) and cell wall loosening (polygalacturonase) were down-regulated.

Tomato is highly sensitive to waterlogging. The proteomic responses of roots (Ahsan et al. 2007a) and leaves (Ahsan et al. 2007b) were recently investigated to identify novel proteins that are regulated by waterlogging stress. In roots, the identified proteins were associated with secondary metabolism (gibberellic acid, phenylpropanoids, flavonoids, brassinosteroids, and porphyrin biosynthesis), suggesting that secondary metabolism is activated as a response to waterlogging stress. In leaves, up-regulation of disease-, stress-, and glycolysis-related proteins was the main feature.

Development of tomato plants is greatly impaired by salt stress. The proteomic response of tomato plants to salt stress was evaluated (Amini et al. 2007). The 2DE of leaf proteins showed that 18 out of 400 detected spots were significantly affected by salt stress. Among them, an epidermal growth factor receptor (EGFR)-like protein was newly synthesized, probably as a consequence of oxidative stress associated with salt stress. Many spots remained to be identified in this work.

Blossom-end rot (BER) is a common physiopathological disorder that occurs on the tomato fruit. BER can be induced by environmental factors but the underlying cause of the disorder is insufficient calcium in the blossom-end of the fruit. A comparison of 2DE protein patterns of healthy green fruit versus BER-affected green fruits identified 26 proteins that were up-regulated in BER-affected fruit (Casado-Vela et al. 2005). Most of the proteins were associated with the pentose phosphate pathway or with antioxidation. This result suggested that, in BER-affected fruits, the capacity of the tissue to produce reducing power increases by activation of the pentose phosphate pathway, in addition to the enhancement of defenses against oxidative damage. No protein function linked to calcium signaling was reported.

13.2.2.3 Biotic Stress

The plant defense reaction against pathogen attack is mediated through the mitogen-activated protein kinase (MAPK) pathway. Constitutive overexpression of *tMEK2* encoding a tomato MAPK kinase enhances the resistance of tomato against the bacterial pathogen, *Pseudomonas syringeae* pv. tomato. Tomato transgenic lines overexpressing a constitutive active form of MAPK kinase, *tMEK2* [MUT], were inoculated with *P. syringeae*, and then up-regulated proteins were identified using MALDI-Qq-TOF-MS (Xing et al. 2003). The pathogenesis-related proteins, β-1,3-glucanase and endochitinase, were amongst the most significantly increased proteins downstream of tMEK2.

To obtain a comprehensive overview of the response of tomato to xylem invasion of the fungal pathogen, *Fusarium oxysporum*, profiles of xylem sap proteins after compatible or incompatible plant-pathogen interaction were analyzed (Rep et al. 2002). Using PMF and ESI-MS/MS, five pathogenesis-related (PR) proteins were unambiguously identified. PR5-x, a 22 kDa protein, was identified as a new member of the PR5 protein family. A precise peptides mass analysis, together with cDNA sequence analysis, revealed that PR5-x underwent deamidation of asparagine residues followed by glycine residues.

13.2.3 Proteomics and QTL Mapping

The richness of tomato germplasm resources and availability of numbers of introgression lines (ILs), (e.g., Eshed and Zamir 1994) provide plant biologists a great opportunity to associate phenotype to genotype. Proteome- and metabolome-analyses facilitate characterization of the so-called 'silent plant phenotypes' (Weckwerth et al. 2004). Thus, proteomic- and metabolomic-phenotyping of tomato varieties, wild species, and ILs aids to identify novel QTL.

13.2.3.1 Pathogen Resistance

With the proteomic characterization of two loci controlling tomato resistance to bacterial canker, the group of D Francis provided the first example of what could be called a 'genetical proteomic approach' (Coaker et al. 2004). Resistance of *Solanum habrochaites* against *Clavibacter michiganensis* was controlled by two QTLs, *Rcm 2.0* and *Rcm 5.1* (Coaker and Francis 2004). Proteomes of two introgression lines from *S. habrochaites* x *Solanum lycopersicum* that contained *Rcm 2.0* and *Rcm 5.1*, respectively, were compared with that of a susceptible control line. In the line containing *Rcm 2.0*, remorin and phospholipid glutathione peroxidase were up-regulated in response to bacterial infection, implying that *Rcm 2.0* was associated with the defense response. On the other hand, responses of the line containing *Rcm 5.1* were characterized by the up-regulation of a protein translation initiator, an RNA-binding protein and *S*-adenosylhomocysteine hydrolase. Thus, *Rcm 5.1* affected proteins related to the transcriptional and translational processes in response to the pathogen.

13.2.3.2 Fruit Quality Traits

Proteomic tools were applied to characterize QTLs that affect organoleptic quality, which is easily perceived by consumers but difficult to quantify by means of instrumentation. Five genomic regions of cherry tomato were identified as carrying QTLs for sugar content, acid content, aroma, fruit weight and texture. The cherry tomato alleles located on chromosomes 1, 2, 4 and 9 were introgressed into a large-fruited recipient genotype to obtain QTL-near isogenic lines (NILs) (Causse et al. 2007). Comparison of the proteomes of these lines showed that the number of varying spots between QTL-NILs and the recurrent parental line was proportional to the size of the introgressed genomic region. Some candidate proteins for sugar and fruit-size QTLs were proposed (Faurobert et al. 2006). The analysis of proteins associated with the texture trait was also studied (D Page, unpublished data).

13.3 Tomato Metabolomics: Techniques, Applications, and QTL Analyses

13.3.1 Metabolomics Tools

13.3.1.1 Metabolite Extraction

The optimum extraction conditions differ for different types of metabolites. To extract carbohydrates, amino acids and fatty acids from tomato tissues for a measurement with gas chromatography-mass spectrometry (GC-MS), an extraction protocol originally developed for potato tuber (Roessner et al. 2000) is widely used. Metabolites from 100–250 mg tomato tissues were extracted with 1.4 ml methanol, which gave final concentrations of methanol ranging from 85% to 93% (Urbanczyk-Wochniak et al. 2003; Schauer et al. 2005; Urbanczyk-Wochniak and Fernie 2005; Carrari et al. 2006). The extracts were then fractionated to polar and non-polar phases by chloroform. A detailed review on extraction conditions for *Arabidopsis* samples for GC-MS is available (Gullberg et al. 2004). To extract secondary metabolites for non-targeted analysis by LC-MS, a methanol extraction using 75% methanol (final concentration) is widely used (Bino et al. 2005; Moco et al. 2006; Iijima et al. 2008). However, more lipophilic extraction (e.g., using chloroform) is required for extracting non-polar secondary metabolites such as carotenoids. To extract volatile metabolites, solid phase micro extraction (Tikunov et al. 2005) or solvent extraction (e.g., propanol, pentane) (Schmelz et al. 2003) were reported.

13.3.1.2 GC-MS

MS has become a primary detection method for plant metabolomic analysis, as well as for proteomics. GC-MS is applicable to volatile metabolites, such as alcohols, monoterpenes, and esters, and also to non-volatile polar metabolites (primary metabolites), such as amino acids, sugars, lipids, and organic acids through derivatization by methylation or trimethylsilylation. The electron ionization (EI) technique that is usually coupled with GC provides reproducible fragmentation patterns. In addition, the retention index procedure (Kovats 1958) compensates the fluctuation of retention times by normalizing them with respect to internal standards, thus giving additional criteria for peak identification. Metabolites can be identified by comparing fragment patterns and retention indices with those of standard compounds in compound databases. Additionally, reproducibility of GC-MS peak detection combined with availability of computer software such as MATLAB (MathWorks, Natick, Massachusetts, USA, http://www.mathworks.com) and GeneMaths (Applied Maths, Sint-Martens-Latem, Belgium, http://www.applied-maths.com) facilitates a reliable comparison between peak profiles of different samples.

13.3.1.3 LC-MS

LC-MS is a versatile technology to analyze non-volatile secondary metabolites. Most of the LC-MS instruments are equipped with ion sources of ESI, atmospheric pressure chemical ionization (APCI), and/or atmospheric pressure photo ionization (APPI). With these ionization methods, LC-MS detects quasi-molecular ions and is suitable for analyzing relatively large metabolites (m/z 500–2,000). Additionally, many LC-MS instruments are equipped with MS/MS that enables the fragmentation of precursor ions, which provide useful information for predicting partial structures of the metabolite. The high-resolution MS systems such as TOF, Fourier transform ion cyclotron resonance (FTICR) and orbitrap Fourier transform (orbitrap FT) coupled to high-precision LC is proving useful in analyzing plant secondary metabolites containing numbers of isomers

(Bino et al. 2005; Iijima et al. 2008; Suzuki et al. 2008). However, in contrast to GC-MS, the lack of a searchable LC-MS database is one of the obstacles in identifying metabolites by LC-MS.

13.3.1.4 CE-MS

Capillary electrophoresis-MS (CE-MS) has been employed to analyze water-soluble metabolites, in particular to analyze organic acids, nucleotides, amino acids, sugars, and sugar phosphates. Although an application of CE-MS to the comprehensive profiling of plant metabolites is relatively limited (Sato et al. 2004; Takahashi et al. 2006; Williams et al. 2007), CE-MS is useful in a broader array of applications due to its sensitivity, ease of sample preparation, and applicability to polar metabolites.

13.3.1.5 NMR

Nuclear magnetic resonance (NMR) is frequently used for metabolic profiling. Unfortunately, the application of NMR to plant metabolomics is currently limited to the profiling of the major metabolites because of its low sensitivity relative to MS-based methods. Ongoing improvements in instrumentation design include increasing resolution by using multi-dimensional NMR (Fan et al. 2001; Kikuchi and Hirayama 2007), and by coupling LC separation to NMR (Bailey et al. 2000). Although NMR remains underused in plant metabolomics, an advantage of NMR has been demonstrated for unbiased metabolic fingerprinting (Ward et al. 2003; Krishnan et al. 2005) and for identification and quantification of metabolites (Fan et al. 2001). Furthermore, NMR is a powerful technology for the analysis of metabolic flux in combination with a stable isotope labeling method (Sekiyama and Kikuchi 2007).

13.3.1.6 Analytical Strategies

The data acquired by MS or NMR are subjected to data mining. Typically, there are two strategies for data mining: metabolic profiling which essentially aims to quantify metabolites present in a given biological sample (Hall 2006), and metabolite annotation, which aims to label peaks with metadata and reliable predictions with respect to the identity and structure of the metabolites (Fiehn et al. 2005; Iijima et al. 2008).

In metabolic profiling, identification of metabolites based on an authentic compound is feasible only for a limited number of metabolites. However, all detected peaks including unknown ones can be quantified and utilized as un-annotated variables for further statistical analysis. Approaches such as principal components analysis (PCA) and hierarchical clustering are most widely used to clarify the differences or similarities between metabolite profiles of multiple samples.

Metabolite annotation involves the collection and integration of information of a given metabolite to provide chemical and structural specifications of the metabolite. Unlike transcripts and proteins, metabolites cannot be identified on the basis of a genomic sequence. Integrated interpretation of mass spectral data are needed for specifying metabolites detected with MS (Iijima et al. 2008). In practice, the procedure includes assignment and quantification of isotopic ions, assignment of in-source derivative ions (adduct ions, multiply charged ions and fragmented ions), prediction of molecular formulae, prediction of structures from MS/MS spectra, and a search of the public metabolite databases. Results of these procedures are collectively attached to the metabolite-representing peaks together with the mass spectral data including retention time, m/z value, MS/MS pattern and absorption spectrum as annotations (Fig. 13-1). Consequently, metabolite annotation facilitates the systematic analysis of unknown metabolites and interpretation of their relationships to known metabolites.

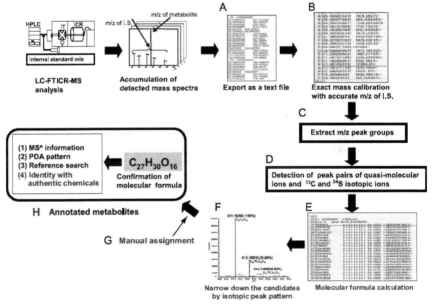

Figure 13-1 Schematic flow of the metabolite annotation procedure. This procedure aims to identify a putative 'metabolite', which is defined as a group of mass signals that are detected in consecutive scans to form a peak group, and are accompanied by isotopic ions. (A) Raw data acquisition. (B) m/z calibration with internal standards. (C) Extraction of peak groups. (D) Isotopic ion assignment. (E) Molecular formula calculation. (F) Molecular formula screening using relative intensity of isotopic ions. (G) Manual curation of isotopic, fragment, and adduct peak assignment. (H) Resultant metabolite annotations.

13.3.2 *Application of Metabolomics to Functional Studies*

13.3.2.1 *Comprehensive Analysis of Tomato Fruit Metabolites*

Tomato metabolites detected using LC-ESI-FTICR-MS were subjected to comprehensive metabolite annotation (Iijima et al. 2008). Metabolite-representing peaks were detected according to the following criteria; (i) whether mass signals formed a chromatographic peak, and (ii) whether the peak was accompanied by isotopic ions. Consequently, 869 metabolites fulfilled these criteria. Metabolite annotations including predicted molecular formulae, predicted structure and database hits were attached to these metabolites. Database searches demonstrated that 494 of 869 metabolites did not match database-registered compounds, suggesting that they were novel metabolites. Based on the m/z values and molecular formulae predictions, it was suggested that modification by amino group, caffeic acid, malonic acid and hexose occurred frequently in the tomato fruit metabolome. Additionally, new metabolic pathways in flavonoid- and glycoalkaloid-groups were hypothesized. Particularly, the presence of a biosynthetic pathway from tomatine to esculeoside A was supported by the finding of intermediate metabolites in the pathway, which had been reported in tomato metabolic profiling studies using LC-Q-TOF-MS (Moco et al. 2007; Mintz-Oron et al. 2008).

13.3.2.2 *Tissue-dependent Accumulation of Metabolites*

Detailed metabolic profiling of tomato fruit tissues at different developmental stages using LC-MS based methods demonstrated spatial and temporal specificity in the accumulation of endogenous metabolites (Moco et al. 2007). LC-photo diode array (PDA)-fluorescence detection (FD) was used to monitor isoprenoids, carotenoids, chlorophylls, xanthophylls, tocopherols and ascorbic acid; LC-PDA-QTOF-MS was used to monitor semi-polar metabolites such as flavonoids and glycoalkaloids. The authors reported that metabolic variation between tissues was more pronounced than developmental variation. For example, carotenoids, flavonoids, glycoalkaloids and ascorbic acid were specifically abundant in the epidermis. Absisic acid-hexose and glycosylated cytokinin were detected in the jelly parenchyma at relatively high levels. The level of total chlorophyll was highest in the vascular attaching region, and lowest in the epidermis. Detailed analysis of tomato fruit metabolites with a specific focus on the peel, together with a comprehensive gene expression analysis, were also reported (Mintz-Oron et al. 2008). These studies provide experimental bases to investigate tissue-specific regulation of metabolic pathways.

13.3.2.3 Metabolome Analysis in Transgenic Tomato Plants

Genetic manipulation of carotenoid pathway genes is of special interest since the pathway produces the predominant pigments of tomato fruit including lycopene and β-carotene (Fraser et al. 2002; Ralley et al. 2004; Tian et al. 2004). *Phytoene synthase-1* (*Psy*-1) in tomato is responsible for carotenoid formation during fruit ripening. To characterize the effect of *Psy*-1 constitutive expression, quantitative, comprehensive metabolite data were obtained in parallel with gene expression data and enzymatic activity data (Fraser et al. 2007). The results demonstrated that constitutive expression of *Psy*-1 enhanced the accumulation of carotenoids (phytoene, β-carotene and intermediates between the two) in green fruits. Early depletion of chlorophyll with increased carotenoid levels in *Psy*-1 fruit suggested that geranylgeranyl diphosphate is preferentially used in carotenoid formation rather than chlorophyll formation. Interestingly, metabolite-induced plastid transition from chloroplast to chromoplast-like plastids was observed. These results showed that overexpression of a metabolic gene could affect the plastid morphology as well as the metabolite profile.

The flavonoid pathway is also a well-studied target of tomato genetic manipulation. Fruits of tomato plants overexpressing *Petunia hybrida chalcone isomerase* gene (*chi-a*) accumulated higher levels of flavonols (quercetin- and kaempferol-glycosides) in the peel (Muir et al. 2001). Overexpression of maize transcription factors *LC* and *C1* in tomato resulted in the increase of kaempferol and naringenin glycosides in the flesh (Bovy et al. 2002; Le Gall et al. 2003a, b). To manipulate the composition of flavonoids in tomato fruits, genetically modified tomato overexpressing foreign flavonoid-related genes *stilbene synthase, chalcone reductase* (together with *chalcone synthase*), and *flavone synthase* (together with *chalcone isomerase*) were produced (Schijlen et al. 2006). In these transgenic tomatoes, flavonoids that are not normally present in tomato fruit including stilbenes (resveratrol and piceid), deoxychalcones (butein and isoliquiritigenin) and flavones (luteolin and luteolin-7-glucoside) were successfully produced. This result shows that the redirection of the flavonoid pathway is possible by genetic engineering.

To manipulate carotenoid and flavonoid pathways simultaneously, Davuluri and coworkers engineered a regulatory gene, *TDET1* (Davuluri et al. 2005). *TDET1*, a tomato ortholog of *Arabidopsis DE-ETIOLATED1* which is proposed to be a negative regulator of light signal transduction, is responsible for the phenotype of the tomato *high pigment-2* (*hp-2*) mutant, in which high accumulation of flavonoids, anthocyanins and carotenoids was observed (Mustilli et al. 1999; Bino et al. 2005). Fruit-specific RNAi-mediated silencing of *TDET1* resulted in elevated levels of lycopene, β-carotene, naringenin chalcone, quercetin derivatives and chlorogenic acid up to 3.5-fold compared with those of wild type fruit (Davuluri et al. 2005).

The entire pathway of carotenoid and flavonoid was affected by *TDET1-*silencing, suggesting that the supply of precursors is not a rate-limiting step in this transgenic tomato.

13.3.3 Metabolomics and QTL Mapping

13.3.3.1 Metabolic Profiling of Wild Species

It is expected that cultivated varieties and wild species of tomato will display large differences in metabolite content. A comparison of metabolite profiles of *S. lycopersicum* and five wild species (*S. pimpinellifolium*, *S. neorickii*, *S. habrochaites*, *S. chmielewskii*, and *S. pennellii*) was recently reported (Schauer et al. 2005). Fruit metabolite profiles showed higher levels of variance than those of leaves. Accumulation levels of sucrose, isocitrate, chlorogenate, and shikimate were higher, while accumulation levels of glucose, fructose, α-tocopherol, and putrescine were lower in the wild species when compared to *S. lycopersicum*.

13.3.3.2 Metabolic Profiling of ILs between S. pennellii and S. lycopersicum

Studies using tomato IL populations have identified a number of QTLs. An IL population between *S. pennellii* and *S. lycopersicum* (M82), consisting of 76 NILs, is one of the most intensively studied IL for agronomic traits (Eshed and Zamir 1995). Some loci influencing tomato metabolism, such as *Brix 9-2-5* (Fridman et al. 2000), *fw2.2* (Alpert and Tanksley 1996) and *B* (Ronen et al. 2000) were mapped using the ILs. A comprehensive metabolic profiling of the ILs using GC-MS has been reported (Schauer et al. 2006). Based on the quantification of 74 metabolites from 76 ILs, the presence of 889 QTLs covering 74 metabolites was demonstrated. The authors reported that the metabolites whose contents were higher in the ILs compared to *S. lycopersicum* were not limited to the metabolites that were more abundant in *S. pennellii* compared to *S. lycopersicum*. The authors combined the analytical results from harvests in 2001, 2003 and 2004, and then estimated the heritability and the mode of inheritance of the metabolic QTLs (Schauer et al. 2008). Semel et al. (2006) demonstrated that levels of sugars, inositol, glycerol and fatty acids were strongly heritable, while levels of stress-related metabolites such as proline, shikimate, aconitate, fumarate and malate exhibited low heritability, suggesting the differing influence of environment on metabolic variation. Estimation of the mode of inheritance (recessive, additive, dominant and overdominant) demonstrated that the majority of metabolic QTLs exhibited additive or dominant modes of inheritance. Limited numbers of metabolites (e.g., glycine, leucine, valine, γ-aminobutylic acid, glycerol, glucose and adenosine-5-phosphate) exhibited recessive QTLs. Overdominant QTLs were even rarer (e.g., cysteine, homoserine, benzoate, citrate, dehydroascorbate, glucose, mannose, trehalose, phosphate and squalene). The increase of metabolite content was associated with a yield penalty (Schauer et al. 2006). But, this negative correlation between metabolite content and yield, observed in lines homozygous for *S. pennellii* introgressions, was not detected in lines heterozygous for the introgressions (Schauer et al. 2008). These results provide many practical implications for crop improvement strategies.

13.4 Application in Phenomics-Assisted Breeding

13.4.1 Integrated -omics

To link proteomic- and metabolomic-data to functional genomics, proteins and metabolites must be assigned to biological pathways. The possibility of linking metabolite profiles with transcript profiles was recently explored (Alba et al. 2005; Carrari et al. 2006). Carrari and co-workers performed a comprehensive parallel analysis of transcript and metabolites during tomato fruit development (Carrari et al. 2006). The authors estimated metabolite-to-metabolite, transcript-to-transcript, and metabolite-to-transcript correlations. A comparison between the metabolite-to-metabolite and the transcript-to-transcript correlation profiles demonstrated that, for metabolite-to-metabolite correlations, metabolites belonging to the same class were regulated in a highly coordinated manner, whereas the transcripts belonging to the same functional group displayed less coordinated behavior. Moreover, metabolite-to-transcript correlation demonstrated that metabolite levels of certain metabolic pathways displayed low correlations with transcript levels responsible for that pathway. These results suggested that metabolism was largely regulated at the post-transcriptional level. However, the authors identified several high correlations between metabolites and transcripts in ripening-associated processes.

In contrast to metabolomics, relatively small numbers of attempts have been reported on the integration of proteomics with other -omics. The main reason for this is that plant proteomics studies have focused more on protein identification rather than quantification. The difficulty of quantification in a complex sample can be overcome by the use of stable isotope labeling methods, such as ICAT™ and iTRAQ™ (Goodlett et al. 2001; Smolka et al. 2001; Ross et al. 2004; Wienkoop and Weckwerth 2006). An alternative approach is label-free quantitative shotgun proteomics based on quantification according to the number of tandem mass spectra collected from a peptide mixture (that is, 'spectral count') (Wienkoop et al. 2008). These approaches will accelerate the generation of quantitative proteomics data, and thus the integration of proteomics data with other -omics data.

Table 13-2 Tomato proteomics and metabolomics databases.

Database and URL	Reference	Description
Tomato Proteomics		
SOLstIS http://www.avignon.inra.fr/solstis	Ferry-Dumazet et al. 2005	Experimental data of 2DE gels and peptide spectra.
Proteomics		
PROTICdb http://cms.moulon.inra.fr/proticdb/web_view/index.php	Ferry-Dumazet et al. 2005	Experimental data of 2DE gels and peptide spectra.
ProMEX http://promex.mpimp-golm.mpg.de/home.shtml	Hummel et al. 2007	Mass spectral library of experimentally obtained tryptic peptides using LC-IT-MS.
World-2DPAGE Repository http://world-2dpage.expasy.org/repository	Gasteiger et al. 2003	Standard-compliant repository for gel-based proteomics data.
MIAPEGelDB http://miapegeldb.expasy.org	Taylor et al. 2007	Repository for MIAPE gel electrophoresis documents.
Tomato Metabolomics		
TOMET http://tomet.bti.cornell.edu	Fei et al. 2006	Metabolite database of *S. pennellii*-derived ILs.
LycoCyc http://solcyc.sgn.cornell.edu/LYCO/server.html	Zhang et al. 2005	Catalog of known and/or predicted biochemical pathways.
MotoDB http://appliedbioinformatics.wur.nl/moto	Moco et al. 2006	Experimental data and database hit information collected by LC-QTOF-MS.
MapMan http://gabi.rzpd.de/projects/MapMan	Urbanczyk-Wochniak et al. 2006	Visualization system for the display of gene expression data on metabolic pathways.
KOMICS http://webs2.kazusa.or.jp/komics	Iijima et al. 2008	Mass spectra and metabolite annotation experimentally obtained by LC-FTICR-MS.
Metabolomics		
CSB.DB (http://csbdb.mpimp-golm.mpg.de/)	Steinhauser et al. 2004	Metabolome module contains custom mass spectral libraries and metabolic profiling experiments.

13.4.2 Databases

With the progress of tomato proteomics and metabolomics, databases are becoming increasingly important not only as repositories of large-scale data sets, but also as tools that allow data interpretation and integration. Standards for presenting and exchanging data were proposed both for proteomics and metabolomics, 'minimum information about a proteomics experiment' (MIAPE) and 'minimum information about a metabolomics experiment' (MIAMET), respectively, so that significant results will be accessible to the greater community (Bino et al. 2004; Taylor et al. 2007). Public databases of the tomato proteome and metabolome are listed in Table 13-2. Access via web services to remotely execute workflows, obtain online information, and analyze data in a fully automated and interoperable way will constitute the next phase of database development. Such development will allow a synergistic integration of all '-omics' tomato data.

From a crop improvement perspective, if the profiling data conforming to the above mentioned standards were to be compiled into one database, it could be used as a phenotypic catalog for screening candidate cultivars and varieties for molecular breeding. Additionally, a comparison of the comprehensive profiling data could illustrate the similarities and differences between newly bred cultivars and varieties versus parental varieties, which is a major responsibility of breeders towards consumers.

References

Aebersold R, Mann M (2003) Mass spectrometry-based proteomics. Nature 422: 198–207.

Ahsan N, Lee D-G, Lee S-H, Lee K-W, Bahk J, Lee B-H (2007a) A proteomic screen and identification of waterlogging-regulated proteins in tomato roots. Plant Soil 295: 37–51.

Ahsan N, Lee DG, Lee SH, Kang KY, Bahk JD, Choi MS, Lee IJ, Renaut J, Lee BH (2007b) A comparative proteomic analysis of tomato leaves in response to waterlogging stress. Physiol Plant 131: 555–570.

Alba R, Payton P, Fei Z, McQuinn R, Debbie P, Martin GB, Tanksley SD, Giovannoni JJ (2005) Transcriptome and selected metabolite analyses reveal multiple points of ethylene control during tomato fruit development. Plant Cell 17: 2954–2965.

Alpert KB, Tanksley SD (1996) High-resolution mapping and isolation of a yeast artificial chromosome contig containing *fw2.2*: a major fruit weight quantitative trait locus in tomato. Proc Natl Acad Sci USA 93: 15503–15507.

America AH, Cordewener JH, van Geffen MH, Lommen A, Vissers JP, Bino RJ, Hall RD (2006) Alignment and statistical difference analysis of complex peptide data sets generated by multidimensional LC-MS. Proteomics 6: 641–653.

Amini F, Ehsanpour AA, Hoang QT, Shin JS (2007) Protein pattern changes in tomato under *in vitro* salt sress. Russ J Plant Physiol 54: 464–471.

Andersson L, Porath J (1986) Isolation of phosphoproteins by immobilized metal (Fe^{3+}) affinity chromatography. Anal Biochem 154: 250–254.

Azarkan M, Huet J, Baeyens-Volant D, Looze Y, Vandenbussche G (2007) Affinity chromatography: a useful tool in proteomics studies. J Chromatogr B Analyt Technol Biomed Life Sci 849: 81–90.

Bailey NJ, Stanley PD, Hadfield ST, Lindon JC, Nicholson JK (2000) Mass spectrometrically detected directly coupled high performance liquid chromatography/nuclear magnetic resonance spectroscopy/mass spectrometry for the identification of xenobiotic metabolites in maize plants. Rapid Commun Mass Spectrom 14: 679–684.

Bino RJ, Hall RD, Fiehn O, Kopka J, Saito K, Draper J, Nikolau BJ, Mendes P, Roessner-Tunali U, Beale MH, Trethewey RN, Lange BM, Wurtele ES, Sumner LW (2004) Potential of metabolomics as a functional genomics tool. Trends Plant Sci 9: 418–425.

Bino RJ, Ric de Vos CH, Lieberman M, Hall RD, Bovy A, Jonker HH, Tikunov Y, Lommen A, Moco S, Levin I (2005) The light-hyperresponsive *high pigment-2*^{dg} mutation of tomato: alterations in the fruit metabolome. New Phytol 166: 427–438.

Bovy A, de Vos R, Kemper M, Schijlen E, Almenar Pertejo M, Muir S, Collins G, Robinson S, Verhoeyen M, Hughes S, Santos-Buelga C, van Tunen A (2002) High-flavonol tomatoes resulting from the heterologous expression of the maize transcription factor genes *LC* and *C1*. Plant Cell 14: 2509–2526.

Brunner E, Gerrits B, Scott M, Roschitzki B (2007) Differential display and protein quantification. In: Baginsky S, Fernie A (eds) Plant Systems Biology, vol 97. Birkäuser, Basel, Switzerland, pp 115–140.

Bykova NV, Rampitsch C, Krokhin O, Standing KG, Ens W (2006) Determination and characterization of site-specific N-glycosylation using MALDI-Qq-TOF tandem mass spectrometry: case study with a plant protease. Anal Chem 78: 1093–1103.

Carrari F, Baxter C, Usadel B, Urbanczyk-Wochniak E, Zanor MI, Nunes-Nesi A, Nikiforova V, Centero D, Ratzka A, Pauly M, Sweetlove LJ, Fernie AR (2006) Integrated analysis of metabolite and transcript levels reveals the metabolic shifts that underlie tomato fruit development and highlight regulatory aspects of metabolic network behavior. Plant Physiol 142: 1380–1396.

Casado-Vela J, Selles S, Bru Martinez R (2005) Proteomic approach to blossom-end rot in tomato fruits (*Lycopersicon esculentum* M.): antioxidant enzymes and the pentose phosphate pathway. Proteomics 5: 2488–2496.

Causse M, Chaib J, Lecomte L, Buret M, Hospital F (2007) Both additivity and epistasis control the genetic variation for fruit quality traits in tomato. Theor Appl Genet 115: 429–442.

Chen S, Harmon AC (2006) Advances in plant proteomics. Proteomics 6: 5504–5516.

Chevalier F, Rofidal V, Rossignol M (2007) Visible and fluorescent staining of two-dimensional gels. In: Thiellement H, Zivy M, Damerval C, Méchin V (eds) Plant Proteomics: methods and protocols, vol 355. Humana Press, Totowa, NJ, USA, pp 145–156.

Chevalier F, Rofidal V, Vanova P, Bergoin A, Rossignol M (2004) Proteomic capacity of recent fluorescent dyes for protein staining. Phytochemistry 65: 1499–1506.

Coaker GL, Francis DM (2004) Mapping, genetic effects, and epistatic interaction of two bacterial canker resistance QTLs from *Lycopersicon hirsutum*. Theor Appl Genet 108: 1047–1055.

Coaker GL, Willard B, Kinter M, Stockinger EJ, Francis DM (2004) Proteomic analysis of resistance mediated by Rcm 2.0 and Rcm 5.1, two loci controlling resistance to bacterial canker of tomato. Mol Plant Microbe Interact 17: 1019–1028.

Collins MO, Yu L, Choudhary JS (2007) Analysis of protein phosphorylation on a proteome-scale. Proteomics 7: 2751–2768.

Cravatt BF, Simon GM, Yates JR (2007) The biological impact of mass-spectrometry-based proteomics. Nature 450: 991–1000.

Davies MJ 2005. The oxidative environment and protein damage. Biochim Biophys Acta 1703: 93–109.

Davuluri GR, van Tuinen A, Fraser PD, Manfredonia A, Newman R, Burgess D, Brummell DA, King SR, Palys J, Uhlig J, Bramley PM, Pennings HM, Bowler C (2005) Fruit-specific RNAi-mediated suppression of *DET1* enhances carotenoid and flavonoid content in tomatoes. Nat Biotechnol 23: 890–895.

Eshed Y, Zamir D (1994) A genomic library of *Lycopersicon pennellii* in *Lycopersicon esculentum* —a tool for fine mapping of genes. Euphytica 79: 175–179.

Eshed Y, Zamir D (1995) An introgression line population of *Lycopersicon pennellii* in the cultivated tomato enables the identification and fine mapping of yield-associated QTL. Genetics 141: 1147–1162.

Fan TW, Lane AN, Shenker M, Bartley JP, Crowley D, Higashi RM (2001) Comprehensive chemical profiling of gramineous plant root exudates using high-resolution NMR and MS Phytochemistry 57: 209–221.

Faurobert M, Barre M, Chaïb J, Causse M (2006) Effect of quality QTL introgression on tomato fruit proteome. PAA/Solanaceae Conference, 23–27 July 2006, Madison, Wisconsin, USA.

Faurobert M, Mihr C, Bertin N, Pawlowski T, Negroni L, Sommerer N, Causse M (2007a) Major proteome variations associated with cherry tomato pericarp development and ripening. Plant Physiol 143: 1327–1346.

Faurobert M, Pelpoir E, Chaib J (2007b) Phenol extraction of proteins for proteomic studies of recalcitrant plant tissues. In: Thiellement H, Zivy M, Damerval C, Méchin V (eds) Plant Proteomics: methods and protocols, vol 355. Humana Press, Totowa, NJ, USA, pp 9–14.

Fei Z, Tang X, Alba R, Giovannoni J (2006) Tomato Expression Database (TED): a suite of data presentation and analysis tools. Nucl Acids Res 34: D766–770.

Ferry-Dumazet H, Houel G, Montalent P, Moreau L, Langella O, Negroni L, Vincent D, Lalanne C, de Daruvar A, Plomion C, Zivy M, Joets J (2005) PROTICdb: a web-based application to store, track, query, and compare plant proteome data. Proteomics 5: 2069–2081.

Fiehn O, Wohlgemuth G, Scholz M (2005) Setup and annotation of metabolomic experiments by integrating biological and mass spectrometric metadata. In: Ludascher B, Raschid L (eds) Data Integration in Life Sciences, Lecture Notes in Computer Science, vol 3615. Springer-Verlag, Berlin, Germany, pp 224–239.

Fraser PD, Enfissi EM, Halket JM, Truesdale MR, Yu D, Gerrish C, Bramley PM (2007) Manipulation of phytoene levels in tomato fruit: effects on isoprenoids, plastids, and intermediary metabolism. Plant Cell 19: 3194–3211.

Fraser PD, Romer S, Shipton CA, Mills PB, Kiano JW, Misawa N, Drake RG, Schuch W, Bramley PM (2002) Evaluation of transgenic tomato plants expressing an additional phytoene synthase in a fruit-specific manner. Proc Natl Acad Sci USA 99: 1092–1097.

Fridman E, Pleban T, Zamir D (2000) A recombination hotspot delimits a wild-species quantitative trait locus for tomato sugar content to 484 bp within an invertase gene. Proc Natl Acad Sci USA 97: 4718–4723.

Fu X, Gharib SA, Green PS, Aitken ML, Frazer DA, Park DR, Vaisar T, Heinecke JW (2008) Spectral index for assessment of differential protein expression in shotgun proteomics. J Proteome Res 7: 845–854.

Gasteiger E, Gattiker A, Hoogland C, Ivanyi I, Appel RD, Bairoch A (2003) ExPASy: The proteomics server for in-depth protein knowledge and analysis. Nucl Acids Res 31: 3784–3788.

Gevaert K, Van Damme P, Ghesquiere B, Impens F, Martens L, Helsens K, Vandekerckhove J (2007) A la carte proteomics with an emphasis on gel-free techniques. Proteomics 7: 2698–2718.

Gibon Y, Blaesing OE, Hannemann J, Carillo P, Hohne M, Hendriks JH, Palacios N, Cross J, Selbig J, Stitt M (2004) A Robot-based platform to measure multiple enzyme activities in *Arabidopsis* using a set of cycling assays: comparison of changes of enzyme activities and transcript levels during diurnal cycles and in prolonged darkness. Plant Cell 16: 3304–3325.

Goodlett DR, Keller A, Watts JD, Newitt R, Yi EC, Purvine S, Eng JK, von Haller P, Aebersold R, Kolker E (2001) Differential stable isotope labeling of peptides for quantitation and de novo sequence derivation. Rapid Commun Mass Spectrom 15: 1214–1221.

Gullberg J, Jonsson P, Nordstrom A, Sjostrom M, Moritz T (2004) Design of experiments: an efficient strategy to identify factors influencing extraction and derivatization of *Arabidopsis thaliana* samples in metabolomic studies with gas chromatography/mass spectrometry. Anal Biochem 331: 283–295.

Hall RD (2006) Plant metabolomics: from holistic hope, to hype, to hot topic. New Phytol 169: 453–468.

Hattrup E, Neilson KA, Breci L, Haynes PA (2007) Proteomic analysis of shade-avoidance response in tomato leaves. J Agric Food Chem 55: 8310–8318.

Higgs RE, Knierman MD, Gelfanova V, Butler JP, Hale JE (2005) Comprehensive label-free method for the relative quantification of proteins from biological samples. J Proteome Res 4: 1442–1450.

Hummel J, Niemann M, Wienkoop S, Schulze W, Steinhauser D, Selbig J, Walther D, Weckwerth W (2007) ProMEX: a mass spectral reference database for proteins and protein phosphorylation sites. BMC Bioinformat 8: 216.

Iijima Y, Nakamura Y, Ogata Y, Tanaka K, Sakurai N, Suda K, Suzuki T, Suzuki H, Okazaki K, Kitayama M, Kanaya S, Aoki K, Shibata D (2008) Metabolite annotations based on the integration of mass spectral information. Plant J 54: 949–962.

Iwahashi Y, Hosoda H (2000) Effect of heat stress on tomato fruit protein expression. Electrophoresis 21: 1766–1771.

Jensen ON (2004) Modification-specific proteomics: characterization of post-translational modifications by mass spectrometry. Curr Opin Chem Biol 8: 33–41.

Kikuchi J, Hirayama T (2007) Practical aspects of uniform stable isotope labeling of higher plants for heteronuclear NMR-based metabolomics. Methods Mol Biol 358: 273–286.

Kovats E (1958) Gas chromatographische charakteriserung organischer verbindungen. I. retentions indices aliphatischer halogenide, alkohole, aldehyde und ketone. Helv Chim Acta 41: 1915–1932.

Krishnan P, Kruger NJ, Ratcliffe RG (2005) Metabolite fingerprinting and profiling in plants using NMR. J Exp Bot 56: 255–265.

Le Gall G, Colquhoun IJ, Davis AL, Collins GJ, Verhoeyen ME (2003a) Metabolite profiling of tomato (*Lycopersicon esculentum*) using ¹H NMR spectroscopy as a tool to detect potential unintended effects following a genetic modification. J Agri Food Chem 51: 2447–2456.

Le Gall G, DuPont MS, Mellon FA, Davis AL, Collins GJ, Verhoeyen ME, Colquhoun IJ (2003b) Characterization and content of flavonoid glycosides in genetically modified tomato (*Lycopersicon esculentum*) fruits. J Agri Food Chem 51: 2438–2446.

Mann M, Hendrickson RC, Pandey A (2001) Analysis of proteins and proteomes by mass spectrometry. Annu Rev Biochem 70: 437–473.

Mintz-Oron S, Mandel T, Rogachev I, Feldberg L, Lotan O, Yativ M, Wang Z, Jetter R, Venger I, Adato A, Aharoni A (2008) Gene expression and metabolism in tomato fruit surface tissues. Plant Physiol 147: 823–851.

Moco S, Bino RJ, Vorst O, Verhoeven HA, de Groot J, van Beek TA, Vervoort J, de Vos CH (2006) A liquid chromatography-mass spectrometry-based metabolome database for tomato. Plant Physiol 141: 1205–1218.

Moco S, Capanoglu E, Tikunov Y, Bino RJ, Boyacioglu D, Hall RD, Vervoort J, De Vos RC (2007) Tissue specialization at the metabolite level is perceived during the development of tomato fruit. J Exp Bot 58: 4131–4146.

Muir SR, Collins GJ, Robinson S, Hughes S, Bovy A, Ric De Vos CH, van Tunen AJ, Verhoeyen ME (2001) Overexpression of petunia chalcone isomerase in tomato results in fruit containing increased levels of flavonols. Nat Biotechnol 19: 470–474.

Mustilli AC, Fenzi F, Ciliento R, Alfano F, Bowler C (1999) Phenotype of the tomato *high pigment-2* mutant is caused by a mutation in the tomato homolog of *DEETIOLATED1*. Plant Cell 11: 145–157.

Navet R, Jarmuszkiewicz W, Douette P, Sluse-Goffart CM, Sluse FE (2004) Mitochondrial respiratory chain complex patterns from *Acanthamoeba castellanii* and *Lycopersicon esculentum*: comparative analysis by BN-PAGE and evidence of protein-protein interaction between alternative oxidase and complex III. J Bioenerg Biomembr 36: 471–479.

Pieroni E, de la Fuente van Bentem S, Mancosu G, Capobianco E, Hirt H, de la Fuente A (2008) Protein networking: insights into global functional organization of proteomes. Proteomics 8: 799–816.

Ralley L, Enfissi EM, Misawa N, Schuch W, Bramley PM, Fraser PD (2004) Metabolic engineering of ketocarotenoid formation in higher plants. Plant J 39: 477–486.

Rep M, Dekker HL, Vossen JH, de Boer AD, Houterman PM, Speijer D, Back JW, de Koster CG, Cornelissen BJ (2002) Mass spectrometric identification of isoforms of PR proteins in xylem sap of fungus-infected tomato. Plant Physiol 130: 904–917.

Rocco M, D'Ambrosio C, Arena S, Faurobert M, Scaloni A, Marra M (2006) Proteomic analysis of tomato fruits from two ecotypes during ripening. Proteomics 6: 3781–3791.

Roessner U, Wagner C, Kopka J, Trethewey RN, Willmitzer L (2000) Technical advance: simultaneous analysis of metabolites in potato tuber by gas chromatography-mass spectrometry. Plant J 23: 131–142.

Ronen G, Carmel-Goren L, Zamir D, Hirschberg J (2000) An alternative pathway to beta-carotene formation in plant chromoplasts discovered by map-based cloning of beta and old-gold color mutations in tomato. Proc Natl Acad Sci USA 97: 11102–11107.

Rose JKC, Bashir S, Giovannoni JJ, Jahn MM, Saravanan RS (2004) Tackling the plant proteome: practical approaches, hurdles and experimental tools. Plant J 39: 715–733.

Ross PL, Huang YN, Marchese JN, Williamson B, Parker K, Hattan S, Khainovski N, Pillai S, Dey S, Daniels S, Purkayastha S, Juhasz P, Martin S, Bartlet-Jones M, He F, Jacobson A, Pappin DJ (2004) Multiplexed protein quantitation in Saccharomyces cerevisiae using amine-reactive isobaric tagging reagents. Mol Cell Proteom 3: 1154–1169.

Sato S, Soga T, Nishioka T, Tomita M (2004) Simultaneous determination of the main metabolites in rice leaves using capillary electrophoresis mass spectrometry and capillary electrophoresis diode array detection. Plant J 40: 151–163.

Schagger H, von Jagow G (1991) Blue native electrophoresis for isolation of membrane protein complexes in enzymatically active form. Anal Biochem 199: 223–231.

Schauer N, Semel Y, Balbo I, Steinfath M, Repsilber D, Selbig J, Pleban T, Zamir D, Fernie AR (2008) Mode of inheritance of primary metabolic traits in tomato. Plant Cell 20: 509–523.

Schauer N, Semel Y, Roessner U, Gur A, Balbo I, Carrari F, Pleban T, Perez-Melis A, Bruedigam C, Kopka J, Willmitzer L, Zamir D, Fernie AR (2006) Comprehensive metabolic profiling and phenotyping of interspecific introgression lines for tomato improvement. Nat Biotechnol 24: 447–454.

Schauer N, Zamir D, Fernie AR (2005) Metabolic profiling of leaves and fruit of wild species tomato: a survey of the *Solanum lycopersicum* complex. J Exp Bot 56: 297–307.

Schijlen E, Ric de Vos CH, Jonker H, van den Broeck H, Molthoff J, van Tunen A, Martens S, Bovy A (2006) Pathway engineering for healthy phytochemicals leading to the production of novel flavonoids in tomato fruit. Plant Biotechnol J 4: 433–444.

Schmelz EA, Engelberth J, Alborn HT, O'Donnell P, Sammons M, Toshima H, Tumlinson JH, (2003) Simultaneous analysis of phytohormones, phytotoxins, and volatile organic compounds in plants. Proc Natl Acad Sci USA 100: 10552–10557.

Sekiyama Y, Kikuchi J (2007) Towards dynamic metabolic network measurements by multi-dimensional NMR-based fluxomics. Phytochemistry 68: 2320–2329.

Semel Y, Nissenbaum J, Menda N, Zinder M, Krieger U, Issman N, Pleban T, Lippman Z, Gur A, Zamir D (2006) Overdominant quantitative trait loci for yield and fitness in tomato. Proc Natl Acad Sci USA 103: 12981–12986.

Sheehan D (2006) Detection of redox-based modification in two-dimensional electrophoresis proteomic separations. Biochem Biophys Res Commun 349: 455–462.

Smolka MB, Zhou H, Purkayastha S, Aebersold R (2001) Optimization of the isotope-coded affinity tag-labeling procedure for quantitative proteome analysis. Anal Biochem 297: 25–31.

Steen H, Mann M (2004) The ABC's (and XYZ's) of peptide sequencing. Nat Rev Mol Cell Biol 5: 699–711.

Steinhauser D, Usadel B, Luedemann A, Thimm O, Kopka J (2004) CSB.DB: a comprehensive systems-biology database. Bioinformatics 20: 3647–3651.

Sugiyama N, Nakagami H, Mochida K, Daudi A, Tomita M, Shirasu K, Ishihama Y (2008) Large-scale phosphorylation mapping reveals the extent of tyrosine phosphorylation in *Arabidopsis*. Mol Syst Biol 4: 193.

Suzuki H, Sasaki R, Ogata Y, Nakamura Y, Sakurai N, Kitajima M, Takayama H, Kanaya S, Aoki K, Shibata D, Saito K (2008) Metabolic profiling of flavonoids in *Lotus japonicus* using liquid chromatography Fourier transform ion cyclotron resonance mass spectrometry. Phytochemistry 69: 99–111.

Takahashi H, Hayashi M, Goto F, Sato S, Soga T, Nishioka T, Tomita M, Kawai-Yamada M, Uchimiya H (2006) Evaluation of metabolic alteration in transgenic rice overexpressing dihydroflavonol-4-reductase. Ann Bot (Lond) 98: 819–825.

Taylor CF, Paton NW, Lilley KS, Binz PA, Julian RK, Jones AR, Zhu W, Apweiler R, Aebersold R, Deutsch EW, Dunn MJ, Heck AJ, Leitner A, Macht M, Mann M, Martens L, Neubert TA, Patterson SD, Ping P, Seymour SL, Souda P, Tsugita A, Vandekerckhove J, Vondriska TM, Whitelegge JP, Wilkins MR, Xenarios I, Yates JR, Hermjakob H (2007) The minimum information about a proteomics experiment (MIAPE). Nat Biotechnol 25: 887–893.

Tian L, Musetti V, Kim J, Magallanes-Lundback M, DellaPenna D (2004) The Arabidopsis *LUT1* locus encodes a member of the cytochrome p450 family that is required for carotenoid epsilon-ring hydroxylation activity. Proc Natl Acad Sci USA 101: 402–407.

Tikunov Y, Lommen A, de Vos CH, Verhoeven HA, Bino RJ, Hall RD, Bovy AG (2005) A novel approach for nontargeted data analysis for metabolomics. Large-scale profiling of tomato fruit volatiles. Plant Physiol 139: 1125–1137.

Tonge R, Shaw J, Middleton B, Rowlinson R, Rayner S, Young J, Pognan F, Hawkins E, Currie I, Davison M (2001) Validation and development of fluorescence two-dimensional differential gel electrophoresis proteomics technology. Proteomics 1: 377–396.

Ünlü M, Morgan ME, Minden JS (1997) Difference gel electrophoresis. A single gel method for detecting changes in protein extracts. Electrophoresis 18: 2071–2077.

Urbanczyk-Wochniak E, Fernie AR (2005) Metabolic profiling reveals altered nitrogen nutrient regimes have diverse effects on the metabolism of hydroponically-grown tomato (*Solanum lycopersicum*) plants. J Exp Bot 56: 309–321.

Urbanczyk-Wochniak E, Luedemann A, Kopka J, Selbig J, Roessner-Tunali U, Willmitzer L, Fernie AR (2003) Parallel analysis of transcript and metabolic profiles: a new approach in systems biology. EMBO Rep 4: 989–993.

Urbanczyk-Wochniak E, Usadel B, Thimm O, Nunes-Nesi A, Carrari F, Davy M, Blasing O, Kowalczyk M, Weicht D, Polinceusz A, Meyer S, Stitt M, Fernie AR (2006) Conversion of MapMan to allow the analysis of transcript data from Solanaceous species: effects of genetic and environmental alterations in energy metabolism in the leaf. Plant Mol Biol 60: 773–792.

Ward JL, Harris C, Lewis J, Beale MH (2003) Assessment of ^{1}H NMR spectroscopy and multivariate analysis as a technique for metabolite fingerprinting of *Arabidopsis thaliana*. Phytochemistry 62: 949–957.

Weckwerth W (2008) Integration of metabolomics and proteomics in molecular plant physiology—coping with the complexity by data-dimensionality reduction. Physiol Plant 132: 176–189.

Weckwerth W, Loureiro ME, Wenzel K, Fiehn O (2004) Differential metabolic networks unravel the effects of silent plant phenotypes. Proc Natl Acad Sci USA 101: 7809–7814.

Wiener MC, Sachs JR, Deyanova EG, Yates NA (2004) Differential mass spectrometry: a label-free LC-MS method for finding significant differences in complex peptide and protein mixtures. Anal Chem 76: 6085–6096.

Wienkoop S, Morgenthal K, Wolschin F, Scholz M, Selbig J, Weckwerth W (2008) Integration of metabolomic and proteomic phenotypes—Analysis of data-covariance dissects starch and RFO metabolism from low- and high-temperature compensation response in *Arabidopsis thaliana*. Mol Cell Proteom 7: 1725–1736.

Wienkoop S, Weckwerth W (2006) Relative and absolute quantitative shotgun proteomics: targeting low-abundance proteins in *Arabidopsis thaliana*. J Exp Bot 57: 1529–1535.

Williams BJ, Cameron CJ, Workman R, Broeckling CD, Sumner LW, Smith JT (2007) Amino acid profiling in plant cell cultures: an inter-laboratory comparison of CE-MS and GC-MS. Electrophoresis 28: 1371–1379.

Wolters DA, Washburn MP, Yates JR (2001) An automated multidimensional protein identification technology for shotgun proteomics. Anal Chem73: 5683–5690.

Xing,T, RampitschC, Miki BL, Mauthe W, Stebbing J-A, Malik K, Jordan M (2003) MALDI-Qq-TOF-MS and transient gene expression analysis indicated co-enhancement of [beta]-1,3-glucanase and endochitinase by tMEK2 and the involvement of divergent pathways. Physiol Mol Plant Pathol 62: 209–217.

Yates JR, Gilchrist A, Howell KE, Bergeron JJ (2005) Proteomics of organelles and large cellular structures. Nat Rev Mol Cell Biol 6: 702–714.

Zhang P, Foerster H, Tissier CP, Mueller L, Paley S, Karp PD, Rhee SY (2005) MetaCyc and AraCyc. Metabolic pathway databases for plant research. Plant Physiol 138: 27–37.

Zivy M (2007) Quantitative analysis of 2D gels. In: Thiellement H, Zivy M, Damerval C, Méchin V (eds) Plant Proteomics: methods and protocols, vol 355. Humana Press, Totowa, NJ, USA, pp 175–194.

Role of Bioinformatics as a Tool in Tomato Research

Catherine M. Ronning

ABSTRACT

With the advent of high-throughput gene sequencing methods, the amount of DNA sequence data in many major crop species, including tomato, has escalated dramatically. For instance, sequencing of expressed sequence tags (ESTs, Adams et al. 1991) allowed researchers to look at the transcribed sequence complement ("transcriptome") for a specific tissue, developmental stage, experimental treatment, etc. Here is provided information on databases and websites relevant to tomato researchers, methods of integrating the various data types to enhance knowledge of tomato biology are addressed, and work in which bioinformatics tools were an integral part of the project is described. This chapter is intended as a guide to the resources available to tomato researchers as of this writing. Obviously, all databases and websites are subject to revision and update, some will be moved, deleted, or no longer maintained, and new sites will be added.

Keywords: database, bioinformatics, genomics, proteomics, metabolomics, transcriptomics

14.1 Introduction

Before initiation of the National Science Foundation (NSF)-funded Tomato EST Sequencing project in 1999, there were six *Solanum lycopersicum* (*Lycopersicon esculentum*) entries in National Center for Biotechnology Information (NCBI) GenBank's sequence tagged sites database (dbSTS) (release 100298, October 1, 1998). One year later the number had grown to

Office of Biological and Environmental Research, SC-23.2/Germantown Building, U.S. Department of Energy, 1000 Independence Ave., SW, Washington, DC 20585 USA.

50,303 (now the dbEST database, release 121799, December 17, 1999), and nearly two years later, to 138,628 (including 8,346 *Solanum pennellii* and 2,504 *Solanum habrochaites* sequences, dbEST release 092801, September 28, 2001). By comparing specific datasets the potential to provide information on genes transcribed under certain conditions now exists on a global level never before possible. As of October 21, 2011, GenBank contained a total of 744,657 nucleotide sequences from tomato (*S. lycopersicum*; including 298,300 ESTs and 418,489 Genome Survey Sequences (GSSs)), 27 peptide structures, 7,204 protein sequences, 1,540 genes, 737 Gene Expression Omnibus (GEO, high-throughput data such as microarrays) datasets and 43,810 GEO expression profiles, 18,346 unigenes, and 4,261 literature citations, all of which are publicly available. Additionally, sequencing of the tomato genome has been initiated, which promises to add greatly to our fundamental knowledge of tomato genetics and physiology. The challenge was and continues to be the integration of the newly acquired genomic and proteomic data with molecular marker data, linkage maps, and phenotypic data.

To meet this challenge a number of resources in the form of databases and websites have been created and are now available to tomato researchers (Yano et al. 2007). The hallmark, "one stop shop" for the tomato bioinformaticist is the SOL Genomics Network (SGN, http://solgenomics.net/; Mueller et al. 2005, 2008; Bombarely et al. 2010). SGN is a comprehensive, clade-oriented database of the Solanaceae family that includes maps and markers, EST and bacterial artificial chromosome (BAC) sequences, bioinformatics tools, and the in-progress tomato genome sequence. Since many of these sites cannot be placed into discrete categories there is considerable overlap between sections. I will not describe the underlying computing environment of these sites, but will outline the types of data, tools, and other features available in each. Their utility will depend on the individual researcher's needs and objectives. Examples from the literature are provided to further illustrate the potential genomics approaches have to offer.

14.2 Gene and Genome Databases

There are many reasons that tomato was chosen from among the Solanaceae for genome sequencing. *S. lycopersicum* is the most widely studied species within this family, particularly within the area of fruit development and ripening, in which tomato is seen as the model organism for all plants. There is also extensive research on disease and stress response. The tomato genome is relatively small (950 Mb, Michaelson et al. 1991), and has an extensive, high-density genetic map (Tanksley et al. 1992; Broun and Tanksley 1996) and BAC-based physical map. Seventy-five percent of the genome is composed of heterochromatin while the remaining 25%

(~220 Mb) consists of gene-rich, contiguous euchromatic regions located primarily in distal regions of the chromosome arms (Arumuganathan et al. 1991; de Jong 1998; Peterson et al. 1998). A fluorescence in situ hybridization (FISH) map is being constructed to locate these regions (see Sect. 14.5.1). The International Tomato Sequencing Project was initiated in 2004 by a consortium composed of members from Korea, China, the United Kingdom, India, the Netherlands, France, Japan, Spain, Italy, and the United States. The mitochondrial genome will be sequenced by Argentina; the chloroplast genome is complete (Kahlau et al. 2006). The goals of the project include 1) sequencing the gene-rich euchromatic arms of each of the 12 chromosomes, 2) processing and annotating the sequence in a manner compatible with that of *Arabidopsis*, rice (*Oryza sativa*), and other plants, and 3) creating a publicly available bioinformatics portal for Solanaceae genomics through which sequence data and information can be shared. That portal is the SOL Genome Network (SGN, see Sect. 14.7.1), hosted at the Boyce Thompson Institute (BTI) located at Cornell University and mirrored at locations throughout the world.

14.2.1 International Tomato Genome Sequencing Project (http://solgenomics.net/)

Initially, a BAC-by-BAC strategy was employed to sequence the tomato genome, beginning with the sequencing of 1,500 "seed" BAC clones that are anchored to the high-density genetic map (*S. lycopersicum* x *S. pennellii* F_2 population) and aided by construction of a minimal tiling path of BAC clones covering the euchromatic regions, a BAC end sequence database, and a fingerprinted contig (FPC) physical map. In 2009 a whole genome shotgun approach was initiated to complement these efforts, and the assemblies are being annotated by the International Tomato Annotation Group (ITAG).

Sequencing progress can be followed at the project's bioinformatics portal, SGN (see Sect. 14.7.1). Numerous resources and tools are available here as well, including gene, phenotype, and quantitative trait locus (QTL) search capabilities. Annotations include gene structures, repeats, RNAs, and Gene Ontology (GO) functional terms. The International Solanaceae Genome Project (SOL) has an innovative approach to annotation by offering a suite of community annotation tools, encouraging contributions from individual researchers with the result that locus and phenotype descriptors will reflect the expertise of the Solanaceae community (Menda et al. 2008). The goal is to develop a uniform protocol for all phases of sequencing and finishing, nomenclature, and structural and functional annotation, facilitating comparative analyses with other sequenced plant genomes such as *Arabidopsis* and rice. Several international SOL sites are also available:

EU-SOL (http://www.eu-sol.net)

By tapping into the natural biodiversity of the Solanaceae for genes associated with valuable traits, the EU-SOL project hopes to develop high-quality tomato and potato (*Solanum tuberosum*) varieties. These genes will then be incorporated into marker-assisted selection (MAS) and genetic engineering breeding strategies.

Italian SOLAnaceae Genomics Resource (ISOL@)
(http://biosrv.cab.unina.it/isola)

ISOL@ seeks to integrate data from the tomato sequencing community into a common publicly available website (Chiusano et al. 2008). The site is divided into two levels, the genome and the transcriptome. Preliminary annotation of the tomato genome is provided, along with analytical tools for mining these data. The extensive tomato EST dataset is the primary constituent of the expression level, together with tomato microarray probe sets.

Lat-SOL (http://cnia.inta.gov.ar/lat-sol)

Lat-SOL provides a network site for the Latin American laboratories dedicated to Solanaceae research.

Solanaceae Research Community UK
(http://www.uk-sol.org/about.htm)

This site provides information on the United Kingdom's participation in the tomato genome project (chromosome 4).

Indian Initiative for Tomato Genome Sequencing
(http://www.nrcpb.org/content/tomato-genome)

The sequencing of tomato chromosome 5 is highlighted here, including methods for physical mapping, shotgun cloning, and DNA sequencing.

14.2.2 *Solanum lycopersicum Plant Genome Database (SlGDB) (http://www.plantgdb.org/SlGDB)*

PlantGDB is a database of annotated sequence data from green plants as well as a resource for genome analysis tools (www.plantgdb.org, Dong et al. 2005; Duvick et al. 2008). All Viridiplantae sequences in GenBank and UniProt are uploaded regularly. PlantGDB-generated Unique Transcripts (PUTs) are then assembled from species with at least 10,000 transcripts, aligned to UniProt entries and assigned GO annotations based on the UniProt match. PlantGDB provides genome browsers for sequenced plant genomes and BAC-based browsers for genomes in progress, including tomato (SlGDB). Here, one can browse both finished and unfinished BACs, and view annotations and EST, cDNA and PUT alignments.

A suite of tools are available within PlantGDB/SlGDB. BLAST searches against tomato microarray probes, cDNAs, ESTs, BACs, and PUTs are possible, as well as against *Arabidopsis* predicted peptides. The existence of alternative splicing in eukaryotes has an impact on correct annotation of genes and gene products; the GeneSeqer tool predicts exon/intron structure and splice sites by aligning expressed transcripts to genomic sequence. Genes can be annotated by the user through Your Gene structure Annotation Tool for Eukaryotes (yrGATE), a web-based tool for community annotation.

To streamline computational processes, PlantGDB is associated with the BioExtract Server (http://www.bioextract.org/). This web interface allows a smooth workflow by integrating database queries, analyses, and data storage in a form that can be saved, revised and shared. PlantGDB/SlGDB provides a comprehensive, single-site resource useful for interspecific comparative research.

14.2.3 Solanaceae Genomics Resource (http://solanaceae.plantbiology.msu.edu)

The Solanaceae Genomics Resource, which integrates genomic sequences from the ongoing tomato, potato, and tobacco (*Nicotiana tabacum*) genome sequencing projects with Solanaceae transcript assemblies, is housed at Michigan State University, having moved from its former location at The Institute for Genomic Research (TIGR, now the J Craig Venter Institute (JCVI)). The genome browser databases contain BAC sequences from the three genome sequencing projects mentioned above as well as assembled Solanaceae transcripts from PlantGDB (PUTs, see Sect. 14.2.2), UniProt-genome sequence alignments, matches to genomes of the model dicots *Arabidopsis*, grape (*Vitis vinifera*), and poplar (*Populus trichocarpa*), and repeat and ncRNA predictions.

A number of comparative analysis databases and tools are available, including mappings of Solanaceae transcripts to *Arabidopsis*, Solanaceae lineage-specific sequences (identified by TBLASTX against non-Solanaceous genomes and transcripts), the Solanaceae ortholog/paralog database, and Solanaceae genetic markers and in silico-generated simple sequence repeats (SSRs). Now that the potato genome sequence is available, all Solanaceae genomic data will be integrated using potato as the reference genome All resources are downloadable, providing a global functional genomics resource for the Solanaceae from a single web portal.

14.2.4 Tomato Genome Project at NCBI
(http://www.ncbi.nlm.nih.gov/genomes/PLANTS/PlantList.html)

The tomato genome project is one of many publicly funded genome sequencing projects featured in NCBI Plant Genomes Central, which provides information on all large-scale, currently active plant sequencing projects that are or will soon contribute to the genome sequence of that particular plant. There are four major sections to this site: 1) completed genome sequencing projects, with links to sequences of each chromosome as well as plastid and mitochondrial sequence (if available) of fully sequenced plants including *Arabidopsis, Medicago truncatula*, rice, and poplar, 2) in-progress genome sequencing projects, which include (among others) tomato and potato, 3) genetic mapping projects, including *Solanum lycopersicoides* (wild nightshade), *Solanum melongena* (eggplant), and *Solanum peruvianum* (Peruvian tomato), in addition to tomato and potato, and 4) EST sequencing projects. This last section provides links from which one can BLAST a sequence of interest against all tomato or all potato EST sequences, or against the entire plant EST database.

A fifth section, Plant-Oriented Resources at NCBI, summarizes and provides links to several additional tools oriented towards plant research, including the Keygene nomenclature for amplified fragment length polymorphism (AFLP) primers, multiple genome Map Viewer, and plant genome or EST BLAST. This site contains available project data including the plastid genome (Kahlau et al. 2006), proteins, ESTs, publications, and trace files. There are also links to the data and the sequencing centers providing the data for BAC ends, plastid genome, nuclear genome, and maps.

14.2.5 C.M. Rick Tomato Genetics Resource Center (TGRC)
(http://tgrc.ucdavis.edu)

Named for the late Dr. Charles M. Rick, the TGRC genebank contains numerous genetic stocks of tomato, including wild relatives and mutants. Located at the University of California at Davis, TGRC is part of the United States Department of Agriculture (USDA) National Plant Germplasm System (NPGS). This extensive database can be queried by any of a number of criteria including accession number, category (e.g., ploidy level, marker, disease resistance, inbred, hybrid, or introgression line), mating system, gene, background genotype, or chromosome number. It is also possible to search for tomato accessions containing specific genes, loci, alleles, marker or mutant types, or phenotype; images of selected genes, phenotypes and accessions are available as well. Seeds of accessions are available to researchers upon request. The database is searchable through SGN.

14.2.6 Genes That Make Tomatoes
(http://zamir.sgn.cornell.edu/mutants/index.html)

Mutant populations provide an invaluable means of exploring gene function. Genes That Make Tomatoes, which is linked to SGN, is a comprehensive library of tomato mutants derived from the inbred processing variety 'M82' treated with ethyl methanesulfonate (EMS) or fast-neutron irradiation to induce mutagenesis (Menda et al. 2004). The current collection contains nearly 3,500 mutations catalogued into 15 primary and 48 secondary categories, including descriptors for seed, plant, leaf, flower and fruit morphology, sterility, and response to disease and stress. This isogenic tomato mutation library is searchable by gene name, keyword, and phenotypic characteristics. Seed is available upon request.

14.2.7 Real-Time QTL (RTQ)
(http://zamir.sgn.cornell.edu/Qtl/Html/home.htm)

Many if not most phenotypic traits of interest to tomato breeders such as yield, flavor, and disease resistance manifest as subtle, continuous phenotypic differences resulting from expression of quantitative trait loci (QTLs) rather single genes. This complexity presents an additional challenge to the researcher trying to mine genomic data for clues to such traits. RTQ integrates QTL data from 76 introgression lines (ILs) derived from the tomato 'M82' and the wild green-fruited species *S. pennellii* with the sequenced-oriented SGN genetic map and sometimes to the genes themselves. Measured traits include yield (e.g., plant weight, brix, total biomass), fruit morphology (size, weight, color), and fruit biochemistry (carotene and lycopene content). After selecting specific traits the user is linked to the SGN chromosome map containing the trait.

14.2.8 Other Plant Resources

UniProt Knowledgebase: HAMAP proteomes: Plastids
(http://hamap.expasy.org/plastid.html)

The complete genome sequence of the tomato plastid genome, with links to its 80 proteins, is contained here.

ChloroplastDB (http://chloroplast.cbio.psu.edu)

This site attempts to provide a unified naming system for chloroplast genes (the plastome, Cui et al. 2006). Chloroplast sequences from over 100 species are currently included in the database, including tomato and potato. Interspecific chloroplast protein families, clustered using the

Markov clustering algorithm TribeMCL (Enright et al. 2003) are available. Another useful tool is a list of RNA edits, results of post-transcriptional, pre-translational processes occurring in chloroplasts of land plants that can cause nonsynonymous nucleotide changes. Such changes result in a protein sequence that differs from that predicted by the DNA sequence, having obvious implications for gene annotation and protein prediction.

Germplasm Resources Information Network (GRIN; http://www.ars-grin.gov)

The USDA Agricultural Research Service (ARS) NPGS was created to maintain genetic diversity within stock collections through acquisition, preservation, evaluation, documentation, and distribution of crop germplasm from throughout the world. This information is contained within the GRIN database, which as of October 21, 2011, contained 15,938 different Solanum accessions, both cultivated and wild, and including 9,207 accessions from *S. lycopersicon*. Information on numerous phenotypic descriptors is provided and is searchable, and includes terms describing plant morphology, disease, insect, nematode, and stress response, growth, phenology and production, collection source and history, economic impact, geographical range, synonyms, and images. Data for individual crops can be downloaded using pcGRIN, allowing offline searching. Germplasm is available to scientific researchers.

14.3 Comparative Genome Databases

Most of the online resources described in this chapter contain the capability for comparative (i.e., intra- and interspecific) genomic analysis. SGN (described in detail in Sect. 14.7.1) includes several Solanaceous species (including cultivated and wild tomato, potato, pepper (*Capsicum annuum*), eggplant, tobacco, petunia (*Petunia hybrida*)) with other fully sequenced genomes (Mueller et al. 2005, 2008). A set of highly conserved, single/low copy number markers (conserved ortholog set, COS) was developed by stringently screening the tomato EST database against the *Arabidopsis* genome (Fulton et al. 2002). Computational screening against the *M. truncatula* genome and hybridization with "garden blots" containing genomic DNA from a wide variety of plant species indicated that at least a subset of these COS markers can be used for comparative studies with more distant species. This set was extended by the identification of single-copy orthologs (COSII) common to the Euasterids and *Arabidopsis* (Wu et al. 2006).

In addition to the Solanaceae, this Asterid I clade-oriented database contains sequences from the Rubiaceae (e.g., coffee (*Coffea* spp.)) and the Plantaginaceae (snapdragon (*Antirrhinum majus*)). In fact, tomato EST sequences were found to be very similar to those of coffee, indicating that Solanaceous species are preferable to *Arabidopsis* as genomic models for other Asterids (Lin et al. 2005).

Another resource for comparative *in silico* analysis is the Dana-Farber Cancer Institute (DFCI) Gene Index Project (formerly the TIGR Gene Indices), which includes the Tomato Gene Index. The DFCI Gene Index project contains annotated unigene sets constructed primarily from EST sequences of 75 plant species, 46 animal species, 15 protists, and 10 fungal species, as well as tools for interspecific comparative analysis. The DFCI Gene Indices are discussed in greater detail in Sect. 14.4.1.

Munich Information Center for Protein Sequence (MIPS) PlantsDB features comparative and integrative analytical tools, with a focus on protein sequences (see Sect. 14.7.4). The Plant Ontology Consortium (POC) database, based upon the GO project, uses a controlled vocabulary to describe genes and gene products, specifically plant structural and developmental stages (see Sect. 14.6.3). This uniformity of terms greatly facilitates cross-species analysis and functional inference.

14.4 Gene Expression Databases

ESTs are partial, single-pass sequence reads from cDNA clones (Adams et al. 1991). Since cDNA libraries can be constructed, e.g., from specific tissues or developmental stages, ESTs capture a "snapshot" of genes transcribed (the "transcriptome"). ESTs provide a venue for gene discovery as well as evidence for gene modeling and alternative splice forms (D'Agostino et al. 2007). One aspect of this methodology is that highly expressed genes are sequenced repeatedly; the deeper a library is sequenced the more redundant the dataset becomes, while very rare transcripts will probably not be detected at all. This feature can be advantageous if the objective is to approximate overall expression levels of the transcriptome (in silico expression analysis); however, when gene discovery is the objective sequencing clones representing the same gene(s) repeatedly is wasteful of resources. This problem can be alleviated somewhat by use of normalized or subtracted libraries, which should increase the chance of discovering new and/or rare transcripts.

Microarray technology allows global gene expression analysis between and within tissues, treatments, or species in a single experiment. Construction of cDNA arrays based on ESTs is usually based on a computationally assembled and thus non-redundant "unigene" dataset (see Sect. 14.4). Oligonucleotide arrays (e.g., Affymetrix) consist of short oligomers that uniquely represent each gene. Since all probes are of the same length, experimental conditions are easier to optimize and cross-hybridization between related genes is minimized. Arrays based on completely sequenced genomes (e.g., the *Arabidopsis* ATH1 array) offer the opportunity to monitor expression of the entire genome.

The advent of global gene expression techniques such as ESTs and microarrays offers analytical potential never before seen, and a number of databases and websites have been constructed to facilitate such analyses. However, before embarking upon such research it is important to understand the limitations as well as the benefits (Alba et al. 2004). First of all, public EST databases contain a certain proportion of errors and contaminants; processing errors occur during assembly and annotation, and human errors occur during clone handling. It is well worth the time to reprocess and recheck quality of these sequences before beginning any extensive analysis or experiment. cDNA microarrays based upon ESTs have these limitations plus the problem of expression artifacts resulting from non-specific hybridization, e.g., chimeric clones or secondary structure. Oligonucleotide arrays are denser, however, they are very expensive. Another important consideration for microarray experimentation is that sufficient quantities of high purity RNA from which to prepare labeled targets are required.

Careful thought should be put into the experimental design of gene expression and transcriptome profiling experiments, including standardization of RNA preparation and sufficient replication. The type of experiment (e.g., time-course, development, inter- or intraspecific comparisons) will influence design and downstream analysis. Data processing pipelines should be thoroughly considered and conform to the resources and expectations of the experiment. Appropriate statistical tests, such as the Students t-test, ANOVA, and non-parametric tests, will depend upon assumptions about the data to be analyzed. Detailed protocols for tomato microarray analysis are available at http://ted.bti.cornell.edu/cgi-bin/TFGD/array/home.cgi.

14.4.1 Tomato Gene Index (at DFCI) (http://compbio.dfci.harvard.edu/tgi)

With the virtual explosion of EST sequences in the late 1990s came the potential for comparative genomic research on a scale never before seen. However, the very attributes that made this technology so powerful, the magnitude and diversity of the data, also made such analysis formidable. Differences between source labs in features such as nomenclature and quality, as well as the inherent redundancy of EST datasets, could easily overwhelm existing data management systems. It was evident that in order to efficiently perform the comparative analyses now possible a means of cataloging the literally hundreds of thousands of newly available sequences was necessary. Thus, bioinformatics researchers at TIGR conceived of and constructed the TIGR Gene Indices (TGI, Quackenbush et al. 2001).

The Gene Index Project seeks to create a non-redundant, high-value set of ESTs by cataloging publicly available EST sequences from organisms having significant numbers of available cDNA sequences. To ensure quality, sequences downloaded from NCBI dbEST are submitted to a cleaning pipeline in which contaminants and low quality sequences are expunged and poly(A/T) tails and vector sequences are trimmed from remaining sequences. The cleaned, trimmed sequences are first clustered under stringent constraints (40 bp minimum match, >94% ID, maximum 30 bp unmatched overhang), after which the clusters are assembled into contigs containing at least two sequences. A single consensus sequence is inferred from the assembled contig (the "tentative consensus," or TC sequence), and the resulting collection of TCs and singletons are autoannotated (see the DFCI Gene Index Project website and accompanying publications for details of these processes).

The Tomato Gene Index (LeGI) was one of the first plant gene indices to come online from this project, made possible as a result of the NSF-funded project "Generation of a Tomato EST Database" (Van der Hoeven et al. 2002). Over 150,000 tomato ESTs from 26 cDNA libraries from various tomato tissues and developmental stages were generated from this project and compiled into LeGI. Other tomato ESTs as well as tomato expressed transcripts (ETs) in dbEST were included. Like the Gene Indices for other species, LeGI supports a number of analytical tools. The Sequence Reports section allows searches by keyword, TC or EST annotation, or library, and the TC Report illustrates the alignment of a given TC with all component sequences. Functional analysis tools provide insight into alternative splice forms, expression analysis, metabolic pathways, and sequence-specific oligomers. The gene expression data page predicts differentially expressed genes based on the R statistic (Stekel et al. 2000) between user-selected libraries.

TGI now resides at DFCI. The current tomato release contains 299,800 ESTs and 2,989 ETs, which collapse into 28,167 TCs and 24,335 singletons for a total of 52,502 unique genes (LeGI release 13.0, April 13, 2010). At present DFCI maintains gene indices for a total of 60 plants (including, in addition to tomato, the Solanaceous species pepper, potato, petunia, tobacco, and *Nicotiana benthamiana*) as well as several species of animals, protists, and fungi. Putative orthologs between tomato Tentative Consensus (TC) sequences and singletons and any other organism in the index can be obtained using the Eukaryotic Gene Orthologues (EGO) tool. If one wishes to construct a "personal" gene index and has access to a Linux or

SunOS system the DFCI Gene Indices software tools are freely available for download to the scientific community. These tools include TGI Clustering (TGICL), the assembly viewer clview, SeqClean for automated sequence cleaning, cdbfasta and cdbyank to index and retrieve FASTA files from flat files, and the genome viewer DAS/XML. The DFCI TGI provides a powerful resource for comparative research for tomato and between tomato and numerous eukaryotic species.

14.4.2 Tomato Functional Genomics Database (TFGD) (http://ted.bti.cornell.edu)

The Tomato Functional Genomics Database (TFGD) is a comprehensive public resource for tomato expression data, metabolite profiles, and small RNA (sRNA) data, as well as numerous computational pipelines and tools to process, query and analyze the increasingly expanding functional genomics data for tomato (Fei et al. 2011). This site expands upon the previously described Tomato Expression Database (TED; Fei et al. 2006) with the inclusion of metabolite and sRNA data and tools that allow the user to explore co-expression and regulatory mechanisms of complex pathways. The site is composed of three integrated components: gene expression data, metabolite profiles, and sRNAs, and is linked to SGN (http://solgenomics. net/; section 14.7.1).

Probes for the three currently available microarray platforms have been annotated, mapped to plant-specific GO terms, and assigned to metabolic pathways (LycoCyc; SGN). Microarray data has been processed to ensure uniformity across experiments, and tools are offered that allow co-expression analysis, pathway analysis, and functional classification. Metabolite and transcriptome profiles can be queried together using a tool that correlates these data sets, allowing the user to identify genes associated with a given pathway. Finally, TFGD is a repository for sRNA sequence data with tools to investigate sRNAs, putative miRNAs and miRNA targets, and small interference RNAs (siRNAs). The TFGD will continue to collect additional data as it becomes available, and plans to incorporate RNA-seq, proteomics and phenotypic data sets are underway.

14.4.3 Kazusa Micro-Tom Database (MiBase) (http://www.pgb.kazusa.or.jp/mibase/)

The dwarf tomato cultivar Micro-Tom is a completely miniaturized version of tomato, reduced not only in height but also in leaves, fruits, and life cycle, thus allowing dense planting and experimentation in artificially controlled environments such as growth chambers (Meissner et al. 1997). Located at the Kazusa DNA Research Institute in Japan, the Micro-Tom database combines publicly availably tomato ESTs with 115,062 ESTs generated from nine different Micro-Tom cDNA libraries and then assembled into an annotated 76,276-element unigene set (KTU3, Kazusa Tomato Unigene ver. 3, Yano et al. 2006). The unigene database can be searched by annotation (tomato or *Arabidopsis*), GO term, name, tissue, strain, pathways, or markers (single nucleotide polymorphisms (SNPs), insertion/deletions (indels), SSRs). Microarray and gene expression networks are searchable on this site, and BLAST and SNP finder tools are available.

14.4.4 Kazusa Full-Length Tomato cDNA Database (KafTom) (http://www.pgb.kazusa.or.jp/kaftom)

Also available at Kazusa is the Micro-Tom Full-Length Tomato cDNA database (KafTom; Aoki et al. 2010). Many of the search capabilities available at MiBase also exist here, as well as a genome sequence search to hunt for BAC clones and predicted protein coding sequences.

14.4.5 Affymetrix GeneChip® Tomato Genome Array (http://www.affymetrix.com)

The "Affy" chip is a 25-mer oligonucleotide array that was developed in collaboration with tomato researchers through the Affymetrix GeneChip® Consortia Program. Based on NCBI Unigene build #20 (2004) and GenBank mRNAs, the chip contains over 10,000 *S. lycopersicum* probe sets, representing 9,200 transcripts plus controls.

14.4.6 RNA-Seq

While DNA microarrays have revolutionized biology by allowing analysis of the transcriptome on a global scale, they are limited in that sequence information is required for transcript analysis and evaluation. The new, next-generation, high-throughput sequencing technologies have now contributed to the development of RNA-Seq (Wang et al. 2009; Wilhelm and Landry 2009), which uses massively parallel sequencing (MPS) to more accurately quantify transcript levels as well as isoforms of the entire transcriptome.

14.4.7 Applications

The work described below represents the various ways gene profiling methods have been used in tomato research. They also illustrate how bioinformatics methods used in conjunction with standard "wet lab" experimentation and classical genetics approaches can provide insights into biological processes such as stress and disease response and metabolism.

Ouyang et al. (2007) used suppression subtractive hybridization (SSH) and microarray analysis to identify early response genes in tomatoes exposed to severe salt stress. Forward and reverse SSH cDNA libraries were constructed from each of two cultivars, LA2711 (salt-tolerant) and ZS-5 (salt sensitive), and 768 random clones were selected from each of the four libraries for spotting onto a cDNA microarray. RNA was isolated from roots at various time points after exposure to high salt, and gene expression profiles measured for each. All cDNAs differentially expressed after 30 min of treatment were sequenced, and the resulting ESTs assembled into 201 unigenes. The unigenes were putatively identified based on BLAST searches against the NCBI nr protein database, and then classified into MIPS functional categories. Transcription-related genes, cell wall genes, and genes associated with other pathways, as well as unknown proteins, were all putatively identified among the differentially expressed genes, indicating a whole plant response and adaptation to salt stress.

Baxter et al. (2005) performed transcriptome analysis of six distinct tomato ILs previously used to identify QTLs associated with fruit quality and yield with the CGEP TOM1 cDNA array. Each line contains a unique, non-overlapping introgressed segment of *S. pennellii* in *S. lycopersicum* 'M82' background that results in altered fruit gene expression. Principal components analysis (PCA) was performed on the complete datasets to illustrate transcriptomic diversity among ILs, successfully distinguishing lines varying in fruit phenotype. Significance Analysis of Microarrays (SAM, http://www.jcvi.org/cms/research/software/) analysis was also performed to identify specific genes that were differentially expressed. The results add to the characterization of primary metabolism (fruit soluble solids and carbohydrates) in these lines.

A cDNA array was constructed by querying the (formerly TIGR) DFCI TGI for root-associated genes to perform root transcriptome analysis of root-knot nematode (RKN) susceptible (*Mi-*, 'Moneymaker') and resistant (*Mi+*, 'Motelle') tomato cultivars (Schaff et al. 2007). GO terms, Hidden Markov Model (HMM) and Interpro protein motifs were also used to identify genes putatively associated with the nematode-plant interaction. Root tissue was collected for RNA extraction and hybridization at various time points after infection, and the data was analyzed by mixed-model ANOVA. Nearly half of the genes were significantly differentially regulated between infected and uninfected roots, indicating that a wide range of genes are affected by RKN infection. Out of the 4,300 genes initially identified, 1,547 were found to be usable, highlighting the importance of checking quality of clones and/or data before proceeding experimentally.

Several papers illustrate the use of gene profiling studies to identify genes involved in the disease response. Transcriptome analysis was performed on *Xanthomonas campestris* pv. *vesicatoria* (bacterial spot) race T1-resistant tomato 'Hawaii 7998' containing the *Rxv, Rxv1, Rxv2* genes, and T3-resistant 'Hawaii 7981' with the *RxvT3* gene, to characterize recognition between the avirulence factor AvrRxv and resistance proteins (Bonshtien et al. 2005). Leaf samples were taken at various time intervals after challenge with isogenic bacterial strains race T3 (AvrRxv-HA) and race T3 [AvrRxv-HA(C244A)] and hybridized to the CGEP TOM1 array. Student t-tests identified 420 genes that were up- or down-regulated. When classified into functional role categories, 32% were unknown, while most of the classified genes (12%) were putative transcription factors. Balaji et al. (2007) extended this work by measuring changes during the resistance response triggered by the avirulence gene *avrXv3* and the resistance gene *RxvT3*. Array results were validated by Northern blot analysis. Affymetrix oligonucleotide arrays were used to identify transcriptional changes upon infection of tomato by *Clavibacter michiganensis* (canker and wilt disease, Balaji et al. 2008). Results were analyzed by Student's t-test and validated by quantitative RT-PCR. One-hundred twenty two genes were found to be differentially expressed at one or more time points.

The tomato chloroplast genome (the "plastome") is a 155,461 bp circular chromosome containing 114 genes and open reading frames (ORFs) (Kahlau et al. 2006). An oligonucleotide array (each element 68–71 nucleotides in length) was constructed to represent all genes and ORFs from the tobacco, tomato, and potato plastomes (Kahlau and Bock 2008). Expression profiling measured differences in transcriptional and translational regulation during tomato fruit development and ripening and during chloroplast to chromoplast differentiation.

S. lycopersicum has glandular trichomes that produce compounds such as terpenes, acyl sugars, and phenylpropanoid-derived metabolites (www.trichome.msu.edu; Schilmiller et al. 2009, 2010; http://bioinfo.bch.msu.edu/trichome_est; Schilmiller et al. 2009). Several previously uncharacterized sesquiterpene synthase cDNAs were identified from stem trichomes of *S. lycopersicum* and the wild species *S. habrochaites* using RNA-seq (Bleeker et al. 2011). Using a combination of high-throughput cDNA sequencing from tomato trichomes with proteomics experiments and metabolite profiling, Schilmiller et al. (2010) was able to identify and functionally characterize a previously unknown sesquiterpene synthase in this tissue.

14.5 Molecular Marker and Genetic Map Databases

Cultivated tomato is genetically highly monomorphic relative to wild tomato species, probably due to domestication (Rick and Fobes 1975; Miller and Tanksley 1990), and so most tomato molecular maps were developed using interspecific crosses. The proliferation of tomato DNA sequence data now coming forth should enhance genome mapping efforts through development of more high resolution intraspecific markers. For an extensive review of genetic maps and markers in tomato and their applicability to tomato breeding see Foolad (2007) and Labate et al. (2009).

14.5.1 Maps

A successful genome sequencing project requires the existence of a high-density genetic map. Twenty tomato maps of various types are available at SGN (http://solgenomics.net/, see Sect. 14.7.1) and are described briefly below.

14.5.1.1 Genetic Maps

The Tomato-EXPEN 1992 map was developed from a composite F_2 population derived from *S. lycopersicum* cv. VF36 x *S. pennellii* LA716 (Tanksley et al. 1992). This map includes 1,005 markers, primarily restriction fragment length polymorphism (RFLP) but also isozyme and morphological markers. The same set of probes was used to construct a high-density linkage map of potato, making it possible to locate the inversion breakpoints that differentiate tomato and potato at the genomic level (Tanksley et al. 1992). The Tomato-EXHIR 1997 map, derived from an interspecific backcross of *S. lycopersicum* TA209 x *S. habrochaites* LA1777, contains a 135-marker subset of EXPEN 1992, and was used to study floral traits and compatibility between these two species (Bernacchi and Tanksley 1997). Tomato-EXPEN 2000 (*S. lycopersicum* LA925 x *S. pennellii* LA716 type F2.2000) is based on 80 F_2 individuals. The 2,506 total markers include a subset from the Tomato-EXPEN 1992 map as well as SSR and COS (Fulton et al. 2002) markers. This map was used to determine synteny between *Arabidopsis* and tomato. The Kazusa F2-2000 map is similar but contains new SSR and intronic polymorphism loci (Fulton et al., 2002; Ohyama et al., 2009; Shirasawa et al., 2010). The Tomato–EXPIMP 2001 map is a consensus map derived from the BC_1, BC_2, and BCIRL populations of *S. lycopersicum* TA209 x *S. pimpinellifolium* LA1589 (Grandillo and Tanksley 1996; Tanksley et al. 1996; Doganlar et al. 2002). It contains 145 RFLP markers, and has been used to identify QTLs associated with several economically important traits. The Tomato-EXPIMP 2008 map (*S. lycopersicum* TA492 x *S. pimpinellifolium* LA1589) contains 181 markers (Gonzalo and van der Knaap, 2008). *S.pimpinellifolium* LA1589 was also used to generate F2 maps with 'Yellow Stuffer' (Yellow Stuffer x LA1589) and 'Sun1642' (Sun1642 x LA1589) (van der Knaap and Tanksley 2003). The two linkage maps were combined to generate an integrated map.

The 294 markers that comprise the Tomato-EXPIMP 2009 map (*S. lycopersicum* (NCEBR-1) x *S. pimpinellifolium* (LA2093)) are putatively associated with disease resistance or defense-related response pathways (Ashrafi et al., 2009). The 91-marker Tomato-LXCHM 2007 is a frame map primarily containing CAPS, SSR and AFLP markers derived from a *S. lycopsersicum* LE777 x *S. chmielewskii* CH6047 F2 population (Jiménez-Gómez et al., 2007).

Finally, two tomato IL maps were constructed from the *S. lycopersicum* Zamir ILs, each having 76 markers based on either the EXPEN1992 or EXPEN2000 map. Each of these near-isogenic lines (NILs), developed through series of backcrosses, have a single chromosome segment from *S. pennellii* in *S. lycopersicum* 'M82' background.

14.5.1.2 Physical Map

The Tomato (*S. lycopersicum*) Physical Map contains anchored BACs mapped relative to the EXPEN2000 map. The BACs are anchored by overgo methodology, by computational matches to BAC end sequences, or through experimentation or literature review. A total of 10,600 markers are located on the 10,501 assigned BACs.

The Tomato FPC (SGN 2011) map contains the positions of 976 contigs comprising the Arizona Genome Institute's SGN2011 physical map. Clicking on a chromosome or on an individual contig leads to additional information.

14.5.1.3 Cytological Map

The Tomato FISH map was constructed by performing FISH on pachytene phase chromosomes with labeled BAC probes. The FISH map is being used to determine the amount of euchromatin present in the genome and to locate the euchromatic/heterochromatic boundaries, information vital to the genome sequencing project. Using FISH and mapping results in combination, the fruit morphology locus *sun* was found to be located within a region of the tomato genome prone to chromosomal rearrangements (van der Knaap et al. 2004).

14.5.1.4 Sequence Progress Map

The Tomato Accessioned Golden Path (AGP) map shows all finished (sequenced) clones and their approximate physical map positions.

14.5.1.5 Other Maps

The Tomato—Kazusa and SolCAP markers mapped to genome has positions of 10,763 of these markers, and the Tomato QTL map contains 201 markers.

14.5.2 Transposon Tagging

Reverse genetics, or the analysis of changes occurring after mutagenesis of a target gene, is a relatively straightforward means of determining gene function. Transposon-induced mutagenesis takes advantage of the fact that the maize (*Zea mays*) transposable elements *Ac* and *Ds* retain most features when integrated into other plant genomes. Closely linked genes can thus be tagged, a feature exploited for high-throughput insertional mutagenesis for functional analysis of tomato (Emmanuel and Levy 2002). The finding that *Ds* preferentially inserts into genes led to the development of 2,932 lines containing approximately 7,500 *Ds* insertions in the miniature tomato cultivar Micro-Tom (Meissner et al. 2000). This Micro-Tom-*Ac/Ds* system allowed for both targeted and non-targeted transposon tagging, useful for the discovery of novel genes and transcriptional regulators (i.e., promoters). A multifunctional plasm13 construct containing a T-DNA-modified *Ds* element and selectable marker genes was fabricated and introduced into the tomato cultivar 'Moneymaker' (Gidoni et al. 2003). Sites for rare-cutting restriction enzymes and site-specific recombinases allow mapping and cloning of regions between the T-DNA and the transposed *Ds* element. Four-hundred five individual tomato lines containing a copy of pJasm13 were constructed.

14.5.3 Targeting Induced Local Lesions IN Genomes (TILLING)

TILLING is another reverse-genetics technique that combines EMS-induced mutagenesis with high-throughput screening and heteroduplex analysis for the detection of point mutations (McCallum et al. 2000; Henikoff et al. 2004). DNA from the M_2 generation is pooled, arrayed, and amplified with gene-specific primers. The PCR products are incubated with CEL I endonuclease, resulting in heteroduplex formation between wild-type and mutant DNA which can be rapidly screened. The location of the mutation can then be determined by screening individual DNA samples from the pool. A TILLING service has been installed at the University of California Davis Genome Center, and has been funded to produce a TILLING service for tomato (http://tilling.ucdavis.edu/index.php/Tomato_Tilling).

14.6 Protein and Metabolome Databases

Proteomics and metabolomics refer to the large-scale study of the entire complement of proteins (structure and function, the "proteome") and of metabolites (small molecules such as metabolic intermediates and products and signaling molecules, the "metabolome"), respectively, within an organism. Proteomics is used in the pharmaceutical industry for novel drug discovery and in the development of diagnostic biomarkers for specific diseases. Metabolomic profiles of mutants with indels in known genes can facilitate prediction of function of unknown genes.

Tomato is the model organism for fruit development and ripening. Rich in metabolites, the compounds comprising flavor volatiles, pigments, and vitamin content, among others, have been well studied in this fruit. As such, tomato is a natural organism to use in metabolomic and proteomic research. The integration of metabolite, biological, and chemical data through efficient database management and bioinformatics tools will lead to a better understanding of overall biological function during processes such as fruit development (Tikunov et al. 2005; Moco et al. 2007).

14.6.1 Tomato Metabolite Data at TFGD (http://ted.bti.cornell.edu/cgi-bin/TFGD/metabolite/home.cgi)

Tomato metabolite profile data from *S. pennellii*- and *S. habrochaites*-derived ILs are housed within this module of the TFGD (section 14.4.2). This site provides a number of tools useful in mining the database for specific compounds or tomato lines. Using the Metabolite Sorter one can select a specific metabolite and find the levels of that metabolite in the various ILs; the Multiple Metabolite Sorter allows searching for ILs containing user-specified levels of metabolites. The Metabolite/Gene Profile Correlation tool allows the user to link a metabolite from a specific metabolite dataset with gene expression data from the TOM1 or TOM2 microarrays, thereby identifying genes with profiles similar to that of the metabolite and allowing correlation of changes in metabolite accumulation with gene expression among ILs.

14.6.2 Secretom (http://solgenomics.net/secretom)

Part of SGN, the Secretom project investigates the cell wall proteome (the "secretome") in tomato, for example, during fruit ripening or the defense response, and catalogs its properties using tomato as a model. Several tools and resources are provided through this project, including computation prediction of signal peptides ("SecreTary") and various datasets related to the cell wall, glycoproteome, and others, available for download.

14.6.3 Other Plant Resources

MetaCyc, BioCyc, SolCyc, AraCyc (http://metacyc.org, http://biocyc.org/, http://solcyc.solgenomics.net, //http://www.arabidopsis.org/biocyc/index.jsp)

MetaCyc, a multi-organism database of experimentally determined, manually curated, nonredundant biochemical pathways, includes more than 600 pathways found in plants, including *S. lycopersicon* (Caspi et al., 2010). The site includes proteins and complexes, compounds, pathways, genes, and reactions, all of which are interrelated. In conjunction with MetaCyc the BioCyc website contains more than 500 Pathway/Genome Databases (PGDBs) describing organism-specific genome and metabolic pathways, including the Solanaceae-specific SolCyc and *Arabidopsis*-specific AraCyc databases. These sites contain manually and computationally predicted annotation and is continuously curated (Mueller et al. 2003; Krieger et al. 2004; Zhang et al. 2005), and together can be useful in discerning function of tomato proteins.

Plant Ontology Consortium (POC)
(http://www.plantontology.org)

The plethora of DNA and protein sequence data made available in the past few years, contributed by innumerable research groups and labs, can have many different terms and definitions, making cross-species comparison quite difficult. The GO project (http://www.geneontology.org/, Ashburner et al. 2000) strives to simplify this by developing a controlled vocabulary to describe genes and gene products which can be applicable to any organism. POC builds upon the GO project by providing uniform descriptions across plant species, specifically defining genetic and phenotypic data associated with structure and developmental stages (The Plant Ontology Consortium 2002; Avraham et al. 2008).

Like GO, the PO browser displays gene product information, but additionally contains annotation for object types such as germplasm and QTLs.

While the POC database has been built primarily upon *Arabidopsis*, rice, and maize, community participation in supplying new descriptive terms from other angiosperms is encouraged and new terms are continuously added. Data from the tomato genome sequencing project should add substantially to the annotations already available in POC, providing a valuable tool for comparative research in Solanaceae.

14.6.4 Applications

An extensive metabolic profiling of various tissues within the tomato fruit and at different stages of post-expansion fruit ripening was conducted using liquid chromatography with quadrupole time-of-flight mass spectrophotometry (LC-QTOF-MS) and with photo-diode array and fluorescence detection (LC-PDA-FD) (Moco et al. 2007). Most of the detected compounds were secondary metabolites, which are responsible for economically important traits such as flavor and aroma. Different metabolites accumulated preferentially in different fruit tissues and at different ripening stages, further defining the tempo-spatial characteristics of metabolic pathways in tomato.

The TOM1 microarray was used to study carotenoid biosynthesis in a tomato mutant that overproduces phytonutrients (Kolotilin et al. 2007). Pericarp tissue from mature green, breaker, and early red fruits of *hp-2^{dg}* and *hp-2^{j}* mutants, which have higher levels of pigmentation, flavonoids and vitamins, were used in the analysis. The authors specifically profiled genes encoding enzymes of the plastid methylerythritol phosphate (MEP) pathway, which leads to carotenogenesis. Quantitative reverse transcription (qRT)-PCR analysis was performed with the few MEP pathway genes that were not represented on the array, and the program MapMan was used to create a visual representation of the microarray as metabolic pathways (Thimm et al. 2004). GO terms assigned to the differentially expressed transcripts demonstrated that chloroplast biogenesis, photosynthesis, and phytonutrient biosynthesis-related transcripts were up-regulated. Correlation of the expression and metabolic profiling data with microscopic analysis of chloroplasts presented a global view of phytonutrient production in tomato fruits.

The final step in the vitamin C (L-ascorbic acid) biosynthetic pathway is catalyzed by L-galactono-1,4-lactone dehydrogenase (L-GalLDH). Expression of this gene was studied in the cherry tomato 'West Virginia 106' using a transcriptomics/metabolomics approach (Alhagdow et al. 2007). One cDNA encoding a tomato *SlGalLDH* was identified by screening the SGN and JCVI/TIGR EST databases. Analysis with the TOM1 microarray and MapMan revealed 1,269 leaf and 92 fruit transcripts that were significantly differentially expressed; functional classifications were assigned according to the MIPS system. *SlGalLDH* expression was correlated with ascorbate content and phenotype in RNAi-silenced transgenic tomato lines. The results suggest an association between ascorbate regulation and plant metabolism.

A comparative multivariate analysis of tomato fruit volatiles from 94 contrasting tomato genotypes based on solid phase microextraction gas chromatography-mass spectrophotometry (SPME-GC-MS) profiles revealed

322 different compounds (Tikunov et al. 2005). Hierarchical cluster analysis (HCA) and PCA disclosed that many of the identifiable compounds could be clustered and characterized based on a few main groups, which were then further divided into subclusters. A high-resolution, unbiased approach to metabolomic data analysis composed of full mass spectral alignment of metabolic profiles, followed by multivariate comparative analysis of individual fragments and mass spectral reconstruction was presented.

14.7 Integration of Different Data Types

Complex biological processes involve a myriad of components and dynamic interactions that form to function in the whole organism. Many of these individual constituents have been identified through traditional reductionist methods. However, thorough understanding of the organism as a whole will require the simultaneous observation and measurement of these intertwining networks (Sauer et al. 2007). This systems biology approach involves the integration of transcriptomic, proteomic, metabolomic, and other "-omic" data with observable phenotype, presenting a data management challenge to the bioinformaticist. To more fully understand the "big picture" it will be incumbent upon the researcher to efficiently quantify and integrate this heterogeneous data into a cohesive, comprehensible view of the entire system.

Many of the previous sections of this chapter overlap with this section, particularly those pertaining to metabolomic/proteomic data (see Sect. 14.6). Additional resources available to the tomato researcher that will facilitate this integration of data, both within and between species, are described below.

14.7.1 The SOL Genomics Network (SGN) (http://solgenomics.net/)

SGN is the main resource for tomato bioinformaticists and other tomato researchers (Mueller et al. 2005, 2008). This comprehensive, clade-oriented database of the Solanaceae family integrates maps and marker data; genes, phenotypes and pathways; bioinformatics tools, including the Breeders Toolbox; and data from the tomato genome sequencing effort. As the bioinformatics center of SOL, SGN strives to integrate this large compendium of bioinformatics resources for tomato, the Solanaceae in general, and related Asterid families, using the tomato genome as the centerpiece. Located at BTI, the site also provides a means of relating Solanaceae data to the other model plant genomes (e.g., rice, *Arabidopsis*), allowing a systems biology approach to tomato research.

SGN contains extensive information and resources for tomato research. The site can be navigated via a toolbar containing links and menus to the various sections, including Search, Maps, Genomes, and Tools.

Another innovative aspect of SGN is its community annotation feature (Menda et al. 2008), especially important for breeders because phenotypes are extensively covered and correlated with genetic loci when known. Each locus has a detail page which can be edited by researchers in the field by requesting "locus editor" privileges.

SGN aims to provide a freely accessible, unrestricted information network that will enable understanding questions such as how the members of the Solanaceae can be so genetically similar and yet so phenotypically diverse.

14.7.2 Solanaceous MapMan (http://mapman.mpimp-golm.mpg.de)

Solanaceaous MapMan allows the visualization of transcript and metabolite profiles by correlating data with pathways and processes (Urbanczyk-Wochniak et al. 2006). Adapted from MapMan, which was developed for the whole genome ATH1 oligomer array (Thimm et al. 2004), Solanaceous MapMan functionally classifies the EST unigene-based tomato cDNA TOM1 array elements (Van der Hoeven et al. 2002). By correlating transcript profiles with phenotype, understanding the interrelationship between experimental results and biological function (e.g., metabolic pathways, cellular processes) is facilitated. Comparisons with wild tomato and *Arabidopsis* can also be made. Software and diagrams can be downloaded at http://gabi.rzpd.de/database/java-bin/MappingDownloader. Solanaceous MapMan is a community resource and will be augmented as annotation from the tomato genome sequencing project becomes available.

14.7.3 Fruit Biology Laboratory at INRA Bordeaux (http://www.bordeaux.inra.fr/umr619/home.htm)

The Fruit Biology lab investigates fleshy fruit development using primarily the model fruits tomato and grape, with the objective of understanding processes underlying the production of quality fruit. The lab focuses on fruit organogenesis, metabolism, and development, integrating aspects of molecular biology, biochemistry, cytology, and physiology. For the bioinformaticist, genomic tools are being developed to identify candidate genes associated with fruit development and composition, including transcriptome and tomato mutant/TILLING platforms (see Sect. 14.5.3).

14.7.4 MIPSPlantsDB
(http://mips.helmholtz-muenchen.de/plant/genomes.jsp)

The Munich Information Center for Protein Sequence Plants Databases (MIPSPlantsDB), featuring genome data from several plant species including tomato, allows comparative and integrative analysis through use of a number of bioinformatics tools (Spannagl et al. 2007). Sequenced contigs, chromosome sequences, markers, clones, and genetic elements (transcripts, pseudogenes, and transposable elements) can be retrieved, and annotations can be obtained using the Generic Genome Browser (Gbrowse). The current MIPS tomato database is currently underdevelopment, and the annotations contained therein should not be considered final.

Determination of putative function as well as estimation of evolutionary relationships depends largely upon the identification of homologs and orthologs from other species. The Similarity Matrix of Proteins (SIMAP) database is a comprehensive, current set of precomputed protein homologies from all publicly available plant, fungal, and mammalian sequences (http://liferay.csb.univie.ac.at/portal/web/simap, Rattei et al. 2008). The similarity matrix of the more than 122 million proteins is precalculated, thus avoiding repeated computations and speeding data retrieval. Homologs of the query protein are identified by sequence and/or domain similarity.

Repeat elements, of which numerous variations exist, make up a large portion of plant genomes and are important elements in understanding genetic diversity as well as phylogenetic relationships. The MIPS Repeat Element Database (mips-REdat) and MIPS Repeat Element Catalog (mips-REcat) contain annotated collections of plant repetitive elements compiled from publicly available datasets together with repeats newly identified from within MIPSPlantsDB. Repeats are classified into SSRs, mobile elements, or high copy number genes.

For researchers interested in the study of regulatory systems, the MotifDB database and the Regulomips *Cis* Regulatory Element Detection Online (CREDO) can be used to identify putative transcription factor binding sites. MotifDB is a browsable database based upon *Arabidopsis* and *Brassica* promoters (Haberer et al. 2006). CREDO integrates results from a number of motif finding programs and provides a visualization of the findings for a given input sequence.

14.7.5 Other Resources

In addition to tomato- and/or plant-specific databases, there are many others that provide useful general bioinformatics tools.

National Center for Biotechnology Information (www.ncbi.nlm.nih.gov)

NCBI provides a comprehensive database of nucleotide, protein, genome, and gene sequences for more than 380,000 organisms derived from all publicly available sequences (GenBank, Benson et al. 2011) as well as an extensive array of retrieval and analytical tools (Wheeler et al. 2008). Files are classified into divisions based on taxonomic group and sequencing strategy (e.g., ESTs, GSSs, high-throughput genotyping (HTG) sequences). A number of database retrieval tools (e.g., Entrez, PubMed, Taxonomy), variations of the BLAST programs, and tools for analysis at the gene and genome levels are available.

SwissProt (http://web.expasy.org/docs/swiss-prot_guideline.html)

SwissProt consists of two sets of data. UniProtKB/SwissProt is a minimally redundant, curated protein sequence database. The high quality annotation contains information on function, domains, post-translational modifications, variants, and other details of note, and is integrated with several other databases. The European Molecular Biology Laboratory (EMBL) sequence translations not integrated into Swiss-Prot are contained within UniProtKB/TrEMBL, the computer-annotated supplement to Swiss-Prot.

European Bioinformatics Institute (EBI, http://www.ebi.ac.uk)

EBI offers an immense collection of databases and tools. Databases include ArrayExpress, a public repository for microarray datasets; the Ensembl collection of annotated, large eukaryotic genomes; the InterPro database of protein families and domains; IntAct, for exploration of protein interactions; and the repository of protein sequence and function, UniProt. Tools are available for sequence searching and alignment, functional and structural analysis, data retrieval, and mining the scientific literature.

14.7.6 Applications

Carrari et al. (2006) correlated metabolite and transcriptome profiles during tomato fruit development. Principal metabolic fluxes for 92 metabolites and transcriptomic changes were analyzed in parallel, providing an overall perspective of gene expression and primary metabolic processes fundamental to development. This work was further developed by Bermudez et al. (2008), who integrated molecular marker analysis, sequencing, allelic variation analysis, gene expression, and metabolite composition traits in order to identify candidate QTLs involved in metabolism.

14.8 Conclusions

Advances in "-omics" technologies together with biotechnology and molecular biology methods have the potential to transform plant breeding during this century. In addition to improving agronomic traits, plant breeders now focus on plants as sources of improved nutrition (nutraceuticals), biofuels and biosynthetics, and living "factories" for the production of therapeutics. Plants contain thousands of interesting and potentially useful compounds (i.e., phytochemicals). Genomics tools, particularly metabolomic profiling, can be useful in identifying and modifying metabolic pathways for the enhancement of crop quality. However, because of the inherent complexity and plasticity of such pathways such modifications can lead to unexpected and unintended pleiotropic effects; the physiology, level, timing, and location of the altered process need to be taken into consideration (Newell-McGloughlin 2008).

Compared to the incredibly genetically diverse wild tomato species, the cultivated tomato *S. lycopersicum* has a narrow genetic base. The tomato genome project should help to better understand domestication-related genes and to identify alleles from wild *Solanum* useful for breeding novel genotypes (Bai and Lindhout 2007). Data from molecular markers and high-throughput genome sequencing projects can be used characterize genetic diversity and thus influence choice of parents in a breeding program as well as increase selection efficiency, but to be fully effective it is important that data management systems be structured so as to effectively integrate "-omics" data with pedigrees, phenotypes, and marker genotypes (Moose and Mumm 2008).

To take full advantage of this new paradigm it will be imperative to understand plant biology and genetics, molecular biology, statistical methods, and field-based breeding in addition to genomics, necessitating a collaborative approach. With proper management and integration of tomato genomics data with classical genetic and phenotypic data, potential advances in tomato breeding in the 21st century are exciting and promising.

Acknowledgements

The author thanks Lukas Mueller, John Quackenbush, and Linda Hannick for their suggestions, and especially Robin Buell for critical reading of the manuscript. Appreciation is also extended to the many authors upon whose work this chapter was based.

References

Adams MD, Kelley JM, Gocayne JD, Dubnick M, Polymeropoulos MH, Xiao H, Merril CR, Wu A, Olde B, Moreno RF, et al. (1991) Complementary DNA sequencing: expressed sequence tags and human genome project. Science 252: 1651–1656.

Alba R, Fei Z, Payton P, Liu Y, Moore SL, Debbie P, Cohn J, D'Ascenzo M, Gordon JS, Rose JK, Martin G, Tanksley SD, Bouzayen M, Jahn MM, Giovannoni J (2004) ESTs, cDNA microarrays, and gene expression profiling: tools for dissecting plant physiology and development. Plant J 39: 697–714.

Alhagdow M, Mounet F, Gilbert L, Nunes-Nesi A, Garcia V, Just D, Petit J, Beauvoit B, Fernie AR, Rothan C, Baldet P (2007) Silencing of the mitochondrial ascorbate synthesizing enzyme L-galactono-1,4-lactone dehydrogenase affects plant and fruit development in tomato. Plant Physiol 145: 1408–1422.

Aoki K, Yano K, Suzuki A, Kawamura S, Sakurai N, Suda K, Kurabayashi A, Suzuki T, Tsugane T, Watanabe M, Ooga K, Torii M, Narita T, Shin-I T, Kohara Y, Yamamoto N, Takahashi H, Watanabe Y, Egusa M, Kodama M, Ichinose Y, Kikuchi M, Fukushima S, Okabe A, Arie T, Sato Y, Yazawa K, Satoh S, Omura T, Ezura H, Shibata D (2010) Large-scale analysis of full-length cDNAs from the tomato (*Solanum lycopersicum*) cultivar Micro-Tom, a reference system for the Solanaceae genomics. *BMC Genomics* 2010, 11: 210 doi:10.1186/1471-2164-11-210.

Arumuganathan K, Slattery JP, Tanksley SD, Earle ED (1991) Preparation and flor cytometric analysis of metaphase chromosomes of tomato. Theor Appl Genet 82: 101–111.

Ashburner M, Ball CA, Blake JA, Botstein D, Butler H, Cherry JM, Davis AP, Dolinski K, Dwight SS, Eppig JT, Harris MA, Hill DP, Issel-Tarver L, Kasarskis A, Lewis S, Matese JC, Richardson JE, Ringwald M, Rubin GM, Sherlock G (2000) Gene ontology: tool for the unification of biology. The Gene Ontology Consortium. Nat Genet 25: 25–29.

Ashrafi H, Kinkade M, Foolad MR (2009) A new genetic linkage map of tomato based on a *Solanum lycopersicum* × *S. pimpinellifolium* RIL population displaying locations of candidate pathogen response genes. Genome 52(11): 935–956.

Avraham S, Tung CW, Ilic K, Jaiswal P, Kellogg EA, McCouch S, Pujar A, Reiser L, Rhee SY, Sachs MM, Schaeffer M, Stein L, Stevens P, Vincent L, Zapata F, Ware D (2008) The Plant Ontology Database: a community resource for plant structure and developmental stages controlled vocabulary and annotations. Nucl Acids Res 36: D449–454.

Bai Y, Lindhout P (2007) Domestication and breeding of tomatoes: what have we gained and what can we gain in the future? Ann Bot (Lond) 100: 1085–1094.

Balaji V, Gibly A, Debbie P, Sessa G (2007) Transcriptional analysis of the tomato resistance response triggered by recognition of the Xanthomonas type III effector AvrXv3. Funct Integr Genom 7: 305–316.

Balaji V, Mayrose M, Sherf O, Jacob-Hirsch J, Eichenlaub R, Iraki N, Manulis-Sasson S, Rechavi G, Barash I, Sessa G (2008) Tomato transcriptional changes in response to Clavibacter michiganensis subsp. michiganensis reveal a role for ethylene in disease development. Plant Physiol 146: 1797–1809.

Baxter CJ, Sabar M, Quick WP, Sweetlove LJ (2005) Comparison of changes in fruit gene expression in tomato introgression lines provides evidence of genome-wide transcriptional changes and reveals links to mapped QTLs and described traits. J Exp Bot 56: 1591–1604.

Benson DA, Karsch-Mizrachi I, Lipman DJ, Ostell J, Sayers EW (2011) GenBank. Nucl Acids Res 39: D32–37.

Bermudez L, Urias U, Milstein D, Kamenetzky L, Asis R, Fernie AR, Van Sluys MA, Carrari F, Rossi M (2008) A candidate gene survey of quantitative trait loci affecting chemical composition in tomato fruit. J Exp Bot 59: 2875–2890.

Bernacchi D, Tanksley SD (1997) An interspecific backcross of Lycopersicon esculentum x L. hirsutum: linkage analysis and a QTL study of sexual compatibility factors and floral traits. Genetics 147: 861–877.

Bleeker PM, Spyropoulou EA, Diergaarde PJ, Volpin H, De Both MTJ, Zerbe P, Bohlmann J, Falara V, Matsuba Y, Pichersky F, Haring MA, Schuurink RC (2011) RNA-seq discovery, functional characterization, and comparison of sesquiterpene synthases from *Solanum lycopersicum* and *Solanum habrochaites* trichomes. Plant Mol Biol 77: 323–336.

Bombarely A, Menda N, Tecle IY, Buels RM, Strickler S, Fischer-York T, Pujar A, Leto J, Gosselin J, Mueller LA (2010) The Sol Genomics Network (solgenomics.net): growing tomatoes using Perl. Nucl Acids Res 39: D1149–D1155.

Bonshtien A, Lev A, Gibly A, Debbie P, Avni A, Sessa G (2005) Molecular properties of the Xanthomonas AvrRxv effector and global transcriptional changes determined by its expression in resistant tomato plants. Mol Plant Microbe Interact 18: 300–310.

Broun P, Tanksley SD (1996) Characterization and genetic mapping of simple repeat sequences in the tomato genome. Mol Gen Genet 250: 39–49.

Carrari F, Baxter C, Usadel B, Urbanczyk-Wochniak E, Zanor MI, Nunes-Nesi A, Nikiforova V, Centero D, Ratzka A, Pauly M, Sweetlove LJ, Fernie AR (2006) Integrated analysis of metabolite and transcript levels reveals the metabolic shifts that underlie tomato fruit development and highlight regulatory aspects of metabolic network behavior. Plant Physiol 142: 1380–1396.

Caspi R, Altman T, Dale JM, Dreher K, Fulcher CA, Gilham F, Kaipa P, Karthikeyan AS, Kothari A, Krummenacker M, Latendresse M, Mueller LA, Paley S, Popescu L, Pujar A, Shearer AG, Zhang P, Karp PD (2010) The MetaCyc Database of metabolic pathways and enzymes and the BioCyc collection of Pathway/Genome Databases. Nucl Acids Res 38: D473–D479.

Chiusano ML, D'Agostino N, Traini A, Licciardello C, Raimondo E, Aversano M, Frusciante L, Monti L (2008) ISOL@: an Italian SOLAnaceae genomics resource. BMC Bioinformatics 9 Suppl 2: S7.

Cui L, Veeraraghavan N, Richter A, Wall K, Jansen RK, Leebens-Mack J, Makalowska I, dePamphilis CW (2006) ChloroplastDB: the Chloroplast Genome Database. Nucl Acids Res 34: D692–696.

D'Agostino N, Traini A, Frusciante L, Chiusano ML (2007) Gene models from ESTs (GeneModelEST): an application on the *Solanum lycopersicum* genome. BMC Bioinformat 8 Suppl 1: S9.

de Jong JH (1998) High resolution FISH reveals the molecular and chromosomal organisation of repetitive sequences in tomato. Cytogenet Cell Genet 81: 104.

Doganlar S, Frary A, Ku HM, Tanksley SD (2002) Mapping quantitative trait loci in inbred backcross lines of *Lycopersicon pimpinellifolium* (LA1589). Genome 45: 1189–1202.

Dong Q, Lawrence CJ, Schlueter SD, Wilkerson MD, Kurtz S, Lushbough C, Brendel V (2005) Comparative plant genomics resources at PlantGDB. Plant Physiol 139: 610–618.

Duvick J, Fu A, Muppirala U, Sabharwal M, Wilkerson MD, Lawrence CJ, Lushbough C, Brendel V (2008) PlantGDB: a resource for comparative plant genomics. Nucl Acids Res 36: D959–965.

Emmanuel E, Levy AA (2002) Tomato mutants as tools for functional genomics. Curr Opin Plant Biol 5: 112–117.

Enright AJ, Kunin V, Ouzounis CA (2003) Protein families and TRIBES in genome sequence space. Nucl Acids Res 31: 4632–4638.

Fei Z, Tang X, Alba R, Giovannoni J (2006) Tomato Expression Database (TED): a suite of data presentation and analysis tools. Nucl Acids Res 34: D766–70.

Fei Z, Joung J-G, Tang X, Zheng Y, Huang M, Lee JM, McQuinn R, Tieman DM, Alba R, Klee HJ, Giovannoni JJ (2011) Tomato Functional Genomics Database: a comprehensive resource and analysis package for tomato functional genomics. Nucl Acids Res 39: D1156–1163.

Foolad MR (2007) Genome mapping and molecular breeding of tomato. Int J Plant Genom 2007: 64358.

Fulton TM, Van der Hoeven R, Eannetta NT, Tanksley SD (2002) Identification, analysis, and utilization of conserved ortholog set markers for comparative genomics in higher plants. Plant Cell 14: 1457–67.

Gidoni D, Fuss E, Burbidge A, Speckmann GJ, James S, Nijkamp D, Mett A, Feiler J, Smoker M, de Vroomen MJ, Leader D, Liharska T, Groenendijk J, Coppoolse E, Smit JJ, Levin I, de Both M, Schuch W, Jones JD, Taylor IB, Theres K, van Haaren MJ (2003) Multi-functional T-DNA/Ds tomato lines designed for gene cloning and molecular and physical dissection of the tomato genome. Plant Mol Biol 51: 83–98.

Gonzalo MJ, van der Knaap E (2008) A comparative analysis into the genetic bases of morphology in tomato varieties exhibiting elongated fruit shape. Theor Appl Genet 116(5): 647–56.

Grandillo S, Tanksley SD (1996) QTL analysis of horticultural traits differentiating the cultivated tomato from the closely related species *Lycopersicon pimpinellifolium*. Theor Appl Genet 92: 935–951.

Haberer G, Mader MT, Kosarev P, Spannagl M, Yang L, Mayer KF (2006) Large-scale cis-element detection by analysis of correlated expression and sequence conservation between Arabidopsis and *Brassica oleracea*. Plant Physiol 142: 1589–602.

Henikoff S, Till BJ, Comai L (2004) TILLING. Traditional mutagenesis meets functional genomics. Plant Physiol 135: 630–6.

Jiménez-Gómez JM, Alonso-Blanco C, Borja A, Anastasio G, Angosto T, Lozano R, Martínez-Zapater JM (2007) Quantitative genetic analysis of flowering time in tomato. Genome 50: 303–315.

Kahlau S, Aspinall S, Gray JC, Bock R (2006) Sequence of the tomato chloroplast DNA and evolutionary comparison of solanaceous plastid genomes. J Mol Evol 63: 194–207.

Kahlau S, Bock R (2008) Plastid transcriptomics and translatomics of tomato fruit development and chloroplast-to-chromoplast differentiation: chromoplast gene expression largely serves the production of a single protein. Plant Cell 20: 856–874.

Kolotilin I, Koltai H, Tadmor Y, Bar-Or C, Reuveni M, Meir A, Nahon S, Shlomo H, Chen L, Levin I (2007) Transcriptional profiling of high pigment-2dg tomato mutant links early fruit plastid biogenesis with its overproduction of phytonutrients. Plant Physiol 145: 389–401.

Krieger CJ, Zhang P, Mueller LA, Wang A, Paley S, Arnaud M, Pick J, Rhee SY, Karp PD (2004) MetaCyc: a multiorganism database of metabolic pathways and enzymes. Nucl Acids Res 32: D438–42.

Labate JA, Robertson LD, Wu F, Tanksley SD, Baldo AM (2009) EST, COSII, and arbitrary gene markers give similar estimates of nucleotide diversity in cultivated tomato (*S. lycopersicum* L.). Theor Appl Genet 118: 1005–1014.

Lin C, Mueller LA, Mc Carthy J, Crouzillat D, Petiard V, Tanksley SD (2005) Coffee and tomato share common gene repertoires as revealed by deep sequencing of seed and cherry transcripts. Theor Appl Genet 112: 114–130.

McCallum CM, Comai L, Greene EA, Henikoff S (2000) Targeting induced local lesions IN genomes (TILLING) for plant functional genomics. Plant Physiol 123: 439–442.

Meissner R, Chague V, Zhu Q, Emmanuel E, Elkind Y, Levy AA (2000) Technical advance: a high throughput system for transposon tagging and promoter trapping in tomato. Plant J 22: 265–274.

Meissner R, Jacobson Y, Melamed S, Levyatuv S, Shalev G, Ashri A, Elkind Y, Levy A (1997) A new model system for tomato genetics. Plant J 12: 1465–1472.

Menda N, Buels RM, Tecle I, Mueller LA (2008) A community-based annotation framework for linking Solanaceae genomes with phenomes. Plant Physiol 147(4): 1788–1799.

Menda N, Semel Y, Peled D, Eshed Y, Zamir D (2004) In silico screening of a saturated mutation library of tomato. Plant J 38: 861–872.

Michaelson MJ, Price HJ, Ellison JR, Johnston JS (1991) Comparison of plant DNA contents determined by feulgen microspectrophotometry and laser flow cytometry. Am J Bot 78: 183–188.

Miller JC, Tanksley SD (1990) RFLP analysis of phylogenetic relationships and genetic variation in the genus *Lycopersicon*. Theor Appl Genet 80: 437–448.

Moco S, Capanoglu E, Tikunov Y, Bino RJ, Boyacioglu D, Hall RD, Vervoort J, De Vos RC (2007) Tissue specialization at the metabolite level is perceived during the development of tomato fruit. J Exp Bot 58: 4131–4146.

Moose SP, Mumm RH (2008) Molecular plant breeding as the foundation for 21st century crop improvement. Plant Physiol 147: 969–977.

Mueller LA, Mills AA, Skwarecki B, Buels RM, Menda N, Tanksley SD (2008) The SGN comparative map viewer. Bioinformatics 24: 422–423.

Mueller LA, Solow TH, Taylor N, Skwarecki B, Buels R, Binns J, Lin C, Wright MH, Ahrens R, Wang Y, Herbst EV, Keyder ER, Menda N, Zamir D, Tanksley SD (2005) The SOL Genomics Network: a comparative resource for Solanaceae biology and beyond. Plant Physiol 138: 1310–1317.

Mueller LA, Zhang P, Rhee SY (2003) AraCyc: a biochemical pathway database for Arabidopsis. Plant Physiol 132: 453–460.

Newell-McGloughlin M (2008) Nutritionally improved agricultural crops. Plant Physiol 147: 939–953.

Ohyama A, Asamizu E, Negoro S, Miyatake K, Yamaguchi H, Tabata S, Fukuoka H (2009) Characterization of tomato SSR markers developed using BAC-end and cDNA sequences from genome databases. Mol Breed 23: 685–691.

Ouyang B, Yang T, Li H, Zhang L, Zhang Y, Zhang J, Fei Z, Ye Z (2007) Identification of early salt stress response genes in tomato root by suppression subtractive hybridization and microarray analysis. J Exp Bot 58: 507–520.

Peterson DG, Pearson WR, Stack SM (1998) Characterization of the tomato (*Lycopersicon esculentum*) genome using *in vitro* and *in situ* DNA reassociation. Genome 41: 346–356.

Quackenbush J, Cho J, Lee D, Liang F, Holt I, Karamycheva S, Parvizi B, Pertea G, Sultana R, White J (2001) The TIGR Gene Indices: analysis of gene transcript sequences in highly sampled eukaryotic species. Nucl Acids Res 29: 159–164.

Rattei T, Tischler P, Arnold R, Hamberger F, Krebs J, Krumsiek J, Wachinger B, Stumpflen V, Mewes W (2008) SIMAP—structuring the network of protein similarities. Nucl Acids Res 36: D289–292.

Rick CM, Fobes JF (1975) Allozyme variation in the cultivated tomato and closely related species. Bull Torry Bot Club 102: 376–384.

Sauer U, Heinemann M, Zamboni N (2007) Genetics. Getting closer to the whole picture. Science 316: 550–551.

Schaff JE, Nielsen DM, Smith CP, Scholl EH, Bird DM (2007) Comprehensive transcriptome profiling in tomato reveals a role for glycosyltransferase in Mi-mediated nematode resistance. Plant Physiol 144: 1079–1092.

Schilmiller AL, Miner DP, Larson M, McDowell E, Gang DR, Wilkerson C, Last RL (2010) Studies of a biochemical factory: tomato trichome deep EST sequencing and proteomics. Plant Physiol 153(3): 1212–1223.

Schilmiller AL, Schauvinhold I, Larson M, Xu R, Charbonneau AL, Schmidt A, Wilkerson C, Last RL, Pichersky E. 2009. Monoterpenes in the glandular trichomes of tomato are synthesized from a neryl diphosphate precursor rather than geranyl diphosphate. PNAS 106(26): 10865–10870.

Shirasawa K, Asamizu E, Fukuoka H, Ohyama A, Sato S, Nakamura Y, Tabata S, Sasamoto S, Wada T, Kishida Y, Tsuruoka H, Fujishiro T, Yamada M, Isobe S (2010) An interspecific linkage map of SSR and intronic polymorphism markers in tomato. Theor Appl Genet 121: 731–739.

Spannagl M, Noubibou O, Haase D, Yang L, Gundlach H, Hindemitt T, Klee K, Haberer G, Schoof H, Mayer KF (2007) MIPSPlantsDB—plant database resource for integrative and comparative plant genome research. Nucl Acids Res 35: D834–840.

Stekel DJ, Git Y, Falciani F (2000) The comparison of gene expression from multiple cDNA libraries. Genome Res 10: 2055–2061.

Tanksley SD, Ganal MW, Prince JP, de Vicente MC, Bonierbale MW, Broun P, Fulton TM, Giovannoni JJ, Grandillo S, Martin GB, et al. (1992) High density molecular linkage maps of the tomato and potato genomes. Genetics 132: 1141–1160.

Tanksley SD, Grandillo S, Fulton TM, Zamir D, Eshed Y, Petiard V, Lopez J, Beck-Bunn T (1996) Advanced backcross QTL analysis in a cross between an elite processing line of tomato and its wild relative *L. pimpinellifolium*. Theor Appl Genet 92: 213–224.

The Plant Ontology Consortium (2002) The plant ontology consortium and plant ontologies. Comp Funct Genom 3: 137–142.

Thimm O, Blasing O, Gibon Y, Nagel A, Meyer S, Kruger P, Selbig J, Muller LA, Rhee SY, Stitt M (2004) MAPMAN: a user-driven tool to display genomics data sets onto diagrams of metabolic pathways and other biological processes. Plant J 37: 914–939.

Tikunov Y, Lommen A, de Vos CH, Verhoeven HA, Bino RJ, Hall RD, Bovy AG (2005) A novel approach for nontargeted data analysis for metabolomics. Large-scale profiling of tomato fruit volatiles. Plant Physiol 139: 1125–1137.

Urbanczyk-Wochniak E, Usadel B, Thimm O, Nunes-Nesi A, Carrari F, Davy M, Blasing O, Kowalczyk M, Weicht D, Polinceusz A, Meyer S, Stitt M, Fernie AR (2006) Conversion of MapMan to allow the analysis of transcript data from Solanaceous species: effects of genetic and environmental alterations in energy metabolism in the leaf. Plant Mol Biol 60: 773–792.

Van der Hoeven R, Ronning C, Giovannoni J, Martin G, Tanksley S (2002) Deductions about the number, organization, and evolution of genes in the tomato genome based on analysis of a large expressed sequence tag collection and selective genomic sequencing. Plant Cell 14: 1441–1156.

van der Knaap E, Sanyal A, Jackson SA, Tanksley SD (2004) High-resolution fine mapping and fluorescence *in situ* hybridization analysis of *sun*, a locus controlling tomato fruit shape, reveals a region of the tomato genome prone to DNA rearrangements. Genetics 168: 2127–2140.

van der Knaap E, Tanksley SD (2003) The making of a bell pepper-shaped tomato fruit: identification of loci controlling fruit morphology in Yellow Stuffer tomato. Theor Appl Genet 107(1): 139–47.

Wang Z, Gerstein M, Snyder M (2009) RNA-Seq: a revolutionary tool for transcriptomics. Nat Rev Genet 10(1): 57–63.

Wheeler DL, Barrett T, Benson DA, Bryant SH, Canese K, Chetvernin V, Church DM, Dicuccio M, Edgar R, Federhen S, Feolo M, Geer LY, Helmberg W, Kapustin Y, Khovayko O, Landsman D, Lipman DJ, Madden TL, Maglott DR, Miller V, Ostell J, Pruitt KD, Schuler GD, Shumway M, Sequeira E, Sherry ST, Sirotkin K, Souvorov A, Starchenko G, Tatusov RL, Tatusova TA, Wagner L, Yaschenko E (2008) Database resources of the National Center for Biotechnology Information. Nucl Acids Res 36: D13–21.

Wilhelm BT, Landry JR (2009) RNA-Seq-quantitative measurement of expression through massively parallel RNA-sequencing. Methods 48(3):249–257.

Wu F, Mueller LA, Crouzillat D, Petiard V, Tanksley SD (2006) Combining bioinformatics and phylogenetics to identify large sets of single-copy orthologous genes (COSII) for comparative, evolutionary and systematic studies: a test case in the euasterid plant clade. Genetics 174: 1407–1420.

Yano K, Aoki K, Shibata D (2007) Genomic databases for tomato. Plant Biotechnol 24: 17–25.

Yano K, Watanabe M, Yamamoto N, Tsugane T, Aoki K, Sakurai N, Shibata D (2006) MiBASE: A database of a miniature tomato cultivar Micro-Tom. Plant Biotechnol 23: 195–198.

Zhang P, Foerster H, Tissier CP, Mueller L, Paley S, Karp PD, Rhee SY (2005) MetaCyc and AraCyc. Metabolic pathway databases for plant research. Plant Physiol 138: 27–37.

Index

Color Plate Section